Chapman & Hall/CRC Mathematical and Computational Biology Series

Bioinformatics

A Practical Approach

CHAPMAN & HALL/CRC
Mathematical and Computational Biology Series

Aims and scope:
This series aims to capture new developments and summarize what is known over the whole spectrum of mathematical and computational biology and medicine. It seeks to encourage the integration of mathematical, statistical and computational methods into biology by publishing a broad range of textbooks, reference works and handbooks. The titles included in the series are meant to appeal to students, researchers and professionals in the mathematical, statistical and computational sciences, fundamental biology and bioengineering, as well as interdisciplinary researchers involved in the field. The inclusion of concrete examples and applications, and programming techniques and examples, is highly encouraged.

Proposals for the series should be submitted to one of the series editors above or directly to:
CRC Press, Taylor & Francis Group
24-25 Blades Court
Deodar Road
London SW15 2NU
UK

Published Titles

Bioinformatics: A Practical Approach
Shui Qing Ye

Cancer Modelling and Simulation
Luigi Preziosi

Computational Biology: A Statistical Mechanics Perspective
Ralf Blossey

Computational Neuroscience: A Comprehensive Approach
Jianfeng Feng

Data Analysis Tools for DNA Microarrays
Sorin Draghici

Differential Equations and Mathematical Biology
D.S. Jones and B.D. Sleeman

Exactly Solvable Models of Biological Invasion
Sergei V. Petrovskii and Bai-Lian Li

Introduction to Bioinformatics
Anna Tramontano

An Introduction to Systems Biology: Design Principles of Biological Circuits
Uri Alon

Knowledge Discovery in Proteomics
Igor Jurisica and Dennis Wigle

Modeling and Simulation of Capsules and Biological Cells
C. Pozrikidis

Niche Modeling: Predictions from Statistical Distributions
David Stockwell

Normal Mode Analysis: Theory and Applications to Biological and Chemical Systems
Qiang Cui and Ivet Bahar

Pattern Discovery in Bioinformatics: Theory & Algorithms
Laxmi Parida

Stochastic Modelling for Systems Biology
Darren J. Wilkinson

The Ten Most Wanted Solutions in Protein Bioinformatics
Anna Tramontano

Chapman & Hall/CRC Mathematical and Computational Biology Series

Bioinformatics

A Practical Approach

Shui Qing Ye

CRC Press
Taylor & Francis Group
Boca Raton London New York

CRC Press is an imprint of the
Taylor & Francis Group, an **informa** business
A CHAPMAN & HALL BOOK

CRC Press
Taylor & Francis Group
6000 Broken Sound Parkway NW, Suite 300
Boca Raton, FL 33487-2742

First issued in paperback 2019

ISBN-13: 978-1-58488-810-9 (hbk)
ISBN-13: 978-0-367-38875-1 (pbk)

Library of Congress Cataloging-in-Publication Data

Ye, Shui Qing, 1954-.
Bioinformatics : a practical approach / Shui Qing Ye.
p. cm. -- (Mathematical and computational biology series)
Includes bibliographical references and index.
ISBN 978-1-58488-810-9 (alk. paper)
1. Bioinformatics. I. Title. II. Series.

QH324.2.Y42 2007
570.285--dc22 2007004464

Visit the Taylor & Francis Web site at
http://www.taylorandfrancis.com

and the CRC Press Web site at
http://www.crcpress.com

Contents

Preface

The idea of this book has been excogitated over the past few years from my own research endeavor to apply an integration of bioinformatics, genomic, and genetic approaches in the identification of genetic and biochemical markers in cardiopulmonary diseases. To this project I have brought my five years as director of the Gene Expression Profiling Core at The Johns Hopkins Center of Translational Respiratory Medicine, my three-year stint as a coordinator of the Affymetrix User Group monthly meeting at The Johns Hopkins University medical institutions and two years as a director of the Molecular Resource Core in the National Heart, Lung, and Blood Institute-funded Program Project Grant on Cytoskeletal Regulation of Lung Endothelial Pathobiology at the University of Chicago, as well as my experience in obtaining my R01 grant award and involvement in a successful Specialized Centers of Clinically Oriented Research application on "Molecular Approaches to Ventilator-Associated Lung Injury," directed by Dr. Joe G.N. Garcia, from the National Institutes of Health. The idea for this book further crystallized by listening to the enlightening advice at the System Biology Symposium held at the University of Chicago in October 2005 from Dr. Phillip A. Sharp, professor of biology from Massachusetts Institute of Technology and a 1993 Nobel laureate (for discovering gene splicing). He pointed out that to be successful in the omics age, every biological or biomedical researcher should know some bioinformatics. I fully concurred with Dr. Sharp's comments and developed a book proposal. Dr. Sunil Nair, a perceptive publisher from Taylor & Francis, was first to hand me a book contract, which I happily accepted.

Bioinformatics is emerging as an ever-evolving new branch of science in which computer tools are applied to collect, store, and analyze biological data to generate new biological information. Over the past few years, major progress in the field of molecular biology, coupled with rapid advances in genomic technologies, has led to an explosive growth in the biological information distributed in a variety of biological databases. Currently, genome resources from a number of species are available at the National Center of Biological Information Website (http://www.ncbi.nlm.nih.gov/Genomes/index.html). This list is being expanded at an unprecedented pace. A challenge facing researchers today is how to piece together and analyze this plethora of data to make new biological discoveries and gain new unifying global biological insights. This led to the absolute requirement for biologists and medical researchers to obtain a reasonable amount of knowledge on computational biology (bioinformatics), i.e., applying computational approaches to facilitate the understanding of various biological processes such as a more global perspective in experimental design and the ability to capitalize on database mining — the process by which testable hypotheses are generated regarding the function or structure of a gene or protein of interest by identifying similar sequences in better characterized organisms. Equally important is that computer gurus need to have some basic understanding of biological problems in order for them to efficiently execute their computer skills in the field of bioinformatics.

Biologists are usually not extensively trained with computer skills, and computer experts rarely have a biology background. One of the goals of this book is to bridge or shorten the knowledge gap between biologists and computer specialists to make better and more efficient application and development of bioinformatics.

This book will cover the most state-of-the-art bioinformatics applications a biologist needs to know. Part I will focus on genome and DNA sequence analysis with chapters on genome analysis, common DNA analysis tools, phylogenetics analysis and SNP and haplotype analysis. Part II will center on transcriptome and RNA sequence analysis, with chapters on microarray, SAGE, regulation of gene expression, miRNA, and siRNA. Part III will present widely applied programs or tools in proteome, protein sequences, protein functions, and functional annotation of proteins in murine models, and Part IV will introduce the most useful basic biocomputing tools in chapter on the application of programming languages in biology, Website and database design, and interchanging data between Microsoft Excel and Access. Abbreviations, a selected glossary, and Websites for selected bioinformatics software are listed in the appendices for reference.

Our goal is to assimilate the most current bioinformatic knowledge and tools relevant to the omics age into one cohesive, concise, self-contained book accessible to biologists, to computer scientists, and to motivated nonspecialists with a strong desire to learn bioinformatics. The twenty-five contributors to this book have been recruited from nine world-renowned institutions in six countries: Johns Hopkins University, University of Chicago, Ohio State University, Georgetown University in the United States; the University of Toronto in Canada; the Swiss Institute of Bioinformatics in Switzerland; Okayama University in Japan; Wuhan University in China; and the International Centre for Genetic Engineering and Biotechnology in India. In each chapter, a theoretical introduction of the subject is followed by the exemplar step-by-step tutorial to help readers both to have a good theoretical understanding and to master a practical application. Although all the chapters were contributed by experts in their respective fields, the book was written to be easily understood. Complex mathematic deductions and jargon were avoided or reduced to a minimum. Any novice, with a little computer knowledge, can learn bioinformatics from this book without difficulty. In the overview of each chapter, several authoritative references were given, so that more experienced readers may explore the subject in depth and find new horizons. The target readers of this book are biologists but computer specialists may find a fruitful, comprehensive, and concise synopsis of biological problems to tackle in each chapter.

This book is the collective efforts of editorial staff and other individuals. I am deeply indebted to all contributing authors for their tremendous efforts to finish their work on time, and for their gracious tolerance of my repeated haggling for revisions. We are grateful to the editorial staff for their tireless and meticulous work on accuracy and detail. We want to thank several other unnamed people who have helped us along the way for their valuable guidance and innumerable improvement suggestions. We apologize to many colleagues whose work was not covered in this book due to space limitations.

We welcome any criticism and suggestions for improvement so that they may be incorporated into the next edition.

Shui Qing Ye

Editor

Dr. Shui Qing Ye, M.D., Ph.D. is professor of surgery, molecular microbiology, and immunology at the University of Missouri, Columbia. He was a research associate professor of medicine and the director of the Molecular Resource Core in a National Institutes of Health (NIH) funded Program Project on Cytoskeletal Regulation of Lung Endothelial Pathobiology at University of Chicago from 2005 to 2006. Before that, he had been an assistant professor of medicine and the director of the Gene Expression Profiling Core at the Center for Translational Respiratory Medicine, Johns Hopkins University, Baltimore, Maryland, for five years.

Dr. Ye obtained his medical education at Wuhan University School of Medicine in China and pursued his Ph.D. degree at the University of Chicago. He has been engaged in biomedical research for more than twenty years, and has authored more than fifty publications, including peer-reviewed research articles, invited reviews, and book chapters. Well versed in most modern cell, molecular, and biochemical research tools and techniques.

Dr. Ye is an expert on current bioinformatics, genomic, and genetic technologies. Using an integrated omics approach, he identified the pre-B-cell colony enhancing factor (PBEF) as a novel biochemical and genetic biomarker in acute lung injury (ALI), which led to his successful R01 application to NIH for the further mechanistic exploration of PBEF in the susceptibility and severity of ALI. Dr. Ye's current interest is to identify candidate genes for human diseases using the combination of bioinformatic tools and experimental approaches.

Editor

Contributors

Brigitte Boeckmann, Ph.D.
Swiss Institute of Bioinformatics
Geneva, Switzerland

George Adrian Calin, M.D., Ph.D.
Molecular Virology, Immunology, and Medical Genetics Department
Comprehensive Cancer Center
The Ohio State University
Columbus, Ohio

Blanca Camoretti-Mercado, Ph.D.
University of Chicago
Chicago, Illinois

Steven Elliot
Johns Hopkins University
Baltimore, Maryland

Dmitry N. Grigoryev, M.D., Ph.D.
Division of Clinical Immunology
Johns Hopkins University
Bayview Medical Center
Baltimore, Maryland

Yurong Guo, Ph.D.
Johns Hopkins University
Bayview Campus
Baltimore, Maryland

Lydie Lane
Swiss Institute of Bioinformatics
Geneva, Switzerland

Chang-Gong Liu, Ph.D.
Molecular Virology, Immunology, and Medical Genetics Department
Comprehensive Cancer Center
The Ohio State University
Columbus, Ohio

Hongfang Liu, Ph.D.
Georgetown University Medical Center
Washington, D.C.

Mingyao Liu, M.D., M.Sc.
School of Graduate Studies
Faculty of Medicine
University of Toronto
Toronto, Ontario
Canada

Xiuping Liu, M.D.
Microarray Shared Resource
Comprehensive Cancer Center
The Ohio State University
Columbus, Ohio

Rodney Lui
Johns Hopkins University
Baltimore, Maryland

Shwu-Fan Ma, Ph.D.
Section of Pulmonary and Critical Care Medicine
University of Chicago School of Medicine
Chicago, Illinois

Claudia C. dos Santos, M.D.
Saint Michael's Hospital
Toronto, Ontario
Canada

Masaharu Seno, Ph. D.
Department of Biomedical Engineering
Graduate School of Natural Science and Technology
Okayama University
Okayama, Japan

Hiroko Tada, Ph.D.
Department of Biomedical Engineering
Graduate School of Natural Science and Technology
Okayama University
Okayama, Japan

Renu Tuteja, Ph.D.
International Centre for Genetic Engineering and Biotechnology
New Delhi, India

Jennifer E. Van Eyk, Ph.D.
Johns Hopkins University
Baltimore, Maryland

Jun Wada, M.D.
Department of Medicine and Clinical Science
Okayama University Graduate School of Medicine
Okayama, Japan

Jerry M. Wright, Ph.D.
Department of Physiology
Johns Hopkins University
School of Medicine
Baltimore, Maryland

Shui Qing Ye, M.D., Ph.D.
Section of Pulmonary and Critical Care Medicine
University of Chicago School of Medicine
Chicago, Illinois

Yum Lina Yip
Swiss Institute of Bioinformatics
Geneva, Switzerland

Li Qin Zhang, M.D.
Section of Pulmonary and Critical Care Medicine
University of Chicago School of Medicine
Chicago, Illinois

Xiao-Lian Zhang, Ph.D.
Department of Immunology
State Key Laboratory of Virology and Hubei Province Key Laboratory of Allergy and Immunology
Wuhan University School of Medicine
Wuhan, China

Fang Zheng, M.D.
Center for Gene Diagnosis Zhongnan Hospital
Wuhan University School of Medicine
Wuhan, China

Abbreviations

Å	Angstrom
AA	African American
AC	Accession number
aCGH	Array comparative genomic hybridization
ANN	Artificial Neural Networks
ANOVA	Analysis of variance
API	Application Programming Interface
ARGs	Androgen-Negulated Genes
ASN1	Abstract Syntactical Notation Version 1
ASP	Alternative Splice Profile
AS-PCR	Allele-Apecific Polymerase Chain Reaction
ATI	Alternative Translational initiation
ATID	Alternative Translational Initiation Database
ATR-X	Alpha-Thalassemia/mental Retardation syndrome
attB and attP	Named for the attachment sites for the phage integrase on the bacterial and phage genomes, respectively.
BED	Browser Extensible Data Format
bFGF	Basic Fibroblast Growth Factor
Bl2seq	Blast 2 Sequences
BLAST	Basic Local Alignment Search Tool
Blat	Blast-Like-Alignment Tool
BLOSUM	BLOcks SUbstitution Matrix
BTC	Betacellulin
CAFASP/EVA	Critical Assessment of Fully Automated Structure Prediction
CAPRI	Critical Assessment of Prediction Interaction
CASP	Community Wide Experiment on the Critical Assessment of Techniques for Protein Structure;
CAST	Clustering affinity search technique
CD	Cluster of differentiation
CD	Conserved domain
CDART	Conserved Domain Architecture Retrieval Tool
CDD	Conserved Domain Database

cDNA	Complementary DNA sequence
CDS	Coding sequence
CEPH	The Centre d'Etude du Polymorphisme Humain
CEU	European ancestry by the Centre d'Etude du Polymorphisme Humain
CFLP	Cleavage fragment length polymorphism
CGAP	Cancer Genome Anatomy Project
CGI	Common Gateway Interface
CHB	Han Chinese individuals from Beijing
ChIP-on-chip	Chromatin immuno-precipitation microarray
CM	Comparative modeling
CPAN	Comprehensive Perl Archive Network
Cre	Cause recombination of the bacteriophage P1 genome
cRNA	Complementary RNA sequence
CSGE	Conformation-sensitive gel electrophoresis
CSS	Cascading Style Sheets
CV/CD	Common variant/common disease
CXCR4	Chemokine receptor chemokine (C–X–C motif) receptor 4
DAG	Directed acyclic graph
DAS	Directly Attached Storage
DBD	Database Driver
DBI	Database Interface
DBMS	Database Management Software
DBMS	Database Management System
dCHIP	DNA chip analyzer
DDBJ	DNA Data Bank of Japan
df	Degree of freedom
DGGE	Denaturing gradient gel electrophoresis
DHPLC	Denaturing high-performance liquid chromatography
DHTML	Dynamic HTML
DIGE	Differential in gel electrophoresis
DLDA	Diagonal linear discriminant analysis
DNA	Deoxyribonucleic acid
DNS	Domain Name Service
DSCAM	The *Drosophila melanogaster* Down Syndrome cell adhesion molecule
DSL	Digital Subscriber Line
dsRNA	Double-stranded ribonucleic acid
EASED	Extended Alternatively Spliced EST Database
EBI	European Bioinformatics Institute
ECRs	Evolutionary Conserved Regions
EGF	Epidermal growth factor
EGFR	Epidermal growth factor receptor
EM	The expectation-maximization algorithms
EMBL	European Molecular Biology Laboratory

EMBL/UniProt	European Molecular Biology Laboratory/ Universal Protein Resource
EMS	Ethylmethanesulfonate
ENCODE	ENCyclopedia of DNA Elements
ENU	*N-ethyl N-nitrosourea*
EPD	Eukaryotic Promoter Database
ER	Estrogen receptor
ERCC	External RNA controls consortium
ES	Embryonic stem cells
ESEs	Exonic splicing enhancers
ESS	Exonic splicing silencers
EST	Expressed Sequence Tag
ET-1 and ET-3	Endothelins-1 and -3
EXIF	Exchangeable Image File format
ExPASY	Expert Protein Analysis System
ExProView	Expression Profile Viewer
FASTA	Fast-All
FDA	U.S. Food and Drug Administration
FDR	False discovery rate
FGF	Fibroblast growth factor
Floxed	Flanked by loxP sites
FOM	Figure of merit
FR	Fold Recognition;
FRET	Fluorescence resonance energy transfer
FRT	Flp recombinase recognition target
FTP	File Transfer Protocol
FWER	Family-wise error rate
GCG/MSF	Genetics Computer Group/ Multiple Sequence Format
GC-RMA	Quinine-cytosine content robust multi-array analysis
GDE	Genetic Data Environment
GEO	Gene expression omnibus
GFF	General Feature Format
GIF	Graphics Interchange Format
GNU	GNUs Not UNIX
GO	Gene Ontology
GOTM	Gene ontology tree machine
GPI	Glycosyl phosphatidylinositol
GRs	Glucocorticoid receptors
GTF	Gene Transfer Format
GUI	Graphical User Interface
GXD	Gene Expression Database
HAMAP	High-quality automated and manual annotation of microbial proteomes
HCA	Hydrophobic Cluster Analysis;

HGNC	HUGO Gene Nomenclature Committee
HLA	Human leukocyte antigen
HMM	Hidden Markov model
hnRNP	Heterogeneous nuclear ribonucleoprotein
HPLC	High performance liquid chromatography
HS_EGFR	Human epidermal growth factor receptor
HSP	High-scoring pairs
HTG	High Throughput Genomic
HTML	Hypertext Markup Language
HUGO	Human Genome Organization
ICANN	Internet Corporation for Assigned Names and Numbers
ICAT	Isotope-coded affinity tag
IDE	Integrated Development Environment
IgBlast	Immunoglobulin Blast
IGTC	International Gene Trap Consortium
IMAC	Immobilized metal-affinity chromatography
IMEX	International molecular exchange consortium
IMSR	International Mouse Strain Resources
INK	Cyclin-dependent kinase inhibitors
InterNIC	Internet Network Information Center
IPTC	International Press Telecommunications Council
ISEs	Intronic splicing enhancers
ISP	Internet Service Provider
ISS	Intronic splicing silencers
iTRAQ	Isobaric tagging for relative and absolute quantitation
IUB	International Union of Biochemistry
IUPAC	International Union of Pure and Applied Chemistry
IUT	Intersection-union testing
JDK	JAVA API development kit
JIT	Just-in-time compiler
JPEG	Joint Photographic Experts Group
JPT	Individuals from Tokyo
JRE	JAVA Runtime Environment
JVM	Java Virtual Machine
Kd	Equilibrium dissociation constant
KDOM	The Knowledge Discovery Object Model
KNN	K-nearest neighbor
Koff	Rate of dissociation
Kon	Rate of association
LD	Linkage disequilibrium
loxP	Locus of X-over in P1
LSID	The Life Science Identifier
MADAM	Microarray data manager

MAF	Minor allele frequency
MAQC	Microarray quality control
MAS	Microarray analysis software
MB	Megabyte
MeV	Multiexperiment viewer
MEV	TIGR MultiExperiment Viewer
MGED	Microarray gene expression data
MGI	Mouse Genome Informatics
MIAME	Minimum information about a microarray experiment
ML	Markup Language
MIAPA	Minimal Information About a Phylogenetic Analysis
MIDAS	Microarray data analysis system
MIM	Mendelian inheritance in man
miRNAs	MicroRNAs
MM	Mismatched probe
mM	Millimol
MMM	Mixture-mode methods
mRNA	Messenger RNA
MS	Mass spectrometry
MS/MS	Tandem mass spectrometry
MSA	Multiple sequence alignment
MSDE	Microsoft Data Engine
MSE	Mean squared error
MTB	Mouse Tumor Biology Database
MudPIT	Multidimensional protein identification technology
MuLV	Murine leukemia virus
MUSCLE	Multiple sequence comparison by log expectation
Mw	Molecular weight
MySQL	Multithreaded, muiltiuser, structured query language
NAE	Alternatively spliced ESTs
NBRF/PIR	The National Biomedical Research Foundation/ Protein Information Resource
NCBI	National Center for Biotechnology Information
NCE	Number of constitutively spliced ESTs
NCICB	National Cancer Institute's Centre for Bioinformatics
ncRNAs	Non-coding RNAs
NEB	New England BioLAbs
NHGRI	National Human Genome Research Institute
NHLBI	National Heart Lung and Blood Institute
NIH	National Institutes of Health
NJ	Neighbor joining
nM	Nanomol
NOS	Nitric oxide synthase

nr	Non-redundant database
NSS	Number of alternative splice sites
OD	Optical density
Oligo	Oligonucleotide
OMIM	Online Mendelian inheritance in man
OMMSA	Open mass spectrometry search algorithm
ORF	Open reading frame
OS	Operation System
PAL	Phylogenetic Analysis Library
PAM	Point Accepted Mutation
PANTHER	Protein ANalysis THrough Evolutionary Relationships;
PBEF	Pre-B-cell colony enhancing factor
PC	Personal Computer
PCA	Principal component analysis
PCGs	Protein coding genes
PCR	Polymerase chain reaction
PDB	Protein data bank
Perl	Practical Extraction Report Language
PFAM	Protein Family
Pfam	Protein Families database
PGA	Programs for Genomic Applications
PHYLIP	PHYLogeny Inference Package
pI	Isoelectric point;
PIR	Protein Information Resource
PLIER	Probe logarithmic intensity error
PLS	Proportional hazard regression model for survival analysis
PM	Perfect match
PMF	Peptide mass fingerprint
PRF	Protein Research Foundation
PRs	Progesterone receptors
PSI	Protein standard initiative
PSI-BLAST	Position-specific iterated Blast
PSL	Process Specification Language
PTGS	Post-transcriptional gene silencing
PTM	Post-translational modification
qRT-PCR	Quantitative real-time PCR
QTL	Quantitative Trait Loci
RACE	Rapid amplification of cDNA ends
RBI	Resampling based inference
RCSB PDB	Research Collaboratory for Structural Bioinformatics Protein Data Bank
REBASE	Restriction Enzyme Database
RFLP	Restriction fragment length polymorphism

RID	Request ID
RMA	Robust multi-array analysis
RNA	Ribonucleic acid
RPS-Blast	Reverse Position-Specific Blast
RSF	Rich Sequence Format
RT	Reverse transcriptase
SADE	SAGE adaptation for downsized extracts
SAGE	Serial Analysis of Gene Expression
SAM	Significant analysis of microarray
SAM	Sequence Alignment and Modeling system
SARS	Severe acute respiratory syndrome
SE	Standard error
SILAC	Stable isotope labeling with amino acids
siRNAs	Small interfering RNAs
SMART	Simple Modular Architecture Research Tool
snoRNAs	Small nucleolar RNAs
SNP	Single nucleotide polymorphism
snRNA	Small nuclear RNAs
snRNP	Small nuclear ribonuclear proteins
SOFT	Simple Omnibus Format in Text
SOM	Self-organizing maps
SQL	Structured Query Language
SSCP	Single-strand conformation polymorphism
SSR	Site-specific recombinase
STS	Sequence Tagged Sites
Taq	Thermus aquaticus
2DE	Two-dimensional gel electrophoresis
TDGS	Two-dimensional gene scanning
TDMD	Tab-delimited, multisample files
TGGE	Temperature gradient gel electrophoresis
Tiff	Tagged image file format
TIGR	Institute of Genome Research
TRC	RNAi Consortium
UCSC	University of California–Santa Cruz
UniProtKB	UniProt Knowledgebase
URL	Uniform Resource Locator
URN	Uniform Resource Name
UTRs	Untranslated Regions
VIGS	Virus-induced gene silencing
VSM	Vector support machines
WIG	Wiggle Track Format
WISIWYG	What You See Is What You Get
WSP	Weighted sum of pairs

XHTML	Extensible Hypertext Markup Language
XML	Extensible Markup Language
YRI	Yoruba people of Ibadan, Nigeria

CHAPTER 1

Genome Analysis

Shwu-Fan Ma

CONTENTS

Since the launch of the International Human Genome Project and official whole-genome sequencing projects from other species, tremendous amounts of sequence data (>~3.3 million) have become available. In addition, large-scale technologies, such as microarray for gene expression detection and genome-wide association studies, also have speeded up the collection of sequences to the public databases. How to effectively display, align, and analyze genomic sequences to harness genomic power therefore becomes crucial in the postgenomic era. This chapter commences with Genome Browser in Section 1 and then introduces the Basic Local Alignment Search Tools (BLAST) in Section 2. In line with the format and style throughout this book, each section starts with a theoretical introduction in Part I, continues with a step-by-step tutorial in Part II, and ends with the presentation of sample data in Part III.

SECTION 1 GENOME STRUCTURE ANALYSIS BY GENOME BROWSER

Part I Introduction

1. What Is Genome Browser?

Genome Browser is a tool that collates all relevant genomic sequence information in one location and provides a rapid, reliable, and simultaneous display of any requested portion of genomes at any scale in a graphical design. In addition, they provide the ability to search for markers and sequences, to extract annotations for specific regions or for the whole genome, and to act as a central starting point for genomic research.

2. Multiple Genome Browser Sites

For the purpose of creating high-resolution graphical interface of specific segments in a known genomic DNA sequence obtained from whole-genome sequencing projects, many genome browsers have been created with some unique and/or overlapping features. They include but are not limited to the following:

1. UCSC genome browser (http://genome.ucsc.edu/): In this browser, data are organized along the genomic sequence backbone and aligned for quick search, data retrieval, and display. All the data are linked out to other databases, Web sites, and literature. In addition, data types, referred to as "annotation tracks," are aligned on the genomic backbone framework. These tracks, including known genes, predicted genes, ESTs, mRNAs, gaps location, chromosomal

bands, comparative genomics, single-nucleotide polymorphisms (SNPs) and other variations, evolutionary conservation, microarray/expression data, etc., are displayed to provide additional information about any given genomic region of interest. In this chapter, we will use specific examples and provide a step-by-step tutorial to explain some of the features in the browser and to retrieve sequences and map regions of interest.

2. Ensembl genome browser (http://www.ensembl.org): It is a joint project between European Molecular Biology Laboratory (EMBL)-European Bioinformatics Institute (EBI) and the Sanger Institute, which provides and maintains automated genome annotation on selected eukaryotic genomes and subsequent visualization. Each species supported by Ensembl has its own home page, which allows you to search the whole Ensembl database of genomic information or categories of information within it. From the species-specific home page, basic release information and statistics and a link to a clickable site map for further information and additional entry points into the Ensembl system are also available.

3. VISTA (http://genome.lbl.gov): It is a comprehensive suite of programs and databases for comparative analysis of genomic sequences. Sequences or alignments can be uploaded as a plain text files in FASTA format, or their GenBank accession numbers, to the VISTA servers for the following analyses:
   a. mVISTA: align and compare sequences (up to 100) from multiple species
   b. rVISTA: combines transcription factor binding sites database search with a comparative sequence analysis
   c. GenomeVISTA: compare your sequences with whole-genome assemblies
   d. Phylo-VISTA: analyze multiple DNA sequence alignments of sequences from different species while considering their phylogenic relationships
   e. Align whole genome: align and compare two finished or draft microbial genome assemblies up to 10 Mb long. VISTA can also be used to examine precomputed alignments of whole-genome assemblies through VISTA Browser.

4. NCBI MapViewer (http://www.ncbi.nlm.nih.gov/mapview/): It allows you to view and search for a subset of organisms in Entrez Genomes including archaea, bacteria, eukaryotae, viruses, viroids, and plasmids. Regions of interest can be retrieved by text queries (e.g., gene or marker name) or by sequence alignment (BLAST). Results can be viewed at the whole-genome level, and chromosome maps displayed and zoomed into progressively greater levels of detail, down to the sequence data for a region of interest. Multiple options exist to configure your display, download data, navigate to related data, and analyze supporting information. The number and types of available maps vary by organism. If multiple maps are available for a chromosome, it displays them aligned next to each other, based on shared marker and gene names, and for the sequence maps, based on a common sequence coordinate system.

5. ECR Browser (http://ecrbrowser.dcode.org): It is a dynamic whole-genome navigation tool for visualizing and studying evolutionary relationships between vertebrate and nonvertebrate genomes. The tool is constantly being updated to include the most recently available sequenced genomes (currently, human, dog, mouse, rat, chicken, frog, Fugu puffer fish, Tetraodon puffer fish, zebra fish, and six fruit flies). Evolutionary conserved regions (ECRs) that have been mapped within alignments of the genomes are presented in the graphical browser, where ECRs in relation to known genes that have been annotated in the base genome are depicted and color-coded. The "Grab ECR" feature allows users to rapidly extract sequences that correspond to any ECR, visualize underlying sequence alignments, and/or identify conserved transcription factor binding sites. In addition to accessing precomputed alignments for the available genomes, the ECR Browser can also be used as an alignment tool. It allows users to map submitted sequences to specific homologous positions within the genomes and to create a detailed alignment using the BlastZ alignment program.

6. Combo (http://www.broad.mit.edu/annotation/argo/): It is a free, downloadable comparative genome browser that provides a dynamic view of whole-genome alignments along with their associated annotations. Combo has two different visualization perspectives: (1) the perpendicular (dot plot) view provides a dot plot of genome alignments synchronized with a display of genome annotations along

each axis and (2) the parallel view displays two genome annotations horizontally, synchronized through a panel displaying local alignments as trapezoids. Users can zoom to any resolution, from whole chromosomes to individual bases. They can select, highlight, and view detailed information from specific alignments and annotations on multiple platforms.

3. *What Can UCSC Genome Browser Do?*

A. Basic Functionality of Genome Browser and BLAT Use The UCSC Genome Bioinformatics home page (http://genome.ucsc.edu/) contains general information and news announcing new features and software or data changes. A list of features available is displayed on the top and on the left column of the navigation bars. From there you can simply start the "Text Search" through "Genomes" or "Genome Browser" or a "Sequence Search" through the Blast-Like Alignment Tool (BLAT) tool (Figure 1.1).

i. Text Search This function featured at UCSC can be gene name, gene symbol, chromosome number, chromosome region, keywords, marker

The UCSC Home page: http://genome.ucsc.edu

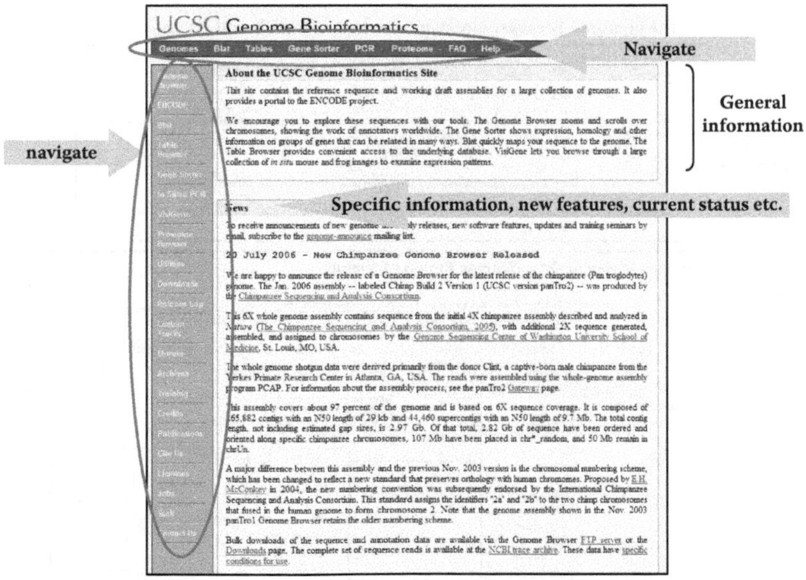
UCSC Genome Bioinformatics
Genomes Blat Tables Gene Sorter PCR Proteome FAQ Help
Navigate
About the UCSC Genome Bioinformatics Site
General information
navigate
Specific information, new features, current status etc.
29 July 2006 - New Chimpanzee Genome Browser Released

FIGURE 1.1. The UCSC Genome Bioinformatics home page. General information, news, software or data changes, as well as a list of available features are displayed.

identification number, GenBank submitter name, and so on. Several options offered with the text search are clade, genome species, date of assembly, and image width. By default, the search is set to vertebrate, human, the most recent assembly (i.e., March 2006), and 620 pixels width, respectively (Figure 1.2A). All the options, including figure image and tracks, can be configured from the pull-down menus (Figure 1.2B). Depending on the text search term used, the results page may appear in a number of different records. Users have to select the one with the correct gene symbol or name from the results page before entering the browser. If there appear to be multiple entries that are likely to be splice variants, the nucleotide range indicated at the end of the link may serve as a reference.

The Genome Viewer section features the diagrammatic representation of the genome, which corresponds to the available annotation tracks, for quick data finding. Depending on the stage of the assembly, the actual track options will increase over time. Currently, UCSC has the following tracks: (1) Mapping and Sequencing Tracks; (2) Phenotype and Disease Associations; (3) Genes and Gene Prediction Tracks; (4) mRNA and EST Tracks; (5) Expression and Regulation; (6) Comparative Genomics, Variation, and Repeats; (7) ENCODE Regions and Genes; (8) ENCODE Transcript Lev-

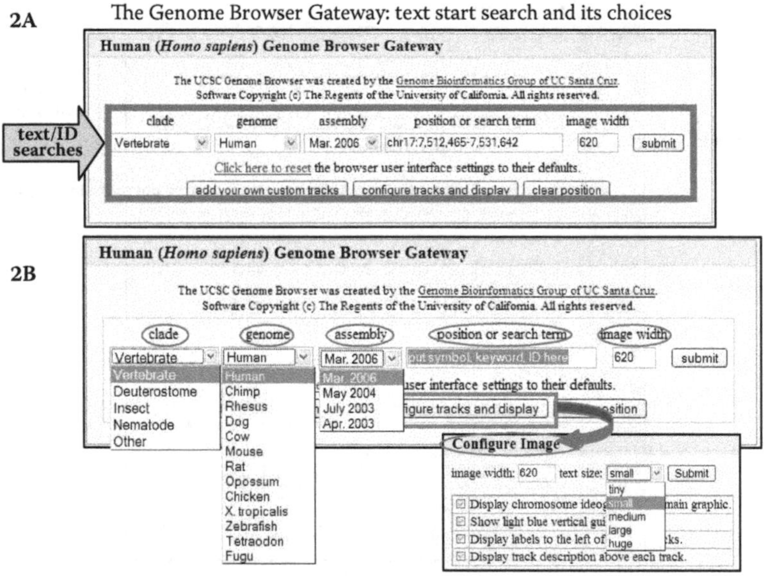
2A
The Genome Browser Gateway: text start search and its choices
Human (Homo sapiens) Genome Browser Gateway
text/ID searches
clade
genome
assembly
position or search term
image width
Vertebrate
Human
Mar. 2006
chr17:7,512,465-7,531,642
620
submit
add your own custom tracks
configure tracks and display
clear position
2B
Human (Homo sapiens) Genome Browser Gateway
Deuterostome
Insect
Nematode
Other
Chimp
Rhesus
Dog
Cow
Mouse
Rat
Opossum
Chicken
X. tropicalis
Zebrafish
Tetraodon
Fugu
May 2004
July 2003
Apr. 2003
Configure Image
image width: 620
text size: small
Submit
tiny
medium
large
huge

FIGURE 1.2. The Genome Browser Gateway: text search and its choices. Text search interface (panel A) and detailed start page choices (panel B) are illustrated.

els; (9) ENCODE Chromatin Immunoprecipitation; (10) ENCODE Chromosome, Chromatin, and DNA Structure; and ENCODE Comparative Genomics and ENCODE Variation (Figure 1.3). When the configuration is performed, the Genome Viewer display needs to be refreshed accordingly. Additionally, UCSC Genome Browser uses the same interface and display for each of the species listed; therefore, the software works similarly as well. However, different species will have different annotation tracks, again depending on the availability of data assembly.

To facilitate data viewing, the Genome Viewer page provides several options to make changes. You can use the buttons with the arrowhead indicators to walk left or right along the chromosome to the region of interest. The number of steps is proportional to the number of arrowheads (i.e., triple arrowheads for big steps and single ones for small steps). You can manipulate the image area up to ten-fold using the Zoom In/Zoom

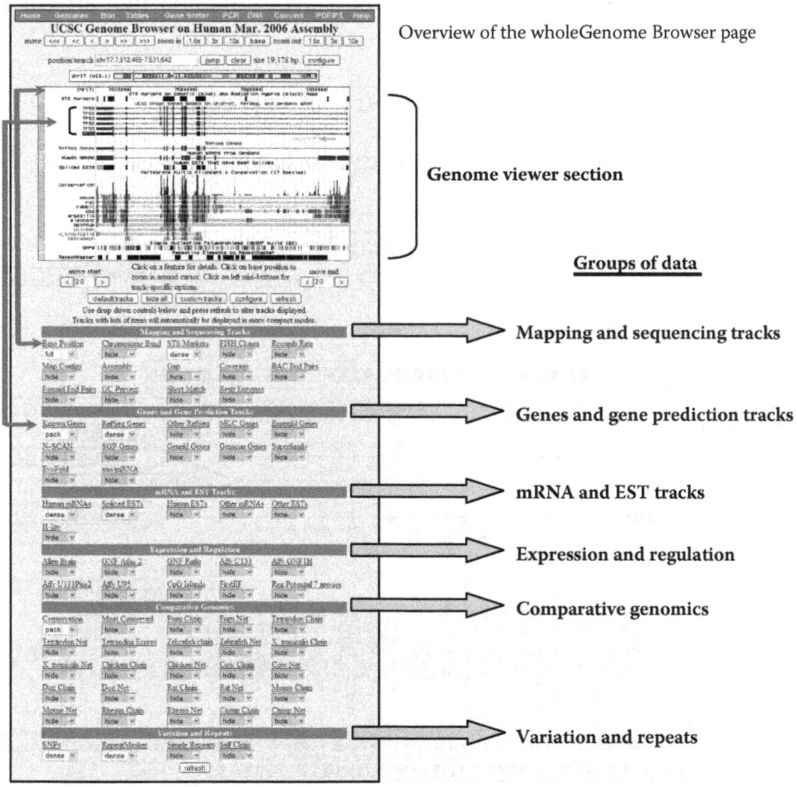
UCSC Genome Browser on Human Mar. 2006 Assembly
Overview of the wholeGenome Browser page
Genome viewer section
Groups of data
Mapping and sequencing tracks
Genes and gene prediction tracks
mRNA and EST tracks
Expression and regulation
Comparative genomics
Variation and repeats

FIGURE 1.3. Overview of the Genome Browser page. Sample diagrammatic representation of the genome corresponding to the available annotation tracks for quick data finding is illustrated.

Out buttons. If you select Base under Zoom In option, it will show you the nucleotides level right away. Alternatively, you can indicate a specific genome coordinate position in the Position box. By subtracting or adding certain base pairs of nucleotides to the coordinate position of the 5' or 3' end, you can retrieve the extra sequences. Alternatively, you can simply use the text search strategy by typing in gene name, gene symbol, or ID, etc. Another handy feature is the Automatic Zoom and Recenter Action — by positioning the mouse over the nucleotide backbone track at the very top, the browser will automatically recenter the image where you clicked, and zoom in threefold (Figure 1.4).

There are some visual cues to help interpret the graphical display on the genome viewer section:

1. Sequence tagged sites (STS) or SNPs are indicated by vertical tick marks.
2. Coding region exons are the tallest (full size) boxes.
3. The half-size boxes are 5' and 3' untranslated regions (UTRs).

Sample Genome Viewer

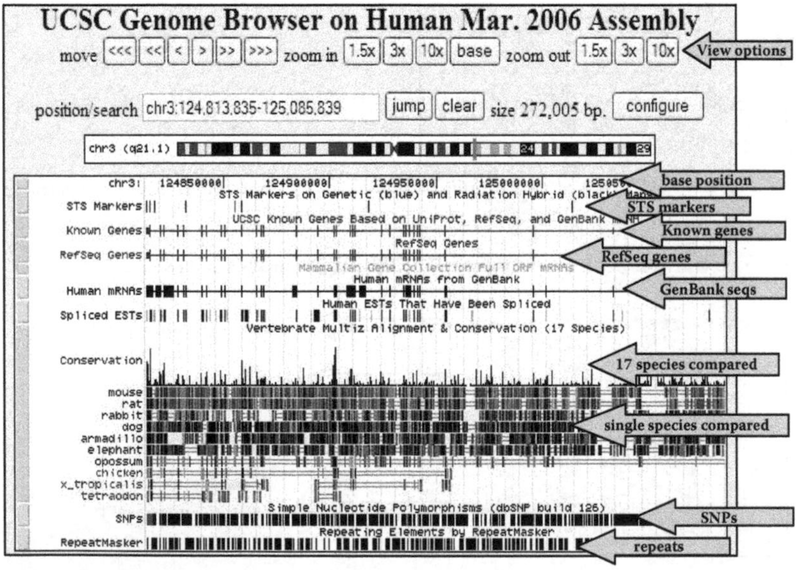
UCSC Genome Browser on Human Mar. 2006 Assembly
move
zoom in
1.5x 3x 10x base
zoom out
1.5x 3x 10x
View options
position/search chr3:124,813,835-125,085,839
jump clear
size 272,005 bp.
configure
base position
STS markers
Known genes
RefSeq genes
GenBank seqs
17 species compared
single species compared
SNPs
repeats
STS Markers
Known Genes
RefSeq Genes
Human mRNAs
Spliced ESTs
Conservation
mouse
rat
rabbit
dog
armadillo
elephant
opossum
chicken
x_tropicalis
tetraodon
RepeatMasker

FIGURE 1.4. Sample Genome Viewer image corresponding to Chr3:124,813,835-125,085,839. View options and annotation features are highlighted by the arrows.

4. Direction of the transcription of this coding unit is indicated by the little arrowheads, i.e., if the arrowheads point to the left, the gene is transcribed from the 5' UTR on the right side to the 3' UTR on the left and vice versa.

5. Degree of conservation is represented by the height of bars. Tall bars indicate the increased likelihood of an evolutionary relationship in that region (Figure 1.5).

For some tracks, colors have important meaning. For example, in the Known Genes track, the color Black indicates that there is a protein data bank (PDB) structure entry for this transcript. Dark blue corresponds to NCBI-Reviewed Sequence and light blue for Provisional Sequences. Types of SNPs are color-coded as well. If you are not sure about any specific representation, you can click on the label (hyperlink) of the track under each annotation for more descriptive information. Understanding these representations will help you to quickly grasp many of the features in any genomic region.

Other than viewing the genome display horizontally, the pull-down menu options for each individual track allow you to see data vertically. Several options, including Hide (completely removes the data from the image); Pack (each item is separate, but efficiently stacked); Squish (keep each item on a separate line, but the graphics are shrunk by 50% of their regular height); Dense (collapse all items into a single line), and Full (one item per line) are available even though by default, some tracks are On and

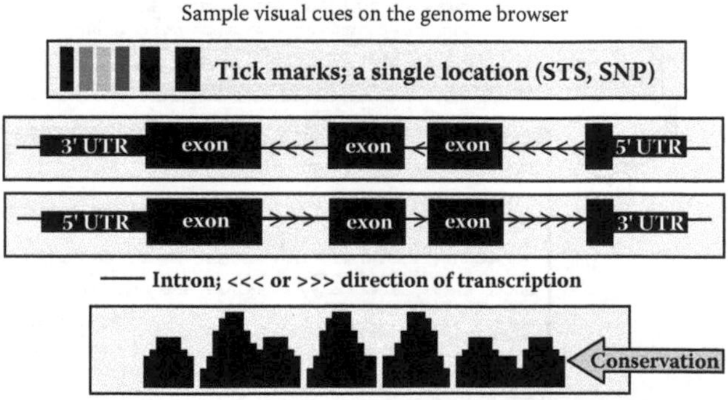
Sample visual cues on the genome browser
Tick marks; a single location (STS, SNP)
3' UTR
exon
exon
exon
5' UTR
5' UTR
exon
exon
exon
3' UTR
Intron; <<< or >>> direction of transcription
Conservation

FIGURE 1.5. Sample visual cues on the Genome Browser. Various data objects representing sequence tagged sites (STS), simple nucleotide polymorphisms (SNPs), exons, 5' and 3' Untranslated Regions (UTR), direction of transcription, and the degree of conservation are illustrated.

others are Hidden. However, whenever any changes are made, you need to click the Refresh button to reload the display. The UCSC Genome Browser retains whatever changes you made until you clear them.

Finally, to learn more about the object (e.g., known genes, conservation, or SNPs, etc.) you are researching, you can position the mouse over that line, click, and a new Web page will appear (Figure 1.6). Many important details, including sequences, microarray data, mRNA secondary structure, protein domain structure, homologues in other species, gene ontology descriptions, mRNA descriptions and pathways, etc., are provided in the Page Index box. Again, not all the genes have the same levels of detail, and not every species has all the information.

ii. Sequence Search This function of the UCSC is called BLAST-Like Alignment Tool or BLAT. BLAT searches require an index of the sequences in the database consisting of all the possible unique 11-oligomer sequences in the genome (or 4-mers for protein sequences). Just as you can quickly scan a book index to find the correct word, BLAT scans the index for matching 11-mers or 4-mers and builds the rest of the match out from there. BLAT works best with high identity and greater similarity (>95% and >21 bp in nucleotide and >80% and >20-mers in protein, respectively).

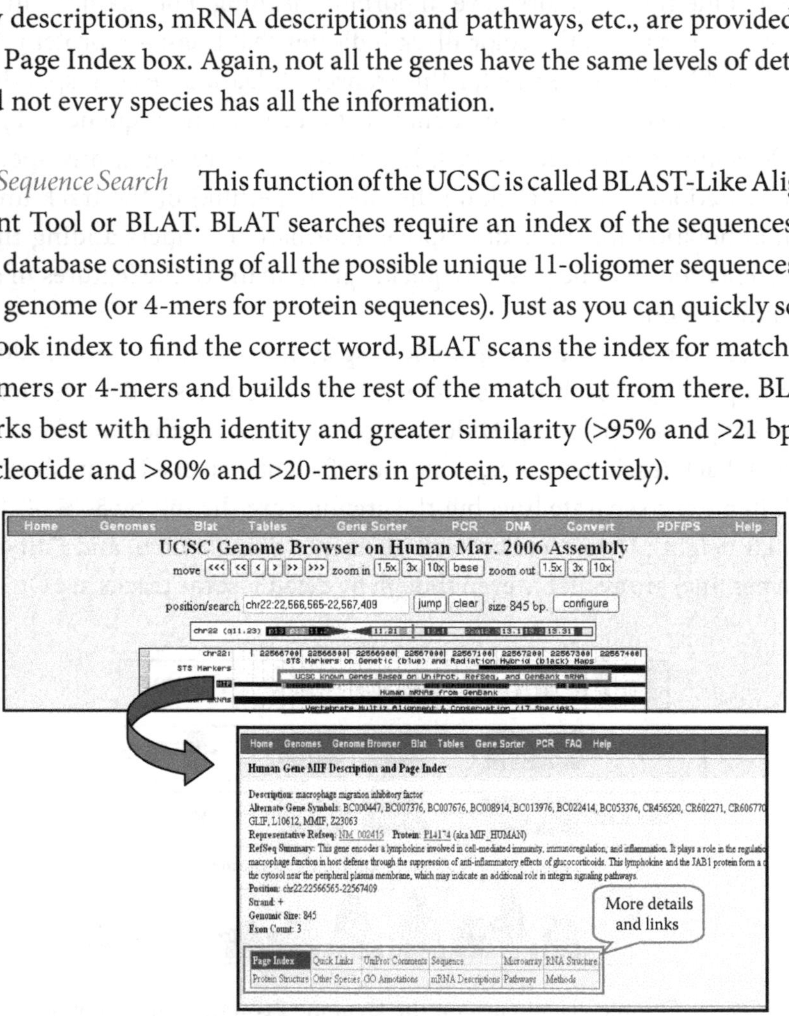
Home Genomes Blat Tables Gene Sorter PCR DNA Convert PDF/PS Help
UCSC Genome Browser on Human Mar. 2006 Assembly
move zoom in 1.5x 3x 10x base zoom out 1.5x 3x 10x
position/search chr22:22,566,565-22,567,409 jump clear size 845 bp. configure
Home Genomes Genome Browser Blat Tables Gene Sorter PCR FAQ Help
Human Gene MIF Description and Page Index
Description: macrophage migration inhibitory factor
Position: chr22:22566565-22567409
Strand: +
Genomic Size: 845
Exon Count: 3
More details and links
Page Index Quick Links UniProt Comments Sequence Microarray RNA Structure
Protein Structure Other Species GO Annotations mRNA Descriptions Pathways Methods

FIGURE 1.6 Sample detailed viewer objects. A new Web page is opened when the object is clicked providing more details and links illustrated in the lower panel.

BLAT can be used to: (1) find the genomic coordinates of mRNA or protein within a given assembly, (2) determine the exon/intron structure of a gene, (3) display a coding region within a full-length gene, (4) isolate an EST of special interest as its own tract, (5) search for gene family members, and (6) find human homologs of a query from another species, etc. Similar to the text search, there are a few parameters (genome, assembly, query type, sort output, and output type) on the BLAT Search Genome page that you can change or specify. If you select BLAT'S Guess under query type, the BLAT tool will automatically guess whether you have entered nucleotides or amino acids, and retrieve data accordingly.

BLAT allows you to paste up to 25,000 bases, 10,000 amino acids, and up to a total of 25 sequences in the common FASTA format (i.e., start with the greater than symbol (>) followed by the gene identification number or reference protein accession number or any name followed by the sequences). Alternatively, you can upload your sequences using the File Upload function. The I'm Feeling Lucky button will take you to the position of your best match right away, in the Genome Viewer (Figure 1.7).

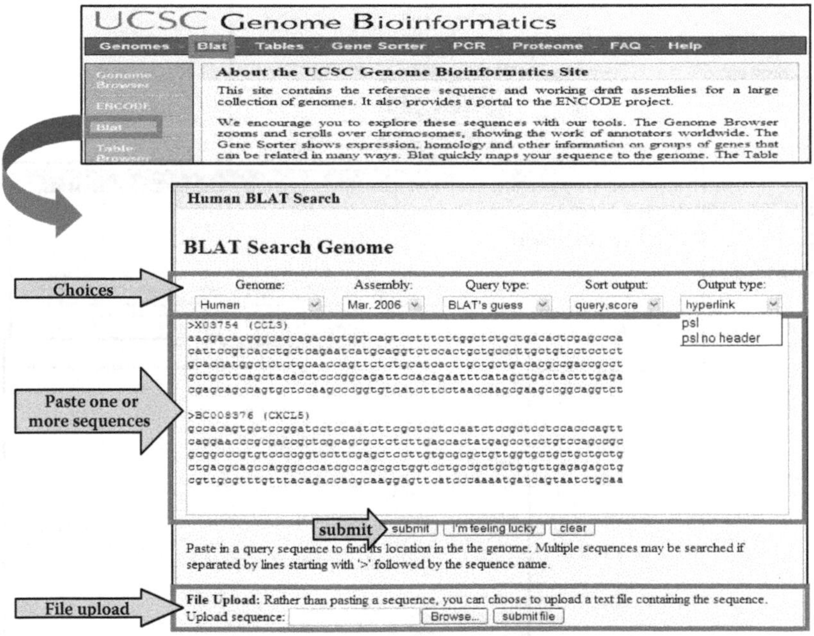
UCSC Genome Bioinformatics
Genomes
Blat
Tables
Gene Sorter
PCR
Proteome
FAQ
Help
About the UCSC Genome Bioinformatics Site
This site contains the reference sequence and working draft assemblies for a large collection of genomes. It also provides a portal to the ENCODE project.
Human BLAT Search
BLAT Search Genome
Choices
Genome:
Assembly:
Query type:
Sort output:
Output type:
Human
Mar. 2006
BLAT's guess
query,score
hyperlink
psl
psl no header
>X03754 (CCL3)
Paste one or more sequences
>BC008376 (CXCL5)
submit
submit
I'm feeling lucky
clear
Paste in a query sequence to find its location in the the genome. Multiple sequences may be searched if separated by lines starting with '>' followed by the sequence name.
File upload
File Upload: Rather than pasting a sequence, you can choose to upload a text file containing the sequence.
Upload sequence:
Browse...
submit file

FIGURE 1.7 BLAT tool overview and interface. The choices for Blat search, sequence input text box and file upload functions are highlighted in arrows. A FASTA format (i.e., start with the greater than symbol [>] followed by gene identification number or reference protein accession number or any name followed by the sequences) is required as shown here.

BLAT results are sorted by query and descending order of score and are displayed either in the browser (hyperlink) or the PSL (ps Layout program) format, which is a differently structured, text-based output. The Browser link will take you to the location of the match in the Genome Viewer. A new line with "YOUR SEQUENCE FROM BLAT SEARCH" will appear on the top, and the name of your query sequence will be highlighted on the genomic viewer. All the other graphical displays are similar to the result of text search as described. The DETAILS link will give you a new page with sequence information. If mRNA sequences are used to Blat genome, color-coded exon/intron structures can be identified, and nucleotide-for-nucleotide alignments displayed (Figure 1.8).

B. Table Browser Use Table Browser is a powerful tool for filtering, manipulating, and downloading data in a very customized and flexible manner, which is not possible with the Genome Browser. The Table browser allows you to:

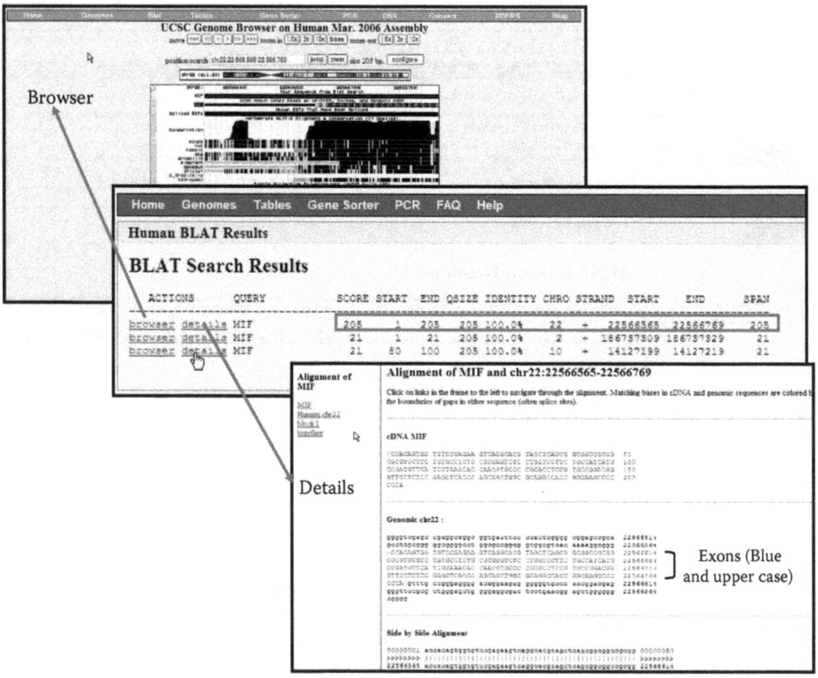
UCSC Genome Browser on Human Mar. 2006 Assembly
Browser
Home Genomes Tables Gene Sorter PCR FAQ Help
Human BLAT Results
BLAT Search Results
ACTIONS QUERY SCORE START END QSIZE IDENTITY CHRO STRAND START END SPAN
browser details MIF 205 1 205 205 100.0% 22 + 22566565 22566769 205
browser details MIF 21 1 21 205 100.0% 2 + 186737309 186737329 21
browser details MIF 21 80 100 205 100.0% 10 + 14127199 14127219 21
Alignment of MIF
Alignment of MIF and chr22:22566565-22566769
cDNA MIF
Genomic chr22 :
Side by Side Alignment
Details
Exons (Blue and upper case)

FIGURE 1.8 Sample BLAT search result and alignment details. BLAT results sorted by query and descending order of score are displayed here. Alignment details are available through the links.

1. Retrieve the DNA sequence data or annotation data underlying Genome Browser tracks for the entire genome, a specified coordinate range, or a set of accessions.
2. Apply a filter to set constraints on field values included in the output.
3. Generate a custom track and automatically add it to your session so that it can be graphically displayed in the Genome Browser.
4. Conduct both structured and free-form SQL queries on the data.
5. Combine queries on multiple tables or custom tracks through an intersection or union and generate a single set of output data.
6. Display basic statistics calculated over a selected dataset.
7. Display the schema for the table and list all other tables in the database connected to the table.
8. Organize the output data into several different formats for use in other applications, spreadsheets, or databases. Tasks such as obtaining a list of all SNPs or nonsynonymous coding variations in a given gene and all the known genes located on a certain chromosome, etc., can be easily performed.

The Table Browser interface is similar to other interfaces in the Genome Browser and related tools. Options, including clade, genome, assembly, and which data table you wish to search in, need to be specified first. A data table consists of three components (group, track, and table). Currently, available options under group are the following: Mapping and Sequencing Tracks; Phenotype and Disease Associations; Genes and Gene Prediction Tracks; Expression and Regulation; mRNA and EST Tracks; Comparative Genomics; Variation and Repeats; ENCODE Regions and Genes; ENCODE Transcription Levels; ENCODE Chromatin Immunoprecipitation; ENCODE Chromosome, Chromatin, and DNA Structure; ENCODE Comparative Genomics; ENCODE Variation; and All Tracks and All Tables. Once the group is specified, the track and table menus automatically change to show the tracks in that group and the tables in that track. The available numbers of tracks and tables are varied within each group. Some may have only one table for a track, and others have multiple tables. In addition, the querying options below each table's menu are varied as well because some output format options and filtering options are appropriate only for certain types of table. If you have any questions about the type of data in the selected table, you can click the Describe

Table Schema button, which leads to a description page that explains the field names, an example value, data type, and definition of each field of the database table. If the database contains other tables that are related to the selected table by a field with shared values, those related tables are listed. If the selected table is the main table for a track, then the track description text is included (Figure 1.9).

In the Table Browser, you can search the entire genome, an ENCODE region, or a specific chromosomal location. ENCODE is a project led by the National Human Genome Research Institute (NGRHI) to identify and characterize all functional elements in human genome sequences. Either with a specific chromosome coordinate range (e.g., chr22:10000000-20000000) in the textbox or with a gene name, the Table Browser will find the location for you. Alternatively, you can copy and paste in a list of names or accession numbers or upload a file.

As mentioned previously, one of the powerful features of the Table Browser is the ability to filter for different parameters of the fields in the table data on various criteria, because it is a form-based SQL query. You can also use the "free form query" to type in your own custom filter. However, knowing SQL is not essential to use this filter form. If there were more tables that the track was based on or related to, those will show up on this page and you can filter on those too. Similar to the Genome Browser, changes and choices made in the Table Browser will be "remembered" until further changes are made or cleared.

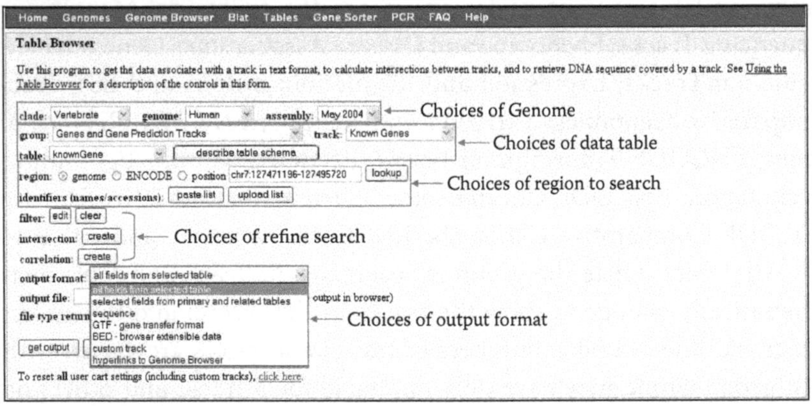
Home Genomes Genome Browser Blat Tables Gene Sorter PCR FAQ Help
Table Browser
Use this program to get the data associated with a track in text format, to calculate intersections between tracks, and to retrieve DNA sequence covered by a track. See Using the Table Browser for a description of the controls in this form.
clade: Vertebrate genome: Human assembly: May 2004
Choices of Genome
group: Genes and Gene Prediction Tracks track: Known Genes
table: knownGene describe table schema
Choices of data table
region: genome ENCODE position chr7:127471196-127495720 lookup
identifiers (names/accessions): paste list upload list
Choices of region to search
filter: edit clear
intersection: create
correlation: create
Choices of refine search
output format: all fields from selected table
output file:
output in browser)
file type return
selected fields from primary and related tables
sequence
GTF - gene transfer format
BED - browser extensible data
custom track
hyperlinks to Genome Browser
Choices of output format
get output
To reset all user cart settings (including custom tracks), click here.

FIGURE 1.9 Sample Table Browser interface. A list of choices for genome, data table, region to search, refine search, and output format are highlighted in boxes.

The intersection function of the Table Browser allows you to find if two datasets have any overlap. For example, you can find out if there is any chromosomal location overlap between the "known genes" dataset and the "simple repeats" dataset. You can specify the group, annotation track, and table that you wish to intersect with the table that you selected on the main page. You can choose complete, none or percentage overlap, and intersection or union. There are also options available regarding how you want to see the data with base-pair-wise intersections and complements. However, the intersection tool can be used only on positional tables containing data associated with specific locations in the genome, such as mRNA alignments, gene predictions, cross-species alignments, and custom annotation tracks. Positional tables can be further subdivided into several categories based on the type of data they describe. For example, alignment data can be best described by using a block structure to represent each element. Other tables require only start and end coordinate data for each element. Some tables specify a translation start and end in addition to the transcription start and end. Some tables contain strand information, and others do not. Most tables, but not all, specify a name for each element. Based on the format of the data described by a table, different query and output formatting options may be offered.

The last feature in the Table Browser is the correlation tool, which was added to the Table Browser in August 2005 and is still under development. It is available for data tables that contain genomic positions and computes a simple linear regression on the scores in two datasets. If a dataset does not contain a score for each base position, then the Table Browser assigns a score of 1 for each position covered by an item in the table, and 0 otherwise. The Table Browser computes the linear regression quickly and then displays several graphs for visualizing the correlation, as well as summary statistics, including the correlation coefficient "r." When datasets and parameters are chosen with some forethought, the correlation feature is a powerful tool to determine, for example, if there is any correlation between GC content and chromosome structure, or between certain types of genes and repeats between two datasets.

The Table Browser offers several choices to output your data. "All fields from selected table" and "Selected fields from primary and related tables" output formats provide you a tab-delineated text file that can be later used in a word processing or spreadsheet program. "Sequence" format provides you the DNA or protein sequence in a FASTA format. Gene Transfer Format (GTF) and the Browser Extensible Data format (BED) provide database formats to be used in other programs and databases. Custom Track output format creates an annotation track of your query in the Genome Browser and in the Table Browser, for further study. This newly created annotation track can be viewed and searched just as any other annotation track. We will focus more on how to create and use the custom annotation tracks in the next section. "Hyperlink to Genome Browser" output provides a list of hyperlinks of the data positions in the Genome Browser. A summary of the specified data is provided via the Summary/Statistics button, which provides a general idea of the number of genes you are working with.

The output file can be saved on your computer when a filename is entered in the Output File textbox; without the filename, the output is displayed in the browser. The exception is the custom track output, which automatically sends you to a separate browser page no matter what is in the textbox.

C. Creating and Using Custom Track The Genome Browser introduced previously provides aligned annotation tracks that have been computed at UCSC or have been provided by outside collaborators. However, custom annotation tracks can be created from the Table Browser searches or your own data, and be viewed and searched as any other standard annotation track. Notably, custom tracks are only persistent for 8 h. Any Table Browser search you have created a custom track from needs to be redone after 8 h if you have not downloaded the file. More information about the custom-annotated track can be provided through a URL as well.

The Custom Track hyperlink on the UCSC home page allows you to create custom tracks of your own data. Detailed instructions on executing the task are provided in Displaying Your Own Annotations in the Genome Browser (http://genome.ucsc.edu/goldenPath/help/customTrack.html). Many custom tracks have been created by members of the scientific community, who have made them available for public viewing and querying from the UCSC Genome home page. The list of submitted data and contributors from custom tracks can be accessed from the following link: (http://genome.ucsc.edu/ goldenPath/customTracks/custTracks.html). We will provide a step-by-step tutorial in Part II of this chapter.

D. Introduction to the Gene Sorter Gene Sorter is a search tool that takes a gene of interest and lists other genes in the genome sorted by a similarity type to a reference gene. Similar to other tools available at UCSC, you are given several choices including genome species, assembly, search term (gene name, accession number, keyword, etc.), and by what criteria you wish to seek for the precalculated similarity to the gene in question. Options include similarity in expression patterns, protein homology (BLASTP), Pfam domain structure, gene distance, name, gene ontology similarity, and others.

Gene Sorter allows you to configure the display to add, subtract, and change data columns even though some columns of data are checked by default. Some data columns can be configured further for more detailed information. For example, under the expression data column, you can configure it to show all tissues or a selected set, or have the values absolute or as a ratio of the mean level of expression, or to change the colors to signify the level of expression. Similarly, you can hide all data columns, which would be useful if you wanted to choose specific columns and eliminate default or other previous choices. You can choose to show all data columns or, if you need to, return to the default columns. At the end, you can save the data column configurations you have chosen for future viewing (Figure 1.10).

A new feature (custom columns) that has been added is the ability to add columns of data that are user generated. A straightforward and simple instruction in the Help section (http://genome.ucsc.edu/goldenPath/help/

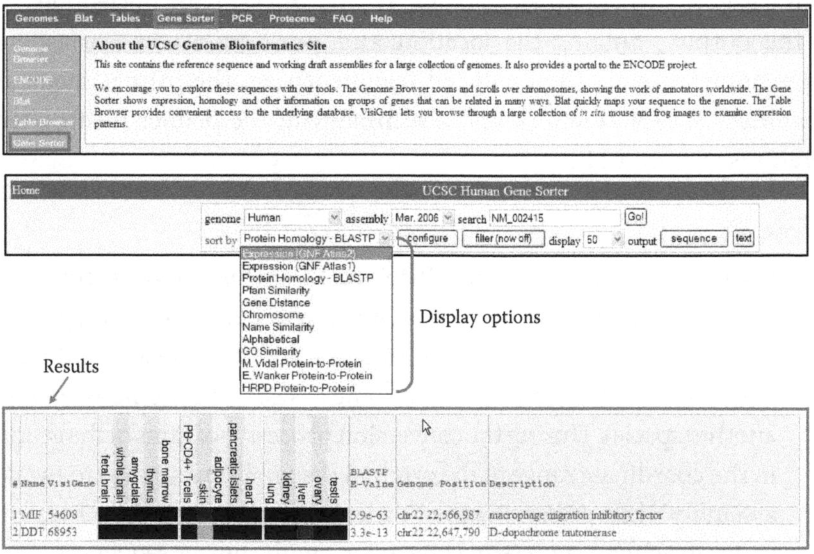
Genomes Blat Tables Gene Sorter PCR Proteome FAQ Help
About the UCSC Genome Bioinformatics Site
This site contains the reference sequence and working draft assemblies for a large collection of genomes. It also provides a portal to the ENCODE project.
UCSC Human Gene Sorter
genome Human assembly Mar. 2006 search NM_002415 Go!
sort by Protein Homology - BLASTP configure filter (now off) display 50 output sequence text
Expression (GNF Atlas1)
Protein Homology - BLASTP
Pfam Similarity
Gene Distance
Chromosome
Name Similarity
Alphabetical
GO Similarity
M. Vidal Protein-to-Protein
E. Wanker Protein-to-Protein
HRPD Protein-to-Protein
Display options
Results
Name VisiGene
fetal brain
whole brain
amygdala
thymus
bone marrow
PB-CD4+ Tcells
skin
adipocyte
pancreatic islets
heart
lung
kidney
liver
ovary
testis
BLASTP E-Value Genome Position Description
1 MIF 54608 5.9e-63 chr22 22,566,987 macrophage migration inhibitory factor
2 DDT 68953 3.3e-13 chr22 22,647,790 D-dopachrome tautomerase

FIGURE 1.10 Sample gene sorter interface. The UCSC homepage with gene sorter navigator bar, displayed options, and sample result are illustrated here.

hgNearHelp.html) will guide you in uploading and formatting your custom column. The Gene Sorter tool allows you to filter every column of data, including those that you have chosen to display in the results and those you have not. For example, you could filter for only genes with gene names that have a specific text string or genes that have a certain minimum or maximum level of expression. For many data column filters, you can paste in or upload a list of filters if you have a large filter list. Once you choose your filters, you can list the resulting genes by name in alphabetical order. Therefore, you can get an idea about how large your results list will be, decide if you need to tighten or loosen your filters, or make sure you are getting the genes you are expecting. All the filters can be saved for further application.

E. In Silico PCR It is a tool used to search a genomic sequence database with a pair of PCR primers. Similar to the BLAT algorithm, it uses an indexing strategy to facilitate the task. To execute the In Silico function, you first need to specify the genome and assembly using the drop-down menus. Currently, not all the genomes are available in this tool, but new ones are added all the time. A minimum of 15 oligonucleotide forward and reverse primers are needed to execute the search. This tool does not handle ambiguous bases, so it is not possible to use "N" to represent any nucleotide.

The output contains the location and position of the corresponding genomic stretch, the predicted fragment size, the primer sequences submitted in capital letters, and a summary of the primer melting temperatures (Figure 1.11).

F. Other Utilities

Batch Coordinate Conversion (liftOver) — This program converts a large number of genome coordinate ranges between assemblies. It is useful for locating the position of a feature of interest in a different release of the same genome or (in some cases) in a genome assembly of another species. During the conversion process, portions of the genome in the coordinate range of the original assembly are aligned to the new assembly while preserving their order and orientation. In general, it is easier to achieve successful conversions with shorter sequences.

DNA Duster — This program removes formatting characters and other non-sequence-related characters from an input sequence. It

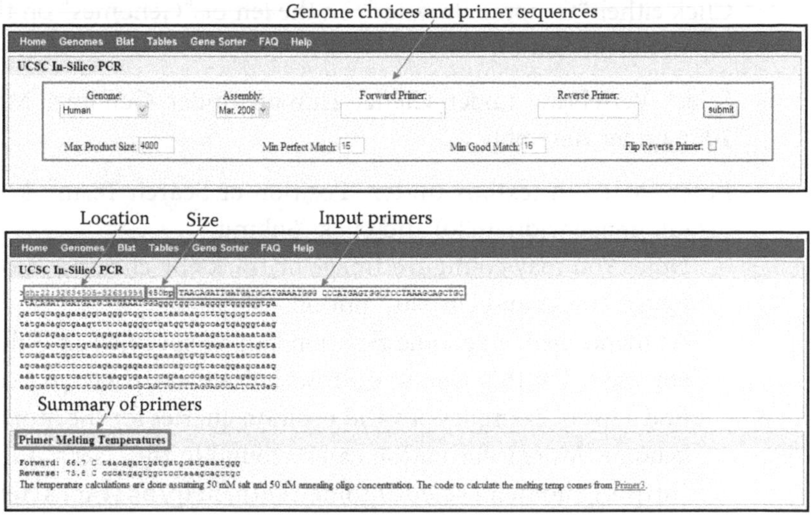
Genome choices and primer sequences
Location
Size
Input primers
Summary of primers
UCSC In-Silico PCR
Primer Melting Temperatures

FIGURE 1.11 Sample view of the "In Silico PCR" interface. Input interface for "In Silico PCR" function is shown on the top panel. A detail description of results with the reverse and forward primers in capital letters, the size, and sequences

offers several configuration options for the output format, including spaces, line breaks, and translated protein, etc.

Protein Duster — Similar to DNA Duster, this program removes formatting characters and other non-sequence-related characters from an input sequence. However, the configuration options for the output format offered are limited.

Phylogenetic Tree Gif Maker — This program creates a gif image from the phylogenetic tree specification given. It offers several configuration options for branch lengths, normalized lengths, branch labels, legend, etc.

Part II Step-By-Step Tutorial

1. Basic Functionality of Genome Browser and BLAT Use

Here, we demonstrate how to perform a basic text and sequence (BLAT) search to retrieve genomic, mRNA, and protein sequences, using the UCSC Genome Browser.

1. Text search

    - Go to the UCSC home page (http://genome.ucsc.edu).

- Click either "Genome Browser" on the left or "Genomes" on the top of the navigation bar.
- Select Vertebrate under Clade; Human under Genome; Mar. 2006 under Assembly.
- Enter "MIF" in textbox under "Position or Search Term"; keep default image width (620) and click "Submit."
  - Note: You may configure image or tracks by clicking "Configure Tracks and Display" button.
  - As mentioned, a genome position can be specified in different ways. On this Genome Browser Gateway page, you can find a list of examples of valid position queries for the human genome. More information can be found in the User's Guide (http://genome.ucsc.edu/goldenPath/help/hgTracksHelp.html).
- Select "MIF (NM_002415) at chr22:22566565-22567409 — macrophage migration inhibitory factor" under Known Gene.
- Click arrows to move left or right; click 1.5x, 3x, or 10x to zoom in or out; click "Base" to zoom in to the nucleotide level.
  - Note: MIF gene is in the forward direction (i.e., arrowheads pointing to the right indicating that the 5' is on the left-hand side and 3' is on the right-hand side).
- Click "DNA" on the top of the navigation bar.
  - A "GET DNA for" page will appear providing options to retrieve regions of sequence.
  - Enter "1000" in both the textboxes to retrieve additional bases from upstream and downstream of the gene.
  - Select All uppercase and Mask repeats to lowercase.
- Click "get DNA."
  - A FASTA output returns with description on the header and sequences followed (total of 2845 bp, Figure 1.12).
- Click the Back arrow on the Web browser to go back to the "Get DNA for" page.
  - Select Extended case/color options.
  - Select Toggle Case and Bold under Known Genes, Bold under SNPs, and Underline under RepeatMasker.
  - Enter "255" under SNPs and Red,

```
>hg18_knownGene_NM_002415 range=chr22:22565565-22568409 5'pad=0 3'pad=0 revComp=FALSE strand=+ repeatMasking=lower
GCCACGAGGGTCCCACAGGCATGGGTGTCCTTCCTATATCACATGGCCTT
CACTGAGACTGGTATATGGATTGCACCTATCAGAGACCAAGGACAGGACC
TCCCTGGAAATCTCTGAGGACCTGGCCTGTGATCCAGTTGCTGCCTTGTC
CTCTTCCTGCTATGTCATGGCTTATCttctttcacccattcattcattca
ttcattcaGCAGTATTAGTCAATGTCTCTTGTATGCCTGGCACCTGCTAG
ATGGTCCCCGAGTTTACCATTAGTGGAAAAGACATTTAAGAAATTCACCA
AGGGCTCTATGAGAGGCCATACACGGTGGACCTGACTAGGGTGTGGCTTC
CCTGAGGAGCTGAAGTTGCCCAGAGGCCCAGAGAAGGGGAGCTGAGCACG
TTTGAACCACTGAACCTGCTCTGGACCTCGCCTCCTTCCCTTCGGTGCCT
CCCAGCATCCTATCCTCTTTAAAGAGCAGGGGTTCAGGGAAGTTCCCTGG
ATGGTGATTCGCAGGGGCAGCTCCCCTCTCACCTGCCGCGATGACTACCC
CGCCCCATCTCAAACACACAAGCTCACGCATGCGGGACTGGAGCCCTTGA
GGACATGTGGCCCAAAGACAGGAGGTACAGGGGCTCAGTGCGTGCAGTGG
AATGAactgggcttcatctctggaagggtaaggggccatcttccgggttc
accgccgcatccccacccccggcacagcgcctcctggcgactaacatcgg
tgacttagtgaaAGGACTAAGAAAGACCCGAGGCGAGGCCGGAACAGGCC
GATTTCTAGCCGCCAAGTGGAGAACAGGTTGGAGCGGTGCGCCGGGCTTA
GCGGCGGTTGCTGGAGGAACGGGCGGAGTCGCCCAGGGTCCTGCCCTGCG
GGGGTCGAGCCGAGGCAGGCGGTGACTTCCCCACTCGGGGCGGAGCCGCA
GCCTCGCGGGGGCGGGGCCTGGCGCCGGCGGTGGCGTCACAAAAGGCGGG
ACCACAGTGGTGTCCGAGAAGTCAGGCACGTAGCTCAGCGGCGGCCGCGG
CGCGTGCGTCTGTGCCTCTGCGCGGGTCTCCTGGTCCTTCTGCCATCATG
CCGATGTTCATCGTAAACACCAACGTGCCCCGCGCCTCCGTGCCGGACGG
GTTCCTCTCCGAGCTCACCCAGCAGCTGGCGCAGGCCACCGGCAAGCCCC
CCCAGGTTTGCCGGGAGGGGACAGGAAGAGGGGGGTGCCCACCGGACGAG
GGGTTCCGCGCTGGGAGCTGGGGAGGCGACTCCTGAACGGAGCTGGGGGG
CGGGGCGGGGGGAGGACGGTGGCTCGGGCCCGAAGTGGACGTTCGGGGCC
CGACGAGGTCGCTGGGGCGGGCTGACCGCGCCCTTTCCTCGCAGTACATC
GCGGTGCACGTGGTCCCGGACCAGCTCATGGCCTTCGGCGGCTCCAGCGA
GCCGTGCGCGCTCTGCAGCCTGCACAGCATCGGCAAGATCGGCGGCGCGC
AGAACCGCTCCTACAGCAAGCTGCTGTGCGGCCTGCTGGCCGAGCGCCTG
CGCATCAGCCCGGACAGGTACGCGGAGTCGCGGAggggcgggggaggggc
ggcggcgcggccaggcccgggACTGAGCCACCCGCTGAGTCCGGCCTC
CTCCCCCCGCAGGGTCTACATCAACTATTACGACATGAACGCGGCCAATG
TGGGCTGGAACAACTCCACCTTCGCCTAAGAGCCGCAGGGACCCACGCTG
TCTGCGCTGGCTCCACCCGGGAACCCGCCGCACGCTGTGTTCTAGGCCCG
CCCACCCCAACCTTCTGGTGGGGAGAAATAAACGGTTTAGAGACTAGGAG
TGCCTCGGGGTTCCTTGGCTTGCGGGAGGAATTGGTGCAGAGCCGGGATA
```

FIGURE 1.12 Sample view of MIF gene search using the BLAT function. A FASTA output with description and sequences are illustrated.

- Click "Submit."
- A FASTA output returns with description on the header and sequences follow (total of 2845 bp). The red color indicates the SNP location; Toggle case and bold indicates the exons' location (total of three exons); and underline for the Repeat-Masker (Figure 1.13).

- Click Back arrow on the Web browser three times to return to the UCSC Genome Browser.
- Click highlighted "MIF" under "UCSC Known Genes based on UniProt, RefSeq, and GeneBank mRNA."
  - The "Human Gene MIF Description and Page Index" page will appear.
  - Select Sequence in the Page Index box.
  - Select "Genomic (chr22:22,566,565-22,567,409)" under sequence. To specify options for sequence retrieval region:
    - Select Promoter/upstream and enter 500 in the Bases box.
    - Select 5' UTR exons, CDS exons, 3' UTR exons, and introns.
    - Select Downstream and enter 500 in the Bases box.

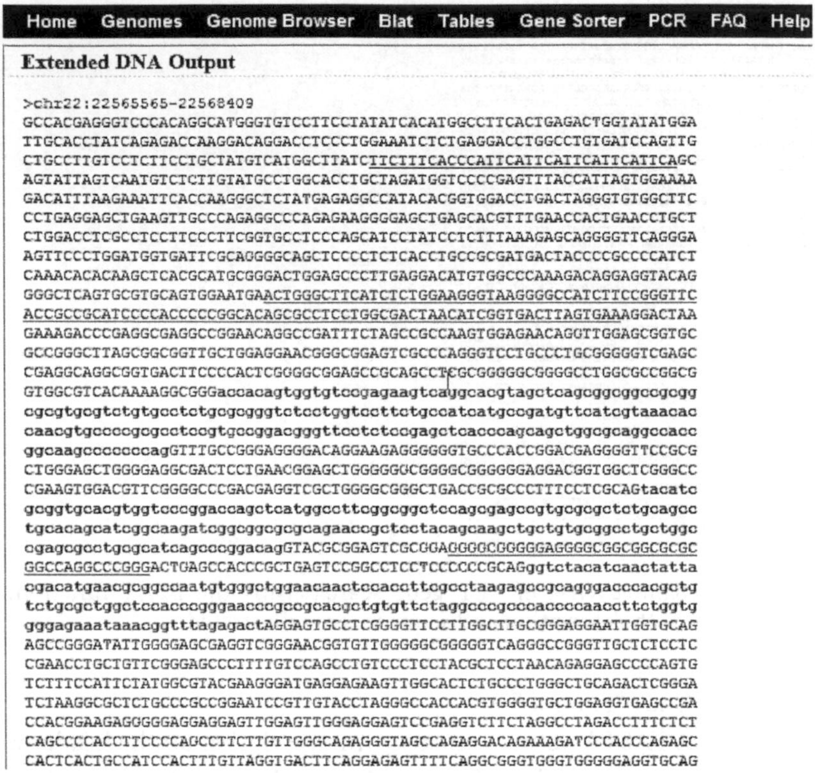
Home Genomes Genome Browser Blat Tables Gene Sorter PCR FAQ Help
Extended DNA Output
>chr22:22565565-22568409

FIGURE 1.13 Sample view of MIF gene search using the BLAT function and Extended case/color options. A FASTA output with sequences where SNPs are highlighted in red, exons are displayed in toggle case and bold, and RepeatMasker underlined.

    - Select One FASTA record per gene.
  - To specify options for sequence formatting:
    - Select Exons in uppercase and everything else in lowercase.
- Click "Submit."
  - A FASTA output returns with description on the header and sequences follow (total of 1845 bp) (Figure 1.14).

2. Sequence (BLAT) Search

- From the UCSC home page, select "Blat" on the navigation bar.
- Select Human under Genome; Mar. 2006 under Assembly; BLAT's guess under Query type; Query, Score under Sort output, and Hyperlink under Output type.

```
>hg18_knownGene_NM_002415 range=chr22:22566065-22567909 5'pad=0 3'pad=0 revComp=FALSE strand=+ repeatMasking=none
atggtgattcgcaggggcagctcccctctcacctgccgcgatgactaccc
cgccccatctcaaacacacaagctcacgcatgcgggactggagcccttga
ggacatgtggcccaaagacaggaggtacaggggctcagtgcgtgcagtgg
aatgaactgggcttcatctctggaagggtaaggggccatcttccgggttc
accgccgcatccccacccccggcacagcgcctcctggcgactaacatcgg
tgacttagtgaaaggactaagaaagacccgaggcgaggccggaacaggcc
gatttctagccgccaagtggagaacaggttggagcggtgcgccgggctta
gcggcggttgctggaggaacgggcggagtcgcccagggtcctgccctgcg
ggggtcgagccgaggcaggcggtgacttccccactcggggcggagccgca
gcctcgcgggggcggggcctggcgccggcggtggcgtcacaaaaggcggg
ACCACAGTGGTGTCCGAGAAGTCAGGCACGTAGCTCAGCGGCGGCCGCGG
CGCGTGCGTCTGTGCCTCTGCGCGGGTCTCCTGGTCCTTCTGCCATCATG
CCGATGTTCATCGTAAACACCAACGTGCCCCGCGCCTCCGTGCCGGACGG
GTTCCTCTCCGAGCTCACCCAGCAGCTGGCGCAGGCCACCGGCAAGCCCC
CCCAGgtttgccgggaggggacaggaagagggggggtgcccaccggacgag
gggttccgcgctgggagctggggaggcgactcctgaacggagctgggggg
cggggcgggggggaggacggtggctcgggcccgaagtggacgttcggggcc
cgacgaggtcgctggggcgggctgaccgcgccctttcctcgcagTACATC
GCGGTGCACGTGGTCCCGGACCAGCTCATGGCCTTCGGCGGCTCCAGCGA
GCCGTGCGCGCTCTGCAGCCTGCACAGCATCGGCAAGATCGGCGGCGCGC
AGAACCGCTCCTACAGCAAGCTGCTGTGCGGCCTGCTGGCCGAGCGCCTG
CGCATCAGCCCGGACAGgtacgcggagtcgcggaggggcgggggaggggc
ggcggcgcgcggccaggcccgggactgagccacccgctgagtccggcctc
ctccccccgcagGGTCTACATCAACTATTACGACATGAACGCGGCCAATG
TGGGCTGGAACAACTCCACCTTCGCCTAAGAGCCGCAGGGACCCACGCTG
TCTGCGCTGGCTCCACCCGGGAACCCGCCGCACGCTGTGTTCTAGGCCCG
CCCACCCCAACCTTCTGGTGGGGAGAAATAAACGGTTTAGAGACTaggag
tgcctcggggttccttggcttgcgggaggaattggtgcagagccgggata
ttggggagcgaggtcgggaacggtgttgggggcgggggtcagggccgggt
tgctctcctccgaacctgctgttcgggagccctttgtccagcctgtccc
tcctacgctcctaacagaggagccccagtgtctttccattctatggcgta
cgaagggatgaggagaagttggcactctgccctgggctgcagactcggga
tctaaggcgctctgcccgccggaatccgttgtacctagggccaccacgtg
gggtgctggaggtgagccgaccacggaagagggggaggaggagttggagt
tgggaggagtccgaggtcttctaggcctagacctttctctcagccccacc
ttccccagccttcttgttgggcagagggtagccagaggacagaaagatcc
cacccagagccactcactgccatccactttgttaggtgacttcag
```

FIGURE 1.14 Sample MIF gene search using the BLAT function and specific sequence retrieval option. A FASTA output of MIF DNA sequences plus 500 bp of 5' UTR is displayed. Uppercase indicates the exons and lowercase for everything else as specified in the query input.

- Paste partial MIF sequence (i.e., exon1; see Sample Data below) to the textbox and click "Submit."
- Select the query output located on chr22 with the highest score (205) indicating the perfect match. At this point, you can:
    1. Select browser
        - A special track in the viewer with a new line says "*YOUR SEQUENCE FROM BLAT SEARCH*" will appear, and the name of your query sequence is highlighted on the left. Click the highlight name to a new page where all the features described in the previous section can be configured. You need to select the refresh button to change the setting.
    2. Select details
        - The "Alignment of Your Seq and chr22:22566565-22566714" page containing the query, genomic match in color cues and letter case (blue and uppercase for exon; black and lower case for intron), and side-by-side alignment, appears at the bottom (Figure 1.8).

2. Table Browser Use

1. Straightforward search for gene of interest.
    - From the UCSC home page, select "Table Browser" or "Tables" on navigation bar.
    - Select Vertebrate under Clade; Human under Genome; Mar. 2006 under Assembly; Gene and gene prediction tracks under Group; Known genes under Tracks; Known genes under Table.
    - Select position under region and enter MIF in the textbox; click lookup.
        - A "Select Position" page with Known Genes listed will appear.
        - Select MIF (NM_002415) at chr22:22566565-22567409 – macrophage migration inhibitory factor.
        - The Table Browser page returns with the chr22:22566565-22567409 filled in the position text box.
    - Select Sequence under Output format. Make sure filter, intersection, and correlation function are Off (i.e., only the "create" button appears).
    - Select "get output."
        - A "Select Sequence Type for Known Genes" page will appear. You can:
            - Select Genomic or protein or mRNA.
            - If you select "Genomic," you have to specify the retrieval sequences from regions of interest. Check boxes for 5' UTR exons, CDS exons, 3' UTR exons, and introns.
            - Select One FASTA record per gene.
            - Select CDs (codings) in upper case and UTR in lowercase, under Sequence Formatting Options.
            - Select "Get Sequence."
            - A FASTA output returns with description on the header and sequences follow (total of 845 bp) (Figure 1.15).
        - Click Back arrow on the Web page three times to return to the Table Browser page.
2. Use filter function to search for specific sequences of gene of interest.
    - Select Genome under Region.

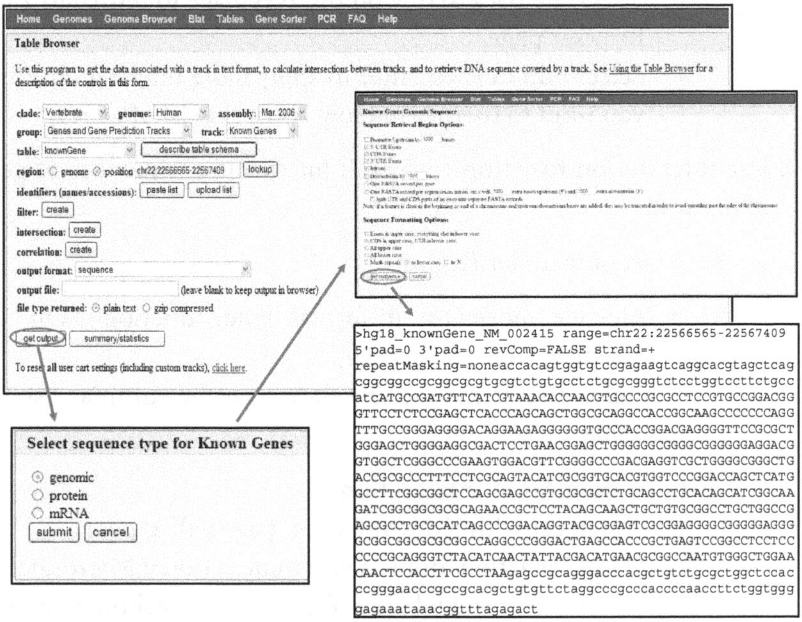
Home Genomes Genome Browser Blat Tables Gene Sorter PCR FAQ Help
Table Browser
clade: Vertebrate genome: Human assembly: Mar. 2006
group: Genes and Gene Prediction Tracks track: Known Genes
table: knownGene describe table schema
region: genome position chr22:22566565-22567409 lookup
identifiers (names/accessions): paste list upload list
filter: create
intersection: create
correlation: create
output format: sequence
output file: (leave blank to keep output in browser)
file type returned: plain text gzip compressed
get output summary/statistics
Select sequence type for Known Genes
genomic
protein
mRNA
submit cancel
>hg18_knownGene_NM_002415 range=chr22:22566565-22567409
5'pad=0 3'pad=0 revComp=FALSE strand=+
repeatMasking=noneaccacagtggtgtccgagaagtcaggcacgtagctcag
cggcggccgcggcgcgtgcgtctgtgcctctgcgcgggtctcctggtccttctgcc
atcATGCCGATGTTCATCGTAAACACCAACGTGCCCCGCGCCTCCGTGCCGGACGG
GTTCCTCTCCGAGCTCACCCAGCAGCTGGCGCAGGCCACCGGCAAGCCCCCCCAGG
TTTGCCGGGAGGGGACAGGAAGAGGGGGGTGCCCACCGGACGAGGGGTTCCGCGCT
GGGAGCTGGGGAGGCGACTCCTGAACGGAGCTGGGGGGCGGGGCGGGGGGAGGACG
GTGGCTCGGGCCCGAAGTGGACGTTCGGGGCCCGACGAGGTCGCTGGGGCGGGCTG
ACCGCGCCCTTTCCTCGCAGTACATCGCGGTGCACGTGGTCCCGGACCAGCTCATG
GCCTTCGGCGGCTCCAGCGAGCCGTGCGCGCTCTGCAGCCTGCACAGCATCGGCAA
GATCGGCGGCGCGCAGAACCGCTCCTACAGCAAGCTGCTGTGCGGCCTGCTGGCCG
AGCGCCTGCGCATCAGCCCGGACAGGTACGCGGAGTCGCGGAGGGGCGGGGGAGGG
GCGGCGGCGCGCGGCCAGGCCCGGGACTGAGCCACCCGCTGAGTCCGGCCTCCTCC
CCCCGCAGGGTCTACATCAACTATTACGACATGAACGCGGCCAATGTGGGCTGGAA
CAACTCCACCTTCGCCTAAgagccgcagggacccacgctgtctgcgctggctccac
ccgggaacccgccgcacgctgtgttctaggcccgcccaccccaaccttctggtggg
gagaaataaacggtttagagact

FIGURE 1.15 Sample view of Table Browser search using the position of the gene. An overview of the query result for chr22:22566565-22567409. Options of sequence outputs and regions of retrieval are illustrated here. A sample of the sequences retrieved is displayed in FASTA format.

- Select Sequence under Output format. Make sure filter, intersection, and correlation function are Off (i.e., only the create button appears).
- Select Summary/statistics.
  - A Known Genes (KnownGene) Summary Statistics table returns showing a total of 39,288 items (genes). Both filter and intersection functions are OFF.
- Select Create next to Filter.
  - A “Filter on Fields from hg18.knownGene” page will appear. Filter for Chr22 by entering 22 in the text box under Chrom match.
  - Select Submit.
  - Select Summary/statistics.
    - A Known Genes (KnownGene) Summary Statistics table returns showing a total of 949 items (genes) with our filtering approach (Figure 1.16).

    - Select the Back arrow on the Web page to return to Table Browser page.
    - Select Edit next to Filter to apply more filtering criteria if you want to further narrow down the list of genes.

3. Use intersection function to search for specific sequences of gene of interest.

   - Keep Genome under Region.
   - Keep Sequence under Output format. Filter function is On (i.e., edit and clear buttons are shown).
     - A total of 949 items (genes) shown on the summary/statistics table.
   - Select Create next to Intersection.
     - An "Intersect with Known Genes" page will appear.
       - Select Variation and Repeats under Group; Microsatellite under Track and tables will automatically show Microsatellite (microsat).
       - Select All Known Genes records that have any overlap with Microsatellite.
       - Click "Submit."
     - Select Summary/statistics.
       - A Known Genes (KnownGene) Summary Statistics table returns showing a total of nine items (genes) with our intersection approach (Figure 1.17).

3. *Creating and Using Custom Track*

   1. To create custom tracks from the Table Browser search queries

      - Continue the setting from previous Table Browser settings (i.e., nine items after filtering and intersection).
      - Select Custom track under the Output format.
      - Select Get output.
        - A new format "Output knownGene as custom track" page will appear.
        - Enter MicrosatGenes under Name; Intersection Microsatellite, KnownGenes under Description.
        - Select Pack from the menu under Visibility.

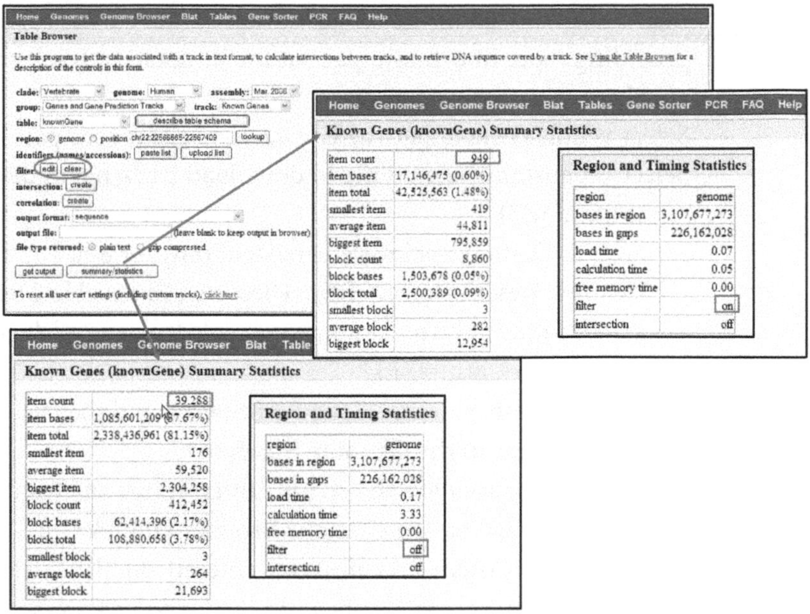

FIGURE 1.16 Sample view of Table Browser search and filter function. The known genes summary statistics tables before and after the filter function are illustrated.

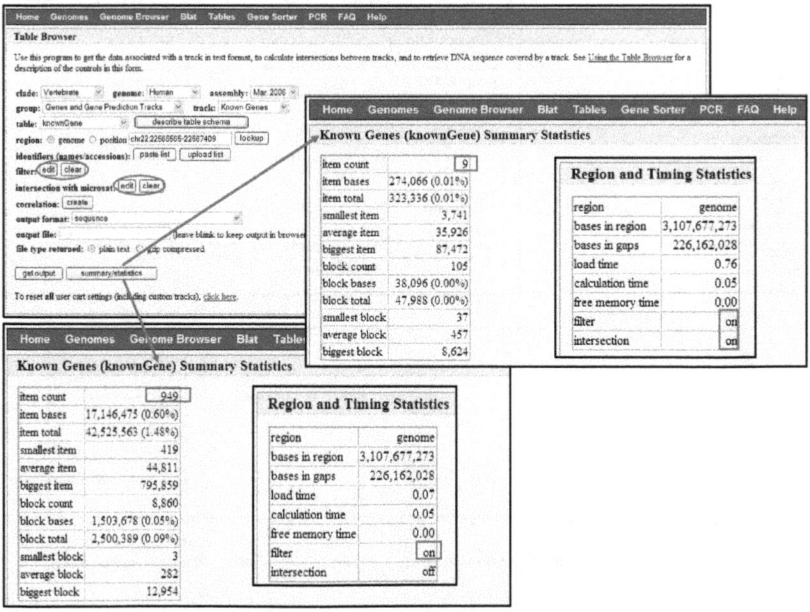

FIGURE 1.17 Sample view of table browser search and intersection function. The known genes summary statistics tables before and after intersection function (while filter function is on) are illustrated.

- Leave the URL blank because we do not have a Web page providing more information about the annotation track created here.
- Select Whole Gene under the Create one BED record per:
- Select get custom track in file to download track file to desktop (Figure 1.18).
  - NOTE: Custom tracks are persistent only for 8 h.
  - Detailed description of BED (Browser Extensible Data) can be found at http://genome.ucsc.edu/goldenPath/help/customTrack.html#BED
- Select Get custom track in Genome Browser, and the custom track will appear in the Genome Browser.
- Position the mouse over the string "Intersection microsatellite KnownGenes" on the Genome Browser and select it.
  - The name (MicrosatGenes) we entered on the custom track header page will appear on the Genome Browser.
- Select Manage custom tracks.
- Click the hyperlink *chr22* under Pos(ition).

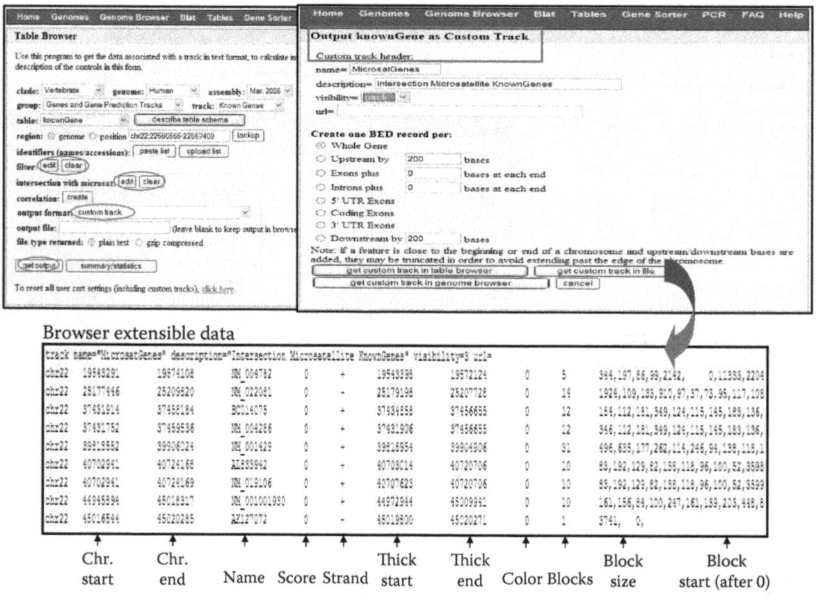
Table Browser
Output knownGene as Custom Track
Create one BED record per:
Browser extensible data
Chr. start
Chr. end
Name
Score
Strand
Thick start
Thick end
Color
Blocks
Block size
Block start (after 0)

FIGURE 1.18 Sample view of Table Browser search using "custom track" as one of the output formats. Configuration interface and display of the custom track in Table Browser using Browser Extensible Data format are illustrated and described in detail.

  - All the nine items (genes) we retrieved are showing on the Browser in ticks spanning Chromosome 22. If you did not see it, double-check the genome region, which should be set at Chr22:19,543,292-46,000,000. A broader genome region ensures the view of all the items being displayed.
  - Hide all the tracks except for custom tracks (Base Position "dense" and MicrosatGenes "dense") and the Known Genes (pack) under Genes and Gene Prediction Tracks.
  - Note: Change dense to full under the MicrosatGenes of the Custom Tracks and see what happened (reveals one gene per data).
- Select NM_004782 under Intersection Microsatellite KnownGenes. A Custom Track: MicrosatGenes page will appear with descriptive information.
- Select Position: chr22:19543292-19574108 will take you to the Genome Browser.
  - Our search (i.e., KnownGene + Microsatellite) identified the SNAP29 gene.
- Positioning the mouse over SNAP29 and clicking will provide you with a more detailed description of this gene.
  - SNAP 29 (synaptosomal-associated protein 29) is a member of the SNAP25 gene family. The protein encoded by this gene binds tightly to multiple syntaxins and is localized to intracellular membrane structures rather than to the plasma membrane. Use of multiple polyadenylation sites has been noted for this gene (Figure 1.19).
  - You can continuously view and compare the rest of genes in our specialized search with other annotated data (i.e., position the mouse over each item and select) in the Genome Browser.

2. To create custom tracks from your own data

- From the UCSC home page select Genome Browser.
- Select Manage custom tracks.
  - On the Manage Custom Tracks, select Add custom tracts.
  - Paste the input file (see Sample Data that follow) in the pasted URL or data text box.

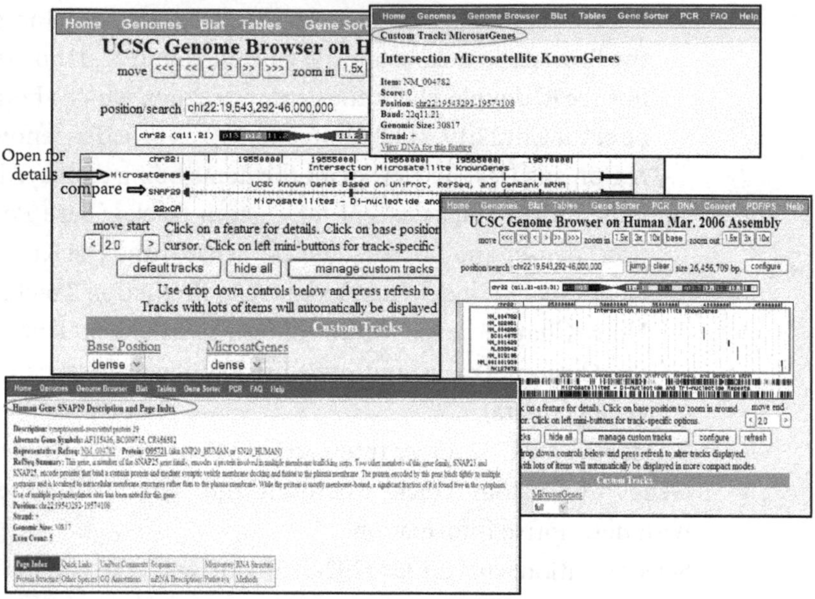
Home Genomes Blat Tables Gene Sorter
UCSC Genome Browser on H
move <<< << < > >> >>> zoom in 1.5x
position/search chr22:19,543,292-46,000,000
Open for details
compare
move start
Click on a feature for details. Click on base position
cursor. Click on left mini-buttons for track-specific
default tracks
hide all
manage custom tracks
Use drop down controls below and press refresh to
Tracks with lots of items will automatically be displayed
Custom Tracks
Base Position
MicrosatGenes
dense
Custom Track: MicrosatGenes
Intersection Microsatellite KnownGenes
UCSC Genome Browser on Human Mar. 2006 Assembly
Human Gene SNAP29 Description and Page Index

FIGURE 1.19 Sample view of the custom track displayed in Genomic Browser format.

- Selecting Submit will take you back to the Manage Custom Tracks page.
- Select either Go to Genome Browser or Go to Table Browser To view your own track that you have just created (Figure 1.20 and Figure 1.21).
  - Note: The input file you created for the custom track can be either pasted to the textbox or uploaded.
  - Formats of Annotation data supported by UCSC include: standard GFF GTF, PSL, BED, or WIG. GFF and GTF files must be tab delimited rather than space delimited to display correctly.
  - Chromosome references must be of the form chrN (the parsing of chromosome names is case sensitive).
  - You may include more than one dataset in your annotation file; these need not be in the same format (see Sample Data 3).
  - More information about how to display your own annotation in the genome browser can be found at http://genome.ucsc.edu/goldenPath/help/ customTrack.html.

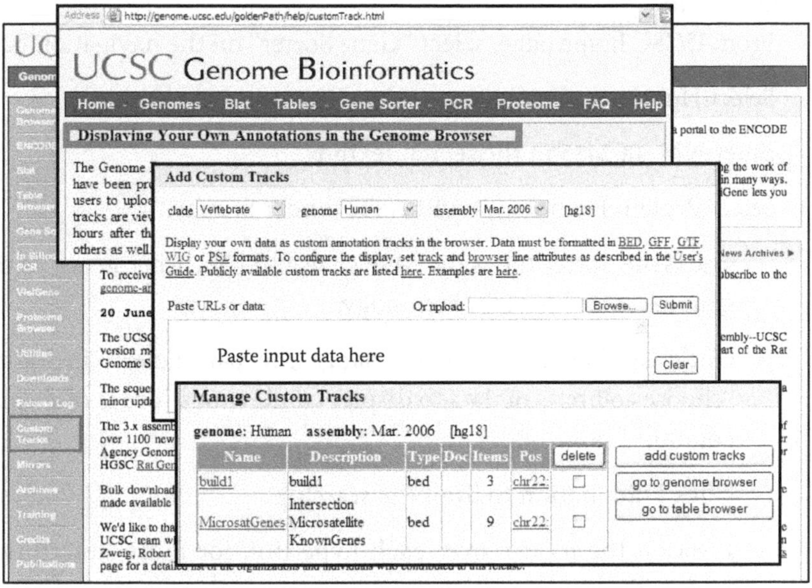
UCSC Genome Bioinformatics
Home Genomes Blat Tables Gene Sorter PCR Proteome FAQ Help
Displaying Your Own Annotations in the Genome Browser
Add Custom Tracks
clade Vertebrate genome Human assembly Mar. 2006 [hg18]
Paste URLs or data:
Or upload:
Browse...
Submit
Paste input data here
Clear
Manage Custom Tracks
genome: Human assembly: Mar. 2006 [hg18]
Name
Description
Type
Doc
Items
Pos
delete
build1
bed
3
chr22:
MicrosatGenes
Intersection Microsatellite KnownGenes
9
add custom tracks
go to genome browser
go to table browser

FIGURE 1.20 Sample view of the "build your own custom tracks."

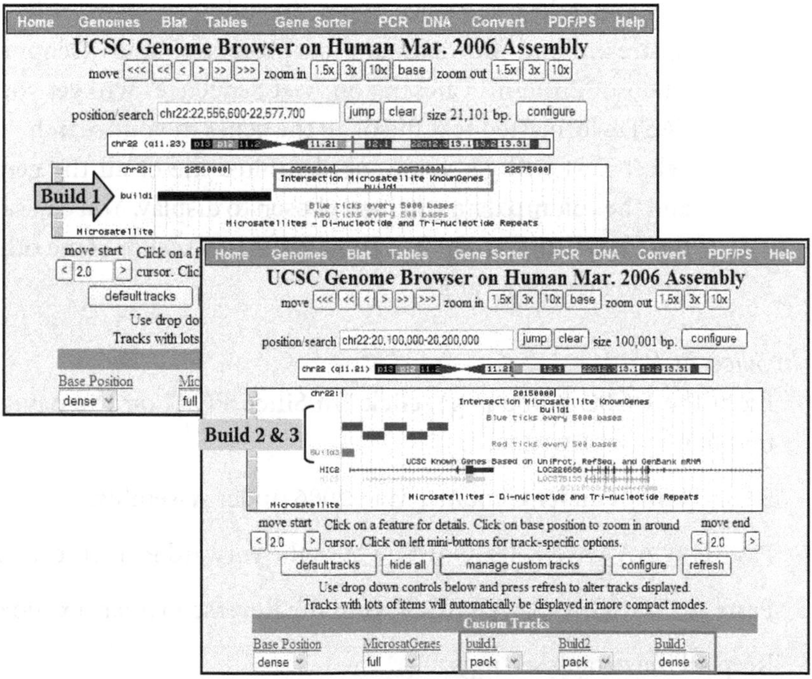
UCSC Genome Browser on Human Mar. 2006 Assembly
position/search chr22:22,556,600-22,577,700
size 21,101 bp.
Build 1
UCSC Genome Browser on Human Mar. 2006 Assembly
position/search chr22:20,100,000-20,200,000
size 100,001 bp.
Build 2 & 3
default tracks
hide all
manage custom tracks
configure
refresh
Custom Tracks
Base Position
MicrosatGenes
build1
Build2
Build3

FIGURE 1.21 Sample view of displaying custom track in Genomic Browser.

4. *Gene Sorter*
   - From UCSC home page, select "Gene Sorter" on the navigation bar.
   - Select Human under Genome; Mar. 2006 under Assembly.
   - Enter NM_002415 in the search textbox.
   - Select Protein Homology — BLASTP under Sort by.
     - The Gene Sorter takes the MIF gene and lists other genes in the genome sorted protein homology.
     - Click Configure button to change the parameters, including "choose columns of data to display" and "choose order of display columns," etc.
     - Click Filter button to limit the search.
     - Position the mouse over each hyperlink for more description or results.
   - You have two output options (sequence and text).
     - Select sequence will take you to a format Get Sequence page. You can select sequence types, including protein, mRNA, upstream and downstream of promoter, and Genomic. Choose "Protein." Clicking on "Get Sequence" will get you a FASTA-formatted text file of all the genes in your search.
     - Select Text will get you a tab-delimited file of all the genes and the column data you have chosen to display. You can save the file on your computer and view it in Excel or some other spreadsheet program.

5. *In Silico PCR*
   - From the UCSC home page, click "In Silico PCR" on the navigation bar.
   - Select Human under Genome; Mar. 2006 under Assembly.
   - Paste CACAAAAGGCGGGACCACA in the Forward Primer text box.
   - Paste ACTGCGAGGAAAGGGCG in the Reverse Primer text box.
   - Keep all the default settings.
   - Select Submit.

    - The search returns a sequence output file in FASTA format containing all sequences in the database that lie between and include the primer pair (Figure 1.11).

6. *Other Utilities (DNA Duster)*
    - From the UCSC home page, click "Utilities" on the navigation bar.
    - Select DNA Duster under UCSC Genome Browser Utilities.
    - Paste the Sample Data 4 to the DNA Duster textbox.
        - The output format settings are 5 for spaces (pull-down menu), 50 for line breaks (pull-down menu), uncheck show numbers option, "unchanged" under case and strand, uncheck translate, start at AUG options, and lowercase in intron checkbox.
    - Select submit and DNA Duster retrieve data as shown in the following text:

```
accac agtgg tgtcc gagaa gtcag gcacg tagct cagcg gcggc cgcgg
cgcgt gcgtc tgtgc ctctg cgcgg gtctc ctggt ccttc tgcca tcatg
ccgat gttca tcgta aacac caacg tgccc cgcgc ctccg tgccg gacgg
gttcc tctcc gagct caccc agcag ctggc gcagg ccacc ggcaa gcccc
cccag tacat cgcgg tgcac gtggt cccgg accag ctcat ggcct tcggc
ggctc cagcg agccg tgcgc gctct gcagc ctgca cagca tcggc aagat
cggcg gcgcg cagaa ccgct cctac agcaa gctgc tgtgc ggcct gctgg
ccgag cgcct gcgca tcagc ccgga caggg tctac atcaa ctatt acgac
atgaa cgcgg ccaat gtggg ctgga acaac tccac cttcg cctaa gagcc
gcagg gaccc acgct gtctg cgctg gctcc acccg ggaac ccgcc gcacg
ctgtg ttcta ggccc gccca cccca acctt ctggt gggga gaaat aaacg
gttta gagac t
```

    - Click Back arrow on the Web page to go back to DNA Duster page (you should still have the input data pasted in the textbox).
    - Change the output format settings to "None" for spaces (pull-down menu) and check Translate and start at AUG options checkbox.
    - Select Submit, and retrieve data as shown in the following text:

```
MPMFIVNTNVPRASVPDGFLSELTQQLAQATGKPPQYIAVHVVPDQLMAF
GGSSEPCALCSLHSIGKIGGAQNRSYSKLLCGLLAERLRISPDRVYINYY
DMNAANVGWNNSTFA
```

Part III Sample Data

1. MIF exon 1 sequence

```
> MIF partial sequence for Blat
ACCACAGTGGTGTCCGAGAAGTCAGGCACGTAGCTCAGCGGCGGCCGCGG
CGCGTGCGTCTGTGCCTCTGCGCGGGTCTCCTGGTCCTTCTGCCATCATG
CCGATGTTCATCGTAAACACCAACGTGCCCCGCGCCTCCGTGCCGGACGG
GTTCCTCTCCGAGCTCACCCAGCAGCTGGCGCAGGCCACCGGCAAGCCCC
CCCAG
```

2. Input file for build 1 (custom track)

```
browser position chr22:22556600-22577700
track name=build1 description="Demo"     visibility=3
chr22 22556700    22556800
chr22 22556800    22556900
chr22 22556900    22562000
```

3. Input file for Build 2 and Build 3 (custom track)

```
browser position chr22:20100000-20200000
track name=Build2 description="Blue ticks every 5000
 bases" color=0,0,255,
chr22 20100000 20105001
chr22 20105000 20110001
chr22 20110000 20115001
chr22 20115000 20120001
chr22 20120000 20125001
track name=Build3 description="Red ticks every 500
 bases" color=255,0,0
chr22 20100000 20100500 Frag 1
chr22 20100501 20101000 Frag 2
chr22 20101001 20101500 Frag 3
chr22 20101501 20102000 Frag 4
chr22 20102001 20102500 Frag 5
chr22 20102501 20103000 Frag 6
```

4. Input file for DNA Duster

```
1 accacagtgg tgtccgagaa gtcaggcacg tagctcagcg gcggccgcgg cgcgtgcgtc
61 tgtgcctctg cgcgggtctc ctggtccttc tgccatcatg ccgatgttca tcgtaaacac
```

```
121 caacgtgccc cgcgcctccg tgccggacgg gttcctctcc gagctcaccc agcagctggc
181 gcaggccacc ggcaagcccc cccagtacat cgcggtgcac gtggtcccgg accagctcat
241 ggccttcggc ggctccagcg agccgtgcgc gctctgcagc ctgcacagca tcggcaagat
301 cggcggcgcg cagaaccgct cctacagcaa gctgctgtgc ggcctgctgg ccgagcgcct
361 gcgcatcagc ccggacaggg tctacatcaa ctattacgac atgaacgcgg ccaatgtggg
421 ctggaacaac tccaccttcg cctaagagcc gcagggaccc acgctgtctg cgctggctcc
481 acccgggaac ccgccgcacg ctgtgttcta ggcccgccca ccccaacctt ctggtgggga
541 gaaataaacg gtttagagac t
```

SECTION 2 SEQUENCE SIMILARITY SEARCHING BY BLAST

Part I Introduction

Because a sequence itself is not informative, finding similarities between sequences by comparative methods against existing databases provides a powerful way to develop novel sequences with previously characterized genes from the same or different organisms (for example, abundant message in a cancer cell line bears similarity to protein phosphatase genes). Based on this, you can infer the function of newly sequenced genes, predict new members of gene families, explore evolutionary relationships, and gain insight into the function and biological importance (for example, phosphorylation and dephosphorylation play an important role in the regulation of cellular transformation in the cancer cell line). In addition, a sequence similarity searching tool can also be used to predict the location and function of protein-coding and transcription regulation regions in genomic DNA.

1. What Is BLAST?

BLAST or Basic Local Alignment Search Tool is a method to ascertain sequence similarity. The core of BLAST services (BLAST 2.0), also known as "Gapped BLAST," is designed to take protein or nucleic acid sequences and compare them against a selection of NCBI databases. The BLAST algorithm introduced in 1990 (Ref Altschul SF) was written to provide better overall speed of searches, while retaining good sensitivity by breaking the query and database sequences into fragments, for distance sequence relationship. Different from other global or multiple sequence alignment programs, BLAST is focused on regions that share only isolated regions of similarity. Therefore, BLAST is a very powerful tool to view sequences aligned with each other, or to find homology. Regions of similarity embedded in otherwise unrelated sequences can be calculated, and statistical significance of matches obtained. Both functional and evolutionary information can be inferred from well-designed queries and alignments.

BLAST comes in variations, for use with different query sequences against different databases. All BLAST applications, as well as information on which BLAST program to use and other help documentation, are listed on the BLAST home page (http://www.ncbi.nlm.nih.gov/BLAST). Recently, NCBI BLAST has implemented several new display features, including: (1) highlight mismatches between similar sequences, (2) show where the query was masked for low-complexity sequence, and (3) integrate information about the database sequences from the NCBI Entrez system into the BLAST display. Additionally, the new report generator has been optimized for databases with large sequences (Jian Ye et al., 2006).

We will briefly introduce the main function of each program and give a step-by-step tutorial with examples in Part II.

2. *What Are the Principles Behind This Software?*

The BLAST algorithm is a heuristic program and is tuned to find functional domains that are repeated within the same protein or across different proteins from different species, as well as shorter stretches of sequence similarity.

To perform a BLAST search, you need to submit sequences (nucleotide or protein) via a BLAST Web page (http://www.ncbi.nlm.nih.gov/BLAST) as a query against all (or a subset of) the public sequence databases. The QBLAST system located on the NCBI BLAST server executes the search, first by making a look-up table of all the "words" (short subsequences) and "neighboring words," i.e., similar words in the query sequence. The sequence database is then scanned for these "hot spots." When a match is identified, it is used to initiate gap-free and gapped extensions of the "word." After the algorithm has looked up all possible "words" from the query sequence, and extended them maximally, it assembles the best alignment for each query sequence pair and writes this information to a SeqAlign data structure, e.g., Abstract Syntax Notation 1 or ASN.1. ASN.1 is an international standard data-representation format used to achieve interoperability between computer platforms. It allows for reliable exchange of data in terms of structure and content by computer and software systems of all types. The BLAST Formatter, which sits on the BLAST server, can then format the results by fetching the ASN.1 and the sequences from the BLAST databases and post them back in a ranked list format to the browser, in the chosen display format. Thus, once a query has been completed, the results can be reformatted without rerunning the search. Alternatively, StandAlone BLAST can be performed locally as a full executable through the command line and can be used to run BLAST searches against private, local databases, or downloaded copies of the NCBI databases. BLAST binaries are provided for Macintosh, Win32 (PC), LINUX, Solaris, IBM AIX, SGI, Compaq OSF, and HP UX systems. StandAlone BLAST executables can be obtained on the NCBI anonymous FTP server (ftp://ftp.ncbi.nih.gov/blast/executables/).

There are many different variations of the BLAST program available to use for different sequence comparisons, e.g., a DNA query to a DNA database, a protein query to a protein database, and a DNA query, translated in all six reading frames, to a protein sequence database. Other adaptations of BLAST, such as Position-Specific Iterated-BLAST (PSI-BLAST) for iterative protein sequence similarity searches and Reversed Position Specific BLAST (RPS-BLAST) for searching for protein domains in the Conserved Domains Database that perform comparisons against sequence profiles, are available as well.

BLAST does not search GenBank flat files (or any subset of GenBank flat files) directly. Rather, sequences are made into BLAST databases. Each entry is split, and two files are formed, one containing just the header information, and the other containing just the sequence information. These are the data that the algorithm uses. If BLAST is to be run in standalone mode, the data file could consist of local, private data, downloaded NCBI BLAST databases, or a combination of the two.

To have some idea of whether the alignment performed by BLAST is "good" and whether it represents a possible biological relationship, or whether the similarity observed is attributable to chance alone, BLAST uses statistical theory to produce a bit score and expect value (E-value) for each alignment pair (query-to-hit). In general terms, this score is calculated from a formula that takes into account the alignment of similar or identical residues, as well as any gaps introduced to align the sequences. The scoring system for nucleotide sequence pairwise alignment is called identity matrix, in which an identity results in a score of +1 and a mismatch results in a score of –3 at that position. However, the scoring system for protein sequence pairwise alignment is based on the selected substitution matrix in which scores for each position are derived from observations of the frequencies of substitutions for any possible pair of residues that are aligned. Scores for an amino acid pair can be positive, zero, or negative depending on how often one amino acid is observed to be substituted for another in nature. Each matrix is tailored to a particular evolutionary distance. The BLOSUM62 matrix (Blocks Substitution Matrix) is the default for most BLAST programs, except for the blastn program, which perform nucleotide–nucleotide comparisons and hence does not use protein-specific matrices and MegaBLAST. In BLOSUM62, the alignment from which scores were derived, was created using sequences sharing no more than 62% identity. Sequences more identical than 62% are represented by a single sequence in the alignment so as to avoid overweighting closely related family members. Other matrices, including BLOSUM80, BLOSUM45, PAM30 (Percent Accepted Mutation), and PAM70, are also available, depending on the type of sequences you are searching with. More information on BLAST substitution matrices can be found at http://www.ncbi.nlm.nih.gov/blast/html/sub_matrix.html. Because the bit scores are normalized, they can be used for different alignment comparisons, even if different scoring matrices were used. Generally, the higher the score, the better the alignment and the greater the degree of similarity.

The E-value gives an indication of the statistical significance of a given pairwise alignment and reflects the size of the database and the scoring system used. For example, a sequence alignment with an E-value of 0.05 means that this similarity has a 5 in 100 chance of occurring by chance alone. Although a statistician might consider this to be significant, it still may not represent a biologically meaningful result, and analysis of the alignments is required to determine biological significance. Generally, the lower the E-value, the more significant the hit is. The default statistical significance threshold for reporting matches against database sequences is 10, such that 10 matches are expected to be found merely by chance. If the statistical significance ascribed to a match is greater than the expected threshold, the match will not be reported. Increasing the E-value forces the program to report less significant matches.

At some positions, where a letter is paired with a null, gaps are introduced. Because a single mutational event may cause the insertion or deletion of more than one residue, the presence of a gap is frequently ascribed more significance than the length of the gap. Hence, the gap is penalized heavily, whereas a lesser penalty is assigned to each subsequent residue in the gap. There is no widely accepted theory for selecting gap costs. Therefore, it is rarely necessary to change gap values from the default.

There are many databases available to compare your query sequences against. A protein database is appropriate for amino acid sequence searches, whereas a nucleic acid database is appropriate for DNA sequence searches.

The exception to this occurs when using programs such as BLASTX and TBLASTN, which perform cross-comparisons between different types of query and database sequences. A commonly used "nr database" is a collection of "non-redundant" sequences from GenBank and other sequence databanks, including EMBL, DDBJ, SwissProt, PIR, PRF, and PDB etc. A detailed list of available subject databases can be accessed from http://www.ncbi.nlm.nih.gov/blast/ blast_databases.shtml.

3. *What Can BLAST Do?*

There are different BLAST programs, which can be distinguished by the type of the query sequence (DNA or protein) and the type of the subject database, to search for various types of homology. Basically, BLAST can tell you: (1) putative identity and function of your query sequence, (2) help to direct experimental design to prove the function, (3) find similar sequences in model organisms (e.g., yeast, *C. elegans*, and mouse), which can be used to further study the gene, and (4) compare complete genomes against each other to identify similarities and differences among organisms. In addition to databases maintained by NCBI, SwissProt and PDB are compiled outside of NCBI, dbEST and month are subsets of the NCBI databases, and other "virtual Databases" can be created using the "Limit by Entrez Query" option. If you are not sure which BLAST program to use, you should check out the "BLAST Program Selection Guide" available at http://www.ncbi.nlm.nih.gov/blast/producttable.shtml.

There are some common features in the BLAST programs:

1. Query sequences should be pasted in the "search" text area.
2. Only FASTA (up to 80 nucleotide bases or amino acids per line), bare sequence, or sequence identifiers (e.g., accession number and Gi) are accepted as input formats.
3. No blank lines are allowed in any input format.
4. Sequences are expected to be represented in the standard IUB/IUPAC amino acid and nucleotide codes.
5. Lowercase letters are accepted and are mapped into uppercase.
6. A single hyphen or dash can be used to represent a gap of indeterminate length.

7. U (for selenocysteine), * (translation stop), and hyphen (gap of indeterminate length) are accepted in amino acid sequences.

The selection of a BLAST program is dependent on: (1) the nature of the query, (2) the purpose of the search, and (3) the database intended as the target of the search and its availability. Following are the brief descriptions of each program (Figure 1.22).

- NUCLEOTIDE
  - **Blastn** compares a nucleotide query sequence (>20 and <3000 bp) against a nucleotide sequence database.
  - **Megablast** uses a greedy algorithm for highly similar sequences (>28 bp) search. This program is optimized for aligning sequences that differ slightly as a result of sequencing or other similar "errors." It also efficiently handles much longer DNA sequences than traditional BLAST algorithms. However, when comparing less conservative sequences (<80%), this approach becomes much less productive than for the higher degree of conservation. Depending on the length of the exact match to start

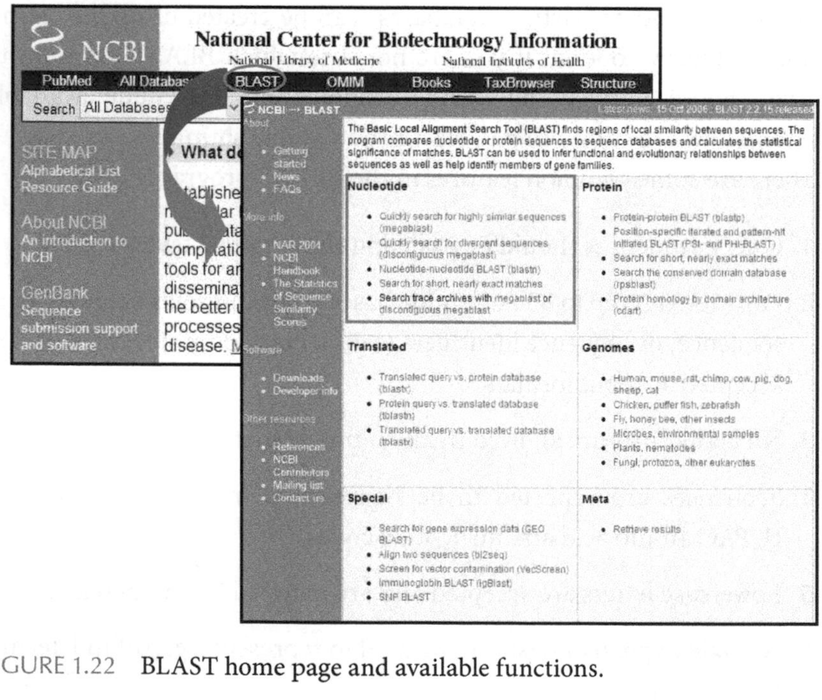
NCBI
National Center for Biotechnology Information
National Library of Medicine
National Institutes of Health
PubMed
BLAST
OMIM
Books
TaxBrowser
Structure
Search
All Databases
SITE MAP
Alphabetical List
Resource Guide
About NCBI
An introduction to NCBI
GenBank
Sequence submission support and software
NCBI → BLAST
The Basic Local Alignment Search Tool (BLAST) finds regions of local similarity between sequences. The program compares nucleotide or protein sequences to sequence databases and calculates the statistical significance of matches. BLAST can be used to infer functional and evolutionary relationships between sequences as well as help identify members of gene families.
Nucleotide
Quickly search for highly similar sequences (megablast)
Quickly search for divergent sequences (discontiguous megablast)
Nucleotide-nucleotide BLAST (blastn)
Search for short, nearly exact matches
Search trace archives with megablast or discontiguous megablast
Protein
Protein-protein BLAST (blastp)
Position-specific iterated and pattern-hit initiated BLAST (PSI- and PHI-BLAST)
Search for short, nearly exact matches
Search the conserved domain database (rpsblast)
Protein homology by domain architecture (cdart)
Translated
Translated query vs. protein database (blastx)
Protein query vs. translated database (tblastn)
Translated query vs. translated database (tblastx)
Genomes
Human, mouse, rat, chimp, cow, pig, dog, sheep, cat
Chicken, puffer fish, zebrafish
Fly, honey bee, other insects
Microbes, environmental samples
Plants, nematodes
Fungi, protozoa, other eukaryotes
Special
Search for gene expression data (GEO BLAST)
Align two sequences (bl2seq)
Screen for vector contamination (VecScreen)
Immunoglobin BLAST (igBlast)
SNP BLAST
Meta
Retrieve results

FIGURE 1.22 BLAST home page and available functions.

the alignments from, it either misses a lot of statistically significant alignments or, on the contrary, finds too many short random alignments. Megablast is the only BLAST Web service that accepts multiple queries in FASTA format. If the query sequences are already present in the Entrez Nucleotide database, the GI or accession numbers can be pasted into the search text box (one identifier per line).

- **Discontiguous megablast** is designed specifically for comparison of diverged sequences, especially sequences from different organisms that have alignments with low degree of identity. In contrast to Megablast and Blastn, which look for exact matches of certain length as the starting points for gapped alignments, discontiguous Megablast finds initial offset pairs, from which the gapped extension is then performed. Additional details on discontiguous Megablast are available at http://www.ncbi.nlm.nih.gov/blast/ discontiguous.html and http://www.ncbi.nlm.nih.gov/Web/Newsltr/FallWinter02/ blastlab.html.
- **Search for short nearly exact matches** is used for short nucleotides (7–20 bp) e.g., primer search. To perform this function, parameter settings for blastn need to be adjusted so that the significance threshold governed by the E-value parameter and default word size parameter are not set too high. Currently, this tool does not work with degenerate bases.

- PROTEIN
  - **Blastp** compares an amino acid query sequence (>15 residues) against a protein sequence database. Generally, blastp is better to use than blastn because the genetic code is degenerate; blastn can often give less specific results as compared to blastp.
  - **PSI-Blast** is designed for position-specific iterated search and can be used to find members of a protein family or build a custom position-specific score matrix.
  - **PHI-Blast** is designed for pattern-hit-initiated Blast and can be used to find proteins similar to the query around a given pattern.
  - **Rpsblast** (Reverse Position-Specific Blast) is used to search conserved domains (i.e., recurring sequence patterns or motifs) of a protein against NCBI's Conserved Domain Database (CDD),

which is a collection of multiple sequence alignments for ancient domains and full-length proteins. Using the Rpsblast algorithm, the query sequence is compared to a position-specific score matrix prepared from the underlying conserved domain alignment. Hits may be displayed as a pairwise alignment of the query sequence with representative domain sequences or as multiple alignments.

  - **Cdart** (Conserved Domain Architecture Retrieval Tool) uses precomputed conserved domain-search results to quickly identify proteins with a set of domains having similar architectures.
  - **Search for short nearly exact matches** is used for short peptide (10 to 15-mer or shorter) search. Similar to the short primer search function mentioned earlier, the parameter settings need to be adjusted for blastp so that the significance threshold governed by the E-value parameter and default word size parameter are not set too high.

- TRANSLATED
  - **Blastx** compares the six-frame conceptual translation products of a nucleotide query sequence (both strands) against a protein sequence database and provides combined significance statistics for hits to different frames. This program is useful for finding: (1) potential translation products of an unknown nucleotide sequence, and (2) homologous proteins to a nucleotide-coding region. Blastx is often the first analysis performed with a newly determined nucleotide sequence and is used extensively in analyzing EST sequences. This search is more sensitive than nucleotide blast because the comparison is performed at the protein level.
  - **Tblastn** compares a protein query sequence against a nucleotide sequence database that is dynamically translated in all six reading frames (both strands). Tblastn can be a very productive way of finding homologous protein-coding regions in unannotated nucleotide sequence databases such as ESTs (expressed sequence tags) and draft genomic records (HTG). Because ESTs and HTGs have no annotated coding sequences, there are no corresponding protein translations in the BLAST protein databases. Therefore, a tblastn search is the only way to search for these potential coding regions at the protein level. Similar to all translating searches, the

tblastn search is especially suited to working with error-prone data similar to ESTs and HTG, because it combines BLAST statistics for hits to multiple reading frames, and thus is robust to frame shifts introduced by sequencing error.

- **Tblastx** compares the six-frame translations of a nucleotide query sequence against the six-frame translations of a nucleotide sequence database. It is useful for identifying novel genes in error-prone nucleotide query sequences. However, this program cannot be used with the nr database on the BLAST Web page, because it is computationally intensive.

- **GENOMES** (20 or 28 bp and above for nucleotide and 15 residues and above for proteins are needed for genome query) You can use this service to map the query sequence, determine the genomic structure, identify novel genes, find homologues, and perform other data mining. Currently, NCBI provides the following genomes for purpose of query:
    - **Blast** human, mouse, rat, chimp, cow, pig, dog, sheep, and cat
    - **Blast** chicken, puffer fish, zebra fish
    - **Blast** fly, honeybee, and other insects
    - **Blast** microbes and environmental samples
    - **Blast** plants and nematodes
    - **Blast** fungi, protozoa, and other eukaryotes
- SPECIAL
    - **GEO Blast** tool enables retrieval of expression profiles on the basis of nucleotide sequence (20 or 28 bp and above) similarity against all GenBank identifiers represented on microarray Platforms or SAGE libraries in GEO. The output resembles conventional BLAST output with each alignment receiving a quality score. Each retrieval has an expression 'E' icon that links directly to corresponding Entrez GEO Profiles.
    - **Bl2seq** (Blast 2 Sequences) is designed to directly align two given sequences (11 bp or above and 15-mer and above) using the BLAST engine. You can Blast nucleotide-to-nucleotide, nucleotide-to-protein, protein-to-nucleotide, or protein-to-protein, as long as

the appropriate program is selected. In the case of aligning short sequence (i.e., primer) to a given sequence, you need to increase the E-value to 1000 or higher, uncheck the filter box, and decrease the word size. The stand-alone executable Bl2seq can be retrieved from NCBI ftp site (ftp://ftp.ncbi.nlm.nih.gov/blast/executables/).

- **VecScreen** is designed for identifying vector sequence (20 or 28 bp and above) contamination in a query sequence against the UniVec database. Currently, UniVec contains a nonredundant set of unique vector segments from a large number of known cloning vectors, and sequences for adapters, linker, stuffers, and primers that are commonly used in the cloning and manipulation of cDNA or genomic DNA. It is recommended to use this service to screen for vector contamination in sequences, before the submission to GenBank. More detailed information about this program can be found at http://www.ncbi.nlm.nih.gov/VecScreen/ UniVec.html).

- **IgBlast** (Immunoglobulin Blast) uses the BLAST search algorithm to facilitate analysis of immunoglobulin V-region sequences in GenBank. This tool has three major functions: (1) reports the three germ line V genes, three D, and two J genes that show the closest match to the query sequence, (2) annotates the immunoglobulin domains (FWR1 through FWR3), (3) and matches the returned hits (for databases other than germ line genes) to the closest germ line V genes, making it easier to identify related sequences. Currently, only human and mouse Ig germ line genes are available.

- **SNP Blast** is linked to NCBI SNP Web page (http://www.ncbi.nlm.nih.gov/SNP/snp_blastByOrg.cgi), where you can perform SNP blast database by organism or blast human SNP database by chromosome. SNP Blast result is displayed in "Pairwise with identify" format, which highlights the mismatches in red. In some cases, you can select "Query anchored with identity" format to get more information.

- META

  - **Retrieve results for an RID** provides multiple accesses to the same result and displays it in various formats. QBLAST issues a unique request ID (RID) for each successfully submitted NCBI BLAST search request, and RID can be used to retrieve the results. However, the server will only make the RID available 24 h after the result is generated (Figure 1.23).

Part II Step-By-Step Tutorial

Here, we give a couple of examples of sequence homology search. As you are more familiar with the NCBI BLAST service, you will notice that the interfaces for both nucleotide and protein search are very similar. In most of the cases, you should keep the default setting for each unique program unless the query you perform did not give you good results.

1. Search for nucleotide sequence homology
   - Using Blastn
     - Go to the NCBI Blast Web page (http://www.ncbi.nlm.nih.gov/BLAST) directly or from the NCBI home page (http://www.ncbi.nlm.nih.gov), and select BLAST from the top of the navigation bar.
     - Select "Nucleotide-nucleotide BLAST (blastn)" under Nucleotide section.
     - Paste the nucleotide sequences in the Search textbox (see Sample Data 1 in the following text).

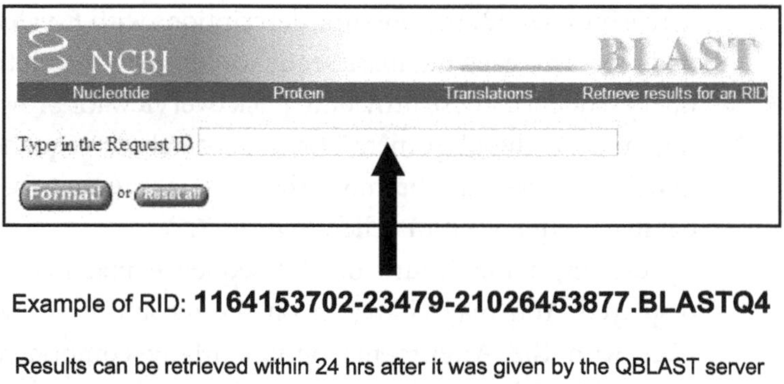
NCBI
BLAST
Nucleotide
Protein
Translations
Retrieve results for an RID
Type in the Request ID
Format!
or
Reset all
Example of RID: 1164153702-23479-21026453877.BLASTQ4
Results can be retrieved within 24 hrs after it was given by the QBLAST server

FIGURE 1.23 Sample view of BLAST interface for retrieving results. An example of the RID (request ID) is illustrated.

- Select "nr" (nonredundant) for the most comprehensive search under the Choose database. Alternatively, you may select other database from the pull-down menu.
  - Note: There are several options for advanced blasting including "Limit by entrez query," "organisms," "Compositional adjustments," "Filter," "Expect value," "Word size," "Matrix," "Gap costs," etc. A hyperlink is provided with each option for a more detailed description.
  - Keep the default E-value to 10 and word size to 11.
- Selecting BLAST will take you to the NCBI formatting BLAST page, with the comment that "Your request has been successfully submitted and put into the Blast Queue." A summary statement about the query is given (in this case, Query = NM_002415 MIF 561 letters). An RID is issued by the QBLAST in the text box (e.g., 1164153702-23479-21026453877.BLASTQ4).
  - Note: The RID number given in the previous paragraph is no longer available for data retrieval. You will be assigned a new RID when you perform the exercises.
- Select Format, which will take you to the NCBI results of BLAST page. Whatever formatting options are being changed, you need to select Format again to apply them.
  - Note: All the BLAST reports consist of four major sections: (1) the header, which contains information about the query sequence and the database searched; (2) interactive browser; (3) the one-line descriptions with E-values and scores for each database sequence found to match the query sequence; these provide a quick overview for browsing; and (4) the alignments for each database sequence matched (there may be more than one alignment for a database sequence it matches, Figure 1.25).
  - By default, a maximum of 500 sequence matches are displayed. You may change it on the advanced BLAST page with the Alignments option. Many components of the BLAST results display via the Internet and are hyperlinked to the same information at different places in the page, to additional information including help documentation, and to the Entrez sequence records of matched sequences. These records provide more infor-

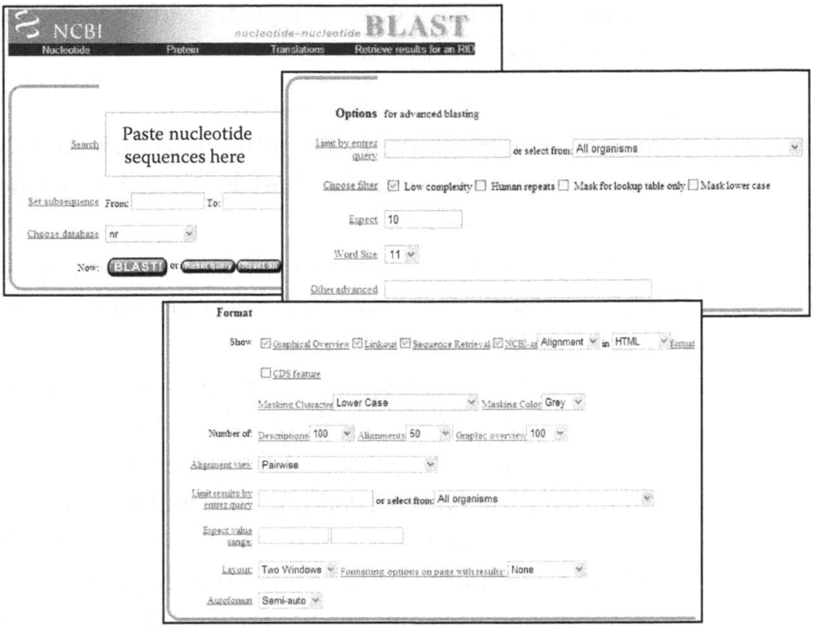
NCBI
nucleotide-nucleotide BLAST
Nucleotide
Protein
Translations
Retrieve results for an RID
Search
Paste nucleotide sequences here
Set subsequence From: To:
Choose database nr
Now: BLAST!
Options for advanced blasting
Limit by entrez query
or select from: All organisms
Choose filter Low complexity Human repeats Mask for lookup table only Mask lower case
Expect 10
Word Size 11
Other advanced
Format
Show Graphical Overview Linkout Sequence Retrieval NCBI-gi Alignment in HTML format
CDS feature
Masking Character Lower Case Masking Color Grey
Number of: Descriptions 100 Alignments 50 Graphic overview 100
Alignment view Pairwise
Limit results by entrez query
or select from: All organisms
Expect value range
Layout Two Windows Formatting options on page with results: None
Autoformat Semi-auto

FIGURE 1.24 Sample view of nucleotide-nucleotide BLAST search and the configuration options.

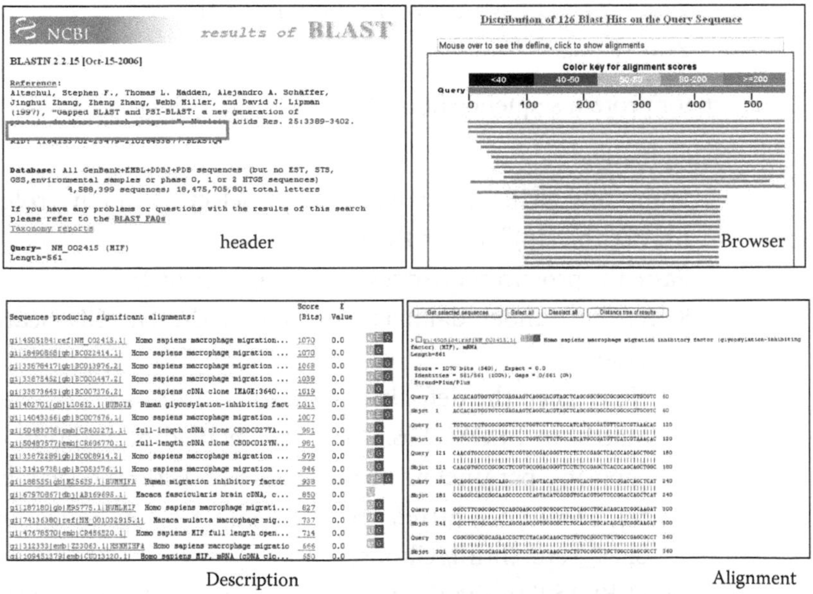
NCBI
results of BLAST
BLASTN 2.2.15 [Oct-15-2006]
header
Distribution of 126 Blast Hits on the Query Sequence
Mouse over to see the define, click to show alignments
Color key for alignment scores
<40
40-50
50-80
80-200
>=200
Query
0
100
200
300
400
500
Browser
Sequences producing significant alignments:
Score (Bits)
E Value
Description
Alignment

FIGURE 1.25 A typical BLAST Search Report. It consists of four major sections: 1) a header (top left); 2) a browser (top right); 3) the one-line descriptions with E-values (lower left); and 4) the alignments for each database sequence matched (lower right).

mation about the sequence, including links to relevant research abstracts in PubMed.

- Using Search for short nearly exact matches:
  - From the NCBI Blast Web page (http://www.ncbi.nlm.nih.gov/BLAST) select "Search for short nearly exact matches" under Nucleotide section.
  - Paste the nucleotide sequences (7–20 bp). We use the excerpted sequence GCGGGTCTCCTGGTCCTTCT from MIF genomic DNA, as an example.
  - Select "nr" (nonredundant) for the most comprehensive search under the Choose database. Alternatively, you may select other database from the pull-down menus.
    - Note: Keep the default E-value to 1000 and word size to 7.
  - Select BLAST, which will take you to the NCBI formatting BLAST page, with the comment that "Your request has been successfully submitted and put into the Blast Queue." A summary statement about the query is given (in this case, Query = 20 letters). An RID is issued by the QBLAST in the textbox.
  - Select Format, which will take you to the NCBI results of BLAST page.

2. Search for protein sequence homology

- Using Blastp:
  - From the NCBI Blast Web page, select "Protein-protein BLAST (blastp)" under **Protein** section.
  - Paste the protein sequences (see Sample Data 2).
  - Select "nr" under database and check the Do CD-Search checkbox.
    - Note: Keep the default E-value to 10, word size to 3, and Matrix to BLOSUM62. A more detailed description on choosing parameters for protein-based BLAST searches can be found at http://www.ncbi.nlm.nih.gov/Education/Blast_setup.html#Choose.
  - Select BLAST to see the results (in this case, Query = NM_002415 MIF 114 letters). Because the Do CD-Search option is selected, putative conserved domains were searched, and in this case was detected with an image. Click on the image

to go to the NCBI Conserved Domains Web page with more detailed descriptions and options.
- Select Format if you wish to view the retrieved results in a different way.
  - Note: By default, a maximum of 100 descriptions, 50 sequence alignments, and 100 graphic overviews are displayed. You may change it on the advanced BLAST page with the Alignments option.

- Using Search for short, nearly exact matches:
  - From the NCBI Blast Web page, select "Search for short nearly exact matches" under Protein section.
  - Paste the nucleotide sequences (10 to 15-mer). We use the extracted sequence PMFIVNTNVPR from MIF protein DNA as an example.
  - Select the "nr" (nonredundant) database.
    - Note: Keep the default E-value to 20000, word size to 2, and Matrix to PAM30.
  - Select BLAST for results.
  - Select Format for a different view.

3. Translated BLAST

- Using Blastx:
  - From the NCBI Blast Web page, select "Translated query vs. protein database (blastx)" under Translated section.
  - Paste the nucleotide sequences (see Sample Data 2).
  - Select the "nr" (nonredundant) database. Keep the default and choose a translation to "TRANSLATED query-PROTEIN database [blastx]" and set Genetic codes to "Standard (1)."
    - Note: Keep the default E-value to 10, word size to 3, and Matrix to BLOSUM62.
  - Select BLAST for results.
  - Note: By default, a maximum of 100 descriptions, 50 sequence alignments, and 100 graphic overviews are displayed. You may change it on the advanced BLAST page with the Alignments option.

- Using tBlastn
  - From the NCBI Blast Web page, select "Protein query vs. translate database (tblastn)" under Translated section.

- Paste the MIF protein sequences (see Sample Data 2).
- Select the "nr" database. Keep default choose a translation to "PROTEIN query-TRANSLATED database [tblastx]" and Genetic codes to "Disabled."
  - Note: Keep the default E-value to 10, word size to 3, and Matrix to BLOSUM62.
- Select BLAST for results (in this case, Query = 115 letters).
- Select Format if you wish.

- Using tBlastn
  - From the NCBI Blast Web page, select "Protein query vs. translate database (tblastn)" under Translated section.
  - Paste the MIF protein sequences as an example.
  - Select the "nr" database. (Keep the default and choose a translation to "PROTEIN query-TRANSLATED database [tblastx]" and set Genetic codes to "Disabled.")
    - Note: Keep the default E-value to 10, word size to 3, and Matrix to BLOSUM62.
  - Select BLAST for result (in this case, Query = 115 letters).
  - Select Format (optional).

4. Genome BLAST

- Blast Human sequences:
  - From the NCBI Blast Web page, select "Human, mouse, rat ...." under Genomes section.
  - Select *"Enter an accession, gi, or a sequence in FASTSA format"* option and enter NM_002415 (accession) or 4505184 (gi) or paste the MIF mRNA sequences in the text box.
  - Select "genome (all assemblies, previous build 35.1)" under Database.
    - Select "cross-species mega-BLAST: Compare nucleotide sequences for other species to this genome" under Program. Keep default values for all the other settings.
  - Select Begin Search for result (in this case, Query = gi|4505184 (561 letters)).
  - Select Format (optional).
  - Selecting ref|NT_011520.10|Hs22_11677 under Alignments will take you to the NCBI Map Viewer with the Blast hit. Select D (highlighted in green) under Alignments will show

you the Definition as "Gi|51476066|ref|NT_011520.10|Hs22_11677 Homo sapiens chromosome 22 genomic contig." Followed by the definition is the sequence including up to 1000 bp flanking sequence.

5. Special BLAST
   - Blast 2 Sequence (Bl2seq):
     1. For nucleotide sequences
        - From the NCBI Blast Web page, select "Align two sequences (bl2seq)" under Special section.
        - Select "blastn" under Program because the input sequences we used here are nucleotide sequences. Keep "Not Applicable" under Matrix.
        - Paste MIF mRNA sequences (Sample Data 1) to sequence 1 textbox.
        - Paste MIF genomic sequences plus 500 bp upstream and downstream of the flanking sequences of the gene (total of 1845 bp) (Sample Data 3) into the sequence 2 textbox.
        - Select Align, which will take you to the NCBI Blast 2 Sequence results page stating the version of the BLASTN (2.2.14) and the date [May-07-2006]. Followed by the heading are the descriptions about the alignment. A brief summary about sequence 1 and sequence 2 is listed.
        - A diagram with three color segments is displayed in sequence matrix where x-axis is sequence 1 and y-axis is sequence 2. Each segment represents an exon aligned to the genomic sequences.
     2. For protein sequences:
        - From the NCBI Blast Web page, select "Align two sequences (bl2seq)" under Special section.
        - Select "blastp" under Program because the input sequences we used here are nucleotide sequences. Keep "BLOSUM62" under Matrix.
        - Paste sequence 1 (RTGVGTHLTSLALPGKAEGVAS-LTSQCSYSSTIVHVGD KKP) to sequence 1 textbox.
        - Paste sequence 2 (RTGVGTHLTSLALPGKAESVAS-LTSQCSYSSTIVHVGD KKP) to sequence 2 textbox.
        - Select Align.

– Results page showing an amino acid mutation (G to S) as shown in the following text.

```
Score = 82.8 bits (203), Expect = 4e - 15
Identities = 40/41 (97%), Positives = 40/41 (97%),
 Gaps = 0/41 (0%)
Query 1 RTGVGTHLTSLALPGKAEGVASLTSQCSYSSTIVHVGDKKP 41
 RTGVGTHLTSLALPGKAE VASLTSQCSYSSTIVHVGDKKP
Sbjct 1 RTGVGTHLTSLALPGKAESVASLTSQCSYSSTIVHVGDKKP 41
```

6. Meta BLAST

- Retrieve results
  - From the NCBI Blast Web page, select "Retrieve results" under Meta section.
  - Type in the Request ID, which was generated within the past 24 h in the textbox.
  - Select Format!
    - Note: In the case where the alignment file is too big, it will be deleted from the server in 30 min.

In addition to the Step-by-Step Tutorial described here, NCBI also provides some BLAST exercises available in the problem set (http://www.ncbi.nlm.nih.gov/Class/FieldGuide/problem_set.html) and exercise page (http://www.ncbi.nlm.nih.gov/Class/MLACourse/ exercises.html#BLAST).

Part III Sample data

1. MIF mRNA sequences (Homo Sapien) for blastn

```
>NM_002415 (MIF)

accacagtggtgtccgagaagtcaggcacgtagctcagcggcggccgcgggcgt
gcgtctgtgcctctgcgcgggtctcctggtccttctgccatcatgcgatgttca
tcgtaaacaccaacgtgccccgcgcctccgtgccggacgggttcctctccgagc
tcacccagcagctggcgcaggccaccggcaagcccccccatacatcgcggtgca
cgtggtcccggaccagctcatggccttcggcggctcagcgagccgtgcgcgctc
tgcagcctgcacagcatcggcaagatcgccagaaccgctcctacagcaagctgc
tgtgcggcctgctggccgagcgcctgcgcacagcccggacagggtctacatcaa
ctattacgacatgaacgcggccaatgtgggctggaacaactccaccttcgccta
agagccgcagggacccacgctgtctgc gctggctccacccgggaacccgccgc
```

```
acgctgtgttctaggcccgcccaccccaaccttctggtggggagaaataaacg
gtttagagact
```

2. MIF protein sequences (Homo Sapien) for blastp

```
>NM_002415 (MIF)

MPMFIVNTNVPRASVPDGFLSELTQQLAQATGKPPQYIAVHVVPDQLMAFGGSSEPCALCS
LHSIGKIGGAQNRSYSKLLCGLLAERLRISPDRVYINYYDMNAANVGWNNSTFA
```

3. MIF genomic DNA sequence

```
>hg18_knownGene_NM_002415
atggtgattcgcaggggcagctcccctctcacctgccgcgatgactacccc
gccccatctcaaacacacaagctcacgcatgcgggactggagcccttgagg
acatgtggcccaaagacaggaggtacaggggctcagtgcgtgcagtggaat
gaactgggcttcatctctggaagggtaaggggccatcttccgggttcaccg
ccgcatccccacccccggcacagcgcctcctggcgactaacatcggtgact
tagtgaaaggactaagaaagacccgaggcgaggccggaacaggccgatttc
tagccgccaagtggagaacaggttggagcggtgcgccgggcttagcggcgg
ttgctggaggaacgggcggagtcgcccagggtcctgccctgcgggggtcga
gccgaggcaggcggtgacttccccactcggggcggagccgcagcctcgcgg
gggcggggcctggcgccggcggtggcgtcacaaaaggcgggACCACAGTGG
TGTCCGAGAAGTCAGGCACGTAGCTCAGCGGCGGCCGCGGCGCGTGCGTCT
GTGCCTCTGCGCGGGTCTCCTGGTCCTTCTGCCATCATGCCGATGTTCATC
GTAAACACCAACGTGCCCCGCGCCTCCGTGCCGGACGGGTTCCTCTCCGAG
CTCACCCAGCAGCTGGCGCAGGCCACCGGCAAGCCCCCCCAGgtttgccgg
gaggggacaggaagagggggtgcccaccggacgaggggttccgcgCtggg
agctggggaggcgactcctgaacggagctgggggcggggcgggggagga
cggtggctcgggcccgaagtggacgttcggggcccgacgaggtcgctgggg
cgggctgaccgcgccctttcctcgcagTACATCGCGGTGCACGTGGTCCCG
GACCAGCTCATGGCCTTCGGCGGCTCCAGCGAGCCGTGCGCGCTCTGCAGC
CTGCACAGCATCGGCAAGATCGGCGGCGCGCAGAACCGCTCCTACAGCAAG
CTGCTGTGCGGCCTGCTGGCCGAGCGCCTGCGCATCAGCCCGGACAGgtac
gcggagtcgcggaggggcgggggaggggcggcggcgcggccaggccggg
actgagccacccgctgagtccggcctcctcccccgcagGGTCTACATCAA
TATTACGACATGAACGCGGCCAATGTGGGCTGGAACAACTCCACCTTCGCC
TAAGAGCCGCAGGGACCCACGCTGTCTGCGCTGGCTCCACCCGGAACCCG
CCGCACGCTGTGTTCTAGGCCCGCCCACCCCAACCTTCTGGTGGGGAGAAA
TAAACGGTTTAGAGACTaggagtgcctcggggttccttggcttgcgggagg
```

```
aattggtgcagAgccgggatattggggagcgaggtcgggaacggtgttgggg
gcgggggtcagggccgggttgctctcctcCgaacctgctgttcgggagccct
tttgtccagcctgtccctcctacgctcctaacagaggagccccagtgtcttt
ccattctatggcgtacgaagggatgaggagaagttggcactctgccctgggc
tgcagactcgggatctaaggcgctctgcccgccggaatccgtttacctaggg
ccaccacgtggggtgctggaggtgagccgaccacggaagagggggaggagga
gttggagttgggaggagtccgaggtcttctaggcctagacctttctctcagc
cccaccttccccagccttcttgttgggcagagggtagccagaggacagaaag
atcccacccagagccactcactgc catccactttgttaggtgacttcag
```

REFERENCES

1. THE HUMAN GENOME: Science Genome Map. 2001. *Science* 291 (5507), 1218.
2. Istrail, S., Sutton, G.G., and Florea, L. et al. Whole-genome shotgun assembly and comparison of human genome assemblies. *Proc Natl Acad Sci USA.* 2004. 101(7):1916-21.
3. Hinrichs, A.S., Karolchik, D., and Baertsch, R. et al. The UCSC Genome Browser Database: update. *Nucleic Acids Res.* 2006. 34: D590-D598.
4. Kent, W.J. BLAT---The BLAST-Like Alignment Tool. *Genome Res.* 2002 12: 656-664
5. McGinnis, S. and Madden, T.L. BLAST: at the core of a powerful and diverse set of sequence analysis tools. *Nucleic Acids Res.* 2004 Jul 1;32 (Web Server issue): W20-5.
6. Ye, J., McGinnis, S., and Madden, T.L. BLAST: improvements for better sequence analysis. *Nucleic Acids Res.* 2006 Jul 1;34 (Web Server issue): W6-9.

CHAPTER 2

Two Common DNA Analysis Tools

Blanca Camoretti-Mercado

CONTENTS

The discovery of restriction enzymes from microorganisms ushered in the era of recombinant DNA technology and the technologies of genetic cloning and genetic engineering during the second half of the twentieth century. Similarly, the polymerase chain reaction (PCR) methodology facilitated unexpected breakthroughs in many life science disciplines, including clinical and basic research. Progress will certainly continue as more robust and faster computational tools emerge and continue to improve. This chapter focuses on the application of bioinformatics tools to restriction mapping and PCR, two of the most common DNA analyses.

SECTION 1 RESTRICTION MAPPING

Part I Introduction

Restriction enzymes are endonucleases that cleave DNA. These proteins are found in bacteria and form part of the defense mechanism against viral and other foreign DNA. Restriction enzymes' names are derived from the organism in which they were isolated (genus, species, strain, and order of discovery). For example, EcoRI refers to the first restriction enzyme isolated from *Escherichia coli*, strain RY13. Bacteria contain over 400 restriction enzymes that recognize and cut more than 100 different DNA sequences.

Restriction enzymes recognize and bind specific oligonucleotide (oligo) sequences of 4 to 12 nucleotides long (restriction sites). Their activity generates DNA fragments with unique ends by making nicks on the double-stranded DNA. Most restriction enzyme sites have dyad symmetry (palindromic) sequences; that is, the sequence on one strand reads the same in the opposite direction *on the complementary strand.* The enzyme recognition sites are written using a single-letter code (see Table 2.1), in 5' to 3' orientation, with the position of the cleavage indicated by a "/" or a "^". For instance, EcoRI recognizes the G/AATTC sequence, and cuts (or digests) each strand of DNA between the G and the A. This leaves "sticky," "cohesive," or "complementary" ends that are single-stranded overhangs of DNA. Other enzymes like SmaI (restriction site, CCC/GGG) make strand incisions just opposite one another, producing DNA fragments with "blunt" ends. Isoschizomers are restriction endonucleases that recognize the same sequence; their cut sites may

TABLE 2.1 Bases Nomenclature

Abbreviation	Base
A	Adenosine
C	Cytosine
G	Guanosine
T	Thymidine
R (purine)	G or A
Y (pyrimidine)	C or T
M	A or C
K	G or T
S (strong)	G or C
W (weak)	A or T
B	not A
D	not C
H	not G
V	not T
N	Any

Note: Mixed bases are also known as degenerate or wobble bases. As there are 11 different possible combinations of 2, 3, or 4 bases, a universal nomenclature has been established, the IUB codes that must be used when specifying nucleic acid content at a mixed-base site.

or may not be identical. For example, HpaII is an isoschizomer of MspI (restriction site, C/CGG), and AccI is an isoschizomer of FblI (restriction site, C/MKAC). DNA fragments produced by restriction enzymes can be reformed by the activity of ligases. For bases' nomenclature, see Table 2.1.

For nonpalindromic enzymes, the point of cleavage is indicated by numbers in parentheses. For instance, GGTCTC(1/5) indicates cleavage at 5'...GGTCTCN/...3' in the top strand and 3'...CCAGANNNNN/...5' on the bottom strand. For enzymes that cleave away from their recognition sequence, the cleavage sites are also indicated in parentheses. For example, for MboII, GAAGA (8/7) indicates cleavage as follows: 5'...GAAGANNNNNNNN/...3' on the top strand and 3'... CTTCTNNNNNNN/...5' on the bottom strand.

The Restriction Enzyme Database REBASE (http://rebase.neb.com) contains information on restriction enzymes and related proteins, their recognition and cleavage sites, published and unpublished references, isoschizomers, commercial availability, methylation sensitivity, as well as crystal and sequence data. REBASE is continuously updated, and each enzyme has its own Web page. Related proteins include homing endonucleases, nicking enzymes, as well as DNA methyltransferases.

1. What Is Restriction Mapping?

The discovery of the restriction enzymes and the elucidation of their mechanism and specificity of action led to the development of recombinant DNA technology. Indeed, restriction enzymes are fundamental means used in many procedures in molecular biology and genetic engineering. They are invaluable tools for dissecting, analyzing, and reconfiguring any genetic information at the molecular level.

The characterization of double-stranded DNA that is based on the location of the restriction endonucleases cleavage sites constitutes the ***restriction map***. A small number of very large DNA fragments (several thousand to million base pairs [bp]) are obtained by digestion with rare-cutter restriction enzymes, which usually have 6-bp recognition sites. Most enzymes, however, cut DNA more often and generate many fragments (of less than a 100 to more than a 1000 bp long). On average, restriction enzymes with 4-base recognition sites will yield fragments of 256 bases, 6-base recognition sites will yield pieces 4,000 bases long, and 8-base recognition sites will yield pieces 64,000 bases long.

2. Why Is Restriction Mapping Useful?

Restriction mapping has been used in several disciplines. In molecular biology, restriction mapping is the first step in many DNA manipulations of recombinant technology. It is a prerequisite for molecular cloning, subcloning of DNA fragments into a variety of vectors, mutagenesis studies, and other interventions that involve description of the DNA. On the other hand, although it is usually desirable to know the sequence of the DNA, in practical terms some DNA manipulations can be performed with no previous knowledge of its sequence.

Restriction digests are performed in genetic fingerprinting and restriction fragment length polymorphism (RFLP) analysis, although recently they have been replaced by faster PCR-based techniques. Restriction mapping is used in DNA profiling in medicine to match potential organ donors; in biological sciences to study wild animal and plant populations; in forensics to analyze DNA from blood, hair, saliva, or semen, and to examine paternity and investigate crimes as well.

In 1978, Werner Arber (Biozentrum der Universität Basel, Switzerland), with Dan Nathans and Hamilton Smith (both from Johns Hopkins

University School of Medicine, Baltimore, MD) shared the Nobel Prize in Physiology or Medicine for the discovery of "restriction enzymes and their application to problems of molecular genetics." Arber discovered restriction enzymes, Smith showed that bacterial restriction enzymes cut DNA in the middle of specific symmetrical sequences, and Nathans pioneered the application of restriction enzymes to genetics. He demonstrated their use for construction of genetic maps and developed and applied new methodologies involving restriction enzymes. As a consequence, these techniques have opened new avenues in the study of the organization and expression of genes of higher animals and in the solution of basic problems in developmental biology. In medicine, increased knowledge in this area has helped in the prevention and treatment of malformations, hereditary diseases, and cancer.

3. Online Restriction Mapping Programs

As stated earlier, restriction maps can be generated with no previous knowledge of the DNA nucleotide sequence by using restriction enzyme digestion with two or more enzymes followed by electrophoresis on agarose or polyacrylamide gels, or by high-performance liquid chromatography (HPLC). The DNA of interest is frequently cloned into a vector (plasmid, phage, virus, cosmid, etc.) that contains a "linker" region which harbors dozens of restriction enzyme recognition sites within a very short segment of artificial DNA. Once the sizes of the fragments generated are calculated against standards run simultaneously, it becomes feasible to deduce where each enzyme cuts. This is an effective, but manual and time-consuming approach of generating restriction maps. However, with the availability of known complete genome sequences from human and other species, and the availability of several computer programs, it is currently easy to search for dozens of restriction enzyme recognition sites as well as to predict and build corresponding maps of any input DNA.

Some of the most popular online programs employed to generate restriction maps are listed in the following text (also see Table 2.2; usually, these programs display graphical as well as text-based maps):

- **NEB Cutter** provided by New England Biolabs (NEB) takes the input DNA sequence (up to 300,000 bases) and finds the sites for all restriction enzymes that cut the sequence only once. This program also detects open reading frames (OFRs). By changing the settings, additional enzymes not available from NEB could be used.

TABLE 2.2 Web Sites of Some Programs That Perform Restriction Enzyme Mapping

Site	Address
NEB Cutter	http://tools.neb.com/NEBcutter2/index.php
Restriction/Mapper	http://arbl.cvmbs.colostate.edu/molkit/mapper/index.html http://www.restrictionmapper.org/
Webcutter	http://www.ccsi.com/firstmarket/cutter/cut2.html
In Silico Restriction	insilico.ehu.es
WatCut	http://watcut.uwaterloo.ca/watcut/watcut/
EnzymeX	http://mekentosj.com/enzymex/
Silent	http://bioweb.pasteur.fr/seqanal/interfaces/silent.html

- **Restriction/Mapper**, available as part of the Molecular Toolkit package, is a restriction mapping software that allows a virtual simultaneous digestion of a given sequence with enzymes of your choice. The database of enzymes used by this program is from the "Prototypes" file at NEB.
- **Webcutter** from BioDirectory, which also uses the NEB database, offers additional features in its new 2.0 version, including highlighted enzymes in color or boldface for easy identification, analysis of sequences containing ambiguous nucleotides such as N, Y, and R, selection of circular or liner sequences, and the ability to find sites that may be introduced by silent mutagenesis (those that do not change the encoded protein sequence).
- **In Silico Restriction** from the University of the Basque Country in Spain allows restriction digests of one or more DNA sequences with commercially available or specific endonucleases. It also compares restriction patterns of multiple prealigned sequences. These are valuable tools for single-nucleotide polymorphism (SNP) and mutation detection.
- **WatCut** from the University of Waterloo, Canada, searches restriction enzyme cleavage sites in the entered DNA, creates new cleavage sites in oligonucleotides using silent mutations, and analyzes SNP-RFLP by introducing restriction sites that will cleave only one variant of the given sequence.
- **EnzymeX,** developed by Dr. Mek and Dr. Tosj, allows restriction analysis and documentation (Mac users). A nice feature provided is that along with the results, the DNA sequence and the protein translation are always displayed at the bottom of the window for a detailed view.

- **Silent**, from Pasteur Institute in France, scans nucleic acid sequences for silent mutation restriction enzyme sites. The program finds positions where a point mutation could be made to introduce a specified restriction enzyme recognition site without changing the translation.

Part II Step-By-Step Tutorial

The following is a guide that can be followed to perform restriction mapping using the NEBcutter2 program. It performs restriction enzyme analysis with parameters set by the user. Other programs share similar characteristics as the NEBCutter.

1. Input Your Genomic or cDNA Sequence

The target sequence to be analyzed can be from a local file (first box), retrieved from GenBank via its accession number (second box), or cut-and-pasted (third box). In this demonstration, the sequence for human SM22alpha was pasted (see Figure 2.1).

Once the DNA sequence has been entered, you should select the options regarding topology of the DNA (linear or circular) and set "Enzymes to use." The default search uses NEB enzymes. NEBCutter also shows all

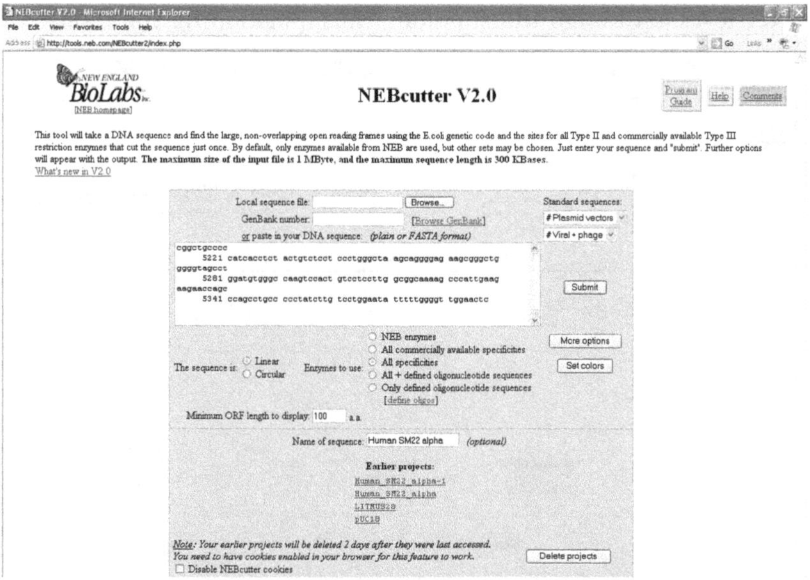
NEBcutter V2.0 - Microsoft Internet Explorer
File Edit View Favorites Tools Help
http://tools.neb.com/NEBcutter2/index.php
NEW ENGLAND BioLabs Inc.
[NEB homepage]
NEBcutter V2.0
Program Guide
Help
Comments
This tool will take a DNA sequence and find the large, non-overlapping open reading frames using the E.coli genetic code and the sites for all Type II and commercially available Type III restriction enzymes that cut the sequence just once. By default, only enzymes available from NEB are used, but other sets may be chosen. Just enter your sequence and "submit". Further options will appear with the output. The maximum size of the input file is 1 MByte, and the maximum sequence length is 300 KBases.
What's new in V2.0
Local sequence file:
Browse...
GenBank number:
[Browse GenBank]
or paste in your DNA sequence: (plain or FASTA format)
Standard sequences:
Plasmid vectors
Viral + phage
cggctgcccc
5221 catcacctct actgtctcct ccctgggcta agcaggggag aagcgggctg
ggggtagcct
5281 ggatgtgggc caagtccact gtcctccttg gcggcaaaag cccattgaag
aagaaccagc
5341 ccagcctgcc ccctatcttg tcctggaata tttttggggt tggaactc
Submit
The sequence is: Linear Circular
Enzymes to use: NEB enzymes
All commercially available specificities
All specificities
All + defined oligonucleotide sequences
Only defined oligonucleotide sequences
[define oligos]
More options
Set colors
Minimum ORF length to display: 100 a.a.
Name of sequence: Human SM22 alpha (optional)
Earlier projects:
Human_SM22_alpha-1
Human_SM22_alpha
LITMUS28
pUC18
Note: Your earlier projects will be deleted 2 days after they were last accessed.
You need to have cookies enabled in your browser for this feature to work.
Disable NEBcutter cookies
Delete projects

FIGURE 2.1 Restriction mapping using NEBcutter2 program. Input of genomic or cDNA sequence and selection of parameters for sequence analysis is shown.

nonoverlapping open reading frames (ORFs) spanning the methionine residue (Met) to the stop codon. The default for ORF size is 100 amino acids, but it can be modified to bigger or smaller values. Giving a name to the sequence (box at the bottom) is useful, especially if several searches are performed. Clicking on the "More options" button will open a window with additional selections such as methylation modifications, usage of type I enzymes, changing the genetic code to one of another species, as well as the possibility of searching a selected region of the input DNA. Projects are saved and can be retrieved for a few days.

2. *Results: Display and Analysis*

After clicking the "Submit" button, results of the search are shown for enzymes that cut once (Figure 2.2). Cleavage codes and enzyme name codes are described at the top in color, below the sequence name. If you select the 2- or 3-cutter enzymes (bottom box) option, a redrawing of the sequence results is displayed. Clicking on "List" shows a table with enzymes (specified by number of cuts) and their cleavage positions. For example, Figure 2.3 shows the list of enzymes that do not cut the DNA. Pointing and/or clicking on the displayed features in the results window retrieves additional information. Thus, positioning the mouse over

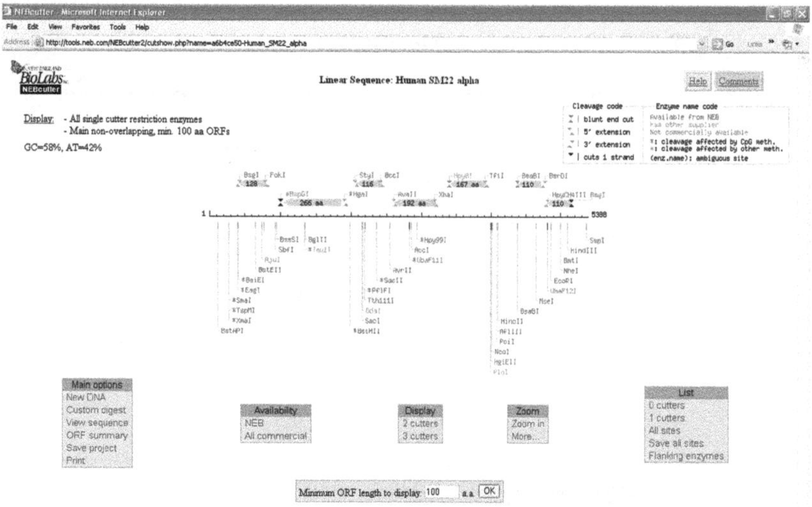

Linear Sequence: Human SM22 alpha
Display: - All single cutter restriction enzymes
- Main non-overlapping, min. 100 aa ORFs
GC=58%, AT=42%
Cleavage code
Enzyme name code
Main options
New DNA
Custom digest
View sequence
ORF summary
Save project
Print
Availability
NEB
All commercial
Display
2 cutters
3 cutters
Zoom
Zoom in
More...
List
0 cutters
1 cutters
All sites
Save all sites
Flanking enzymes
Minimum ORF length to display 100 a.a. OK

FIGURE 2.2 Results display using NEBcutter2 program. Sequence is represented in linear form, and sites for enzymes that cut only once (default) are shown (for details, see text).

the name of the restriction enzymes pops up the recognition site and the position of cutting. Clicking on the name of the enzymes instead, opens a new window with complete information about that enzyme, including information on isoschizomers and enzymes that (unambiguously) generate compatible cohesive sites.

All long ORFs are denoted by filled arrows, flanked by the closer restriction enzymes that could be used in a complete digest to excise each

NEBcutter - Microsoft Internet Explorer
File Edit View Favorites Tools Help
Address http://tools.neb.com/NEBcutter2/listbycuts.php?name=a6b4ce50-Human_SM
Go
Links
NEW ENGLAND BioLabs Inc.
NEBcutter
Enzymes that don't cut
Human SM22 alpha
Help
Comments
[Back to main display]
Number of cuts = 0
OK
Save as text file

#	Enzyme	Specificity
1	AatII	G▲ACGT▼C
2	AclI	AA▼CG▲TT
3	AfeI	AGC▼▲GCT
4	AflIII	C▼TTAA▲G
5	AgeI	A▼CCGG▲T
6	AlfI	▲NN▼(N)$_{10}$GCA(N)$_{6}$TGC(N)$_{10}$▲NN▼
7	ApaLI	G▼TGCA▲C
8	AscI	GG▼CGCG▲CC
9	AseI	AT▼TA▲AT
10	AsiSI	GCG▲AT▼CGC
11	BaeI	▲(N)$_{5}$▼(N)$_{10}$ACNNNNGTAYC(N)$_{7}$▲(N)$_{5}$▼
12	BamHI	G▼GATC▲C
13	BceAI	ACGGC(N)$_{12}$▼NN▲
14	BcgI	▲NN▼(N)$_{10}$CGA(N)$_{6}$TGC(N)$_{10}$▲NN▼
15	BclI	T▼GATC▲A
16	BmgBI	CAC▼▲GTC
17	BsaAI	YAC▼▲GTR
18	BsaWI	W▼CCGG▲W
19	BsiWI	C▼GTAC▲G
20	BsmI	GAATG▲CN▼
21	BspDI	AT▼CG▲AT
22	BspEI	T▼CCGG▲A
23	BspHI	T▼CATG▲A
24	BsrGI	T▼GTAC▲A
25	BstBI	TT▼CG▲AA
26	BstZ17I	GTA▼▲TAC
27	BtgZI	GCGATG(N)$_{10}$▼NNNN▲
28	ClaI	AT▼CG▲AT
29	CspCI	▲NN▼(N)$_{11}$CAA(N)$_{5}$GTGG(N)$_{10}$▲NN▼

FIGURE 2.3 Analysis of restriction mapping by selected features. An example of an option offered in "List" box displayed in Figure 2.2 is shown. The table lists enzymes (and their recognition sites) that do not cut the selected sequence.

ORF. The length of each ORF is also shown. Clicking on an ORF opens a new window (ORF Sequence, Figure 2.4) that displays the coding region coordinates and the deduced protein sequence in a one-letter code. Additionally, further options exist to locate flanking restriction enzyme sites or to BLAST the ORF sequence against GenBank. Other possibilities for analysis of the selected ORF are "Edit" and "Delete." The "Locate multiple cutters that excise this ORF" function displays flanking sites for restriction enzymes (Figure 2.5) and their number; the "Silent Mutagenesis" option shows unique sites within the ORF that can be created by mutating one or more bases without changing the protein sequence (Figure 2.6). In our example, it is composed of two pages; the first one is shown in Figure 2.6. Clicking on "List all sites" displays a table that contains detailed information on the candidate mutated bases (shown in red, underlined and highlighted) and the protein sequence (in a three-letter code) underneath. The enzyme's name, restriction site position, and specificity are also displayed. The "Help" box gives additional valuable information.

Back to the initial display (Figure 2.2): A red mark can be made by clicking once (or twice in order to select a region bounded by the two marks) on the horizontal line that represents the DNA sequence. The selected position or region can be zoomed to visualize it in more detail. Finally, the "Main options" box includes a "Custom digest" selection to create maps with enzymes of your choice and with the ability of displaying the digests in a digital gel. Options include any set of enzymes that have sites within the DNA, enzymes with compatible ends or buffers, enzymes producing particular kinds of termini, and others. The "View sequence"

ORF Sequence
Human SM22 alpha
Coding region: complement(2621..3199)
[Edit] - [Delete] - [Add new ORF] - [Locate multiple cutters that excise this ORF] - [Silent Mutagenesis]
Protein sequence:
Blast this sequence at NCBI

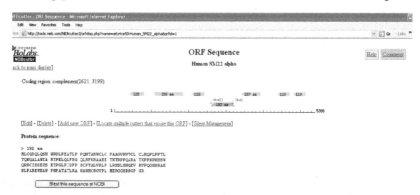

FIGURE 2.4 Predicted ORF sequence using NEBcutter2 program (for details, see text).

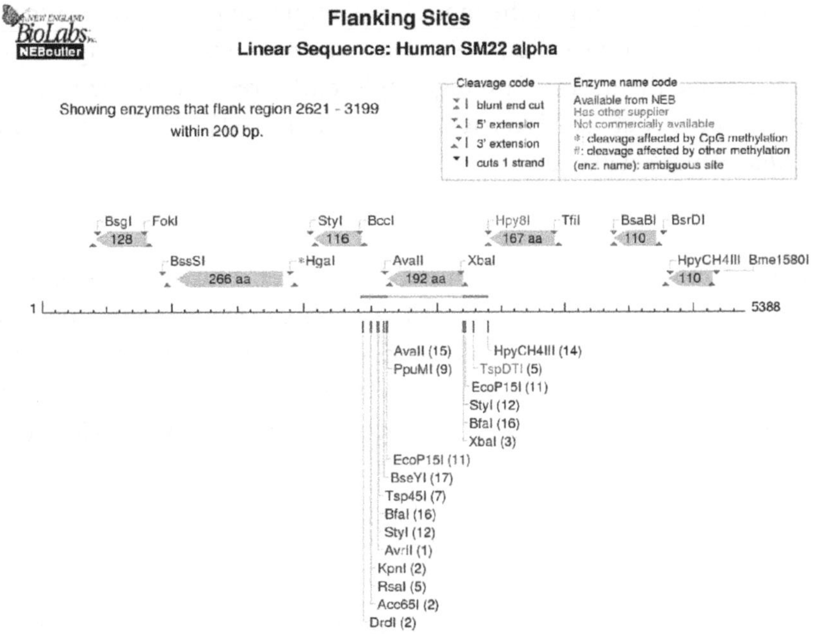
NEW ENGLAND BioLabs Inc.
NEBcutter
Flanking Sites
Linear Sequence: Human SM22 alpha
Showing enzymes that flank region 2621 - 3199 within 200 bp.
Cleavage code
blunt end cut
5' extension
3' extension
cuts 1 strand
Enzyme name code
Available from NEB
Has other supplier
Not commercially available
*: cleavage affected by CpG methylation
#: cleavage affected by other methylation
(enz. name): ambiguous site
BsgI
FokI
128
StyI
BccI
116
Hpy8I
TfiI
167 aa
BsaBI
BsrDI
110
BssSI
266 aa
*HgaI
AvaII
192 aa
XbaI
HpyCH4III
Bme1580I
110
1
5388
AvaII (15)
PpuMI (9)
HpyCH4III (14)
TspDTI (5)
EcoP15I (11)
StyI (12)
BfaI (16)
XbaI (3)
EcoP15I (11)
BseYI (17)
Tsp45I (7)
BfaI (16)
StyI (12)
AvrII (1)
KpnI (2)
RsaI (5)
Acc65I (2)
DrdI (2)

FIGURE 2.5 Restriction enzyme sites flanking ORFs using NEBcutter2 program. This option displays the enzymes that potentially release a selected ORF and their cleavage sites for downstream manipulations such as mutagenesis and expression purposes.

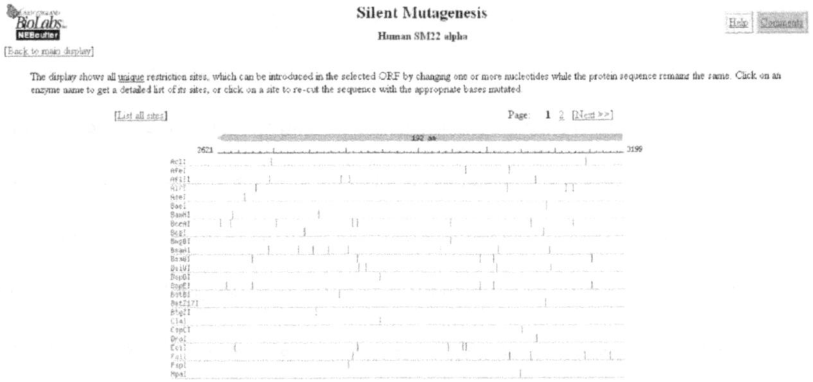
Silent Mutagenesis
Human SM22 alpha
NEBcutter
192 aa
2621
3199

FIGURE 2.6 Silent mutagenesis predicted by NEBcutter2 program. This option displays unique restriction enzyme sites that can be created by nucleotide substitutions without changing the ORF (for details, see text).

option allows browsing of the input DNA sequence file. "ORF summary" provides a table of the genes, their coordinates, polypeptide lengths, protein IDs at GenBank, and flanking, single-cutter restriction enzymes.

Importantly, published restriction maps generated by NEBcutter should be cited (Vincze et al., 2003).

Part III Sample Data

The first 1500 bases of the genomic DNA sequence used in the previous demo process (for which 5388 bases was used) are shown in the following text. It is the human SM22 alpha gene starting at the initiation of transcription. It has the accession number AF013711, for *Homo sapiens* 22-kDa actin-binding protein (SM22) gene, complete cds (Camoretti-Mercado et al., 1998).

```
gi|2501853|gb|AF013711.1|HUMSM22S2[2501853]

1 atcctgtctg tccgaaccca gacacaagtc ttcactcctt cctgcgagcc ctgaggaagc

61 cttgtgagtg cattggctgg ggcttggagg gaagttgggc tggagctgga caggagcagt

121 gggtgcattt caggcaggct ctcctgaggt cccaggcgcc agctccagct ccctggctag

181 ggaaacccac cctctcagtc agcatggggg cccaagctcc aggcagggtg ggctggatca

241 ctagcgtcct ggatctctct cagactgggc agccccgggc tcattgaaat gccccggatg

301 acttggctag tgcagaggaa ttgatggaaa ccaccggggt gagagggagg ctccccatct

361 cagccagcca catccacaag gtgtgtgtaa gggtgcaggc gccggccggt taggccaagg

421 ctctactgtc tgttgcccct ccaggagaac ttccaaggag gtgagtgtgt gaacgcacct

481 gtgttggagc acggtgtccc actctggcgg gtccccaggg cctgagcagc agcgatagcc

541 ctgtgacaat gtgaagggcc acagaactct tgtattccag tcaggggcaa agagtggaga

601 cggaaggcct gtccttctga caagcagccc cttccactgt ctgacagtgg gaggtcaccc

661 ccactgtagc agaggggtgg ggggcgggta ctgccaagga ggagctttgg agtgataggc

721 ggggcaggcc tgggaccctt ggtctttccc aaagggtggt tccctttcaa agttgctatt

781 ccaaaggtag cagtgaggat ggcacaaatt tatgtagcca ggccactcct gtttgtccgt

841 ggagtggaga ggagccaccc tcctgccctc tcagaggtcc aggtgtgctc actcctattt

901 gggaagagag aaggggcaca gcagccctgc aggggcccct ggctctgctc gtgtttctgg

961 gcttctgctg gtagggggtg cgaccttccc gtttgaccct ggactttctt tctccactgc
```

```
1021 ccactttcac ccactgggca gctcagggga ggggtcctgg caggagccac agggcaggag

1081 ggtccatgcc tgtccatgct gtcctgggag tgtcttgaga tgcccctggg ggggccgccc

1141 tgtaatcatc ctcctctccc cctccccact ggctcgaagg caggactcgg gaggcttcat

1201 cagaggactc taaaggattt ctggggattc tccacttttc gaccctgacc caggaggagg

1261 aaggggaagg atggtggtgc tgggtgggag tggggatggt gtgtgcttca tcccccctctg

1321 accgaaatcc taatcttgtc tctagatctg ggggctgcag tgttgtgtac ctgtcacccct

1381 tagccttgct gctttgacct gtattgtctg ttctgaccct cctgagactg gaagctgggg

1441 gtaggggaca cactctcctt ccatcctgtt cctcaggagc ccagcagggg tgcagaaggg
```

SECTION 2 PCR APPLICATION

Part I Introduction

Polymerase chain reaction (PCR) is an enzymatic *in vitro* procedure by which a desired DNA sequence of any origin (viral, bacterial, animal, plant, or human) can be selectively amplified several orders of magnitude (hundreds of millions of times) in a short time (from a few hours down to 5 to 10 min with new heating and cooling gas-based instruments). PCR was first described in 1985 by Kary B. Mullis, who won the Nobel Prize in Chemistry in 1993 for this invention. PCR allows amplification from minute amounts of DNA, a task that would require days with standard recombinant methodologies. PCR technology has greatly impacted many areas of science, including basic research, clinical medicine, genetic disease diagnostics, forensic, and evolutionary biology. Justly, it has been said that PCR is to genes what Gutenberg's printing press was to the written word (Johannes Gensfleisch zur Laden zum Gutenberg, 1398–1468, was a German goldsmith and inventor credited with inventing movable type printing in Europe in 1450).

1. How PCR Works

PCR is based on the utilization of specialized thermostable DNA polymerases. It requires, therefore, a DNA template that harbors the target sequence, all four dNTPs, and two specific DNA primers (each about 20 bases long) that flank the sequence to be amplified. To synthesize DNA fragments (amplicons) by PCR, repeated heating and cooling cycles are generated automatically using a thermocycler, which alters the temperature of the reaction mixture to certain preestablished values and duration. The mixture is first heated to separate the strands of double-stranded DNA (*denaturation*), and then is cooled to an optimal temperature to allow the primers to bind to their complementary target sequences (*annealing*). In a third step, the temperature is modified again to let the enzyme extend the primers into new complementary strands (*extension*). Highly specific amplification of the target DNA occurs exponentially (by a factor of 2^n, where n = number of cycles) because each new double-stranded DNA is separated in the next cycle to become two templates for further synthesis. In theory, every cycle will duplicate the amount of DNA present and, for example, after 20 PCR cycles, the target DNA will be amplified by a millionfold.

2. Quantitative Real-Time PCR

Quantitative real-time PCR (q-RT-PCR) has revolutionized our ability to measure nucleic acid abundance. Real-time PCR instruments quantify the amount of PCR product at *each step of the reaction* (in real time), enhancing the accuracy, reproducibility, and ease of quantification compared to classic PCR-based methods. These instruments support chemistries for template detection that include SYBR green dye intercalation (the most popular method), hybridization probes, hydrolysis probes, and molecular beacons. Primer sets for q-RT-PCR may be designed using standard primer design algorithms (see following text).

PCR amplicons are produced exponentially, but because it takes a number of cycles to readily detect enough product (either from the intercalated SYBR green or by using fluorescent primers), the plot of fluorescence vs. cycle number is sigmoidal. At later cycles, the substrates become depleted and the curve starts to flatten. The point on the curve in which the fluorescence begins to increase rapidly is termed the threshold cycle (Ct value). The plot of Ct vs. template amounts is linear and allows comparison of Ct values between multiple reactions. Therefore, calculation of the concentration of the target can be performed from a standard curve, or relative to control genes. The slope of this line provides a measure of PCR efficiency.

3. PCR Applications

PCR has impacted wide areas across scientific disciplines and clinical specialties, including applications in molecular biology, and in the diagnosis and research of cancers, infectious and genetic diseases, and other medical conditions. PCR technology has increased the speed with which studies can be performed, which directly affects the capacity for simultaneous sample handling. PCR is a highly sensitive and reliable process, easily adapted for automation. However, great care should be taken to prevent, detect, and remediate potential contamination.

PCR is a valuable laboratory tool used routinely for tasks that can be done more easily and faster than with other methods. Numerous applications based on PCR technology exist, with dozen of variations of published methods available. Moreover, multiplex PCR offers the ability to amplify not just one region of the DNA or RNA but multiple portions, facilitating the interrogation and survey of several regions simultaneously in a single reaction. In the following text we briefly describe two general applications of PCR as representative examples: DNA subcloning and PCR-mediated *in vitro* mutagenesis. Subcloning of DNA targets using PCR allows cell-free cloning of DNAs into plasmid vectors when no convenient restriction enzymes sites are available to use classic approaches. Three strategies are commonly employed to subclone PCR products: (1) T/A cloning, (2) addition of restriction sites, and (3) blunt-end ligation. T/A cloning takes advantage of the property of Taq DNA polymerase of adding a single A residue to the 3' end of amplified DNA fragments. These products are easily ligated with vectors containing overhanging T residues in their cloning polylinker region. Addition of restriction sites is another common method in which PCR primer pairs incorporate restriction enzyme recognition sites into their 5' ends, which are subsequently used for subcloning. For blunt-end cloning of PCR products, "polishing" enzymes such as T4 polymerase or Pfu polymerase are used to remove overhanging single nucleotides and facilitate cloning into vectors digested with an appropriate enzyme such as SmaI.

PCR-mediated *in vitro* mutagenesis exploits the elevated error rate of Taq polymerase in the presence of $MnCl_2$ and high $MgCl_2$ concentrations to incorporate random mutations. The advantage over chemical mutagenesis lies in the selectivity of targeting the random mutations to a defined segment of DNA flanked by the PCR primers. In addition, PCR carried out with designed nested primers allows generation of a series of DNA fragments with progressively smaller regions either at the 5' or 3' end of the external amplicon.

PCR is widely used in gene expression studies using cDNA templates generated by reverse transcription from messenger RNA (RT-PCR). PCR is particularly useful for amplification of differentially expressed gene sequences because standard differential screening and subtraction hybridization require large amounts of RNA to synthesize sufficient quantities of an enriched cDNA probe for library screening. PCR allowed the development of Serial Analysis of Gene Expression (SAGE), a powerful quantitative technique for profiling gene expression.

In the clinical arena, there are three main diagnostic applications of PCR: detection of pathogens, screening specific genes for unknown mutations, and genotyping. For instance, PCR permits mutation detection and early identification of deadly infective agents like SARS, small pox, HIV, and influenza. PCR can be applied for tumor cell detection as well, mixed chimerism after bone marrow transplantation, and noninvasive prenatal screening of fetal DNA in the maternal circulation. Single-strand conformational polymorphism (SSCP) is one of the most widely used methods for detecting single base pair mutations. SSCP is based on the sequence-dependent intramolecular folding that is responsible for the differential migration of two single-stranded DNA molecules of identical length but dissimilar sequence on nondenaturing acrylamide gel. Finally, efficient extraction of RNA and DNA from formaldehyde fixed tissues was recently achieved and, importantly, it was shown that it is possible to perform PCR on these often-damaged short DNAs. This accomplishment gives researchers the exciting opportunity of utilizing PCR to obtain information from archival materials collected many years ago, thus contributing to the understanding of common complex, chronic diseases.

Applications of real-time PCR include, among others, gene expression quantification and detection, validation of gene expression data obtained by microarray analysis, measurement of DNA copy number, detection and quantification of viral particles and potentially lethal microorganisms, allelic discrimination, and SNP detection.

4. PCR Primer Design

Successful PCR relies on the utilization of suitable primers. Human-designed PCR primers often fail due to low amplification efficiency, nonspecific amplification, and primer-dimers formation. Several programs are available that help design the most appropriate set of primers for a given application; many of these tools are freely available online. Several companies that synthesize custom oligos usually offer either in-house, commercial, or academic sites for primer design. With these computational tools, primer pairs are computed from user-selected target regions and then screened against a series of parameters to maximize priming efficiency for trouble-free PCR. The following attributes are included in primer design: target sequence, amplicon length, cross homology with related genes and pseudogenes (if present), amplicon location (distance from 3' end), primer Tm (melting temperature), primer length, primer deltaG 3' end, intron spanning, GC content, and primer hairpin structure. For qRT-PCR, added parameters include perfect probe/template annealing Tm, probe 5' and 3' extensions, probe melting point, probe length, and probe GC content. Getting these characteristics right is critical to achieving sensitive, specific, and reproducible PCR results.

In the following text is a list of some online sites for the design PCR primers (see also Table 2.3). Several sites include the development of repositories for primer sets, reaction conditions, and the primers themselves that would benefit investigators interested in contributing to and taking advantage of the information available:

- **Primer3** designs PCR primers from DNA sequences according to thermodynamic, primer size, and product size characteristics (Rozen and Skaletsky, 2000). Primer3 software can check existing primers and design hybridization probes as well. This can be useful, for example, for spotted arrays for mRNA expression profiling. Primer3 was developed at the Whitehead Institute and the Howard Hughes Medical Institute. The Primer3 Web site is also funded by the National Institutes of Health.
- **AutoPrime** designs primers for real-time PCR for eukaryotic gene expression. Primer pairs are selected in such a way that at least one of them matches an exon-exon border sequence that is present in the cDNA but not in the genomic sequence. Alternatively, the pair is designed by placing each primer in a different exon so that a genomic product would include a long intronic sequence unlikely to be amplified under the running PCR conditions.
- **RTPrimerDB** is a public database for primer and probe sequences used in real-time PCR that employs popular chemistries such as SYBR Green I, Taqman, hybridization probes, and molecular

TABLE 2.3 Web Sites of Some Programs That Perform PCR Primer Design

Site	Address
Primer3	http://frodo.wi.mit.edu/cgi-bin/primer3/primer3_www.cgi
AutoPrime	http://www.autoprime.de/AutoPrimeWeb
RTPrimerDB	http://medgen.ugent.be/rtprimerdb/index.php
PrimerBank	http://pga.mgh.harvard.edu/primerbank/index.html
QPPD	http://web.ncifcrf.gov/rtp/gel/primerdb/

beacons. This site encourages users to submit their validated primer and probe sequences, so that other users can benefit from their experience. The goals of this site are twofold: to prevent time-consuming primer design and experimental optimization (Pattyn et al., 2006), and to introduce a certain level of uniformity and standardization among different laboratories.

- **PrimerBank** is another public resource for PCR primers for gene expression detection or quantification (real-time PCR). It contains about 180,000 primers covering most known human and mouse genes. The company claims that primer design algorithm has been extensively tested by real-time PCR experiments for PCR specificity and efficiency.
- **QPPD** (Quantitative PCR Primer Database) provides information about primers and probes gathered from published articles cited in PubMed. Primers are used to quantify human and mouse mRNA.

Part II Step-By-Step Tutorial

The following is a guide to designing primers using the Primer3 program.

1. Sequence Input and Parameters Selection

The first step is to cut and paste the sequence in the input window (Figure 2.7). Note that the entire window has been divided into two figures, (Figure 2.7 and Figure 2.8). Sequences should be devoid of cloning artifacts or chimeric sequences and should not contain repetitive elements. Low-quality bases should be changed to N's or be made part of "Excluded Regions." "Sequence Id" is an assigned identifier that is reproduced in the output to enable you to identify the chosen primers. For standard PCR, only the "left" and "right" primer options should be selected. In the "Excluded Regions" box, primers will not overlap any region specified in this tag. Enter the value as a space-separated list of *start,length* pairs, where *start* is the index of the first base of the excluded region, and *length* is its length. This feature is useful for excluding regions of low sequence quality or rich in repetitive elements such as ALUs.

Figure 2.8 shows commonly used settings for primer design. "Product Size Range" displays a list of target size ranges (100 bp in the example shown). Primer3 attempts to pick primers starting with the first range, then goes to the next range and tries again. The program continues in this way until the last range is screened or until it has picked all necessary

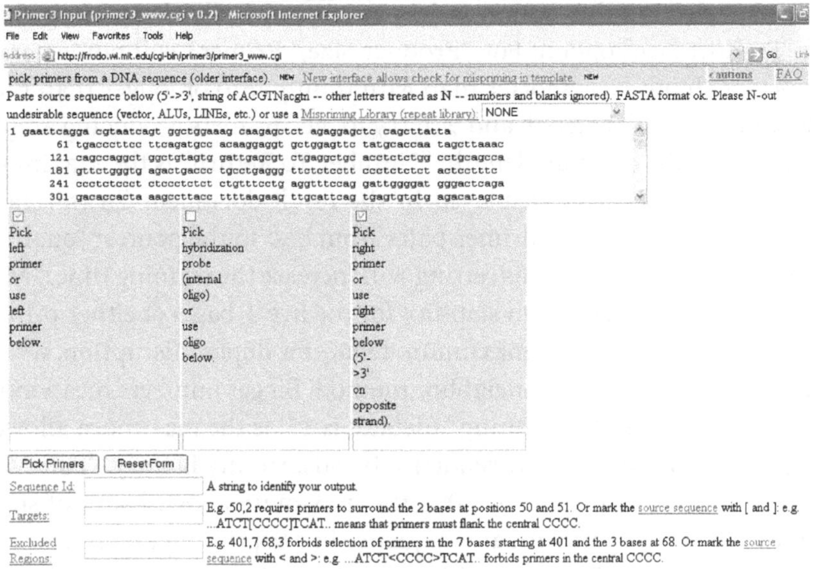
Primer3 Input (primer3_www.cgi v 0.2) - Microsoft Internet Explorer
File Edit View Favorites Tools Help
http://frodo.wi.mit.edu/cgi-bin/primer3/primer3_www.cgi
pick primers from a DNA sequence (older interface). NEW New interface allows check for mispriming in template NEW
cautions FAQ
Paste source sequence below (5'->3', string of ACGTNacgtn -- other letters treated as N -- numbers and blanks ignored). FASTA format ok. Please N-out undesirable sequence (vector, ALUs, LINEs, etc.) or use a Mispriming Library (repeat library): NONE
1 gaattcagga cgtaatcagt ggctggaaag caagagctct agaggagctc cagcttatta
61 tgacccttcc ttcagatgcc acaaggaggt gctggagttc tatgcaccaa tagcttaaac
121 cagccaggct ggctgtagtg gattgagcgt ctgaggctgc acctctctgg cctgcagcca
181 gttctgggtg agactgaccc tgcctgaggg ttctctcctt ccctctctct actcctttc
241 ccctctccct ctccctctct ctgtttcctg aggtttccag gattggggat gggactcaga
301 gacaccacta aagccttacc ttttaagaag ttgcattcag tgagtgtgtg agacatagca
Pick left primer or use left primer below.
Pick hybridization probe (internal oligo) or use oligo below.
Pick right primer or use right primer below (5'->3' on opposite strand).
Pick Primers
Reset Form
Sequence Id:
A string to identify your output.
Targets:
E.g. 50,2 requires primers to surround the 2 bases at positions 50 and 51. Or mark the source sequence with [and]: e.g. ...ATCT[CCCC]TCAT.. means that primers must flank the central CCCC.
Excluded Regions:
E.g. 401,7 68,3 forbids selection of primers in the 7 bases starting at 401 and the 3 bases at 68. Or mark the source sequence with < and >: e.g. ...ATCT<CCCC>TCAT.. forbids primers in the central CCCC.

FIGURE 2.7 Primer design using Primer3 program. The sequence to be used for primer selection is entered in the open box. (For details of sequence properties, see text). Primer design for standard PCR is shown. If real-time PCR is to be performed using Taqman, the hybridization probe can be designed by also selecting the middle box.

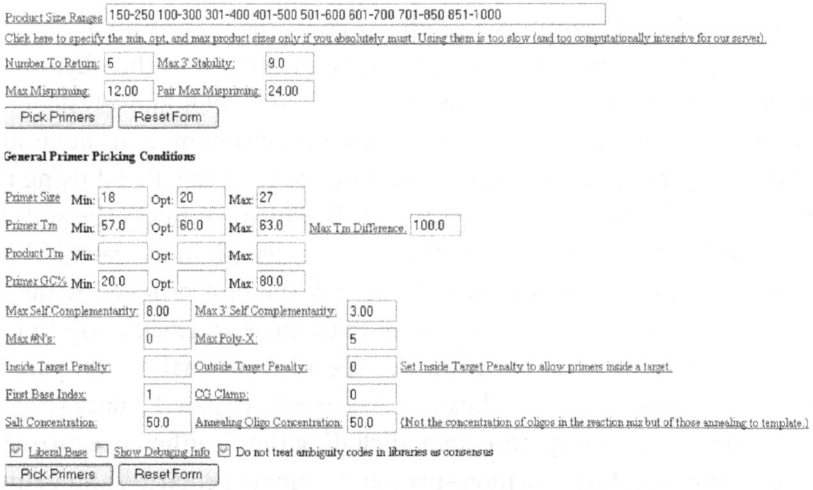
Product Size Ranges
150-250 100-300 301-400 401-500 501-600 601-700 701-850 851-1000
Click here to specify the min, opt, and max product sizes only if you absolutely must. Using them is too slow (and too computationally intensive for our server).
Number To Return: 5
Max 3' Stability: 9.0
Max Mispriming: 12.00
Pair Max Mispriming: 24.00
Pick Primers
Reset Form
General Primer Picking Conditions
Primer Size Min: 18 Opt: 20 Max: 27
Primer Tm Min: 57.0 Opt: 60.0 Max: 63.0
Max Tm Difference: 100.0
Product Tm Min: Opt: Max:
Primer GC% Min: 20.0 Opt: Max: 80.0
Max Self Complementarity: 8.00
Max 3' Self Complementarity: 3.00
Max #N's: 0
Max Poly-X: 5
Inside Target Penalty:
Outside Target Penalty: 0
Set Inside Target Penalty to allow primers inside a target.
First Base Index: 1
CG Clamp: 0
Salt Concentration: 50.0
Annealing Oligo Concentration: 50.0
(Not the concentration of oligos in the reaction mix but of those annealing to template.)
Liberal Base
Show Debuging Info
Do not treat ambiguity codes in libraries as consensus
Pick Primers
Reset Form

FIGURE 2.8 Primer design using Primer3 program (continuation). Commonly used settings for primer design are shown (for details, see text).

primers. Selecting this option is less computationally demanding than the "Product Size" option. For "Product Size," Primer3 will not generate primers with products shorter than Min (minimum) or longer than Max (maximum) length entered, and will try to pick primers producing amplicons close to the optimum length. "Number To Return" is the maximum number of primer pairs displayed in the result document (in this case, 5). The program sorts the primer pairs from best to the poorest "quality." Choosing a large value in this setting will increase the running time. "Max 3' Stability" is the maximum stability for the five 3' bases of either primer measured as a value of the maximum deltaG for duplex disruption, which is calculated by the nearest-neighbor method. Bigger numbers mean more stable 3' ends. "Max Mispriming" (default is 12) is the maximum allowed weighted similarity with any sequence in Mispriming Library. "Pair Max Mispriming" (default value is 24) is the maximum allowed sum of similarities of a primer pair with any sequence in Mispriming Library.

For the "General Primers Picking Conditions" section, enter values in "Primer Size" for minimum, optimum, and maximum primer lengths. Primer3 will attempt to pick primers close to Opt and not shorter than Min or longer than Max (which cannot be larger than 36 bases). The "Primer T_m" option sets the minimum and maximum melting temperatures (in Celsius) for a primer. Primer3 will try to pick primers with melting temperatures ("Primer T_m") close to Opt and not smaller than Min or larger than Max. "Maximum T_m Difference" is the maximum acceptable difference between the "Primer T_m" values. Primer3 uses the oligo melting temperature formula given by Rychlik et al. (1990) and Breslauer et al. (1986). The "Product T_m" is the minimum, optimum, and maximum melting temperature of the amplicon. Product T_m is calculated using the formula from Bolton and McCarthy (1962), which considers the sodium concentration, %GC content, and the sequence length in the calculations.

Primer3 uses penalty components to pick the best primers by taking into consideration values less or greater than those specified as optimum. The score indicates deviation from the specified optimal design parameters; a lower penalty score indicates a better primer pair. Penalty weights are governed by various parameters, including the T_m difference, primer–primer complementarity, primer–primer 3' complementarity, and primer pair mispriming similarity. Position Penalty Weight determines the overall weight of the position penalty in calculating the penalty for a primer. Deviations from the optimum primer size and T_m have a large influence on the penalty score. "Max Complementarity" considers the tendency of the primer to anneal to itself or to each other. The scoring system gives 1.00 for complementary bases, −0.25 for a match of any base with an N, −1.00 for a mismatch, and −2.00 for a gap. Only single base pair gaps are allowed. A score of 0.00 indicates that there is no reasonable local alignment between two oligos. "Max 3' Complementarity" tests for complementarity between left and right primers. A score of 0.00 indicates that there is no reasonable 3'-anchored global alignment between two oligos. The score of a local alignment will always be at least as great as the score of a global alignment. "Max Poly-X" refers to the length of mononucleotide repeat.

In "CG Clamp" box, the number of consecutive G's and C's at the 3' end of both the left and right primer are specified. When "Liberal base" is checked, Primer3 accepts IUB/IUPAC codes for ambiguous bases (see Table 2.1).

2. Results

After clicking the "Pick Primers" button, a window with a text document (Primer3 output) appears (Figure 2.9 and Figure 2.10).

The window displays a table with the first (i.e., best) left and right primer pair sequences shown on the right (always in 5' to 3') along with their start positions, lengths, melting temperatures, and percentage of G or C bases. Their self- and 3' self-complementarity scores are also displayed. Below the table, the predicted product size is shown as well as the DNA sequence and the positions where the left and right primers indicated by arrowheads map (in the figure, the forward left primer is boxed). Other programs provide extinction coefficient, molecular weight, µg/OD, nmol/OD, predicted secondary structures (hairpins), and potential duplexes when the oligo can anneal to any target sequence.

At the bottom of the window (Figure 2.10), similar information is provided for the remaining four additional primer pairs (note that "Number To Return" was set at 5 [see Figure 2.8]). Some statistics are given regarding the number of considered and unacceptable primers.

Some programs offer the possibility of sending your input sequence to NCBI Blast search for short, nearly exact matches. When working with software with multiplex capability, it may be necessary to try several values for melting temperature and %GC content before finding a multiplex primer set for your sequences. The time spent designing the primers should be worthwhile as it will reduce the time in the multiplex PCR optimization step.

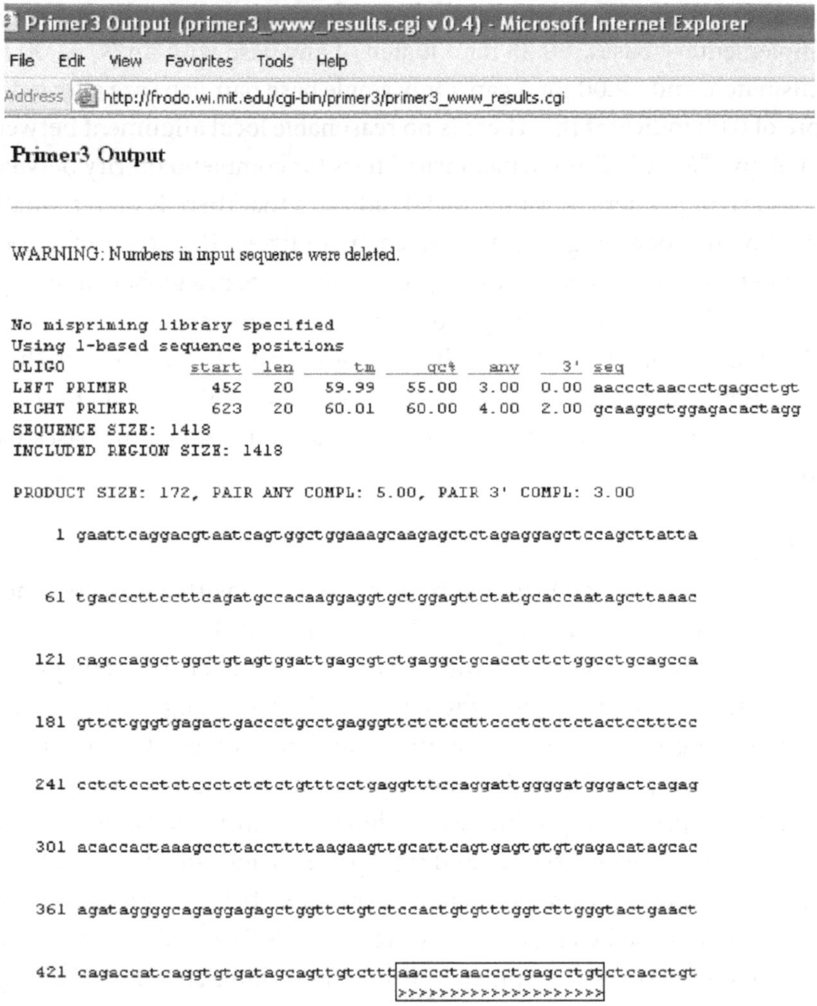
Primer3 Output (primer3_www_results.cgi v 0.4) - Microsoft Internet Explorer
File Edit View Favorites Tools Help
Address http://frodo.wi.mit.edu/cgi-bin/primer3/primer3_www_results.cgi
Primer3 Output
WARNING: Numbers in input sequence were deleted.
No mispriming library specified
Using 1-based sequence positions
OLIGO start len tm gc% any 3' seq
LEFT PRIMER 452 20 59.99 55.00 3.00 0.00 aaccctaaccctgagcctgt
RIGHT PRIMER 623 20 60.01 60.00 4.00 2.00 gcaaggctggagacactagg
SEQUENCE SIZE: 1418
INCLUDED REGION SIZE: 1418
PRODUCT SIZE: 172, PAIR ANY COMPL: 5.00, PAIR 3' COMPL: 3.00
1 gaattcaggacgtaatcagtggctggaaagcaagagctctagaggagctccagcttatta
61 tgacccttccttcagatgccacaaggaggtgctggagttctatgcaccaatagcttaaac
121 cagccaggctggctgtagtggattgagcgtctgaggctgcacctctctggcctgcagcca
181 gttctgggtgagactgaccctgcctgagggttctctccttccctctctctactcctttcc
241 cctctccctctccctctctctgtttcctgaggtttccaggattggggatgggactcagag
301 acaccactaaagccttacctttaagaagttgcattcagtgagtgtgtgagacatagcac
361 agataggggcagaggagagctggttctgtctccactgtgtttggtcttgggtactgaact
421 cagaccatcaggtgtgatagcagttgtctttaaccctaaccctgagcctgtctcacctgt
>>>>>>>>>>>>>>>>>>>>
481 cccttcccaagaccactgaagctaggtgcaagataagtggggaccctttctgaggtggta

FIGURE 2.9 Results displayed using Primer3 program. The output window lists, among others, the primer pair sequences and their properties (start positions, lengths, melting temperatures, and percentage of G or C bases), as well as the predicted amplification product size and their position within the sequence.

Part III Sample Data

The sequence used in the previous demo procedure is shown in the following text. It is the mouse SM22alpha gene flanking sequence, accession number L41161.1, GI:793751, for *Mus musculus* SM22 alpha gene (Solway et al., 1995). Position 1 in exon 1 at nucleotide 1341 is shown underlined in boldface.

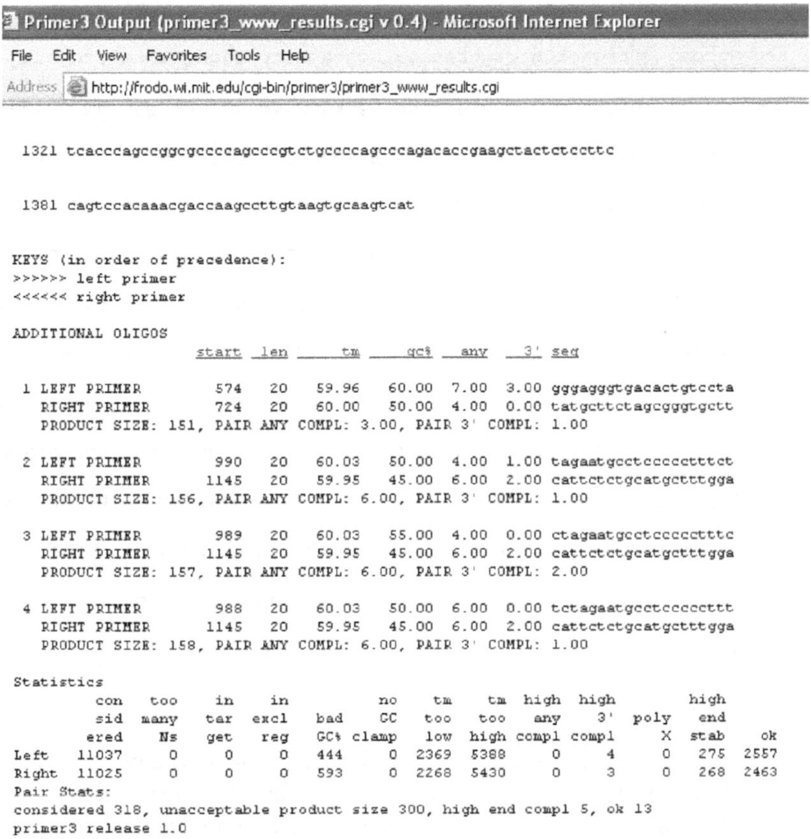

```
Primer3 Output (primer3_www_results.cgi v 0.4) - Microsoft Internet Explorer
File  Edit  View  Favorites  Tools  Help
Address  http://frodo.wi.mit.edu/cgi-bin/primer3/primer3_www_results.cgi

 1321 tcacccagccggcgccccagcccgtctgccccagcccagacaccgaagctactctccttc


 1381 cagtccacaaacgaccaagccttgtaagtgcaagtcat


KEYS (in order of precedence):
>>>>>> left primer
<<<<<< right primer

ADDITIONAL OLIGOS
                     start  len      tm     gc%   any    3' seq

 1 LEFT PRIMER         574   20   59.96   60.00  7.00  3.00 gggagggtgacactgtccta
   RIGHT PRIMER        724   20   60.00   50.00  4.00  0.00 tatgcttctagcgggtgctt
   PRODUCT SIZE: 151, PAIR ANY COMPL: 3.00, PAIR 3' COMPL: 1.00

 2 LEFT PRIMER         990   20   60.03   50.00  4.00  1.00 tagaatgcctccccctttct
   RIGHT PRIMER       1145   20   59.95   45.00  6.00  2.00 cattctctgcatgctttgga
   PRODUCT SIZE: 156, PAIR ANY COMPL: 6.00, PAIR 3' COMPL: 1.00

 3 LEFT PRIMER         989   20   60.03   55.00  4.00  0.00 ctagaatgcctccccctttc
   RIGHT PRIMER       1145   20   59.95   45.00  6.00  2.00 cattctctgcatgctttgga
   PRODUCT SIZE: 157, PAIR ANY COMPL: 6.00, PAIR 3' COMPL: 2.00

 4 LEFT PRIMER         988   20   60.03   50.00  6.00  0.00 tctagaatgcctcccccttt
   RIGHT PRIMER       1145   20   59.95   45.00  6.00  2.00 cattctctgcatgctttgga
   PRODUCT SIZE: 158, PAIR ANY COMPL: 6.00, PAIR 3' COMPL: 1.00

Statistics
         con   too    in    in          no    tm    tm  high  high        high
         sid  many   tar  excl   bad    GC   too   too   any    3'  poly   end
        ered    Ns   get   reg   GC% clamp   low  high compl compl     X  stab    ok
Left   11037     0     0     0   444     0  2369  5388     0     4     0   275  2557
Right  11025     0     0     0   593     0  2268  5430     0     3     0   268  2463
Pair Stats:
considered 318, unacceptable product size 300, high end compl 5, ok 13
primer3 release 1.0
```

FIGURE 2.10 Results displayed using Primer3 program (continuation). Additional primer pairs are displayed with some of their properties.

```
1 gaattcagga cgtaatcagt ggctggaaag caagagctct agaggagctc cagcttatta

61 tgacccttcc ttcagatgcc acaaggaggt gctggagttc tatgcaccaa tagcttaaac

121 cagccaggct ggctgtagtg gattgagcgt ctgaggctgc acctctctgg cctgcagcca

181 gttctgggtg agactgaccc tgcctgaggg ttctctcctt ccctctctct actcctttc

241 ccctctccct ctccctctct ctgtttcctg aggtttccag gattggggat gggactcaga

301 gacaccacta aagccttacc ttttaagaag ttgcattcag tgagtgtgtg agacatagca

361 cagataggggg cagaggagag ctggttctgt ctccactgtg tttggtcttg ggtactgaac

421 tcagaccatc aggtgtgata gcagttgtct ttaaccctaa ccctgagcct gtctcacctg

481 tcccttccca agaccactga agctaggtgc aagataagtg gggaccсttt ctgaggtggt

541 aggatctttc acgataagga ctattttgaa gggagggagg gtgacactgt cctagtcctc
```

```
601 ttaccctagt gtctccagcc ttgccaggcc ttaaacatcc gcccattgtc accgctctag
661 aaggggccag ggttgacttg ctgctaaaca aggcactccc tagagaagca cccgctagaa
721 gcataccata cctgtgggca ggatgaccca tgttctgcca cgcacttggt agccttggaa
781 aggccacttt gaacctcaat tttctcaact gttaaatggg gtggtaactg ctatctcata
841 ataaagggga acgtgaaagg aaggcgtttg catagtgcct ggttgtgcag ccaggctgca
901 gtcaagacta gttcccacca actcgatttt aaagccttgc aagaaggtgg cttgtttgtc
961 ccttgcaggt tcctttgtcg ggccaaactc tagaatgcct ccccctttct ttctcattga
1021 agagcagacc caagtccggg taacaaggaa gggtttcagg gtcctgccca taaaaggttt
1081 ttcccggccg ccctcagcac cgccccgccc cgacccccgc agcatctcca aagcatgcag
1141 agaatgtctc cggctgcccc cgacagactg ctccaacttg gtgtctttcc ccaaatatgg
1201 agcctgtgtg gagtgagtgg ggcggcccgg ggtggtgagc caagcagact tccatgggca
1261 gggaggggcg ccagcggacg gcagaggggt gacatcactg cctaggcggc ctttaaaccc
1321 ctcacccagc cggcgcccca gcccgtctgc cccagcccag acaccgaagc tactctcctt
1381 ccagtccaca aacgaccaag ccttgtaagt gcaagtcat
```

REFERENCES

1. Press release on the Nobel Prize in Physiology or Medicine 1978. http://nobelprize.org/nobel_prizes/medicine/laureates/1978/press.html.
2. Vincze, T., Posfai, J., and Roberts, R.J. NEBcutter: a program to cleave DNA with restriction enzymes. *Nucl Acids Res* 31: 3688–3691, 2003.
3. Camoretti-Mercado, B., Forsythe, S.M., LeBeau, M.M., Espinosa, R. III, Vieira, J.E., Halayko, A.J., Willadsen, S., Kurtz, B., Ober, C., Evans, G.A., Thweatt, R., Shapiro, S., Niu, Q., Qin, Y., Padrid, P.A., and Solway, J. Expression and cytogenetic localization of the human SM22 gene (TAGLN). *Genomics* 49(3), 452–457, 1998.
4. Bustin, S.A., Benes, V., Nolan, T., and Pfaffl, M.W. Quantitative real-time RT-PCR — a perspective. *J Mol Endocrinol* 34: 597–601, 2005.
5. Hajek, A.E., Jensen, A.B., Thomsen, L., Hodge, K.T., and Eilenberg, J. Molecular evolution and systematics: PCR-RFLP is used to investigate relations among species in the entomopathogenic genera *Eryniopsis* and *Entomophaga*. *Mycologia* 95: 262, 2003.
6. Kocher, T.D. PCR, direct sequencing, and the comparative approach. *Genome Res* 1: 217, 1992.
7. Breslaur, K.J., Frank, R., Blocker, H., and Marky, L.A. Predicting DNA duplex stability from the base sequence. *PNAS* 83: 3746–3750, 1986.
8. Rychlik, W., Spencer, W.J., and Rhoads, R.E. Optimization of the annealing temperature for DNA amplification *in vitro*. *Nucl Acids Res* 18(21): 6409–6412, 1990.

9. Solway, J., Seltzer, J., Samaha, F.F., Kim, S., Alger, L.E., Niu, Q., Morrisey, E.E., Ip, H.S., and Parmacek, M.S. Structure and expression of a smooth muscle cell-specific gene, SM22 alpha. *J Biol Chem* 270(22): 13460–13469, 1995.
10. Bolton, E.T., and McCarthy, B.J. A general method for the isolation of RNA complementary to DNA. *PNAS* 48: 1390–1397, 1962.
11. Pattyn, F., Robbrecht, P., Speleman, F., De Paepe, A., and Vandesompele, J. RTPrimerDB: the real-time PCR primer and probe database, major update 2006. *J Nucl Acids Res* 34(Database issue): D684–D688, 2006.
12. Rozen, S. and Skaletsky, H. Primer3 on the WWW for general users and for biologist programmers. *Methods Mol Biol* 132: 365–386, 2000.

ACKNOWLEDGMENTS

ALA, Blowitz Ridgeway Foundation, ATS, LAM Foundation, NIH

CHAPTER 3

Phylogenetic Analysis

Shui Qing Ye

CONTENTS

Phylogenetics is the field of biology that deals with identifying and understanding the relationships between the many different kinds of life on earth. In the past, phylogenetic analysis was heavily based on morphological data. With the sequencing of the human genome now complete and a number of other animal and plant genome sequences on the horizon, well-resolved molecular trees based on molecular data (nuclear DNA and derived protein data) have become an important tool in phylogenetical analysis. Molecular trees can serve as scaffolds for investigating evolutionary relationships. We are inching closer to understanding the organization and structure of the ancestral mammalian and plant genome, as well as the differences that make each species unique. This chapter commences with multiple sequence alignments in Section 1 and then introduces phylogenetical analysis in Section 2. In line with the format and style throughout this book, each section starts with a theoretical introduction in Part I, continues with a step-by-step tutorial in Part II, and ends with the presentation of sample data in Part III.

SECTION 1 MULTIPLE SEQUENCE ALIGNMENTS

Part I Introduction

1. Why Are Multiple Sequence Alignments Needed?

Multiple sequence alignments are an important tool in studying sequences, by aligning more than two DNA or protein sequences. Sequence alignment is the poster child of bioinformatics. Alignment is the most basic component of biological sequence manipulation, and it has diverse applications in sequence annotation, structural and functional predictions for genes and proteins, phylogeny and evolutionary analysis. The basic information multiple sequence alignment can provide is identification of conserved sequence regions. This is very useful in designing experiments to test and modify the function of specific proteins, in predicting the function and structure of proteins, and in identifying new members of protein families.

The completion of human genome sequencing has dramatically speeded up the discovery of new proteins. The number of newly available protein sequences far outpaces the limited number of determined protein three-dimensional structures, and multiple alignments of protein sequences remains the main method to infer protein structure, function, active sites, and evolutionary history.

2. What Is Involved in Multiple Sequence Alignments?

Four basic steps are involved in multiple sequence alignment: selecting DNA or protein sequences, inputting into an automatic multiple sequence alignment Program such as ClustalW, editing alignments, and interpreting the alignments.

Selecting DNA or protein sequences. Aligning protein sequences is the better choice because protein sequences are three times shorter than the corresponding DNA sequences, and protein sequences use a more informative alphabet of twenty amino acids in contrast to only four nucleotides in DNA sequences. Inputting protein or DNA sequences need to be converted to appropriate formats for an automatic multiple sequence alignment program such as ClustalW. Sequence formats are simply the way in which the amino acid or DNA sequence is recorded in a computer file. Different programs expect different formats, so if you are to submit a job successfully, it is important to understand what the various formats are. There are at least a couple of dozen sequence formats in existence at the moment. The program ClustalW accepts sequences in the following formats: Pearson (FASTA), Align/ClustalW (ALN/ClustalW), The National Biomedical Research Foundation/Protein Information Resource (NBRF/PIR), The European Molecular Biology Laboratory/Universal Protein Resource (EMBL/UniProt), Genetic Data Environment (GDE), Genetics Computer Group/ Multiple Sequence Format (GCG/MSF), and Rich Sequence Format (RSF). You can submit protein or DNA sequences in any of these formats to ClustalW. An example of the partial human pre-B-cell colony-enhancing factor (PBEF) in the FASTA format is provided in Figure 3.1. The FASTA sequence format is a widely accepted format. It starts with the greater than symbol (>), gene identification number, its reference protein accession number, and its name followed by the sequence.

```
>gi|5031977|ref|NP_005737.1| pre-B-cell colony enhancing factor 1 isoforma [Homo sapiens]
MNPAAEAEFNILLATDSYKVTHYKQYPPNTSKVYSYFECREKKTENSKLRKVKYEETVFYGLQYILNKYL
KGKVVTKEKIQEAKDVYKEHFQDDVFNEKGWNYILEKYDGHLPIEIKAVPEGFVIPRGNVLFTVENTDPE
CYWLTNWIETILVQSWYPITVATNSREQKKILAKYLLETSGNLDGLEYKLHDFGYRGVSSQETAGIGASA
HLVNFKGTDTVAGLALIKKYYGTKDPVPGYSVPAAEHSTITAWGKDHEKDAFEHIV......
```

FIGURE 3.1 Partial human PBEF protein sequence in the FASTA format.

Multiple sequence alignment in alignment programs such as ClustalW. The sequences can either be pasted into the Web form or uploaded from a file to the Web form in the ClustalW program. Sequences can be aligned across their entire length (global alignment) or only in certain regions (local alignment). This is true for pairwise and multiple alignments. Global alignments need to use gaps (representing insertions or deletions) although local alignments can avoid them, by aligning regions between gaps. The standard computational formulation of the pairwise problem is to identify the alignment that maximizes protein-sequence similarity, which is typically defined as the sum of substitution matrix scores for each aligned pair of residues, minus some penalties for gaps. This approach is generalized to the multiple sequence case by seeking an alignment that maximizes the sum of similarities for all pairs of sequences. A substitution matrix describes the likelihood that two residue types would mutate to each other in evolutionary time. Understanding theories underlying a given matrix can aid in making proper choice. ClustalW can adopt four types of matrices: Point Accepted Mutation (PAM), Blocks Substitution Matrix (BLOSUM), GONNET, and DNA Identity Matrix (Unitary Matrix). PAM matrices are traditionally amino acid scoring matrices, which refer to various degrees of sensitivity, depending on the evolutionary distance between sequence pairs. In this manner, PAM40 is most sensitive for sequences 40 PAMs apart. PAM250 is for more distantly related sequences and is considered a good general matrix for protein database searching. PAM40 and PAM250 mean 40 and 250 mutations per 100 amino acids of sequence, respectively. The BLOSUM matrices, also used for protein database search scoring (the default in blastp), are divided into statistical significance degrees, which in a way are reminiscent of PAM distances. The BLOSUM 45 matrix means a sequence blocks clustered at the 45% identity level. The BLOSUM matrices are most sensitive for local alignment of related sequences and are therefore ideal for identifying an unknown nucleotide sequence. The GONNET matrix is a different method to measure differences among amino acids; it was developed by Gonnet et al. using exhaustive pairwise alignments of the protein databases as they existed at that time. They used classical distance measures to estimate an alignment of the proteins. They then employed these data to estimate a new distance matrix. This was useful to refine the alignment, estimate a new distance matrix, and so on, iteratively. In the DNA Identity Matrix, you only get a positive score for a match, and a score of –10000 for a mismatch. As such a high penalty is given for a mismatch, no substitution should be allowed, although a gap may be permitted. A penalty is subtracted for each gap introduced into an alignment because the gap increases uncertainty into an alignment. A gap is a maximal consecutive run of spaces in a single string of a given alignment. It corresponds to an atomic insertion or deletion of a substring. A single mutation can create a gap (very common). There are several causes for gaps. Unequal crossover in meiosis can lead to insertion or deletion of strings of bases. DNA slippage in the replication procedure can result in the repetition of a string. Retrovirus insertions and translocations of DNA between chromosomes can also create gaps.

Editing alignments. When you generated a multiple sequence alignment with any kind of automatic program, you probably need to edit it manually or in an editing program before presenting or publishing the alignment. Multiple Align Show (http://bioinformatics.org/sms/index.html) and Jalview (http://www.jalview.org/) are two such multiple alignment editors. Multiple Align Show will be employed as a demo in the next part. Multiple Align Show accepts a group of aligned sequences (in FASTA or GDE format) and formats the alignment to your specifications. You can specify the number of residues per line, the font size and font style, and colors for different regions. You can also set a consensus level, which specifies the fraction of residues that need to be identical or similar in a column of the alignment for highlighting to be added. Multiple Align Show is one of many such programs that can be used to enhance the output of sequence alignment programs.

Interpreting the result of multiple sequence alignment. You want to identify important positions or motifs in your protein from your multiple alignment, which are conserved even when aligning distantly related proteins. When you find a good alignment with too many conserved positions, you need to add a few distantly related sequences one by one and check the effect of these sequences on the overall alignment quality. Those sequences that BLAST reported as marginal hits when you first scanned SWISS-PROT for homologous sequences can be construed as distantly related sequences. Initially aligned conserved patterns, if they survive after aligning with distantly related sequences, may be true conserved patterns. You can also resort to other pattern-identifying tools such as ScanProsite and MotifScan hyperlinked in ExPASy Proteomics Server Site (http://ca.expasy.org/) to verify these conserved patterns, revealed by the multiple sequence alignment.

TABLE 3.1 Several Useful URLS of Multiple Sequence Alignment Programs

3D-Coffee	http://igs-server.cnrs-mrs.fr/Tcoffee/tcoffee_cgi/index.cgi
MUSCLE	http://www.drive5.com/muscle/
PROBCONS	http://probcons.stanford.edu/
MAFFT	http://timpani.genome.ad.jp/Art was here, but was deleted.mafft/server/

3. New Development in Multiple Sequence Alignment Programs

In recent years, protein multiple sequence alignment tools have improved rapidly in both scalability and accuracy. Table 3.1 lists useful uniform resource locators (URLS) of several newly developed multiple sequence alignment (MSA) programs. Some features of these programs are briefly discussed as follows.

3D-Coffee. 3D-Coffee is designed to create protein sequence alignments that incorporate three-dimensional structural information, when appropriate structures exist. In principle, utilizing three-dimensional structures facilitates the alignment of distantly related sequences. Structural elements are generally more conserved than primary sequences, retaining their align ability well into the twilight zone (≤25% sequence identity). 3D-Coffee is a fast, simple, and accurate method for incorporating heterogeneous structural data into an alignment and to improve its accuracy, even when only one or two structures are available. The drawback is that using this information can be complicated.

MUSCLE (multiple sequence comparison by log-expectation). MUSCLE is a new progressive alignment package that is extremely fast and accurate. The first step in MUSCLE is to rapidly generate a rough draft of the alignment, using a very crude guide tree. The next stage in the process is to refine the rough draft by generating a more accurate guide tree, which is based on the initial alignment. A second progressive alignment is generated using this improved tree. The speed of MUSCLE was also demonstrated, by aligning 5000 sequences on a PC in 7 min. The latest version of MUSCLE, version 6, is a collaboration between the developers of MUSCLE and PROBCONS and uses a new refinement strategy based on the PROBCONS algorithm. It gives a significant increase in accuracy at a modest computational cost.

PROBCONS. PROBCONS is currently the most accurate multiple alignment method. Initially, all the sequences are aligned with each other using a pair-HMM (hidden Markov model) generated with the maximum expected accuracy objective function. Next, a consistency transformation is applied. The multiple alignment is then generated by using a progressive alignment scheme. The final alignment is then subjected to an iterative refinement protocol.

MAFFT. MAFFT uses a fast Fourier transform to quickly generate a guide tree for progressive alignment. A fast tree-based iteration strategy is then used to refine the alignment by optimizing the weighted sum of pairs (WSP) objective function. This protocol results in very accurate and very fast alignments.

Multiple alignments are so widely used that any further improvements to software or algorithms can have a significant impact on the scientific community. Future improvements of multiple sequence alignment programs are likely to come by combining sequence alignment with other information, such as known structures of some of the proteins being aligned or homology to a larger pool of proteins. Parameter selection for alignment tools remains an important problem, as demonstrated by the sensitivity of RNA benchmarking results to parameter choice. Algorithmically, consideration of all sequences at once as an alternative to progressive alignment (consistency-based methods are a step in this direction) has been shown to be an effective strategy. Finally, better utilization of phylogenetic relationships and incorporation of models of protein sequence evolution also hold promise for improved alignment performance.

Part II Step-By-Step Tutorial

The ClustalW program is used for the demonstration purpose because it is by far the most commonly used program for making multiple sequence alignments. ClustalW is a fully automatic program for global multiple alignment of DNA and protein sequences. The alignment is progressive and considers sequence redundancy. Trees can also be calculated from multiple alignments. The program has some adjustable parameters with reasonable defaults.

The Multiple Align Show program is used for the demonstration of editing and publishing multiple aligned sequences because it is user-friendly with a suite of molecular tools of multiple capability.

1. Align Multiple Protein Sequences

The demo example is to align human, mouse, and rat Pre-B-cell Colony-Enhancing Factor (PBEF) protein sequences:

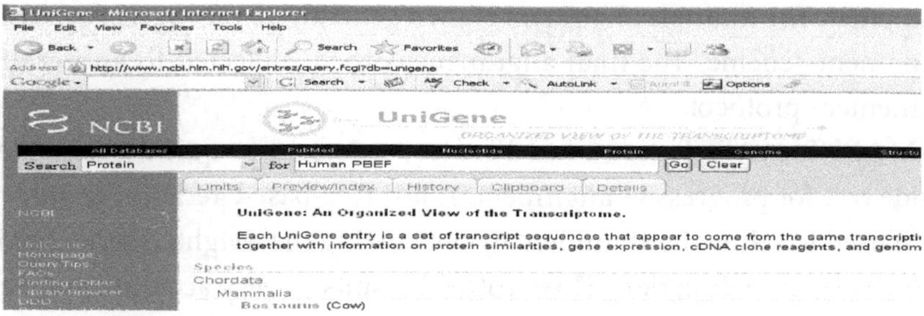
File Edit View Favorites Tools Help
Back Search Favorites
http://www.ncbi.nlm.nih.gov/entrez/query.fcgi?db=unigene
Search Check AutoLink Options
NCBI
UniGene
Search Protein for Human PBEF Go Clear
Limits Preview/Index History Clipboard Details
UniGene: An Organized View of the Transcriptome.
Each UniGene entry is a set of transcript sequences that appear to come from the same transcripti
together with information on protein similarities, gene expression, cDNA clone reagents, and genom
Chordata
Mammalia
(Cow)

FIGURE 3.2 Search for human PBEF protein sequence from the NCBI Unigene database. This screenshot displays how to initiate a retrieval of a protein sequence from the NCBI Unigene database using a protein name symbol.

1. Fetch human, mouse, and rat PBEF protein sequences from the NCBI Web site.
   a. As demonstrated in Figure 3.2, type the address (http://www.ncbi.nlm.nih.gov/entrez/query.fcgi?db=unigene) in your browser (Internet Explorer or others). In "Search," select "Protein" category. In the "for," type "human PBEF."
   b. Click "Go" in your browser or press "Enter" on your keyboard. The accession number of the human PBEF protein sequence will be displayed as in the Figure 3.3.
   c. Click "NP_005737" to display all human PBEF protein sequence information. Then in the Display, select "FASTA" format, and you will have the human PBEF protein sequence in the FASTA format (Figure 3.4).

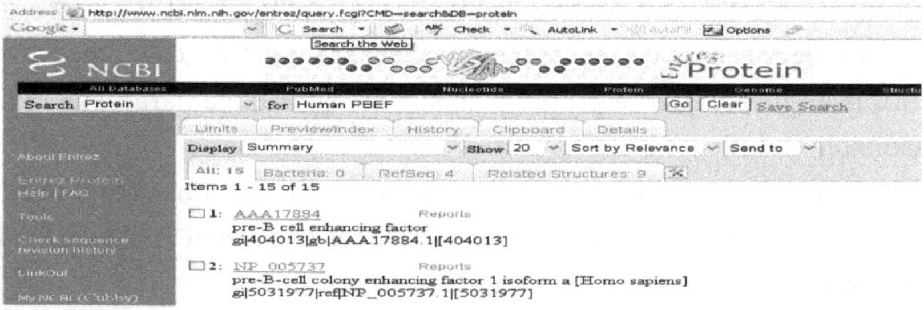
http://www.ncbi.nlm.nih.gov/entrez/query.fcgi?CMD=search&DB=protein
Search Check AutoLink Options
Search the Web
NCBI
Protein
Search Protein for Human PBEF Go Clear Save Search
Limits Preview/Index History Clipboard Details
Display Summary Show 20 Sort by Relevance Send to
All: 15 Bacteria: 0 RefSeq: 4 Related Structures: 9
Items 1 - 15 of 15
1: AAA17884 Reports
pre-B cell enhancing factor
gi|404013|gb|AAA17884.1|[404013]
2: NP_005737 Reports
pre-B-cell colony enhancing factor 1 isoform a [Homo sapiens]
gi|5031977|ref|NP_005737.1|[5031977]

FIGURE 3.3 Display the accession number of human PBEF protein sequence. Item 2 (NP_005737) in the screenshot is the reference accession number of human PBEF protein sequence.

NCBI
Protein
PubMed
Nucleotide
Protein
Genome
Structure
PMC
Search Protein
for
Go
Clear
Limits
Preview/Index
History
Display FASTA
Show 5
Send to
Range: from begin
to end
Refresh
1: NP_005737. Reports pre-B-cell colony...[gi:5031977]
>gi|5031977|ref|NP_005737.1| pre-B-cell colony enhancing factor 1 isoform a [Homo sapiens]
MNPAAEAEFNILLATDSYKVTHYKQYPPNTSKVYSYFECREKKTENSKLRKVKYEETVFYGLQYILNKYL
KGKVVTKEKIQEAKDVYKEHFQDDVFNEKGWNYILEKYDGHLPIEIKAVPEGFVIPRGNVLFTVENTDPE
CYWLTNWIETILVQSWYPITVATNSREQKKILAKYLLETSGNLDGLEYKLHDFGYRGVSSQETAGIGASA
HLVNFKGTDTVAGLALIKKYYGTKDPVPGYSVPAAEHSTITAWGKDHEKDAFEHIVTQFSSVPVSVVSDS
YDIYNACEKIWGEDLRHLIVSRSTQAPLIIRPDSGNPLDTVLKVLEILGKKFPVTENSKGYKLLPPYLRV
IQGDGVDINTLQEIVEGMKQKMWSIENIAFGSGGGLLQKLTRDLLNCSFKCSYVVTNGLGINVFKDPVAD
PNKRSKKGRLSLHRTPAGNFVTLEEGKGDLEEYGQDLLHTVFKNGKVTKSYSFDEIRKNAQLNIELEAAH
H

FIGURE 3.4 Human PBEF protein sequence in the FASTA format.

d. Repeat step a to step c to fetch mouse and rat PBEF protein sequences, respectively, in the FASTA format.

2. Load a group of protein sequences into the ClustalW program.
   a. In your browser (Internet Explorer or others), type the address (http://www.ebi.ac.uk/clustalw/) to get access to the ClustalW program (Figure 3.5).
   b. Change the first line of each sequence into human, mouse, and rat after sign >, respectively. Paste human PBEF, mouse PBEF, and rat PBEF protein sequence one by one into the sequence window of the ClustalW program (Figure 3.6). Keep the default setting, and click "Run."
   c. Results come in three sections: Scores Table, Alignment, and Guide Tree. Score Table lists pairwise scores. Alignment contains the detailed alignment file. The guide tree contains the tree used to guide its progressive alignment strategy, which will be

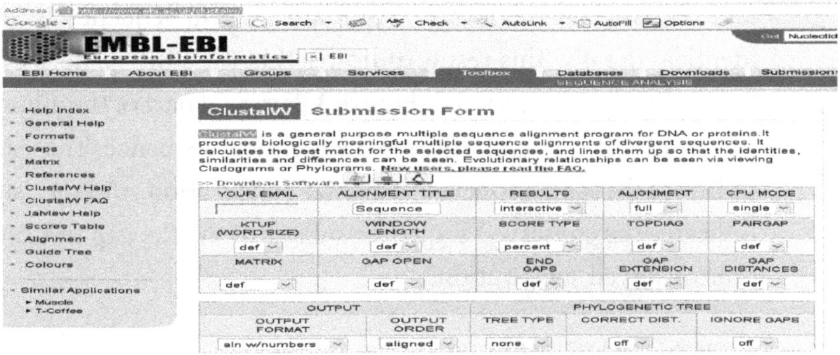
EMBL-EBI
European Bioinformatics
EBI Home
About EBI
Groups
Services
Toolbox
Databases
Downloads
Submissions
SEQUENCE ANALYSIS
Help Index
General Help
Formats
Gaps
Matrix
References
ClustalW Help
ClustalW FAQ
Jalview Help
Scores Table
Alignment
Guide Tree
Colours
Similar Applications
Muscle
T-Coffee
ClustalW Submission Form
YOUR EMAIL
ALIGNMENT TITLE
RESULTS
ALIGNMENT
CPU MODE
Sequence
interactive
full
single
KTUP (WORD SIZE)
WINDOW LENGTH
SCORE TYPE
TOPDIAG
PAIRGAP
def
percent
MATRIX
GAP OPEN
END GAPS
GAP EXTENSION
GAP DISTANCES
OUTPUT
PHYLOGENETIC TREE
OUTPUT FORMAT
OUTPUT ORDER
TREE TYPE
CORRECT DIST.
IGNORE GAPS
aln w/numbers
aligned
none
off

FIGURE 3.5 The EBI ClustalW server. This screenshot displays the parameter-setting window in the ClustalW program.

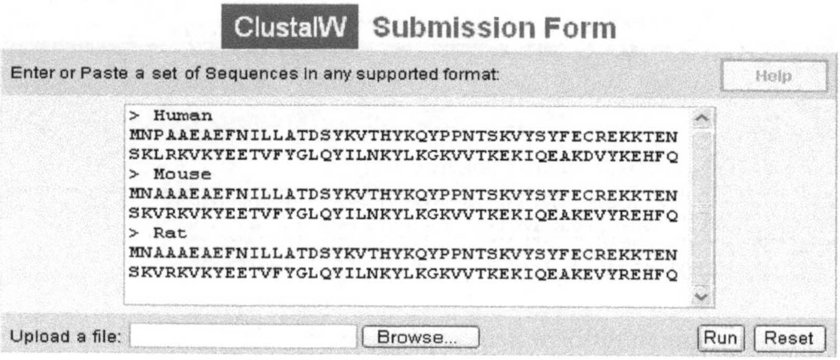
ClustalW Submission Form
Enter or Paste a set of Sequences in any supported format:
Help
> Human
MNPAAEAEFNILLATDSYKVTHYKQYPPNTSKVYSYFECREKKTEN
SKLRKVKYEETVFYGLQYILNKYLKGKVVTKEKIQEAKDVYKEHFQ
> Mouse
MNAAAEAEFNILLATDSYKVTHYKQYPPNTSKVYSYFECREKKTEN
SKVRKVKYEETVFYGLQYILNKYLKGKVVTKEKIQEAKEVYREHFQ
> Rat
MNAAAEAEFNILLATDSYKVTHYKQYPPNTSKVYSYFECREKKTEN
SKVRKVKYEETVFYGLQYILNKYLKGKVVTKEKIQEAKEVYREHFQ
Upload a file:
Browse...
Run
Reset

FIGURE 3.6 Submit 3 PBEF protein sequences in FASTA formats.

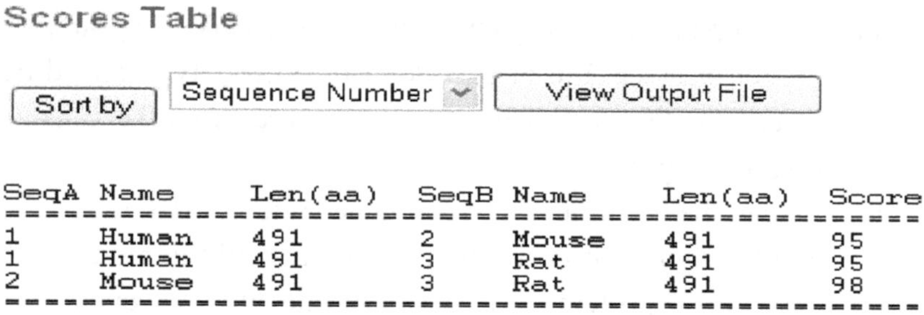
Scores Table

Sort by | Sequence Number | View Output File

SeqA	Name	Len(aa)	SeqB	Name	Len(aa)	Score
1	Human	491	2	Mouse	491	95
1	Human	491	3	Rat	491	95
2	Mouse	491	3	Rat	491	98

FIGURE 3.7 Report of pairwise score.

dealt with in detail in the next section. In Scores Table, users can sort the scores by Alignment Score, Sequence Number, Sequence Name, and Sequence Length. Figure 3.7 shows that the Human PBEF amino acid sequence has 95% identity to those of mouse and rat, whereas the mouse PBEF amino acid sequence is 98% identical to rat's. This result indicates that the PBEF gene is evolutionally highly conserved. Figure 3.8 present part of the aligned output file. The * sign indicates the identical sequence. The . sign indicates the semiconserved sequence substitution, and the : sign indicates the conserved sequence substitution. The gap indicates nonconserved substitution.

2. Edit and Publish Aligned Multiple Protein Sequences

Human PBEF, mouse PBEF, and rat PBEF protein sequences are aligned as In the preceding text except the output file set in the GDE format.

```
Mouse     MNAAAEAEFNILLATDSYKVTHYKQYPPNTSKVYSYFECREKKTENSKVRKVKYEETVFY 60
Rat       MNAAAEAEFNILLATDSYKVTHYKQYPPNTSKVYSYFECREKKTENSKVRKVKYEETVFY 60
Human     MNPAAEAEFNILLATDSYKVTHYKQYPPNTSKVYSYFECREKKTENSKLRKVKYEETVFY 60
          **.*********************************************:***********

Mouse     GLQYILNKYLKGKVVTKEKIQEAKEVYREHFQDDVFNERGWNYILEKYDGHLPIEVKAVP 12C
Rat       GLQYILNKYLKGKVVTKEKIQEAKEVYREHFQDDVFNERGWNYILEKYDGHLPIEVKAVP 12C
Human     GLQYILNKYLKGKVVTKEKIQEAKDVYKEHFQDDVFNEKGWNYILEKYDGHLPIEIKAVP 12C
          ************************:**:*********:*****************:****

Mouse     EGSVIPRGNVLFTVENTDPECYWLTNWIETILVQSWYPITVATNSREQKRILAKYLLETS 18C
Rat       EGSVIPRGNVLFTVENTDPECYWLTNWIETILVQSWYPITVATNSREQKKILAKYLLETS 18C
Human     EGFVIPRGNVLFTVENTDPECYWLTNWIETILVQSWYPITVATNSREQKKILAKYLLETS 18C
          ** **********************************************:**********
```

FIGURE 3.8 ClustalW(ver.1.83) multiple sequence alignments. The left column contains names of different species of the same protein. The middle column shows the sequence alignments. The number on the right indicates the position of an amino acid.

1. Load the aligned sequences into the Multiple Align Show program.
   a. Go to the Web site: http://cgat.ukm.my/tools/sms/multi_align.html
   b. Paste the aligned sequences in the GDE format into the Multiple Align Show program (Figure 3.9). For the demo purpose, only limited amino acid sequences of each species are shown.
2. Run the program.

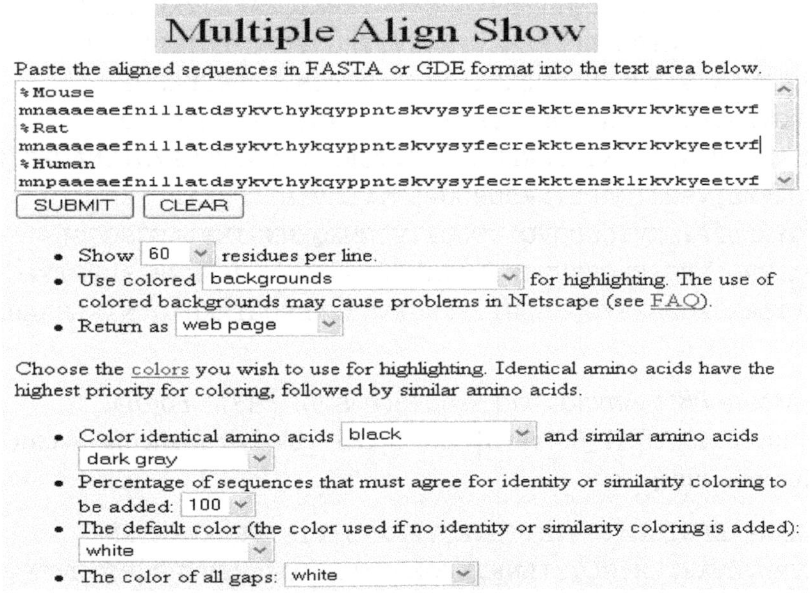

Multiple Align Show
Paste the aligned sequences in FASTA or GDE format into the text area below.
%Mouse
mnaaaeaefnillatdsykvthykqyppntskvysyfecrekktenskvrkvkyeetvf
%Rat
mnaaaeaefnillatdsykvthykqyppntskvysyfecrekktenskvrkvkyeetvf
%Human
mnpaaeaefnillatdsykvthykqyppntskvysyfecrekktensklrkvkyeetvf
SUBMIT
CLEAR
Show 60 residues per line.
Use colored backgrounds for highlighting. The use of colored backgrounds may cause problems in Netscape (see FAQ).
Return as web page
Choose the colors you wish to use for highlighting. Identical amino acids have the highest priority for coloring, followed by similar amino acids.
Color identical amino acids black and similar amino acids dark gray
Percentage of sequences that must agree for identity or similarity coloring to be added: 100
The default color (the color used if no identity or similarity coloring is added): white
The color of all gaps: white

FIGURE 3.9 Paste the aligned sequences in the GDE format into Multiple Align Show program.

Mouse mnaaaeaefnillatdsykvthykqyppntskvysyfecrekktenskvrkvkyeetvfy 60
Rat mnaaaeaefnillatdsykvthykqyppntskvysyfecrekktenskvrkvkyeetvfy 60
Human mnpaaeaefnillatdsykvthykqyppntskvysyfecrekktensklrkvkyeetvfy 60
Mouse glqyilnkylkgkvvtkekiqeakevyrehfqddvfnergwnyilekydghlpievkavp 120
Rat glqyilnkylkgkvvtkekiqeakevyrehfqddvfnergwnyilekydghlpievkavp 120
Human glqyilnkylkgkvvtkekiqeakdvykehfqddvfnekgwnyilekydghlpieikavp 120
Mouse egsviprgnvlftventdpecywltnwietilvqswypitvatnsreqkrilakylleets 180
Rat egsviprgnvlftventdpecywltnwietilvqswypitvatnsreqkkilakylleets 180
Human egfviprgnvlftventdpecywltnwietilvqswypitvatnsreqkkilakylleets 180

FIGURE 3.10 **Sequence Manipulation Suite: Multiple Align Show.**

After selecting the desired parameters (sixty residues per line; identical residues as black; conserved substitution as dark gray and nonconserved substitution as white), click "SUBMIT." The partial result is shown in Figure 3.10.

Part III Sample Data

1. Human PBEF Amino Acid Sequence in the FASTA Format

>gi|5031977|ref|NP_005737.1| pre-B-cell colony enhancing factor 1 isoform a [Homo sapiens]

```
MNPAAEAEFNILLATDSYKVTHYKQYPPNTSKVYSYFECREKKTEN-
SKLRKVKYEETVFYGLQYILNKYLKGKVVTKEKIQEAKDVYKEHFQDDVF-
NEKGWNYILEKYDGHLPIEIKAVPEGFVIPRGNVLFTVENTDPECYWLTNWI-
ETILVQSWYPITVATNSREQKKILAKYLLETSGNLDGLEYKLHDF-
GYRGVSSQETAGIGASAHLVNFKGTDTVAGLALIKKYYGTKDPVP-
GYSVPAAEHSTITAWGKDHEKDAFEHIVTQFSSVPVSVVSDSYDIYNACEKIW-
GEDLRHLIVSRSTQAPLIIRPDSGNPLDTVLKVLEILGKKFPVTEN-
SKGYKLLPPYLRVIQGDGVDINTLQEIVEGMKQKMWSIENIAFGSGGGL-
LQKLTRDLLNCSFKCSYVVTNGLGINVFKDPVADPNKRSKKGRLSLHRTPAG-
NFVTLEEGKGDLEEYGQDLLHTVFKNGKVTKSYSFDEIRKNAQLNIELEAAHH
```

2. Mouse PBEF Amino Acid Sequence in the FASTA Format

>gi|10946948|ref|NP_067499.1| pre-B-cell colony-enhancing factor 1 [Mus musculus]

```
MNAAAEAEFNILLATDSYKVTHYKQYPPNTSKVYSYFECREKKTEN-
SKVRKVKYEETVFYGLQYILNKYLKGKVVTKEKIQEAKEVYREHFQDDVFN-
ERGWNYILEKYDGHLPIEVKAVPEGSVIPRGNVLFTVENTDPECYWLTNWI-
ETILVQSWYPITVATNSREQKRILAKYLLETSGNLDGLEYKLHDS-
GYRGVSSQETAGIGASAHLVNLKGTDTVAGIALIKKYYGTKDPVP-
```

```
GYSVPAAEHSTITAWGKDHEKDAFEHIVTQFSSVPVSVVSDSYDIYNACEKIW-
GEDLRHLIVSRSTEAPLIIRPDSGNPLDTVLKVLDILGKKFPVTEN-
SKGYKLLPPYLRVIQGDGVDINTLQEIVEGMKQKKWSIENVSFGSGGAL-
LQKLTRDLLNCSFKCSYVVTNGLGVNVFKDPVADPNKRSKKGRLSLHRTPAG-
NFVTLEEGKGDLEEYGHDLLHTVFKNGKVTKSYSFDEVRKNAQLNIEQDVAPH
```

3. Rat PBEF Amino Acid Sequence in the FASTA Format

>gi|29293813|ref|NP_808789.1| pre-B-cell colony enhancing factor 1 [Rattus norvegicus]

```
MNAAAEAEFNILLATDSYKVTHYKQYPPNTSKVYSYFECREKKTEN-
SKVRKVKYEETVFYGLQYILNKYLKGKVVTKEKIQEAKEVYREHFQDDVFN-
ERGWNYILEKYDGHLPIEVKAVPEGSVIPRGNVLFTVENTDPECYWLTNWI-
ETILVQSWYPITVATNSREQKKILAKYLLETSGNLDGLEYKLHDF-
GYRGVSSQETAGIGASAHLVNFKGTDTVAGIALIKKYYGTKDPVP-
GYSVPAAEHSTITAWGKDHEKDAFEHIVTQFSSVPVSVVSDSYDIYNACEKIW-
GEDLRHLIVSRSTEAPLIIRPDSGNPLDTVLKVLDILGKKFPVSEN-
SKGYKLLPPYLRVIQGDGVDINTLQEIVEGMKQKKWSIENVSFGSGGAL-
LQKLTRDLLNCSFKCSYVVTNGLGVNVFKDPVADPNKRSKKGRLSLHRTPAGT-
FVTLEEGKGDLEEYGHDLLHTVFKNGKVTKSYSFDEVRKNAQLNMEQDVAPH
```

SECTION 2 BUILDING A PHYLOGENETIC TREE

Part I Introduction

1. What Is Phylogenetics?

The word *phylogenetics* is derived from the Greek words, *phylon*, which means tribe or race, and *genetikos*, which means birth. Phylogenetics, also known as phylogenetic systematics, studies evolutionary relatedness among various groups of organisms. Phylogenetics is a special kind of phylogeny that studies the origin and evolution of a set of organisms.

2. Why Is Phylogenetic Analysis Needed?

1. To classify living species of organisms: Unlike traditional classification rooted in the work of Carolus Linnaeus, who grouped species according to shared physical characteristics, modern classification is based on molecular phylogenetic analysis in which the characters are aligned nucleotide or amino acid sequences. Every living organism contains DNA, RNA, and proteins. Closely related organisms generally have a high degree of agreement in the molecular structure of

these substances, whereas the molecules of organisms distantly related usually show a pattern of dissimilarity. Molecular phylogeny uses such data to build a "relationship tree" that shows the probable evolution of various organisms, which is the basis of species classification.

2. To apply to genetic testing and forensics: Phylogenetics has been applied to the very limited field of human genetics, such as genetic testing to determine a child's paternity as well as the criminal forensics focused on genetic evidence.

3. To infer functions of new genes: One can use phylogenetic analysis to examine whether a new gene is orthologous to another well-characterized gene in another species to infer the potential functions of that new gene. Phylogenetic analyses are increasingly being performed on a genomic scale to predict gene and protein functions, especially in the functional genomic era after the completion of human genome sequencing.

3. What Activities Are Involved in Phylogenetic Analyses?

Collecting data. It is critical that the data subjected to the phylogenetic analysis are homologous, that is, related by evolutionary descent. Among many methods is the sequence-similarity searching method, which can be used to retrieve homologous sequences. This tool can be accessed at sites such as the National Center of Biological Information (http://www.ncbi.nlm.nih.gov), the Japanese GenomicNet server (http://www.blast/genome.ad.jp), and the European Bioinformatics Institute (http://www.ebi.ac.uk). The starting point of a phylogenetic analysis is usually a set of related proteins because it is more informative to work with proteins; more distant relationships can be analyzed.

Multiple sequence alignments. Once a set of sequences to be subjected to the phylogenetic analysis is collected, the next step is to perform a multiple sequence alignment. Readers can refer to Section 1 of this chapter for the details.

Building the phylogenetic tree. Once the sequences have been aligned, the multiple alignment file becomes the input for a phylogenetic analysis program. Numerous phylogenetic methods have been proposed, because no single method performs well in all situations. Complete discussion of all these methods is beyond the scope of this section. There are three major types of methods: distance matrix, maximum parsimony, and maximum likelihood. In distance matrix methods, the number of nucleotide or amino acid substitutions between sequences is treated as a distance and computed for all pairs of taxa. The phylogenetic tree is constructed based on these distances. Neighbor-joining, one of commonly applied distance methods, is statistically consistent under many models of evolution, and hence capable of reconstructing the true tree with high probability. Distance methods are quick to compute. A central idea of the maximum parsimony is that the preferred evolutionary tree requires the smallest number of evolutionary changes to explain the differences observed among the taxa under study. Although parsimony makes no explicit assumptions, there is the critical assumption of the parsimony criterion that a tree requiring fewer substitutions or changes is better than a tree requiring more. This can be contrasted with likelihood methods, which make explicit assumptions about the rate of evolution and patterns of nucleotide substitution. Maximum parsimony is a very simple approach, and is popular for this reason. However, it is not statistically consistent; that is, it is not guaranteed to produce the true tree with high probability, given sufficient data. Maximum parsimony methods take longer to compute. *Maximum likelihood method* is a popular statistical method used to make inferences about parameters of the underlying probability distribution of a given data set. Maximum likelihood methods are very slow, typically computer intensive, and have not been implemented on the Internet as widely as other methods.

Phylogenetics programs are available for both desktop computers and mainframes. A combination of three popular and user-friendly programs, NCBI Protein-protein BLAST (blastp), ClustalW, and the *PHYL*ogeny *I*nference *P*ackage (PHYLIP), will be used as demo in next section to perform phylogenetic analyses. Blastp (http://www.ncbi.nlm.nih.gov/BLAST/) will be employed to search for and collect homologous protein amino acid sequences. ClustalW (http://www.ebi.ac.uk/clustalw/) will be used to carry out multiple protein sequence alignments and build the phylogenetic tree. PHYLIP will be employed to build the phylogenetic tree. Expert Protein Analysis System (ExPASY) proteomics tools (http://ca.expasy.org/tools/#phylo) has, under the heading "Phylogenetic Analysis," a fairly comprehensive collection of phylogenetic analysis programs. There, nearly 300 phylogenetic analysis programs and servers are listed. Many of those listed programs are available on the Web, including both free and nonfree ones. The software programs are conveniently categorized by methods available, by computer systems on which they work, and the particular types of data analyzed. This site is periodically updated, though users should be cautioned that the host does not make any attempt to exclude programs that do not meet some standard of quality or importance. It is important that users for particular software programs should have a good understanding of their underlying methods and potential pitfalls.

Tree evaluation. After a phylogenetic tree is built, it is necessary to evaluate it for robustness. The most common method for doing this is bootstrap analysis, which essentially involves resampling the database and then analyzing the resampled data. Those robust results obtained from initial phylogenetic analysis will tolerate variations introduced by the resampling process, whereas nonrobust results will be altered and yield different trees when small changes in data are made through resampling.

Tree visualization. After robust trees are created, the final step is to best visualize trees and produce publication-quality printouts of the results.

4. New Developments in Phylogenetic Analyses

Phylogenomic analyses. In recent years, phylogenetic analyses have been performed on a genomic scale to address issues ranging from the prediction of gene and protein function to organismal relationships, to the influence of polyploid and horizontal gene transfer on genome content and structure, and to the reconstruction of ancestral genome characteristics. Thus, the term *phylogenomic* has emerged. Phylogenomic analyses are broadly defined as the integration of phylogenetic and genomic analysis, which places genome sequence, gene expression, and functional data in a historical context and thereby helps elucidate the processes shaping the structure and function of genes, genetic systems, and whole organisms. The development and refinement of searchable phylogeny databases such as TreeBase or gene tree databases is an important step in the advancement of phylogenomics. A group of scientists in the field of phylogenetics recently has taken the initiative to develop a Minimal Information About a Phylogenetic Analysis (MIAPA) standard to bring phylogenetic analyses more fully into the informatics age. This will have many beneficial effects on the utility and impact of phylogenomics. A starting proposal on MIAPA includes: (1) the objective of the phylogenetic analysis, (2) the description of raw data, (3) sample voucher information, (4) procedures for establishing character homology, (5) the sequence alignment or other character matrix, (6) the detailed description of the phylogenetic analysis, and (7) the phylogenies, including branch lengths and support values.

New software developments. Widespread recognition of the importance of phylogenetics to genome biology comes at a time when the availability of whole-genome sequences is increasing at an unprecedented rate. It has spurred a rapid expansion and refinements of new and existing programs in phylogenetic analyses. The following are a few examples. Polar and Interactive Tree (PolnTree, http:/geneproject.altervista.org) is an application that allows building, visualizing, and customizing phylogenetic trees in a polar interactive and highly flexible view. It takes as input a FASTA file or multiple alignment formats. Phylogenetic tree calculation is based on a sequence distance method and utilizes the Neighbor Joining (NJ) algorithm. It also allows displaying precalculated trees of the major protein families based on Pfam classification. In PolnTree, nodes can be dynamically opened and closed, and distances between genes are graphically represented. Tree roots can be centered on a selected leaf. The text search mechanism, color-coding, and labeling display are integrated. The visualizer can be connected to an Oracle database containing information on sequences and other biological data, helping guide their interpretation within a given protein family across multiple species. DNA assembly with gaps (Dawg, http://scit.us/dawg/) simulates phylogenetic evolution of DNA sequences in continuous time, using the robust general-time reversible model with gamma and invariant rate heterogeneity and a novel length-dependent model of indel formation. On completion, Dawg produces the true alignment of the simulated sequences. It can be used to parametrically bootstrap an estimation of the rate of indel formation for the phylogeny. Because Dawg can assist in parametric bootstrapping of sequence data, its usefulness extends beyond phylogenetics, such as studying alignment algorithms or parameters of molecular evolution. Bio++, a set of object-oriented libraries written in C++, has been established to bundle bioinformatics applications in the fields of biosequence analysis, molecular evolution, and population genetics. Bio++ enables easy extension and new methods development. It contains a defined general hierarchy of classes that allow developers to implement their own algorithms, which remain compatible with the rest of the libraries. Bio++ source code is distributed free of charge under the CeCILL general public license from its Web site http://kimura.univ-montp2.fr/BioPP.

Part II Step-By-Step Tutorial

As described in the Introduction, a combination of three popular and user-friendly programs, NCBI Protein-protein BLAST (blastp), ClustalW, and PHYLIP, will be used to demonstrate phylogenetic analyses in this section. The BLAST program was described in Chapter 1. The ClustalW program was demonstrated in the first section of this chapter. PHYLIP is maintained by Joe Felsenstein in the Department of Genome Sciences and the Department of Biology, University of Washington, Seattle, WA, U.S. It consists of thirty-five programs. It is distributed as source code, documentation files, and a number of different types of executables. It is available free over the Internet, and is designed to work on as many different kinds of computer systems as possible. Methods that are available in the package include parsimony, distance matrix, and likelihood methods, including bootstrapping and consensus trees. Data types that can be handled include molecular sequences, gene frequencies, restriction sites and fragments, distance matrices, and discrete characters. The programs are controlled through a menu, which asks the users which options they want to set, and allows them to start the computation. PHYLIP is the most widely distributed phylogenetic analysis package since 1980.

1. Collecting a Set of Homologous Protein Sequences

Here again, we use the PBEF protein as an example:

1. Fetch the human PBEF protein sequence from the NCBI Web site.
   a. As demonstrated in Figure 3.2, type the following address (http://www.ncbi.nlm.nih.gov/entrez/query.fcgi?db=unigene) in your browser (Internet Explorer or others). In "Search," select "Protein" category. In the "for," type "human PBEF."
   b. Click "Go" in your browser or press "Enter" on your keyboard. The accession number of human PBEF protein sequence will be displayed as in the Figure 3.3.
   c. Click "NP_005737" to display all human PBEF protein sequence information. Then in the Display, select "FASTA" format, and you will have the human PBEF protein sequence in the FASTA format (Figure 3.4).
2. Obtain PBEF homologous sequences from NCBI Blastp site.

Go to http://www.ncbi.nlm.nih.gov/BLAST/, click "Protein-protein BLAST" (blastp) under Protein, then paste human PBEF protein sequence into the Search window before clicking "BLAST!" followed by clicking "Format!" The results of BLASTp search will be graphically displayed online on the Distribution of Blast Hits on the Query Sequence as well as sequences producing the significant alignments. A partial list of homologous protein sequences to the human PBEF is presented in Figure 3.11.

NCBI *results of* **BLAST**

Sequences producing significant alignments:	Score (Bits)	E Value	
gi|5031977|ref|NP_005737.1| visfatin precursor [Homo sapiens]...	974	0.0	G
gi|110590440|pdb|2GVJ|B Chain B, Crystal Structure Of Human N...	966	0.0	S
gi|72535204|ref|NP_001026963.1| visfatin [Sus scrofa] >gi|682...	947	0.0	U G
gi|76615241|ref|XP_883602.1| PREDICTED: similar to Nicotinami...	942	0.0	G
gi|50293167|gb|AAT72933.1| nicotinamide phosphoribosyltransfe...	940	0.0	U G
gi|89030888|ref|XP_934340.1| PREDICTED: similar to Nicotinami...	939	0.0	U G
gi|112490980|pdb|2GVL|B Chain B, Crystal Structure Of Murine ...	937	0.0	S
gi|89031848|ref|XP_945766.1| PREDICTED: similar to Nicotinami...	936	0.0	G
gi|74147564|dbj|BAE38671.1| unnamed protein product [Mus musculu	936	0.0	U G
gi|55716092|gb|AAH85681.1| Pre-B-cell colony enhancing factor...	936	0.0	U G

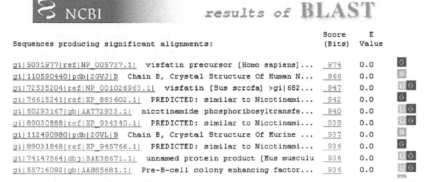

FIGURE 3.11 Partial list of homologous proteins to human PBEF protein. This screenshot shows the partial results of BLAST search, those sequences producing significant alignments to human PBEF protein.

2. *Multiple Sequence Alignments*

1. Edit the input sequences in a text editor or in Microsoft Word. Organize all homologous sequences into one file and keep only species or strain names in the first line of each sequence after sign >, respectively (see Part III titled "Sample Data"). Paste all sequences into the sequence window of the ClustalW program. Keep the default setting, and click "Run" as described in Section 1.

2. Result delivery: As described in Section 1, results come in three sections: Scores Table, Alignment, and Guide Tree. For the following phylogenetic tree construction, only the alignment file as presented in Figure 3.12 is needed.

```
CLUSTAL W (1.83) multiple sequence alignment


Murine          MNAAAEAEFNILLATDSYKVTHYKQYPPNTSKVYSYFECREKKTENSKVRKVKYEETVFY 60
Rat             MNAAAEAEFNILLATDSYKVTHYKQYPPNTSKVYSYFECREKKTENSKVRKVKYEETVFY 60
Bovine          MNAAAEAEFNILLATDSYKVTHYKQYPPNTSKVYSYFECREKKTENSKIRKVKYEETVFY 60
Pig             MNAAAEAEFNILLATDSYKVTHYKQYPPNTSKVYSYFECREKKTENSKIRKVKYEETVFY 60
Human           MNPAAEAEFNILLATDSYKVTHYKQYPPNTSKVYSYFECREKKTENSKLRKVKYEETVFY 60
Xenopus         ------------------------------------------------------------
Puffyfish       -MEHRDSDFNILLATDSYKVTHYKQYPPNTSKVYSYFECREKRTDPSKSRKVTYDKTVFY 59
Sponge          ------MDRNILLETDSYKVTHHLQYPPGAEHVYSYFESRGG----------KFPETVFF 44
Proteobacteria  ----MHYLDNLLLNTDSYKASHWLQYPPGTDASFFYVESRGG----------LYEQTVFF 46


Murine          GLQYILNKYLKGKVVTKEKIQEAKEVYREHFQDDVFNERGWNYILEKYDGHLPIEVKAVP 120
Rat             GLQYILNKYLKGKVVTKEKIQEAKEVYREHFQDDVFNERGWNYILEKYDGHLPIEVKAVP 120
Bovine          GLQYILHKYLKGKVVTKEKIQEAKEVYREHFQDDVFNEKGWNYILEKYDGHLPIEVKAVP 120
Pig             GLQYILNKYLKGKVVTKEKIQEAKEVYKEHFQDDVFNEKGWNYILEKYDGHLPIEVKAVP 120
Human           GLQYILNKYLKGKVVTKEKIQEAKDVYKEHFQDDVFNEKGWNYILEKYDGHLPIEIKAVP 120
Xenopus         -----------GRVVTKQKIQEAKEVYREHFQDDVFNEKGWNYILEKYDGHLPIEIKAVP 49
Puffyfish       GLQYILHKYLKGKVVTPEKIQEAKDVYREHFQDDVFNEKGWTYILEKYNGHLPIEIKAVP 119
Sponge          GLQYILKKSLVGKVVTREKIEEAAAVFDAHLGPGLFNKEGWMYILEKHGGKLPVRIKAVA 104
Proteobacteria  GLQSILKEAIN-RPVTHADIDDAKELLAAHGEP--FNEAGWRDIVDRLGGQLPIRIRAVP 103
                   : **  .*::*  :   *      **: **  *:::  .*:**:.::**.
```

FIGURE 3.12 Output file of multiple sequence alignments (partial presentation).

3. Building the Phylogenetic Tree

1. Load the alignment file into a Phylip server. Go to the Phylip server sponsored by the Pasteur Institute (http://bioweb.pasteur.fr/seqanal/phylogeny/phylip-uk.html), select "advanced form" in "protdist" under Programs for molecular sequence data "sequence.doc," then paste the alignment file from the ClustalW program into the sequence window as shown in Figure 3.13.

2. To run the program, Enter your e-mail address required by the server, select Bootstrap options with an odd random number seed such as 5 and 100 repeats before clicking "Run protdist." The analysis result will be produced as shown in Figure 3.14. Select Neighbor, which is a good distance matrix program, and click "Run the selected program" on outfile. The Web page is then displayed as in Figure 3.15.

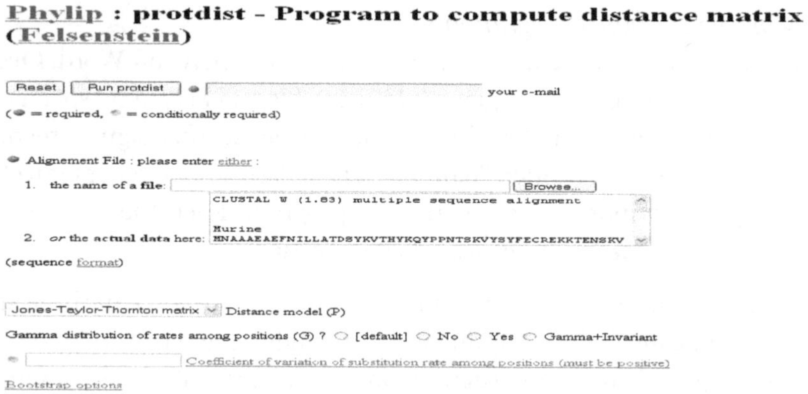
Phylip : protdist - Program to compute distance matrix (Felsenstein)
Reset
Run protdist
your e-mail
(= required, = conditionally required)
Alignement File : please enter either :
1. the name of a file:
Browse...
CLUSTAL W (1.83) multiple sequence alignment
Murine
2. or the actual data here:
MNAAAEAEFNILLATDSYKVTHYKQYPPNTSKVYSYFECREKKTENSKV
(sequence format)
Jones-Taylor-Thornton matrix
Distance model (P)
Gamma distribution of rates among positions (G) ? [default] No Yes Gamma+Invariant
Coefficient of variation of substitution rate among positions (must be positive)
Bootstrap options

FIGURE 3.13 Paste the aligned file into the sequence window of the Phylip program.

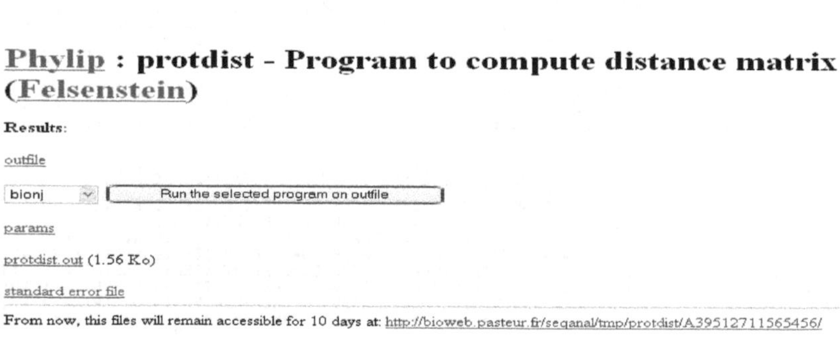
Phylip : protdist - Program to compute distance matrix (Felsenstein)
Results:
outfile
bionj
Run the selected program on outfile
params
protdist.out (1.56 Ko)
standard error file
From now, this files will remain accessible for 10 days at: http://bioweb.pasteur.fr/seqanal/tmp/protdist/A39512711565456/

FIGURE 3.14 Result file.

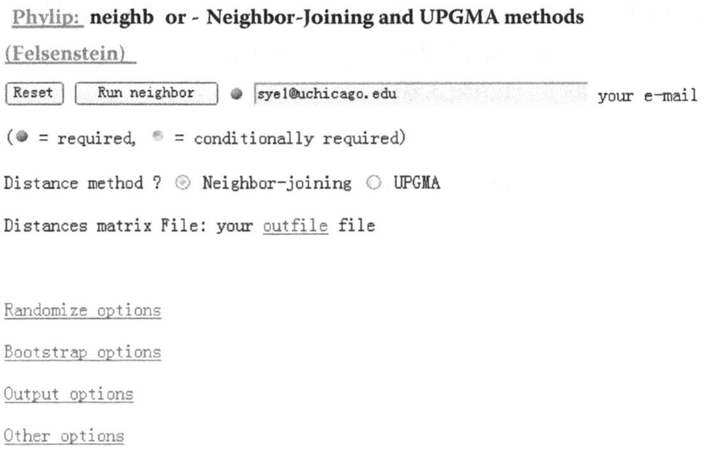
Phylip: neighb or - Neighbor-Joining and UPGMA methods
(Felsenstein)
Reset
Run neighbor
syel@uchicago.edu
your e-mail
(● = required, ● = conditionally required)
Distance method ? Neighbor-joining UPGMA
Distances matrix File: your outfile file
Randomize options
Bootstrap options
Output options
Other options

FIGURE 3.15 Select neighbor-joining matrix.

4. *Tree Evaluation*

1. Bootstrap options: Click "bootstrap options," and select the parameters as shown in Figure 3.16.

2. Result display: Click the "Run" button, and the result will be displayed as in Figure 3.17. The top two files are consensus files. The bottom two files are normal files. Outfiles contains text versions of phylogenetic trees, and outtree files are the Newick formats of the same phylogenetic trees.

5. *Tree Visualization*

Phylogentic trees can be displayed online or with other tree display programs after saving the output files. The following is the example using the online drawtree program. After selecting drawtree, click "Run the selected program" on outtree.consensus, and the Web page will be dis-

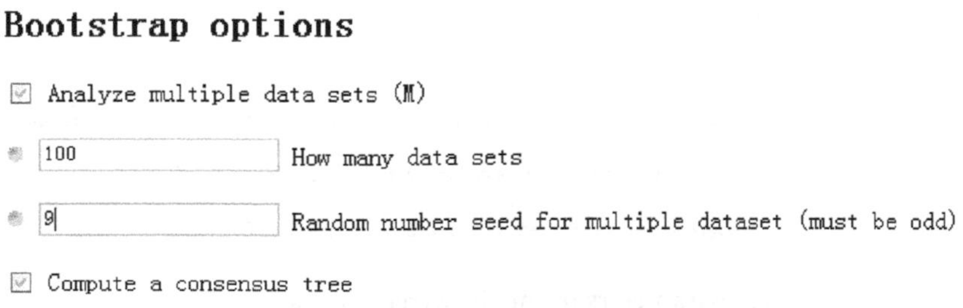
Bootstrap options
Analyze multiple data sets (M)
100
How many data sets
9
Random number seed for multiple dataset (must be odd)
Compute a consensus tree

FIGURE 3.16 Select parameters in bootstrap options.

played as in Figure 3.18. Click "Run drawtree," and the result will be as shown in Figure 3.19. Click "plotfile.ps," and the final phylogenetic tree will be displayed as in Figure 3.20.

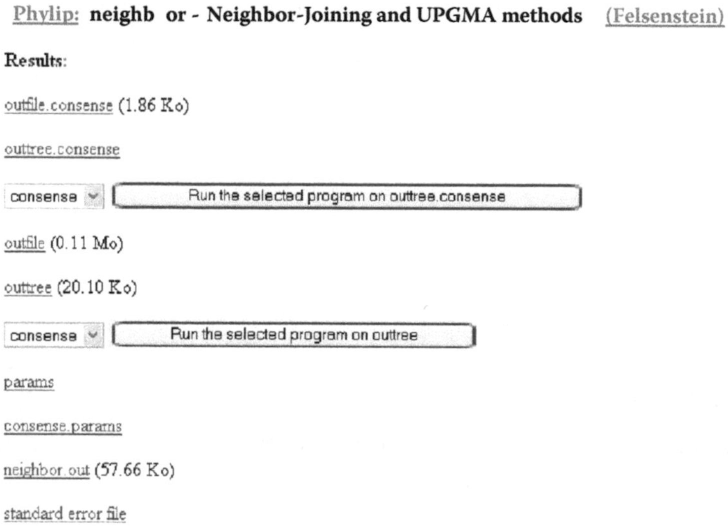
Phylip: neighb or - Neighbor-Joining and UPGMA methods (Felsenstein)
Results:
outfile.consense (1.86 Ko)
outtree.consense
consense
Run the selected program on outtree.consense
outfile (0.11 Mo)
outtree (20.10 Ko)
consense
Run the selected program on outtree
params
consense.params
neighbor.out (57.66 Ko)
standard error file

FIGURE 3.17 Files of Phylogenetic analyses. The upper part lists files of outfile. consense and outtree.consense. The lower part lists files of outfile and outtree.

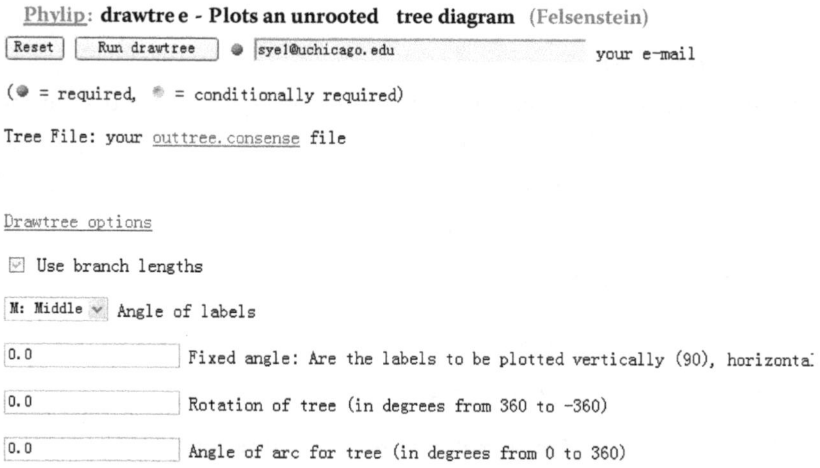
Phylip: drawtre e - Plots an unrooted tree diagram (Felsenstein)
Reset
Run drawtree
syel@uchicago.edu
your e-mail
(● = required, ● = conditionally required)
Tree File: your outtree.consense file
Drawtree options
Use branch lengths
M: Middle
Angle of labels
0.0
Fixed angle: Are the labels to be plotted vertically (90), horizonta
0.0
Rotation of tree (in degrees from 360 to -360)
0.0
Angle of arc for tree (in degrees from 0 to 360)

FIGURE 3.18 Select parameters in the drawtree program.

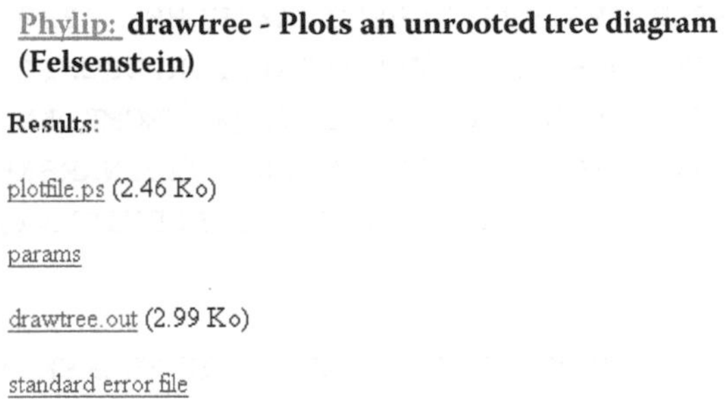
Phylip: **drawtree - Plots an unrooted tree diagram (Felsenstein)**

Results:

plotfile.ps (2.46 Ko)

params

drawtree.out (2.99 Ko)

standard error file

FIGURE 3.19 Plotfile by the drawtree program.

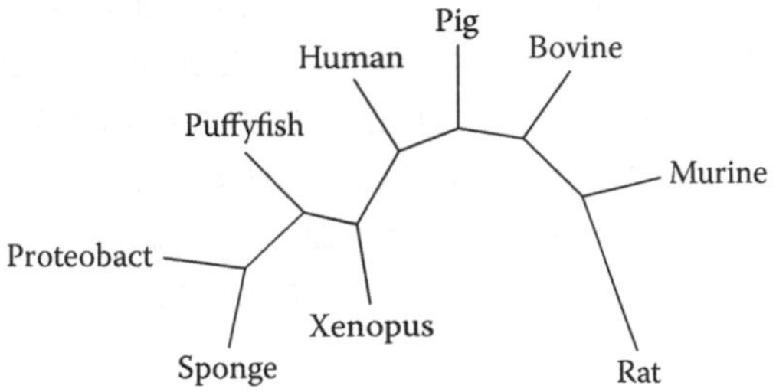
Pig
Human
Bovine
Puffyfish
Murine
Proteobact
Xenopus
Sponge
Rat

FIGURE 3.20 Output by the drawtree program in the plottree.ps program.

Part III Sample Data (Partial Listing)

>Homo sapiens

```
MNPAAEAEFNILLATDSYKVTHYKQYPPNTSKVYSYFERKKTENSKLRKVKYEE
TVFYGLQYILNKYLKGKVVTKEKIQEAKDVYKEHFQDDVFNEKGWNYILEKYDG
HLPIEIKAVPEGFVIPRGNVLFTVENTDPECYWLTNWIETILVQSWYPITVATN
SREQKKILAKYLLETSGNLDGLEYKLHDFGYRGVSSQETAGIGASAHLVNFKGT
DTVAGLALIKKYYGTKDPVPGYSVPAAEHSTITAWGKDHEKDAFEHIVTQFSSV
PVSVVSDSYDIYNACEKIWGEDLRHLIVSRSTQAPLIIRPDSGNPLDTVLKVLE
ILGKKFPVTENSKGYKLLPPYLRVIQGDGVDINTLQEIVEGMKQKMWSIENIAF
GSGGGLLQKLTRDLLNCSFKCSYVVTNGLGINVFKDPVADPNKRSKKGRLSLHR
TPAGNFVTLEEGKGDLEEYGQDLLHTVFKNGKVTKSYSFDEIRKNAQLNIELEA
AHH
```

>Chimpanzee

```
MKAKSDHPQYLIVQVTHYKQYPPNTSKVYSYFECREKKTENSKLRKVKYEETVF
YGLQYILNKYLKGKVVTKEKIQEAKDIYKEHFQDDVFNEKGWNYILEKYDGHLP
IEIKAVPEGFVIPRGNVLFTVENTDPECXWLTNWIETILVQSWYPITVATNSRE
QKKILAKYLLETSGNLDGLEYKLHDFGYRGVSSQETAGIGASAHLVNFKGTDTV
```

AGLALIKKYYGTKDPVPGYSVPAAEHSTITAWGKDHEKDAFEHIVTQFSSVPVS
VVSDSYDIYNACEKIWGEDLRHLIVSRSTQAPLIIRPDSGNPLDTVLKVLEILG
KKFPVTENSKGYKLLPPYLRVIQGDGVDINTLQEIVEGMKQKMWSIENIAFGSG
GGLLQKLTRDLLNCSFKCSYVVTNGLGINVFKDPVADPNKRSKKGRLSLHRTPA
GNFVTLEEGKGDLEEYGQDLLHTVFKNGKVTKSYSFDEIRKNAQLNIELEAAHH

>Rhesus monkey

MNAAAEAEFNILLATDSYKVTHYKQYPPNTSKVYSYFECREKKTENSKLRKVKY
EETVFYGLQYILNKYLKGKVVTKEKIQEAKEVYKEHFQDDVFNEKGWNYILEKY
DGHLPIEVKAVPEGSVIPXXXILIKMQSTDQCWSMPCLIQTILVQSWYPITVAT
NSREQKKILAKYLLETSGNLDGLEYKLHDFGYRGVSSQETAGIGASAHLVNFKG
TDTVAGIALIKKYYGTKDPVPGYSVPAAEHSTITAWGKDHEKDAFEHIVTQFSS
VPVSVVSDSYDIYNACEKIWGEDLRHLIVSRSTEAPLIIRPDSGNPLDTVLKVL
EILGKKFPVTENSKGYKLLPPYLRVIQGDGVDINTLQEIVEGMKQKKWSIENVS
FGSGGALLQKLTRDLLNCSFKCSYVVTNGLGINVFKDPVADPNKRSKKGRLSLH
RTPAGNFVTLEEGKGDLEEYGHDLLHTVFKNGKVTKSYSFDEVRKNAQLNIELE
AAPH

>*Mus musculus*

MNAAAEAEFNILLATDSYKVTHYKQYPPNTSKVYSYFECREKKTENSKVRKVKY
EETVFYGLQYILNKYLKGKVVTKEKIQEAKEVYREHFQDDVFNERGWNYILEKY
DGHLPIEVKAVPEGSVIPRGNVLFTVENTDPECYWLTNWIETILVQSWYPITVA
TNSREQKRILAKYLLETSGNLDGLEYKLHDSGYRGVSSQETAGIGASAHLVNLK
GTDTVAGIALIKKYYGTKDPVPGYSVPAAEHSTITAWGKDHEKDAFEHIVTQFS
SVPVSVVSDSYDIYNACEKIWGEDLRHLIVSRSTEAPLIIRPDSGNPLDTVLKV
LDILGKKFPVTENSKGYKLLPPYLRVIQGDGVDINTLQEIVEGMKQKKWSIENV
SFGSGGALLQKLTRDLLNCSFKCSYVVTNGLGVNVFKDPVADPNKRSKKGRLSL
HRTPAGNFVTLEEGKGDLEEYGHDLLHTVFKNGKVTKSYSFDEVRKNAQLNIEQ
DVAPH

>*Rattus norvegicus*

MNAAAEAEFNILLATDSYKVTHYKQYPPNTSKVYSYFECREKKTENSKVRKVKY
EETVFYGLQYILNKYLKGKVVTKEKIQEAKEVYREHFQDDVFNERGWNYILEKY
DGHLPIEVKAVPEGSVIPRGNVLFTVENTDPECYWLTNWIETILVQSWYPITVA
TNSREQKKILAKYLLETSGNLDGLEYKLHDFGYRGVSSQETAGIGASAHLVNFK
GTDTVAGIALIKKYYGTKDPVPGYSVPAAEHSTITAWGKDHEKDAFEHIVTQFS
SVPVSVVSDSYDIYNACEKIWGEDLRHLIVSRSTEAPLIIRPDSGNPLDTVLKV
LDILGKKFPVSENSKGYKLLPPYLRVIQGDGVDINTLQEIVEGMKQKKWSIENV
SFGSGGALLQKLTRDLLNCSFKCSYVVTNGLGVNVFKDPVADPNKRSKKGRLSL

HRTPAGTFVTLEEGKGDLEEYGHDLLHTVFKNGKVTKSYSFDEVRKNAQLNMEQ
DVAPH

>*Sus scrofa*

MNAAAEAEFNILLATDSYKVTHYKQYPPNTSKVYSYFECREKKTENSKIRKVKY
EETVFYGLQYILNKYLKGKVVTKEKIQEAKEVYKEHFQDDVFNEKGWNYILEKY
DGHLPIEVKAVPEGSVIPRGNVLFTVENTDPECYWLTNWIETILVQSWYPITVA
TNSREQKKILAKYLLETSGNLDGLEYKLHDGYRGVSSQETAGIGASAHLVNFKG
TDTVAGIALIKKYYGTKDPVPGYSVPAAEHSTITAWGKDHEKDAFEHIVTQFSS
VPVSVVSDSYDIYNACEKIWGEDLRHLIVSRSTEAPLIIRPDSGNPLDTVLKVL
DILGKKFPVTENSKGYKLLPPYLRVIQGDGVDINTLQEIVEGMKQKKWSIENIA
FGSGGALLQKLTRDLLNCSFKCSYVVTNGLGINVFKDPVADPNKRSKKGRLSLH
RTPGGNFVTLEEGKGDLEEYGHDLLHTVFKNGKVTKSYSFDEVRKNAQLNIELE
AAPH

REFERENCES

1. Wallace, I.M., Blackshields, G., and Higgins, D.G. Multiple sequence alignments. *Curr Opin Struct Biol* 15(3): 261–266, 2005.
2. Edgar, R.C. and Batzoglou, S. Multiple sequence alignment. *Curr Opin Struct Biol* 16(3): 368–373, 2006.
3. Batzoglou, S. The many faces of sequence alignment. *Brief Bioinf* 6(1): 6–22, 2005.
4. Chenna, R., Sugawara, H., Koike, T., Lopez, R., Gibson, T.J., Higgins, D.G., and Thompson, J.D. Multiple sequence alignment with the Clustal series of programs. *Nucl Acids Res* 31(13): 3497–3500, July 1, 2003.
5. Stothard, P. The sequence manipulation suite: JavaScript programs for analyzing and formatting protein and DNA sequences. *Biotechniques* 28(6): 1102, 1104, 2000.
6. Gonnet, G.H., Cohen, M.A., and Benner, S.A. Exhaustive Matching of the Entire Protein Sequence Database. *Science* 256(5062), 1992, pp. 1443–1445.
7. Felsenstein, J. *Inferring Phylogenies* Sinauer Associates, Sunderland, MA, 2004.
8. Littlejohn, T. Phylogenetics-a web of trees. In *Biocomputing: Computer Tools for Biologists*. Ed. Brown, S.M., Eaton Publishing, Westborough, MA, 2003, pp. 357–360.
9. Leebens-Mack, J., Vision, T., Brenner, E., Bowers, J.E., Cannon, S., Clement, M.J., Cunningham, C.W., dePamphilis, C., deSalle, R., Doyle, J.J., Eisen, J.A., Gu, X., Harshman, J., Jansen, R.K., Kellogg, E.A., Koonin, E.V., Mishler, B.D., Philippe, H., Pires, J.C., Qiu, Y.L., Rhee, S.Y., Sjolander, K., Soltis, D.E., Soltis, P.S., Stevenson, D.W., Wall, K., Warnow, T., Zmasek, C. Taking the first steps towards a standard for reporting on phylogenies: minimum information about a phylogenetic analysis (MIAPA). *OMICS* 10(2): 231–237, 2006.

10. Marco, C., Eleonora, G., Luca, S., Edward, P.S., Antonella, I., and Roberta, B. PolnTree: a polar and interactive phylogenetic tree. *Gen Prot Bioinf* 31: 58–60, 2005.
11. Cartwright, R.A. DNA assembly with gaps (Dawg): simulating sequence evolution. *Bioinformatics* 21(Suppl. 3): iii31–iii38, November 1, 2005.
12. Dutheil, J., Gaillard, S., Bazin, E., Glemin, S., Ranwez, V., Galtier, N., and Belkhir, K. Bio++: a set of C++ libraries for sequence analysis, phylogenetics, molecular evolution and population genetics. *BMC Bioinf* 7: 188, 2006.
13. Felsenstein, J. PHYLIP (Phylogeny Inference Package) version 3.6. Distributed by the author. Department of Genome Sciences, University of Washington, Seattle, WA, 2005.

CHAPTER 4

SNP and Haplotype Analyses

Shui Qing Ye

CONTENTS

SECTION 1 SNP ANALYSIS

Part I Introduction

1. What Is SNP?

SNP, pronounced "snip," stands for single-nucleotide polymorphism, which represents a substitution of one base for another, e.g., C to T or A to G. SNP is the most common variation in the human genome and occurs approximately once every 100 to 300 bases. SNP is terminologically distinguished from mutation based on an arbitrary population frequency cutoff value: 1%, with SNP > 1% and mutation < 1%. A key aspect of research in genetics is associating sequence variations with heritable phenotypes. Because SNPs are expected to facilitate large-scale association genetics studies, there has been an increasing interest in SNP discovery and detection.

2. SNP Discovery and Assay

Much effort has been devoted to developing reliable, efficient, and cost-effective modalities in the discovery and genotyping of SNPs across the human genome. A variety of technologies have been available. Here, a few representative low-throughput methods on the SNP identification and genotyping will be introduced first. This is not only because these methods have contributed significantly to the discovery of SNPs in human population in the past, but also because they are still valuable and economical in any individual lab for a low-to-mid-throughput genotyping need. Then, several high-throughput platforms on the SNP identification and genotyping will be described.

Low-throughput methods. Here, three major methods for SNP discovery will be introduced:

1. **Sequencing-based:** The most reliable strategy for identifying SNPs in a population of interest is the direct sequencing of the genomic DNA region in each individual. Direct DNA sequencing has been adopted as the gold standard. The sample size of the population being resequenced is important. In general, larger sample sizes are

needed to identify SNPs with low minor allele frequency. Smaller sizes are required to identify SNPs with high minor allele frequency. The Direct DNA sequencing method is still not widely applied to the large population survey of SNPs because the cost of DNA sequencing is still quite high. As the costs associated with sequencing are projected to decrease over time, direct sequencing may become the main method to identify and assay SNPs in populations so that the "$1000 per genome" scenario could come true.

2. **Conformation-based:** Single-strand conformation polymorphism (SSCP), cleavage fragment length polymorphism (CFLP), and conformation-sensitive gel electrophoresis (CSGE) are three popular conformation-based methods. SSCP is based on the fact that a single-stranded DNA molecule will adopt a unique conformation due to the formation of intrastrand base pairing under nondenaturing conditions. That conformation is strictly dependent on its sequence context. A single base change can result in conformation change, which can be detected by an alteration in electrophoretic mobility. Conventional SSCP analysis requires denaturation of the double-stranded PCR product by heating, immediate chilling on ice, followed by gel electrophoresis under nondenaturing conditions. Although SSCP is one of the widely used methods, it is a little cumbersome to perform because it requires more than one electrophoretic condition to be run to observe all possible conformational changes. Its assay cannot detect all point mutational differences. Recently, SSCP analysis has been coupled with automated capillary array sequencers for high-throughput SNP identification and genotyping. CFLP is based on the fact that single-stranded DNAs form reproducible hairpin duplexes during self-annealing. In CFLP, the hairpins are cleaved by endonuclease cleavage I (cleavage 1), a structure-specific endonuclease, at the 5' side of the junctions between the single-stranded and the duplex region. The cleavage products show sequence-specific patterns of bands on an electrophoretic gel. Compared with SSCP, CFLP is more rapid, more accurate, and permits the analysis of larger DNA fragments. However, its reproducibility is a concern. CSGE is based on differences in conformation between homoduplex and heteroduplex double-stranded DNA fragments. Heteroduplexes are generated by heat denaturation and reannealing of a mixture of wild-type and mutant DNA molecules. The resulting homoduplexes

and heteroduplexes exhibit either distinct electrophoretic mobility or distinct cleavage patterns under appropriate conditions. Only when combined with SSCP can the mutation-detection rate of CSGE approach 100%.

3. **Melting-based:** Denaturing high-performance liquid chromatography (DHPLC), denaturing gradient gel electrophoresis (DGGE), and two-dimensional gene scanning (TDGS) are widely used melting-based methods. DHPLC uses ion-pair reverse phase liquid chromatography to detect DNA heteroduplexes. Under partially denaturing conditions, heteroduplexes denature more readily and display reduced column retention time compared to their homoduplex counterparts. DHPLC is an automated, fast method with the capacity to analyze fragment sizes from 200 up to 700 bp, but its current format is not suitable for large-scale testing. DGGE is based on the fact that single-nucleotide differences are sufficient to alter melting behavior. When a DNA fragment reaches a denaturant concentration or temperature equivalent to the melting temperature of its lowest-melting domain, partial strand separation, branching, and reduction in electrophoretic mobility occur in a gradient gel of increasing denaturant concentration (DGGE) or increasing temperature (temperature gradient gel electrophoresis, TGGE). DGGE is generally considered the most accurate method for detecting DNA sequence variation. TDGS is based on DGGE in a two-dimensional format, enabling analysis of an entire gene for all possible sequence variants in one gel under one set of conditions. With the recent introduction of high-speed 2D electrophoresis with multicolor fluorescent detection, TDGS has become at least an order of magnitude more cost-effective than nucleotide sequencing at equal accuracy.

High-throughput methods. Recently, the advent of newer technologies has increased the throughput of SNP genotyping while simultaneously decreasing the cost. Newer technologies have allowed the evaluation of SNPs, not just at a single locus, but on a genomewide level at densities that were previously thought to be unobtainable. Here, three technologies are briefly introduced: TaqMan assay (Applied Biosystems), GeneChip human mapping assays (Affymetrix), and Infinium genotyping assay (illumina).

1. **TaqMan assay:** TaqMan assays are based on a 5' nuclease assay in which two probes hybridize to a given polymorphic sequence in an allele-specific manner in a single PCR reaction. Each probe carries a 5' end fluorescent reporter and a 3' end dye quencher. The quencher suppresses the fluorescent reporter until a probe is hybridized to the appropriate SNP allele during the PCR reaction. When the quencher is cleaved by the 5' nuclease activity of Taq polymerase, a fluorescent signal from the released reporter is produced. The fluorescent signal for each of the two allele-specific reporter dyes will report the genotype of that SNP. The TaqMan assay combines the PCR amplification and genotyping assay into a single step, greatly reducing sample processing. Its 380-sample plate format facilitates the midthroughput though the instrument and associated reagents for the TaqMan assay is relatively high for any individual lab.
2. **GeneChip human mapping assays**: Affymetrix has utilized an adaptor-PCR to create a reduced-complexity genome for SNP analysis. Genomic DNA is digested with a restriction enzyme (e.g., XbaI), and universal adaptor sequences are annealed to the sticky ends. The fragments are then amplified by PCR, using a single primer recognizing the adaptor-ligated sequences. The reaction conditions are such that only a subset of the fragments is amplified. As a result, a reproducible subset of the genome is isolated. The fragments are labeled and hybridized to a chip containing oligos that allow genotyping of the SNPs present within the reduced genome. Owing to the nature of the genome-complexity reduction approach, it can be scaled to higher levels of SNP genotyping by simply altering the restriction enzyme or PCR conditions to create a higher-complexity subset and a greater number of amplified SNPs. This approach has been used in the creation of the GeneChip human mapping 100 K and 500 K products for the analysis of over 100,000 and 500,000 SNPs, respectively. This platform has its disadvantages. The assay requires high-quality DNA. The instrument cost is high and mainly suitable for the service core lab. The procedure involves multiple steps and takes days to finish.
3. **Infinium genotyping assay:** The illumina platform combines its BeadChip arrays with an allele-specific extension reaction, which can analyze over 100,000 SNPs in an exon-centric manner across the genome. In the assay, genomic DNA is amplified by a non-PCR

approach in an isothermal reaction. Amplified DNA is fragmented and hybridized to a mixture of bead types. Two bead types are dedicated for the genotyping of each SNP site, resulting in over 200,000 bead types for the analysis of over 100,000 SNPs. To each pair of beads a ~50-mer locus-specific oligo is attached in which the last nucleotide is positioned at the polymorphic base. Upon hybridization to the amplified DNA, the bead-bound primers undergo an extension reaction in the presence of unlabeled and labeled nucleotides. Primers with a mismatch at the SNP site will fail to extend and not be labeled, whereas those with a perfect match will extend and become labeled. The reporter signal associated with each bead is determined by imaging each of the 200,000 bead types. The relative signal observed for each pair of beads is used to determine the genotype of each SNP. Products have recently been released for the analysis of over 250,000 SNPs. The Infinium assay also involves multiple steps and requires days to complete. One significant advantage of the Infinium assay over the Affymetrix system is that it can utilize DNA that is partially degraded; thus, it can be applied to assay older DNA samples.

3. SNP and Human Disease

The most abundant source of genetic variation in the human genome is represented by SNPs, which can account for heritable interindividual differences in complex phenotypes. Identification of SNPs that contribute to susceptibility to common diseases will provide highly accurate diagnostic information that will facilitate early diagnosis, prevention, and treatment of human diseases. Common SNPs, ranging from a minor allele frequency of 5 to >20%, are of interest because it has been argued that common genetic variation can explain a proportion of common human disease — the common variant/common disease (CV/CD) hypothesis. SNPs occur in the coding region, intron region, 5′ and 3′ untranslated region, promoter region, and intragenic region. There are two types of coding SNPs: nonsynonymous SNPs and synonymous SNPs. nonsynonymous SNPs result in changes in amino acids, whereas synonymous SNPs do not change amino acids. Because nonsynonymous SNPs directly affect protein function, many investigators focus on the genotyping of coding SNPs in genetic association studies; this strategy is known as a "direct" approach. The challenge of this approach lies in predicting or determining *a priori* which SNPs are likely to be causative or predicting the phenotype of interest. Nonsynonymous SNPs, though obvious suspects in causing a proportion of human disease, do not account for all SNPs that can cause disease or susceptibility to disease. The "indirect" approach to genetic association studies differs from the direct approach in that the causal SNP is not assayed directly. The assumption is that the assayed or genotyped SNPs will be in linkage disequilibrium or associated with the causative SNP; thus, the assayed SNP would be overrepresented among cases compared with controls because it is highly correlated with the disease-causing SNP. Besides nonsynonymous SNPs, other functional SNPs, located in promoters, introns, splice sites, and intragenic regions, are implicated in human diseases or susceptibility to diseases. Furthermore, even synonymous (or "silent") SNPs have been implicated as having functional consequences via unknown mechanisms. We are still in the early stages of fully discerning effects of DNA polymorphisms in relation to human disease.

4. SNP Databases

A number of SNP databases exist. Here, two selective SNP resources are briefly described.

dbSNP: dbSNP (http://www.ncbi.nlm.nih.gov/projects/SNP/get_html.cgi?whichHtml=overview) was established by The National Center for Biotechnology Information (NCBI), U.S. in collaboration with the National Human Genome Research Institute, U.S. to serve as a central repository for both single-base nucleotide substitutions and short deletion and insertion polymorphisms. The data in dbSNP are integrated with other NCBI genomic data. The data in dbSNP are freely available to the scientific community and made available in a variety of forms.

dbSNP takes the looser "variation" definition for SNPs, so there is no requirement or assumption about minimum allele frequency. dbSNP records all submitted SNP data regardless of any allele frequency in populations. Thus, many SNPs in the database are not validated, nor is any allele frequency data percent.

Scientific users of dbSNP should be aware of the current status in dbSNP to effectively harness the power of this resource. As of November 1, 2006, dbSNP has collected SNPs from 35 organisms. Among them, *Homo sapiens*, *Mus musculus*, *Canis familiaris*, *Gallus gallus*, *Pan troglodytes*, *Oryza sativa*, and *Anopheles gambiae* have had at least 1 million SNPs curated. In dbSNP build 126 for human SNPs, there are 27,846,394 submitted SNPs and 11,961,761 reference SNPs with 47.20% of them validated; 43.42% of those reference SNPs occur within the gene. For those submitted SNPs, only about 20% of them have genotype data and <3% of them have frequency data.

SeattleSNPs: Seattle SNPs (http://pga.gs.washington.edu/) has been funded as part of the National Heart Lung and Blood Institute's (NHLBI) Programs for Genomic Applications (PGA). The SeattleSNPs PGA has been focused on identifying, genotyping, and modeling the associations between single-nucleotide polymorphisms (SNPs) in candidate genes and pathways that underlie inflammatory responses in humans. As of November 2, 2006, they have sequenced 286 genes and identified 28,312 SNPs from 24 African-American (AA) subjects and 17,560 SNPs from 23 European (CEPH) subjects. The Web sites provides useful sequence, genotype, software, and educational resources as well as linking to other relevant SNP databases.

Part II Step-By-Step Tutorial

In this part, pre-B-cell colony-enhancing factor (PBEF1) gene will be used to demonstrate how to search for its SNP information from dbSNP and display all SNP data of the IL-10 gene in VG2 from SeattleSNP.

1. Search for SNP Information of PBEF 1 Gene from dbSNP

1. Go to the SNP home page (http://www.ncbi.nlm.nih.gov/SNP/). There are six different search options: entrez SNP, ID numbers, Submission info, Batch, Locus info, and between markers. Here, entrezSNP is demonstrated. Type gene symbol PBEF1 in the textbox at the top of the page, as displayed in Figure 4.1.

2. Click "GO," and the graphic summary on how many SNPs collected from each species as well as each SNP with reference SNP number (rs#) is shown as in Figure 4.2. Users can have 31 display options to present different formats of the data, and sort the data in 6 different ways. Furthermore, users can go to each particular SNP by simply clicking the relevant linking bar to get detailed information such as its chromosomal location, gene view, sequence view, and genotypes.

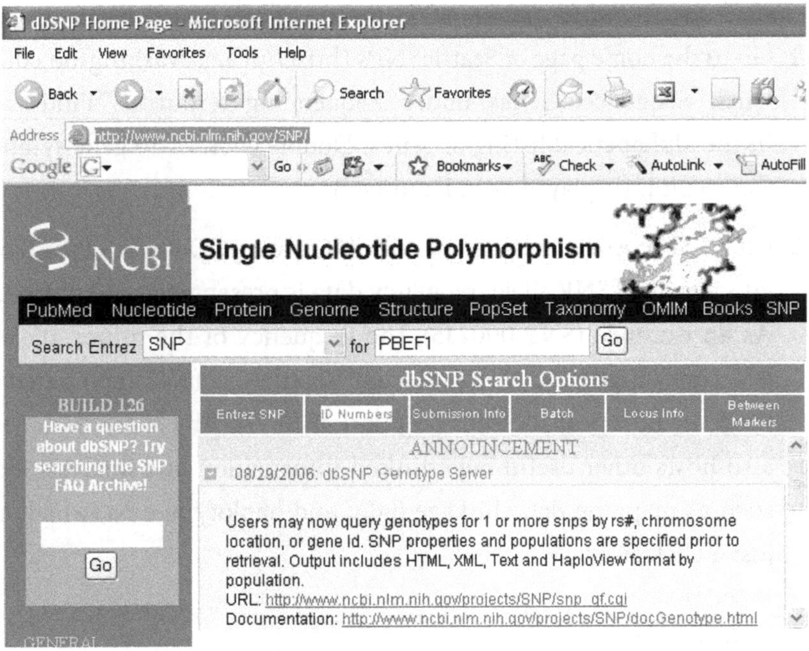
dbSNP Home Page - Microsoft Internet Explorer
File Edit View Favorites Tools Help
Back Search Favorites
Address http://www.ncbi.nlm.nih.gov/SNP/
Go Bookmarks Check AutoLink AutoFill
NCBI
Single Nucleotide Polymorphism
PubMed Nucleotide Protein Genome Structure PopSet Taxonomy OMIM Books SNP
Search Entrez SNP for PBEF1 Go
dbSNP Search Options
BUILD 126
Have a question about dbSNP? Try searching the SNP FAQ Archive!
Go
Entrez SNP ID Numbers Submission Info Batch Locus Info Between Markers
ANNOUNCEMENT
08/29/2006: dbSNP Genotype Server
Users may now query genotypes for 1 or more snps by rs#, chromosome location, or gene Id. SNP properties and populations are specified prior to retrieval. Output includes HTML, XML, Text and HaploView format by population.
URL: http://www.ncbi.nlm.nih.gov/projects/SNP/snp_gf.cgi
Documentation: http://www.ncbi.nlm.nih.gov/projects/SNP/docGenotype.html

FIGURE 4.1 Search for SNP information of PBEF 1 gene in dbSNP. This screenshot displays the search for PBEF1 SNP information from dbSNP by typing the gene symbol PBEF1 in the text box.

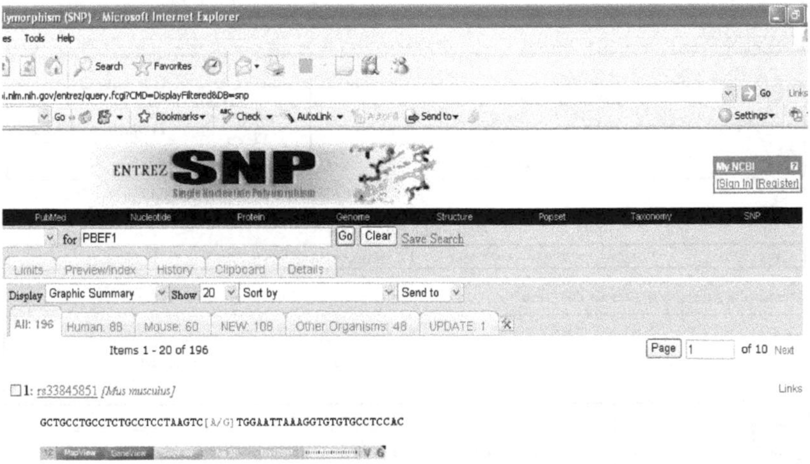
ENTREZ SNP
for PBEF1 Go Clear Save Search
Limits Preview/Index History Clipboard Details
Display Graphic Summary Show 20 Sort by Send to
All: 196 Human: 88 Mouse: 60 NEW: 108 Other Organisms: 48 UPDATE: 1
Items 1 - 20 of 196
Page 1 of 10 Next
1: rs33845851 [Mus musculus]
Links
GCTGCCTGCCTCTGCCTCCTAAGTC[A/G]TGGAATTAAAGGTGTGTGCCTCCAC

FIGURE 4.2 The graphic summary of PBEF1 SNPs. This screenshot shows how many PBEF1 SNPs were collected from each species as well as each SNP with reference SNP number (rs#).

2. Search for SNP Information of IL-10 Gene from SeattleSNPs

1. Go to the home page of SeattleSNPs (http://pga.gs.washington.edu/). Click "Sequenced Genes" under "Sequencing Resources." Find IL-10 in the alphabetical listing of genes. Double click IL-10 gene symbol, the result is displayed as in Figure 4.3.

2. Click "SNP Allele Frequency" under "Genotype Data," and the tabular view of all SNP allele frequency data is presented as in Figure 4.4. As an example (SNP 000245), the frequency of the minor allele T is 2 times higher in Africa-descent population (0.47) compared to those in European-descent population (0.23). The same Web page also hosts other useful hyperlinked information and analysis tools such as mapping data, linkage data, and haplotype data, which are just a click away.

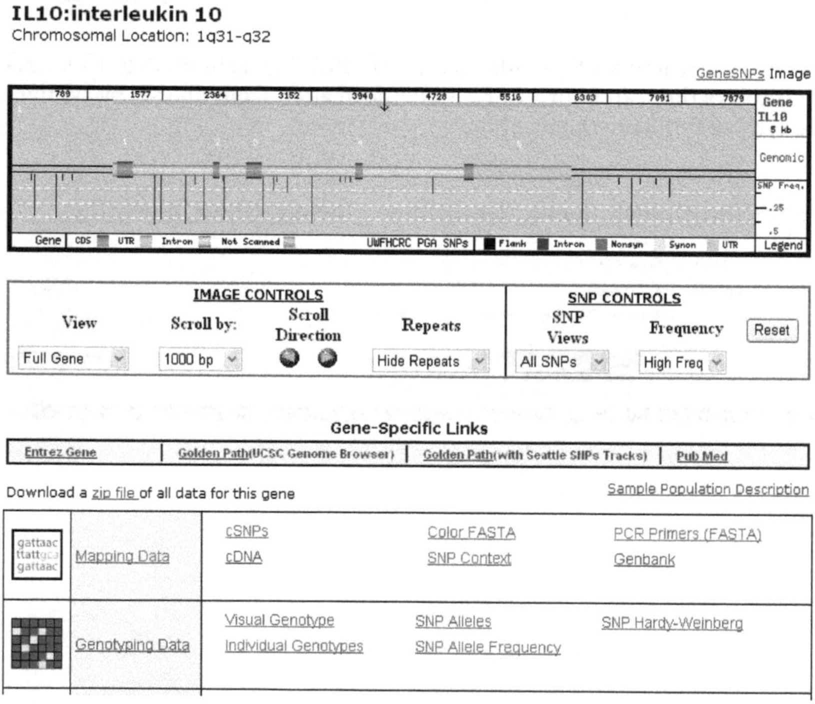
IL10:interleukin 10
Chromosomal Location: 1q31-q32
GeneSNPs Image
Gene
IL10
5 kb
Genomic
Legend
IMAGE CONTROLS
View
Scroll by:
Scroll Direction
Repeats
Full Gene
1000 bp
Hide Repeats
SNP CONTROLS
SNP Views
Frequency
Reset
All SNPs
High Freq
Gene-Specific Links
Entrez Gene
Golden Path(UCSC Genome Browser)
Golden Path(with Seattle SNPs Tracks)
Pub Med
Download a zip file of all data for this gene
Sample Population Description
Mapping Data
cSNPs
cDNA
Color FASTA
SNP Context
PCR Primers (FASTA)
Genbank
Genotyping Data
Visual Genotype
Individual Genotypes
SNP Alleles
SNP Allele Frequency
SNP Hardy-Weinberg

FIGURE 4.3 Search for SNP information of IL-10 gene from SeattleSNPs. This screenshot shows that various types of information on a particular gene, such as the IL-10 gene, is just a click away.

Site	Allele1	AD-pop	ED-pop	Total-pop	Allele2	AD-pop	ED-pop	Total-pop	AD-hz	ED-hz	Total-hz
000213	A	0.00	0.07	0.04	G	1.00	0.93	0.96	0.00	0.13	0.07
000245	T	0.47	0.23	0.35	C	0.53	0.77	0.65	0.50	0.35	0.45
000472	A	0.50	0.24	0.37	C	0.50	0.76	0.63	0.50	0.36	0.47
000532	G	0.02	0.00	0.01	A	0.98	1.00	0.99	0.04	0.00	0.02
000635	T	0.05	0.00	0.02	G	0.95	1.00	0.98	0.09	0.00	0.05
001498	T	0.39	0.46	0.42	C	0.61	0.54	0.58	0.47	0.50	0.49
001568	T	0.50	0.26	0.38	G	0.50	0.74	0.62	0.50	0.39	0.47
001833	T	0.36	0.45	0.41	G	0.64	0.55	0.59	0.46	0.49	0.48
001927	C	0.06	0.10	0.08	A	0.94	0.90	0.92	0.10	0.18	0.15
002018	T	0.50	0.26	0.38	G	0.50	0.74	0.62	0.50	0.39	0.47
002220	C	0.00	0.02	0.01	T	1.00	0.98	0.99	0.00	0.05	0.02
002234	A	0.50	0.24	0.38	G	0.50	0.76	0.62	0.50	0.36	0.47
002646	T	0.48	0.22	0.35	C	0.52	0.78	0.65	0.50	0.34	0.46
002767	T	0.04	0.20	0.12	A	0.96	0.80	0.88	0.08	0.31	0.21
002802	A	0.00	0.02	0.01	G	1.00	0.98	0.99	0.00	0.04	0.02
002911	T	0.06	0.22	0.14	G	0.94	0.78	0.86	0.12	0.34	0.24
003166	G	0.35	0.46	0.40	C	0.65	0.54	0.60	0.46	0.50	0.48
003469	G	0.00	0.02	0.01	T	1.00	0.98	0.99	0.00	0.04	0.02
003528	A	0.11	0.00	0.05	G	0.89	1.00	0.95	0.19	0.00	0.10
003582	C	0.00	0.07	0.03	T	1.00	0.93	0.97	0.00	0.13	0.06
004467	A	0.06	0.22	0.14	G	0.94	0.78	0.86	0.12	0.34	0.24
005016	C	0.35	0.48	0.41	T	0.65	0.52	0.59	0.46	0.50	0.49
005090	G	0.02	0.00	0.01	A	0.98	1.00	0.99	0.04	0.00	0.02
005244	T	0.00	0.02	0.01	A	1.00	0.98	0.99	0.00	0.04	0.02
005351	G	0.15	0.19	0.17	A	0.85	0.81	0.83	0.26	0.31	0.28
005851	G	0.02	0.00	0.01	A	0.98	1.00	0.99	0.05	0.00	0.02
006049	C	0.36	0.47	0.41	T	0.64	0.53	0.59	0.46	0.50	0.49
006433	A	0.02	0.00	0.01	G	0.98	1.00	0.99	0.04	0.00	0.02
006570	A	0.33	0.50	0.41	G	0.67	0.50	0.59	0.44	0.50	0.48
006659	G	0.05	0.00	0.02	A	0.95	1.00	0.98	0.09	0.00	0.05
006831	T	0.02	0.00	0.01	C	0.98	1.00	0.99	0.05	0.00	0.02
006976	T	0.09	0.23	0.16	C	0.91	0.77	0.84	0.16	0.35	0.26

Note:

AD: African_Descent
ED: European_Descent
Total: All_population
pop : population

FIGURE 4.4 Tabular view of all SNP allele frequency data in the IL-10 gene. The tabular view presents SNP site, allele frequencies, and heterozygosities in both African- and European-descent populations of each SNP in the IL-10 gene.

Part III Sample Data

1. Gene symbol: PBEF 1

2. Gene symbol: IL-10 (Interleukin 10)

SECTION 2 HAPLOTYPE ANALYSIS

Part I Introduction

1. What Is a Haplotype?

The term *haplotype* is a contraction of "haploid genotype." Haplotypes are a combination of alleles at different markers along the same chromosome that are inherited as a unit. Although each marker can be analyzed independently of other marker, it is much more informative to analyze markers in a region of interest simultaneously. There is a growing interest in understanding haplotypes structures in the human genome using identified genetic markers because: (1) haplotype structure may provide critical information on human evolutionary history and the identification of genetic variants underlying various human traits; and (2) molecular technologies now make it possible to study hundreds of thousands of genetic polymorphisms in population samples of reasonable sizes. One major aspect of haplotype analysis is to identify linkage disequilibrium (LD), or allelic association, patterns in different regions and different populations because the very existence of LD among markers makes it possible to infer population histories and localize genetic variants underlying complex traits.

2. Methods of Haplotype Analysis

Currently, there are two broad categories of tools that can unambiguously determine haplotypes: directly genotyping pedigrees and using molecular methods in combination with genotyping for individual samples that do not have pedigree information. The pedigree or family-based method relies on the fact that different loci on the same chromosome (haplotype) will be inherited as a unit unless they are separated by a recombination event. The probability of a recombination depends partly on the physical distance between the markers. Markers that are closer in physical distance have a greater chance of being "linked," which means that their alleles are transmitted from parent to offspring as a haplotype. For population-based data, molecular or experimental methods have become the "gold standard" method for constructing haplotypes. Several molecular methods are available to construct unambiguous haplotypes. Two widely used molecular methods include allele-specific polymerase chain reaction (AS-PCR) and somatic cell hybrids. These molecular methods distinguish which allele is on which chromosome, a step generally not required by family-based studies because this information can be extracted from knowledge of the alleles transmitted by the parents to the offspring. A common PCR reaction on an individual sample without pedigree information will tell the investigator which two alleles are present in an individual sample, but an allele-specific PCR will reveal which allele is present in the context of another allele on the same chromosome. The somatic cell hybrid is a technique that physically separates the maternal and paternal chromosomes of an individual before genotyping. Both AS-PCR and somatic cell hybrids have been used to unambiguously determine haplotypes in relatively small- to moderate-sized population surveys.

Although pedigrees and molecular methods are more reliable in assigning haplotypes, both tools are costly and tedious relative to statistical inference. A number of the statistical-inference software packages are currently available. These haplotype inference programs can be categorized into three broad groups: parsimony, maximum likelihood, and Bayesian. Each group has advantages and disadvantages. The Clark algorithm, based on parsimony, first determines the known haplotypes and then searches for genotype combinations that are congruent with the known haplotypes. The Clark algorithm is easily understandable but cannot be applied to all datasets. The expectation-maximization (EM) algorithms can assign all alleles to haplotypes with a high probability. However, the EM algorithms do not make assumptions about recombination or mutation, assume that the data are in Hardy Weinberg equilibrium, and cannot handle large datasets efficiently. Bayesian approaches incorporate assumptions or prior information as a guide for the inference of haplotypes not previously observed. The Bayesian algorithms perform better than the parsimony and maximum-likelihood algorithms. Like other algorithms, Bayesian approaches do not account for recurrent mutation or gene conversion. The drawback of nearly every statistical-inference package is that not all inferred haplotypes are correct.

3. Linkage Disequilibrium, Haplotype Block, and Haplotype Tagging

Linkage disequilibrium. Linkage disequilibrium (LD) is a term used in the study of population genetics for the nonrandom association of alleles at two or more loci. A study of haplotype consisting of a short tandem repeat polymorphism and an Alu deletion polymorphism at the CD4 locus in forty-two worldwide populations first demonstrated that the LD patterns between these two polymorphisms could provide evidence of a common and recent African origin for all non-African populations. In addition to its use in inferring population history, the extent of LD is a critical factor in identifying disease-associated genetic variants and designing efficient studies to detect disease gene associations. Various measures have been proposed for characterizing the statistical association that arises between alleles at different loci. The most commonly used ones are D' and r^2. Both D' and r^2 range between 0 and 1.

D' is a measure of linkage disequilibrium between two genetic markers. A value of $D' = 1$ (complete LD) indicates that two SNPs have not been separated by recombination, whereas values of $D' < 1$ (incomplete LD) indicate that the ancestral LD was disrupted during the history of the population. Only D' values near one are a reliable measure of LD extent; lower D' values are usually difficult to interpret as the magnitude of D' strongly depends on sample size. To calculate this value, see the article by Lewontin (1988) for the details.

r^2 is a measure of linkage disequilibrium between two genetic markers. For SNPs that have not been separated by recombination or have the same allele frequencies (perfect LD), $r^2 = 1$. In such case, the SNPs are said to be redundant. Lower r^2 values indicate less degree of LD. One useful property of r^2 for association studies is that its inverse value, $1/r^2$, provides a practical estimate of the magnitude by which the sample size must be increased in a study design to detect association between the disease and a marker locus, when compared with the size required for detecting association with the susceptibility locus itself. To calculate this value, see the article by Pritchard and Przeworski (2001) for the details. It should be pointed out that based on the analysis of many studies, LD is both locus and population specific. Although LD between two markers tends to decrease as their physical distance increases, the variation is so great that is not possible to predict LD between two polymorphisms reliably, based only on their physical distance. The amount of LD differs among different populations, and LD is usually weaker among Africans than other populations. Therefore, only with a systematic empirical study that covers the genome and involves many human populations could the full spectrum of LD patterns in the human genome be understood.

Haplotype block. Some recent studies have found that chromosomes are structured such that each chromosome can be divided into many blocks, i.e., haplotype blocks, within which there is limited haplotype diversity. The concept of "blocks" arose from an initial report that the 500-kb region of 5q31 genotyped in a European-descent population had discrete regions of low haplotype diversity. The regions, termed *blocks*, were up to 100 kb long and generally consisted of 2–4 haplotypes, which accounted for >90% of the chromosomes surveyed. Within the blocks, there was little to no recombination and between the blocks, there was a clustering of recombination events. Afterward, haplotype blocks have been demonstrated across the human genome in several populations. It should be cautioned that there is no universally accepted definition of haplotype blocks; thus, the block structures identified in each study depend strongly on the definition used, and there has been no systematic comparison of haplotype blocks identified under various definitions.

The main advantage of low haplotype diversity or haplotype blocks is that only a few markers or SNPs need to be genotyped to represent haplotypes within a block in a whole-genome association study. Fueled by the prospect that the human genome can be described in terms of haplotype blocks, the National Human Genome Research Institute (NHGRI) at the National Institutes of Health (NIH) initiated the International "HapMap" Project.

The International HapMap Project is an organization whose goal is to develop a haplotype map of the human genome (the HapMap) that will describe the common patterns of human genetic variation and make these data available for researchers interested in whole-genome association studies. The project is a collaboration among researchers at academic centers, nonprofit biomedical research groups, and private companies in Canada, China, Japan, Nigeria, the U.K., and the U.S. Four populations were selected for inclusion in the HapMap: 30 adult-and-both-parents trios from Ibadan, Nigeria (YRI), 30 trios of U.S. residents of northern and western European ancestry (CEU), 44 unrelated individuals from Tokyo, Japan (JPT), and 45 unrelated Han Chinese individuals from Beijing, China (CHB). All the data generated by the project, including SNP frequencies, genotypes, and haplotypes, have been placed in the public domain at http://www.hapmap.org. International HapMap Project becomes a useful public tool in the search for disease-causing genes and loci important in public health.

Haplotype tagging. Haplotype tagging refers to methods of selecting minimal number of SNPs that uniquely identify common haplotypes (>5% in frequency). The reason for haplotype tagging or tagging SNP selection is that the number of markers now available per gene makes genotyping all markers very expensive for the average research budget, although the costs associated with genotyping have decreased. Because of this, much interest has been devoted to choosing a set of markers to best represent the genetic variation of the candidate gene.

There are two principal uses of tagging. The first is to select a "good" subset of SNPs to be typed in all the study individuals from an extensive SNP set that has been typed in just a few individuals. Until recently, this was frequently a laborious step in study design, but the International HapMap Project and related projects now allow selection of tag SNPs on the basis of publicly available data.

A secondary use for tagging is to select for analysis a subset of SNPs that have already been typed in all the study individuals. Although it is undesirable to discard available information, the amount of information lost might be small, and reducing the SNP set in this way can simplify analyses and lead to more statistical power by reducing the degrees of freedom (df) of a test.

Many current approaches to haplotype tagging have limitations. Many algorithms require haplotypes but do not account for incorrectly inferred haplotypes or variously defined haplotype blocks. The population that underlies a particular study will typically differ from the populations for which public data are available, and a set of tag SNPs that have been selected in one population might perform poorly in another. Investigators should be aware that population stratification inflates estimates of linkage disequilibrium; therefore, populations of combined race or ethnicity (e.g., the Polymorphism Discovery Resource Panel) are not ideal for choosing tagSNPs to be genotyped in larger populations.

4. Medical Applications of Haplotyping

Candidate gene search. In the research setting, haplotypes are commonly used to localize a disease-conferring gene or locus. Currently, much interest surrounds the use of genetic association studies because this study design is suggested to be more powerful than linkage studies in localizing susceptibility loci for common diseases (e.g., heart disease, asthma, diabetes, autoimmune disease, or cancer) that have moderate risk. Similar to the linkage study design, an association study design genotypes markers in affected and unaffected individuals, and it is expected that markers which co-occur or are associated with the disease phenotype either contribute to the phenotype or are associated with the disease susceptibility locus. In a candidate gene association study, a gene is chosen for study based on an educated guess of the location (usually by a linkage study), genetic studies in model organisms, or the biology of the disease locus in relation to the disease phenotype. The specific role of haplotypes in a candidate gene association study depends on the hypothesis being tested. For example, haplotypes can represent a combined effect of several sites along the same chromosome (cis-acting loci) that cannot be detected when these sites are tested one by one.

More commonly, investigators rely on haplotypes to serve as proxies for ungenotyped SNPs. In this case, for a traditional case-control gene association study design, a statistical test is performed to determine if an allele or genotype of a SNP in a particular candidate gene is overrepresented among cases compared with controls. If a SNP allele is associated with a disease phenotype, the allele is either contributing to the disease phenotype or is in linkage disequilibrium with the SNP allele that contributes to the phenotype.

Diagnosis. Human leukocyte antigen (HLA) matching is a clear example of how haplotypes can be used in the clinic to improve outcome. In this scenario, transplant recipients and donors are genotyped at several markers along the major histocompatibility complex. The HLA haplotypes are then determined by ordering the alleles along the chromosomes.

Patients who match the donor haplotypes closely are predicted to have a better transplant outcome than those who do not. The development of HLA haplotype matching has proved to be crucial in making transplantation between unrelated patients and donors a success. There are instances in which haplotypes rather than genotypes at a single locus can predict severity of disease. For example, some research suggests that a specific β-globin locus haplotype is associated with less severe sickle cell disease phenotypes. More recently, a promoter region haplotype in IL-10 was associated with a lower incidence of graft-versus-host disease and death compared with other haplotypes among patients receiving hematopoietic-cell transplants.

Pharmacogenomics and phamacogenetics. *Pharmacogenomics* and *phamacogenetics* are terms used to describe the study of genetic variants and how these variants relate to interindividual response to drug therapy. Both terms can be used to describe how genetic variation affects key pathways for drug metabolism, delivery excretion, sight of action, and toxicity. However, pharmacogenomics emphasizes a larger, genome-wide approach that considers not only single-gene effects, but also mutagen interactions and pathways. Pharmacokinetics will usually be used to describe a single-gene approach to understand the effects of genetic variation on drug response. Many examples of varied clinical response to medication based on heritable differences have been described, and it has been estimated that 20 to 90% of drug effects and efficacy may be caused by heritable differences. PharmGKB (http://www.pharmgkb.org/) is curating information that establishes knowledge about the relationships among drugs, diseases and genes, including their variations and gene products. PharmGKB now contains Applied Biosystems' variant data from 4 human populations on over 200 drug-metabolizing genes. Incorporation of haplotypes into studies of pharmacogenomics and pharmacogenetics will certainly increase in the near future and provide a more complete picture of the sites that are relevant, either alone or in concert, in the practice of "genetic medicine" at the population or individual level.

Part II Step-By-Step Tutorial

In this part, we describe how to find information on all genotyped SNPs of PBEF1 and select tagSNP from the HapMap, and download them into the Haploview program to perform LD select and haplotype analysis.

1. Find All SNPs in the Human PBEF1 Gene

1. Go to the HapMap home page, select Browse Project Data under Project Data, type "PBEF1" in the Search window under Landmark or Region and click "Search," and the result is shown as in Figure 4.5.
2. Click hyperlink chr7:105.5..105.5 Mbp (34.7 kbp) by the NM_005746 pre-B-cell colony enhancing factor 1 isoform a [PBEF1], all 36 genotyped SNPs within the PBEF1 gene region in 4 different human populations (as of November 6, 2006) will be shown as in Figure 4.6. Reference # SNP is followed by (+/–) strand information. Blue color represents reference allele in the human genome assembly, and red color represents alternative allele. In the pie chart, blue and red displays allele frequency in each population, and white color means that data are not available. Double-clicking each reference SNP will lead to the detailed SNP information, including its genomic location, frequency report, and available assays.

2. Select tagSNP

1. Under Report and Analysis, select Annotate tagSNP Picker from its drop-down menu, click "Configure" … and its configuring page is shown as in Figure 4.7. Several parameters such as population, r^2 value, and MAF cutoff can be adjusted at the discretion of investigators.

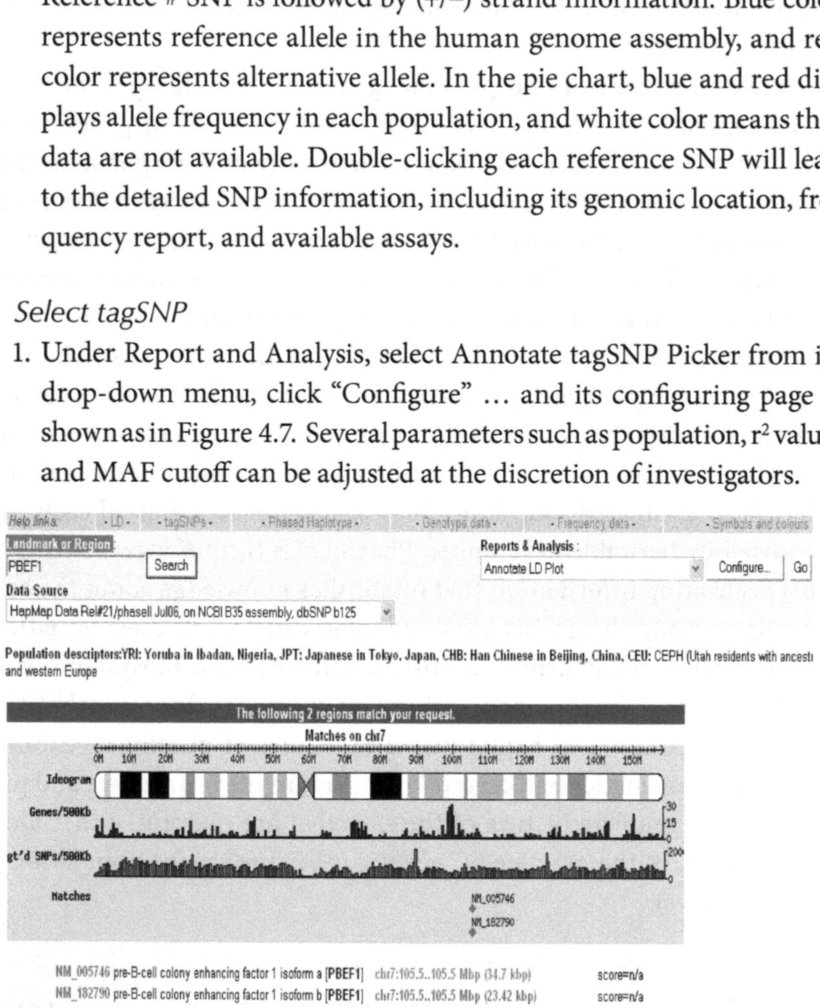
Help links
LD
tagSNPs
Phased Haplotype
Genotype data
Frequency data
Symbols and colours
Landmark or Region
Reports & Analysis:
PBEF1
Search
Annotate LD Plot
Configure...
Go
Data Source
HapMap Data Rel#21/phaseII Jul06, on NCBI B35 assembly, dbSNP b125
Population descriptors:YRI: Yoruba in Ibadan, Nigeria, JPT: Japanese in Tokyo, Japan, CHB: Han Chinese in Beijing, China, CEU: CEPH (Utah residents with ancestı
and western Europe
The following 2 regions match your request.
Matches on chr7
Ideogram
Genes/500Kb
gt'd SNPs/500Kb
Matches
NM_005746
NM_182790
NM_005746 pre-B-cell colony enhancing factor 1 isoform a [PBEF1] chr7:105.5..105.5 Mbp (34.7 kbp) score=n/a
NM_182790 pre-B-cell colony enhancing factor 1 isoform b [PBEF1] chr7:105.5..105.5 Mbp (23.42 kbp) score=n/a

FIGURE 4.5 Find all SNPs in the human PBEF1 gene from the HapMap site. This screenshot displays two hyperlinked results of both PBEF1 and PBEF2 SNPs that matched the search request.

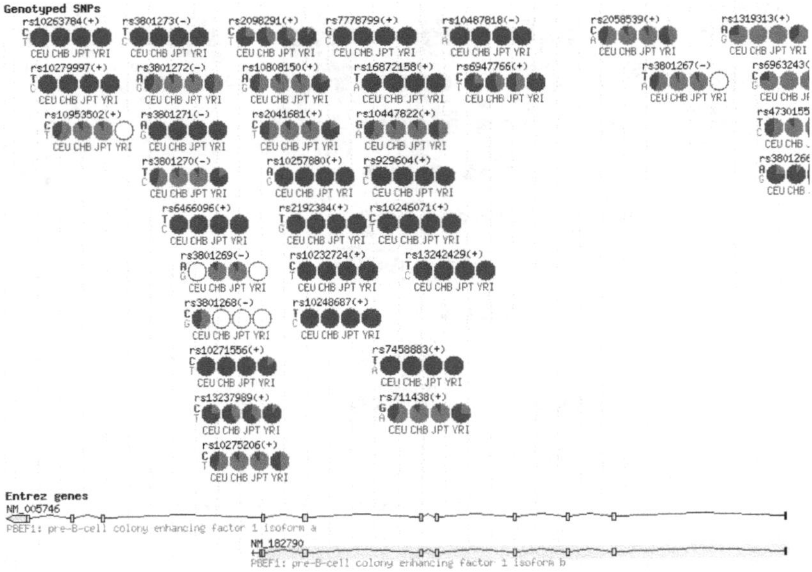
Genotyped SNPs
Entrez genes
NM_005746
PBEF1: pre-B-cell colony enhancing factor 1 isoform a
NM_182790
PBEF1: pre-B-cell colony enhancing factor 1 isoform b

FIGURE 4.6 All 36 genotyped SNPs within the PBEF1 gene region in 4 different human populations. Top panel displays all genetyped SNPs of the PBEF1 gene. Lower panel illustrates two isoforms of the PBEF gene.

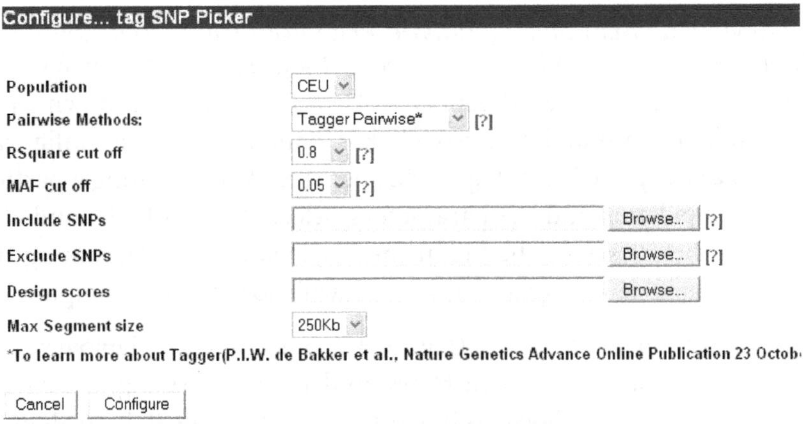
Configure... tag SNP Picker
Population
CEU
Pairwise Methods:
Tagger Pairwise*
[?]
RSquare cut off
0.8
[?]
MAF cut off
0.05
[?]
Include SNPs
Browse...
[?]
Exclude SNPs
Browse...
[?]
Design scores
Browse...
Max Segment size
250Kb
*To learn more about Tagger(P.I.W. de Bakker et al., Nature Genetics Advance Online Publication 23 Octob
Cancel
Configure

FIGURE 4.7 Configure PBEF1 gene tagSNP picker.

2. Click "Configure," and six tagSNPs in the human PBEF gene are shown as in the lower part of Figure 4.8.

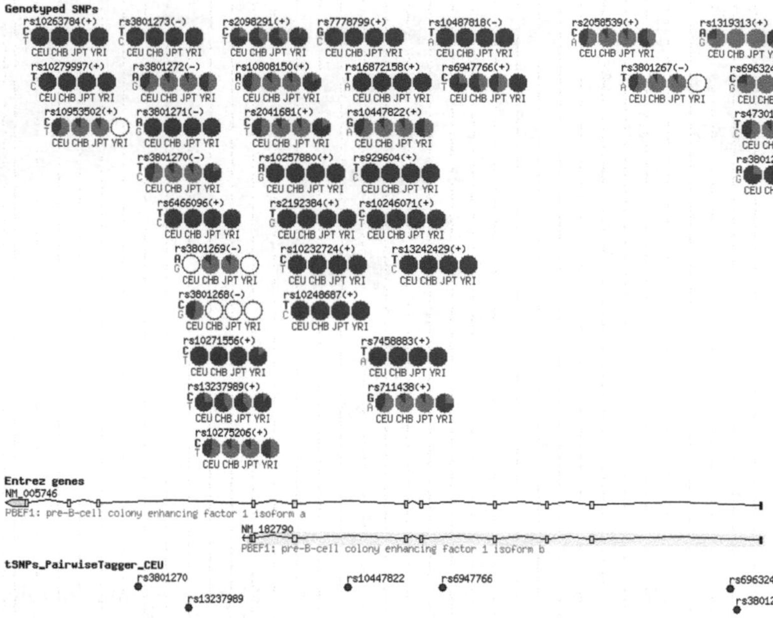
Genotyped SNPs
Entrez genes
NM_005746
PBEF1: pre-B-cell colony enhancing factor 1 isoform a
NM_182790
PBEF1: pre-B-cell colony enhancing factor 1 isoform b
tSNPs_PairwiseTagger_CEU
rs3801270
rs13237989
rs10447822
rs6947766
rs6963243

FIGURE 4.8 Output of PBEF1 gene tagSNPs. Top panel displays all genetyped SNPs of the PBEF1 gene. The middle panel illustrates two isoforms of the PBEF gene. The bottom panel in the figure reports six PBEF1 gene tagSNPs from the European population.

3. Download PBEF1 SNP Genotype Data from HapMap to the Haploview Program to Perform LD Select and Haplotype Analysis

1. Select Download SNP genotype Data from the drop-down menu Under Reports and Analysis, click "Configure" … . The configuring page is displayed as in Figure 4.9, panel A. After choosing a particular population and the parameters, select "Save to Disk" and click "Go" before saving the file in the Haploview folder (assuming that the Haploview program has been downloaded to local computer).

2. Select Load HapMap data (Figure 4.9, panel B) in the Haploview program to load downloaded HapMap data. After loading, the page of Check Marker is shown as in Figure 4.10. Out of 19 SNPs, 6 SNPs are unchecked because their minor allele frequencies (MAF) are all zero.

3. Click "Haplotypes" in the upper left to display Haplotypes as in Figure 4.11. Click "LD plot" to display the LD plot as in Figure 4.12. Under Display options, click "LD zoom" to view annotated details.

Part III Sample Data

1. PBEF1

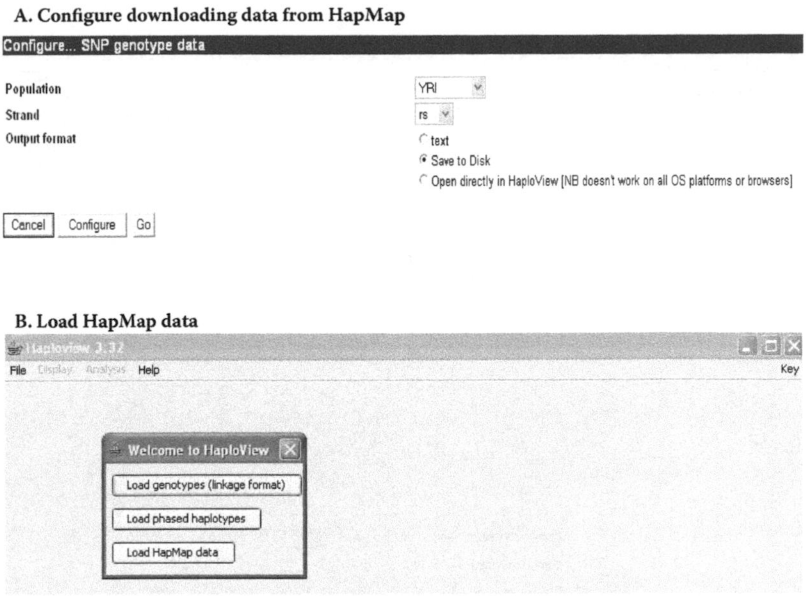
A. Configure downloading data from HapMap
Configure... SNP genotype data
Population
Strand
Output format
YRI
rs
text
Save to Disk
Open directly in HaploView [NB doesn't work on all OS platforms or browsers]
Cancel
Configure
Go
B. Load HapMap data
Haploview 3.32
File
Display
Analysis
Help
Key
Welcome to HaploView
Load genotypes (linkage format)
Load phased haplotypes
Load HapMap data

FIGURE 4.9 Download PBEF1 SNP genotype data from HapMap to the Haploview program. Panel A shows configuration of downloading data from the HapMap. Panel B shows Load HapMap data.

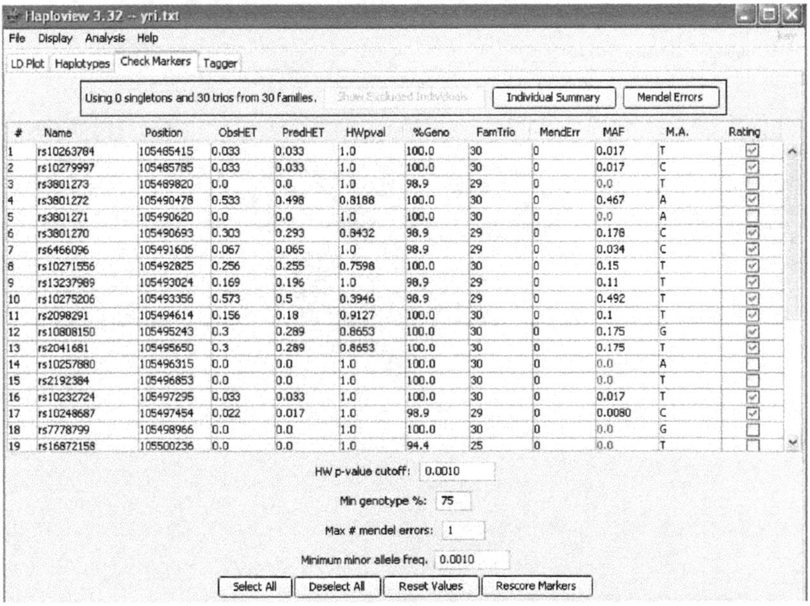
Haploview 3.32 -- yri.txt
File
Display
Analysis
Help
LD Plot
Haplotypes
Check Markers
Tagger
Using 0 singletons and 30 trios from 30 families.
Individual Summary
Mendel Errors
HW p-value cutoff: 0.0010
Min genotype %: 75
Max # mendel errors: 1
Minimum minor allele Freq. 0.0010
Select All
Deselect All
Reset Values
Rescore Markers

#	Name	Position	ObsHET	PredHET	HWpval	%Geno	FamTrio	MendErr	MAF	M.A.	Rating
1	rs10263784	105485415	0.033	0.033	1.0	100.0	30	0	0.017	T	✓
2	rs10279997	105485785	0.033	0.033	1.0	100.0	30	0	0.017	C	✓
3	rs3801273	105489820	0.0	0.0	1.0	98.9	29	0	0.0	T	
4	rs3801272	105490478	0.533	0.498	0.8188	100.0	30	0	0.467	A	✓
5	rs3801271	105490620	0.0	0.0	1.0	100.0	30	0	0.0	A	
6	rs3801270	105490693	0.303	0.293	0.8432	98.9	29	0	0.178	C	✓
7	rs6466096	105491606	0.067	0.065	1.0	98.9	29	0	0.034	C	✓
8	rs10271556	105492825	0.256	0.255	0.7598	100.0	30	0	0.15	T	✓
9	rs13237989	105493024	0.169	0.196	1.0	98.9	29	0	0.11	T	✓
10	rs10275206	105493356	0.573	0.5	0.3946	98.9	29	0	0.492	T	✓
11	rs2098291	105494614	0.156	0.18	0.9127	100.0	30	0	0.1	T	✓
12	rs10808150	105495243	0.3	0.289	0.8653	100.0	30	0	0.175	G	✓
13	rs2041681	105495650	0.3	0.289	0.8653	100.0	30	0	0.175	T	✓
14	rs10257880	105496315	0.0	0.0	1.0	100.0	30	0	0.0	A	
15	rs2192384	105496853	0.0	0.0	1.0	100.0	30	0	0.0	T	
16	rs10232724	105497295	0.033	0.033	1.0	100.0	30	0	0.017	T	✓
17	rs10248687	105497454	0.022	0.017	1.0	98.9	29	0	0.0080	C	✓
18	rs7778799	105498966	0.0	0.0	1.0	100.0	30	0	0.0	G	
19	rs16872158	105500236	0.0	0.0	1.0	94.4	25	0	0.0	T	

FIGURE 4.10 Table view of 19 PBEF1 gene SNPs in the Check Markers Page of the Haploview program.

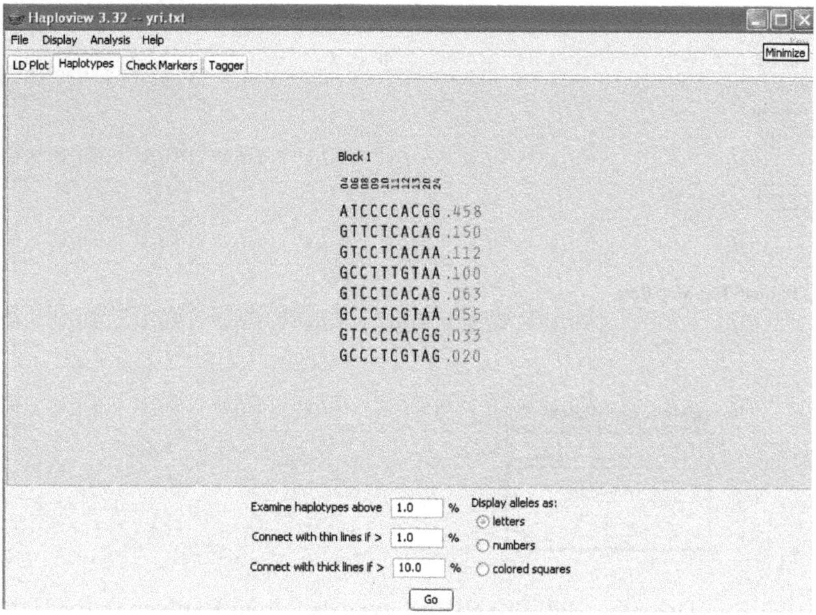
Haploview 3.32 -- yri.txt
File Display Analysis Help
Minimize
LD Plot Haplotypes Check Markers Tagger
Block 1
ATCCCCACGG .458
GTTCTCACAG .150
GTCCTCACAA .112
GCCTTTGTAA .100
GTCCTCACAG .063
GCCCTCGTAA .055
GTCCCCACGG .033
GCCCTCGTAG .020
Examine haplotypes above 1.0 %
Connect with thin lines if > 1.0 %
Connect with thick lines if > 10.0 %
Display alleles as:
letters
numbers
colored squares
Go

FIGURE 4.11 Haplotype display of PBEF gene SNPs.

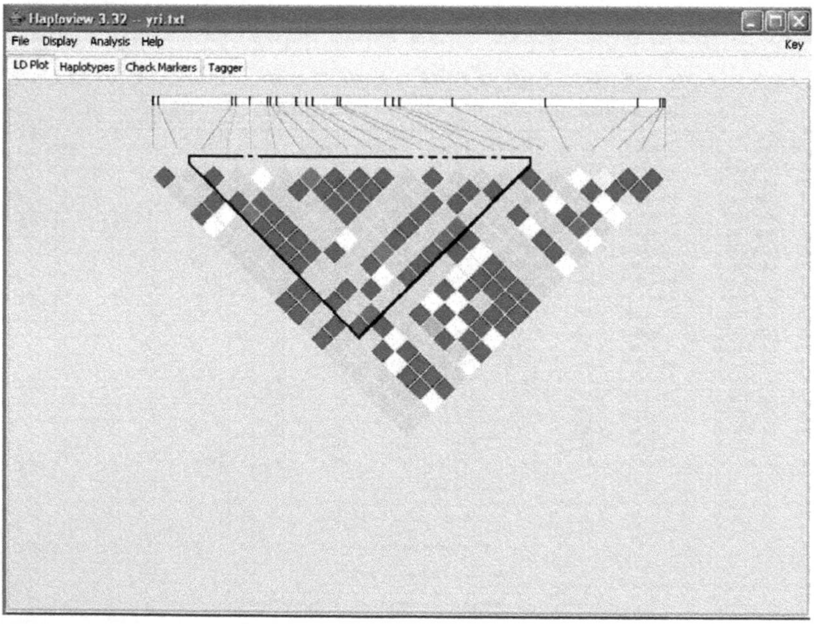
Haploview 3.32 -- yri.txt
File Display Analysis Help
Key
LD Plot Haplotypes Check Markers Tagger

FIGURE 4.12 LD plot display.

REFERENCES

1. Crawford, D.C. and Nickerson, D.A. Definition and clinical importance of haplotypes. *Annu Rev Med* 56: 303–320, 2005.
2. Suh, Y. and Vijg, J. SNP discovery in associating genetic variation with human disease phenotypes. *Mutat Res* 573(1–2): 41–53, 2005.
3. Engle, L.J., Simpson, C.L., and Landers, J.E. Using high-throughput SNP technologies to study cancer. *Oncogene* 25(11): 1594–1601, 2006.
4. Crawford, D.C., Akey, D.T., and Nickerson, D.A. The patterns of natural variation in human genes. *Annu Rev Genomics Hum Genet* 6: 287–312, 2005.
5. Zhao, H., Pfeiffer, R., and Gail, M.H. Haplotype analysis in population genetics and association studies. *Pharmacogenomics* 4(2): 171–178, 2003.
6. Lewontin, R.C. On measures of gametic disequilibrium. *Genetics* 120(3): 849–852, 1988.
7. Pritchard, J.K. and Przeworski, M. Linkage disequilibrium in humans: models and data. *Am J Hum Genet* 69(1): 1–14, 2001.
8. Balding, D.J. A tutorial on statistical methods for population association studies. *Nat Rev Genet* 7(10): 781–791, 2006.
9. Hawkins, G.A., Weiss, S.T., and Bleecker, E.R. Asthma pharmacogenomics. *Immunol Allerg Clin North Am* 25(4): 723–742, 2005.

CHAPTER 5

Gene Expression Profiling by Microarray

Claudia C. dos Santos and Mingyao Liu

CONTENTS

INTRODUCTION

Completion of the human genome project signaled a new beginning for modern biology, one in which the majority of biological and biomedical research will be conducted in a sequence-based fashion. The need to generate, analyze, and integrate large and complex sets of molecular data has led to the development of whole-genome approaches, such as microarray technology, to expedite the process of translating molecular data to biologically meaningful information. The paradigm shift from traditional single-molecule studies to whole-genome approaches requires not only standard statistical modeling and algorithms but also high-level, hybrid, computational/statistical automated learning systems for improving the understanding of complex traits.

Microarray experiments provide unprecedented quantities of genome-wide data on gene-expression patterns. The implementation of a successful uniform program of expression analysis requires the development of various laboratory protocols, as well as the development of database and software tools for efficient data collection and analysis. Although detailed laboratory protocols have been published, the computational tools necessary to analyze the data are rapidly evolving, and no clear consensus exists as to the best method for revealing patterns of gene expression. Consequently, choosing the appropriate algorithms for analysis is a crucial element of the experimental design.

The purpose of this chapter is to provide a general overview of some existing approaches for gene expression analysis. This is not comprehensive, but instead represents a tutorial on some of the more basic tools. The focus here is on basic analysis principles that are generally applicable to expression data generated using spotted arrays (Agilent), SAGE (Serial Analysis of Gene Expression), or oligo arrays (Affymetrix and illumina), provided the data is presented in an appropriate format. Figure 5.1 shows a general schematic framework for microarray analysis. A unique aspect of this book is the integration of practical tutorials with didactic information. In the tutorial, the reader is encouraged to explore multiple steps in the analysis process and become familiar with various software and analysis tools. Many of the topics covered have been recently reviewed (From Hoheisel, J.D. *Nat Genet* 7(3): 200–210, March 2006), and pertinent recent references are provided for further reading at the end of the chapter.

SECTION 1 EXPERIMENTAL DESIGN

The underlying theory behind expression microarray technology is that gene-specific probes, representing thousands of transcribed sequences, are

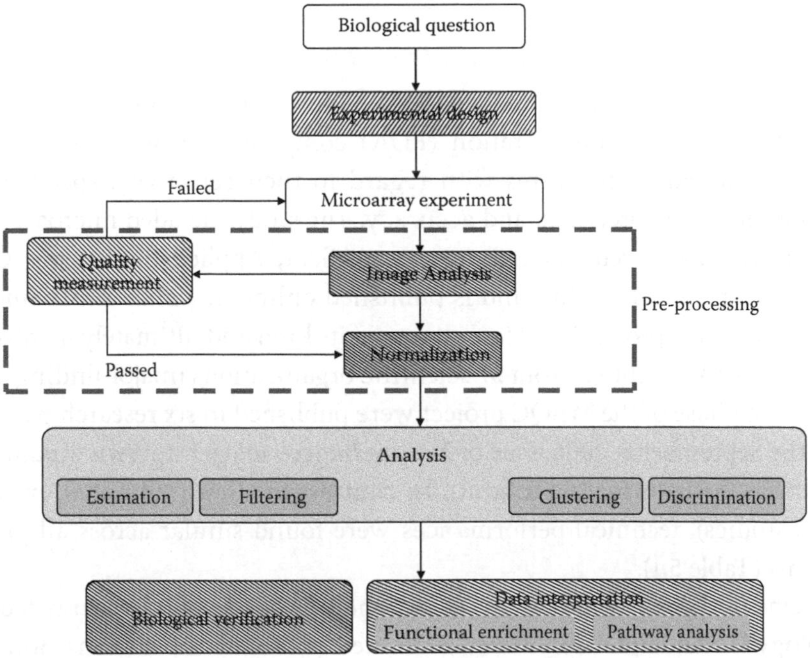
Biological question
Experimental design
Failed
Microarray experiment
Quality measurement
Image Analysis
Pre-processing
Normalization
Passed
Analysis
Estimation
Filtering
Clustering
Discrimination
Biological verification
Data interpretation
Functional enrichment
Pathway analysis

FIGURE 5.1. Schematic of microarray analysis.

arrayed on a fixed substrate and used to assay the levels of gene expression in a target biological sample. RNA is extracted from a source of interest (tissue, cells, or other materials), labeled with a detectable marker (typically, fluorescent dye) and allowed to hybridize to the arrays with individual messages hybridizing to their complementary gene-specific probes on the array. Stoichiometry dictates that the relative abundance of nucleic acid bound to any probe is a function of concentration. A more intense signal corresponds to a higher degree of hybridization, implying higher levels of expression. Once hybridization is complete, samples are washed and imaged using a confocal laser scanner. The relative fluorescence intensity for each gene is extracted and transformed to a numeric value. The actual value reported depends on the microarray technology platform and experimental design.

Currently, multiple microarray platforms can be exploited to analyze global gene expression. Regardless of the approach, subsequent data analyses are the expression measures for each gene in each experiment. Summary of all microarray commercial products, including software for analysis is available at: http://www.nature.com/nature/journal/v442/n7106/pdf/4421071a.pdf.

1. Gene Expression Microarray Platforms

The MicroArray Quality Control (MAQC) project sponsored by the U.S. Food and Drug Administration (FDA) compared performance of different microarray platforms with regard to their sensitivity, specificity, dynamic range, precision, and accuracy. The study included microarrays from five major vendors: Affymetrix, Agilent, Applied Biosystems, GE Healthcare, and Illumina, and is published online in *Nature BioTechnology* 2006. This project is the largest one of its kind and ultimately involved a total of 137 scientists from 51 scientific organizations (major findings of the first phase of the MAQC project were published in six research papers in the September 8, 2006 issue of *Nature Biotechnolog*: http://www.nature.com/nbt/focus/maqc/index.html). In contrast to a few (but not all previous studies), technical performances were found similar across all platforms (Table 5.1).

One of the major limitations of expanding the use of microarray technology is that expression values generated on different platforms cannot be directly compared because of the unique labeling methods and probe sequences used. To assess relative accuracy, expression detected on Taq-

Man quantitative real-time RT-PCR (qRT-PCR) was compared to microarray assays. Affymetrix, Agilent, and illumina platforms displayed high correlation values of 0.9 or higher, based on comparison of approximately 450–550 genes amplified by qRT-PCR. GE Healthcare and NCI platforms had a reduced average correlation of 0.84, but identified almost 30% more genes in the data comparison, reflecting perhaps better lower level of detection of the technology and consequently greater variation. The results of the MAQC consortium demonstrated that most major commercial platforms can be selected with confidence. Moreover, the MAQC data set offers scientists a highly validated reference for future work. The MAQC data are available through GEO (series accession number: GSE5350), Array Express (accession number: E-TABM-132), ArrayTrack (http://www.fda.gov/nctr/science/centers/toxicoinformatics/ArrayTrack/), and the MAQC website (http://edkb.fda.gov.MAQC/).

2. Sources of Variability and Experimental Design

Gene expression microarrays are powerful, but variability arising throughout the measurement process can obscure the biological signals of interest. The many sources of variation in a microarray experiment have, from its inception, been considered in four separate categories: (1) manufacturing of arrays, (2) generation of biological sample (the experimental unit), (3) technical variation (preprocessing), and (4) processing of samples (obtaining image). Variability due to errors introduced in the manufacturing of the arrays is specific to the technology and, for all intents and purposes, beyond the scope of this chapter. Consequently, the three remaining factors comprise the three layers where variability can be introduced.

The first layer corresponds to variation due to the biological component of the experiment. This is intrinsic to all biological systems and includes features influenced by genetic and environmental factors, as well as by whether the samples are pooled or individual. It is often difficult to distribute systemic errors introduced because of the "biology" of the experiment equally to prevent bias; this is mostly achieved through appropriate randomization (see following text). Technical variation surrounding the preprocessing of the samples can arise at any stage of the operation (RNA extraction, labeling, etc.). Protocolization of the processing steps improves the quality and reproducibility of the technique. Measurement error is associated with reading the fluorescent signal, which may be affected by factors such as dust on the arrays.

TABLE 5.1 Microarray Platforms Comparison

	CodeLink	**Affymetrix**	**Agilent**	**Applied Biosystems**	**Illumina**
Array	30-mer	25-mer	60-mer	60-mer	50-mer
Starting RNA requirements (Total RNA)	200 ng-2μg	One-Cycle: 1-15 μg Two - Cycle Target Labeling Assays: 10-100 ng of total RNA	Fluorescent linear amplification: 5 μg Direct Labeling 10μ Low RNA input: 50 ng	RT-IVT labeling: 0.5μg RT-labeling: 40 μg	RT-IVT 50-100 ng
Amount of cRNA/ cDNA hybridized per array	10 μg cDNA	15 μg cRNA	0.75 μg Cy3 cRNA and 0.75 Cy5 per 22,000 array, Combined Cy3 and Cy5 fluorescence cRNA	10 μg cDNA or cFNA	100 ng cRNA
Detection Method	Streptavidin-Alexa Fluor® 647	Streptavidin-phycoerythrin	Cy3 and Cy5	Digoxygenin	Cy3 and Cy5
Secondary detection method	None	Biotinylated anti-streptavidin	None	None	Yes
Sensitivity	1:900,000	1:100,000	1:1,000,000 Multi-pak 1.69M	1:600,000	
Specificity	<1 transcript copies/cell		High	~0.5 transcript copies/cell	<1:250K
Limit of detection (LOD)		8 pM	100 pM	50 fM	0.25 pM
Precision Fold Change	<1.5	<1.2	<1.2	<1.6	≥ .3 fold
Dynamic Range	>3 logs	>3 logs	New Multi-pack Gene Expression Array 6 logs	>3 logs	3 logs

TABLE 5.1 Microarray Platforms Comparison

	CodeLink	**Affymetrix**	**Agilent**	**Applied Biosystems**	**Illumina**
Array-to-Array variability	<10%	<10%	7-12%	10-11%	<10%
Probe redundancy	6 fold	22 probes used for each expression measurement - replicated also included	Prevents redundancy in gene coverage	Prevents redundancy in gene coverage	30 fold (~30 beds coated with thousands of copies of probe-array
Advantages	Sensitivity: 3D surface; liquid hybridization kinetics; can be utilizes with any microarray scanner; customization	Reproducibility; content; mature platform; customization	Reproducibility; content; mature platform; sensitivity; customization	Sensitivity; less background with chemiluminescence	Sensivitiy and specificity, small amount of RNA requirement, scans 3 micron features assembled into the highest-density arrays available for genetic analysis; much lower price
Disadvantages	Non-contact printing - printing related isse, such as poor spot morphology	Short oligonucleotides - less sensitive	Two-color dye bias and ozone related degradation	Currently only available for human and more recently for mice studies	Currently only available for human and mice studies

Valid statistical tests for differential expression of a gene across the samples can be constructed on the basis of any of these variance components, but there are important distinctions in how the different types of tests should be interpreted. There are two broad types of replicate experiments: "*biological replicates*" refer broadly to analysis of RNA of the same type but from different subjects (e.g., blood samples from various different patients treated with the same drug). "*Technical replicates*" refer to multiple array analyses performed with the same RNA (e.g., one blood sample from the same individual analyzed many times). If we are interested in determining how the treatments affect different biological populations represented in our samples, statistical tests should be based on the biological variance. If our interest is to detect variations within treatment groups, the tests should be based on technical variation.

Lastly, identifying the independent units in an experiment is a prerequisite for a proper statistical analysis. Details of how individual animals and samples were handled throughout the course of an experiment can be important to identify which biological samples and technical replicates are "independent." In general, two measurements may be regarded as independent only if the experimental materials on which the measurements were obtained could have received different treatments, and if the materials were handled separately at all stages of the experiment, where the variation might have been introduced.

3. Sample Size and Replication

The precision of estimated quantities depends on the variability of the experimental material, the number of experimental units, the number of repeated observations per unit, and the accuracy of the primary measurements. The basis for drawing inferential conclusions is the residual error (or mean-squared error, MSE), which quantifies the precision of estimates and thus allows one to determine whether estimated quantities are significantly different in the statistical sense. There are only two types of mistakes to be considered: (1) type I error or α error — erroneously concluding that there is a real biological effect when in reality there is not; and (2) type II error or β error — concluding that the treatment had no effect when in reality it did.

A simple way to assess the adequacy of a design is to determine the degrees of freedom (df). This is done by counting the number of independent units and subtracting from it the number of distinct treatments (count all combinations that occur if there are multiple treatment factors). If there are no df left, there may be no information available to estimate the biological variance. The statistical tests will rely on technical variance alone, and the scope of the conclusions will be limited to the samples in hand. If there are 5 or more df, then the analysis may be considered adequate. Although it is generally recommended to have no fewer than 5 residual df, it is quite common to see fewer in microarray experiments, even to the point of having no residual df at all. In the latter case, some strong (that is, questionable) assumptions about the variability in the experiment must be made in order to draw conclusions that can be generalized. Replication and/or repetition of measurements at various levels in the experiment can increase precision. The most direct method to achieve this is to increase the number of experimental units. The MSE decreases in proportion to the square root of the sample size. It is also possible to increase precision by taking measurements on multiple technical replicates obtained from the experimental units. However, this approach cannot reduce the biological variance component, and the gain achieved by taking repeated measurements of single RNA samples will be limited.

For large studies involving highly heterogeneous populations, such as human studies, the problem of calculating the number of independent observations required in a microarray experiment is similar to that of sample size/power calculations in clinical trials and other experiment designs. In general, the required sample size depends on several factors: the true magnitude of the change of gene expression, the desired statistical power (that is, probability) to detect the change, and the specified type I error rate. Any method for sample size/power calculations has to depend on the specific statistical test to be used in data analysis. Free software to calculate sample size is now available from various sites (Table 5.2) to assist in determining how many samples are needed to achieve a specified power for a test of whether a gene is differentially expressed, and the reverse, to determine the power of a given sample size. Because the calculation of sample size depends largely on the statistical approach used to subsequently analyze the data, there are two fundamental functions for two types of experimental design that need to be addressed to determine sample size: (1) completely randomized treatment-control design (where each measurement is considered independent) and (2) a matched-pairs design (where the observations are not independent; instead, pairs of samples are related so that *n* pairs of matched samples are created). For the calculation of power, there are three functions for four

TABLE 5.2 Online Tools for Power Analysis

From	Web site
UCLA Department of Statistics	http://calculators.stat.ucla.edu/powercalc/
U of Iowa Department of Statistics	http://ww.stat.uiowa.edu/[1]rlenth/Power/
York U Department of Math	http://www.math.yorku.ca/SCS/Online/power
Bioconductor	http://bioconductor.org/packages/1.9/bioc/html/sizepower.html

types of experimental designs: (1) a completely randomized treatment-control design, (2) a matched-pairs design, (3) a multiple treatment design having an independent treatment effect, and (4) a randomized block design (where related groups of samples are analyzed together as a single group; http://www.biostat.harvard.edu/people/faculty/mltlee/web-front-r.html).

It is also important to remember that longitudinal designs (the use of multiple samples from the same subject) provide considerably greater power at lower numbers of replicates. They best control for interindividual variability because each subject serves as their own control. When measurement is expensive and/or the individual measurements are very precise, it is preferable to add experimental units rather than technical replicates. Pilot experiments should always be considered to make accurate estimates of effect size (http://discover.nci.nih.gov/microarrayAnalysis/Experimental.Design.jsp). When the variability of measurements is greater than the variability between experimental units, technical replication and repeated measurements will effectively increase precision.

4. Pooling

There is considerable disagreement about whether to pool individual samples, among practitioners and also among statisticians. Sometimes, the amount of sample from any one individual sample is insufficient for hybridization, and in that case pooling is a practical necessity. In theory, if the variation of a gene among different individuals is approximately normally distributed, then pooling independent samples would result in a reduction of variance.

Pooling can reduce the biological component of variation, but it cannot reduce the variability due to sample handling or measurement error. In such cases, the variation can be reduced further by making replicates of the pool, and hybridizing to replicate arrays. Because technical variation is usually less than (roughly half of) individual variation, this strategy would in theory give more accurate estimates of the group means for each gene.

In practice, the distribution of expression levels of many genes among individuals are not roughly normal; often, there are more very high values (outliers) than the normal distribution. This can be because of many factors unrelated to the experimental treatment: for example, individual animals or subjects may be infected, or some tissue samples may be anoxic for long periods before preservation, which allows cells to respond to stress (Pritchard, C.C. Hsu, L., Delrow, J. and Nelson, P.S. PNAS 98(23) 13266–13271, November 6, 2001). In some studies, in which the same samples are analyzed by pooled and unpooled designs, the majority of genes that are identified as differently expressed between two groups turn out to represent extremes in only one individual. Also, if one pools samples, there is no way to estimate variation between individuals. More importantly, by pooling samples we also eliminate all independent replication and jeopardize the df allowed for inferential analysis.

If pooling is inevitable, then a commitment to confirming altered gene expression using an alternative technique, such as qRT-PCR, is imperative. This can be done by retesting samples from individual samples used in the pool; a biologically more robust way is to test the genes of interest in different samples not included in the pool.

5. Randomization

The importance of randomization cannot be overlooked; it forms the physical basis of the validity of statistical tests. It is most crucial to apply randomization or random sampling at the stage of treatment assignment. If the treatment is something that can be applied to the units (e.g., injection of a drug), then a carefully randomized experiment will enable inferences regarding cause and effect to be made during the analysis steps. In contrast, if a treatment assignment is attached to the units (e.g., sex and strain), then conclusions are limited to associations. In this particular case, the valid scope of the conclusions is contingent upon how well the population of interest (both in its mean behavior and its diversity) is represented by the sample of experimental units in the experiment. True random sampling of populations is an ideal that is difficult to achieve, but often a good representative sample can be obtained.

Randomization should also be considered at other stages of the microarray experiment to help minimize or avoid hidden bias. Steps amenable to randomization include dye assignment, order and position of slide printing and scanning, selecting which slide will be used to hybridize which sample and, finally, the arrangement of spots on an array.

6. Designs for Two-Color Arrays

Multiple software programs are available to assist in experimental design. An easy-to-use and freely available program is TIGR_Madam 4.0 (http://www.tm4.org/madam.html). Experiment designer will assist in the planning and development of microarray experiments (Figure 5.2) (Churchill, G.A. *Nat Genet* 32 suppl: 490–495, December, 2005). Multiple experimental designs have been considered for analysis of two-color (competitively hybridized spotted) arrays.

The ability to make direct comparisons between two samples on the same microarray slide is a unique and powerful feature of the two-color microarray system (Figure 5.3a and Figure 5.3b). By pairing samples, we can account for variation in spot size that would otherwise contribute to the error. One complication in two-color arrays is that the two dyes do not get taken up equally well, so that the amount of label per amount of RNA differs (dye bias). An early approach to compensate for dye bias was to

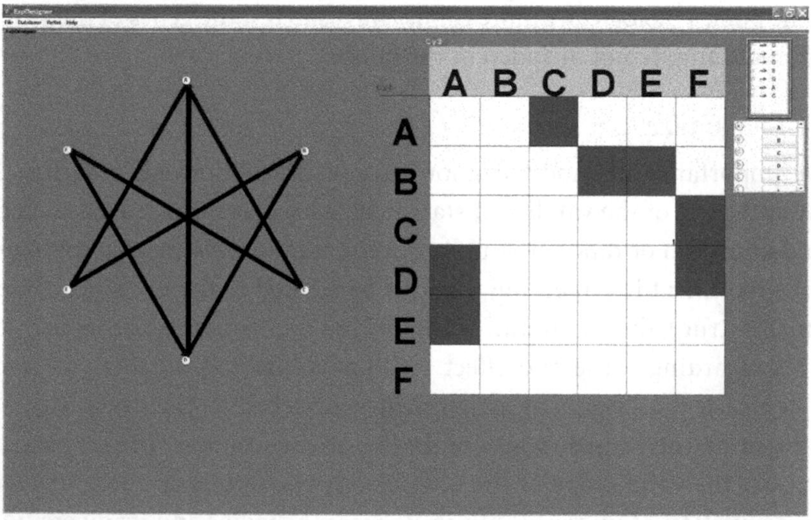
A
B
C
D
E
F
A
B
C
D
E
F

FIGURE 5.2 Experimental designer. This software program can be downloaded from http://www.tm4.org/madam.html, and is used to design two-colored microarray experiments in a systematic fashion. In conjunction with a program to calculate sample size and number of replicates, this is a powerful adjunct to microarray experimental design. (Reporduced from T_m4 Tutorial, with permission).

make duplicate hybridizations with the same samples, using the opposite labeling scheme (Figure 5.3c). However, dye bias is not consistent, and in practice, the ratios in dye-swap experiments do not precisely compensate each other. Robust normalization strategies are currently preferred to dye-swap experiments (e.g., doping controls for microarray; http://genome-www.stanford.edu/turnover/supplement.shtml).

The most common design is the "reference design": each experimental sample is hybridized against a common reference sample (Figure 5.3e and Figure 5.3f). The reference design has several practical advantages: (1) it extends easily to other experiments, if the common reference is preserved; (2) it is robust to multiple chip failures; (3) reduces incidence of laboratory

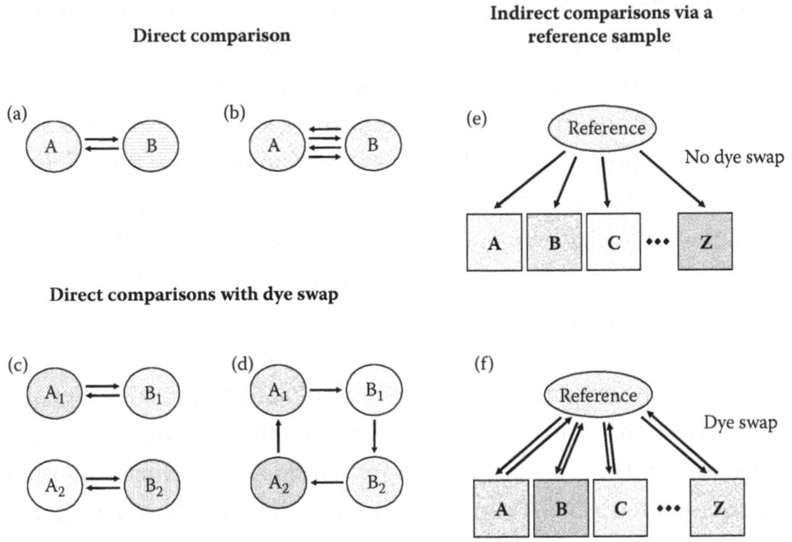

Direct comparison
Indirect comparisons via a reference sample
(a)
A
B
(b)
A
B
(e)
Reference
No dye swap
A
B
C
Z
Direct comparisons with dye swap
(c)
A_1
B_1
A_2
B_2
(d)
A_1
B_1
A_2
B_2
(f)
Reference
Dye swap
A
B
C
Z

FIGURE 5.3 Two-sample comparison. Circles, representing mRNA samples, are labeled as varieties A or B. Subscripts indicate the number of independent biological replicates of the same treatment. Arrows represent hybridizations between the mRNA samples and the microarray. The sample at the tail of the arrow is labeled with red (Cy5) dye, and the sample at the head of the arrow is labeled with green (Cy3) dye. Direct comparison: This figure shows (a) a dye swap, (b) a repeated dye swap, (c) a replicated dye swap, and (d) a simple loop design. For example, in (a), sample A (labeled red) and sample B (labeled green) are hybridized to one array; then sample A (labeled green) and sample B (labeled red) are hybridized to another. Indirect comparisons via a reference sample: (e) boxes represent RNA samples, and arrows represent microarrays, as in Figure 5.4a, the standard reference design uses a single array to compare each test sample (A, B, C, and so on) to the reference RNA. (f) Shows variation, uses a dye swap for each comparison. (From Churchill, G.A. *Nat Genet* 32 Suppl.: 490–495, December 2002. With permission.)

mistakes, because each sample is handled the same way; and (4) provides built-in technical replication. The loop (Figure 5.3d) is a simple alternative to the reference design. This works well for small numbers, but becomes inefficient for larger experiments.

In reference designs, the path connecting any two samples is never longer (or shorter) than two steps; thus, all comparisons are made with equal efficiency. Methods for identifying genes that are differentially expressed across two experimental conditions can be extended to more general settings in which multiple conditions (samples) are considered. Extensions may include time-course experiments, in which two conditions are considered; factorial design, in which the effects of multiple factors and their interactions are explored simultaneously; and so forth. Moreover, reference designs can be extended (as long as the reference sample is available) to assay large numbers of samples that are collected over a period of time. From a practical perspective, every new sample in a reference experiment is handled in the same way. This reduces the possibility of laboratory error and increases the efficiency of sample handling in large projects. Standardizing practices for spotted cDNA arrays is especially problematic, because the manufacture of the arrays varies considerably from place to place. In addition, all spotted arrays use cohybridization of a test RNA sample labeled with one color fluorophore, with a control RNA labeled with a different color to which the test is compared on the same spot. The output is in the form of a ratio of hybridization signals that is comparable to other experiments only if the same control RNA is always used. Consequently, the most important consideration when choosing an appropriate reference RNA sample is that it is plentiful, homogeneous, and stable over time. These are the objectives of the External RNA Controls Consortium (ERCC). It aims to identify and help make commercially available a collection of RNA "spike-in" controls that can be included in any microarray experiment to assess variables such as labeling and hybridization efficiency.

It has been argued that the reference samples are not necessary and that the practice of making all comparisons to a reference sample can lead to inefficient experiments. Half of the measurements in a reference experiment are made on the reference sample, which is presumably of little or no interest. As a consequence, technical variation is inflated four times relative to the level that can be achieved with direct comparisons. Designs that interweave two or more loops together or combine loops with reference designs improve efficiency and robustness by creating multiple links among the samples (Figure 5.4a and Figure 5.4b). The difficulty presented by loop designs is that the deconvolution

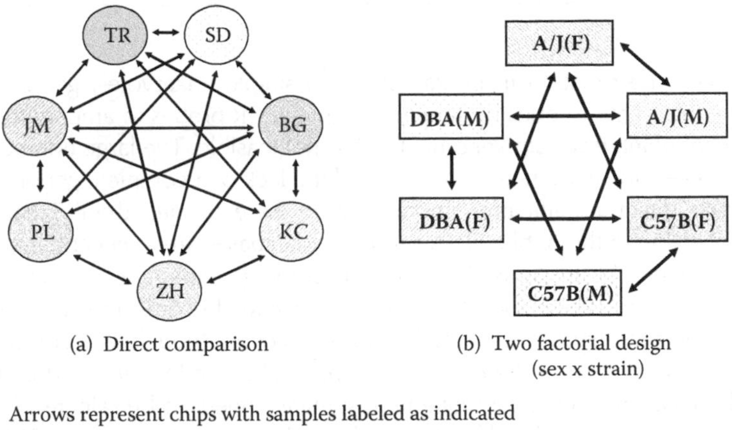
TR
SD
JM
BG
PL
KC
ZH
(a) Direct comparison
A/J(F)
DBA(M)
A/J(M)
DBA(F)
C57B(F)
C57B(M)
(b) Two factorial design
(sex x strain)
Arrows represent chips with samples labeled as indicated

FIGURE 5.4 Multiple sample comparison. The ability to make direct comparisons between two samples on the same microarray slide is a unique and powerful feature of the two-color microarray system. Novel design strategies explore obviating the use of reference hybridizations. (a) and (b) show examples for how to explore this possibility for both multiple sample comparisons and more complex factorial designs. (From Ness, S.A. *Methods Mol Biol* 316: 13–33, 2006. With permission.)

of relative expression values is not always intuitive. However, the availability of software tools that can analyze general designs reduces this concern.

SECTION 2 DATA ACQUISITION AND PREPROCESSING

In this section, we will focus on key features of data extraction and preprocessing, with a special emphasis on quality control and standardization of microarray data. For the purpose of analysis, the first step begins with understanding the characterization and annotation of the probes used to generate microarray chips. The relevant software is presented in Box 5.1.

1. Image Analysis

In terms of image analysis, how to appropriately quantify spots on microarrays is a topic of vigorous inquiry. The millions of sequence strings created at each probe location produce individual hybridization signals that become partitioned into neighborhoods, called pixels, by an optical scanner according to its resolution. A typical scanner resolution scale is at most 10 μm per pixel length. Each square micrometer contains many tens of thousands of oligonucleotide strings. This digital record typically takes the form of a pair of 16-bit TIFF (Tagged Image File Format) images, one for each channel, which records the intensities at each of a large number of

BOX 5.1 | ARRAY PROBES

Unlike the situation in prokaryotes and simple eukaryotes, generating microarray probes by simply designing PCR probes to amplify the genes of interest from genomic DNA is not feasible. The large number of genes, the existence of introns, and the lack of a complete genome make direct amplification impractical. In these species, the EST data collection in the public DNA sequence databases are a valuable representation of the transcribed portion of the genome, and the cDNA clones from which the ESTs are derived have become the primary reagents for expression analysis. Clone selection, though, is a considerable challenge — there are over 3 million human ESTs in the dbEST database, from which a single representative needs to be selected for each gene included in the array. Each database attempts to group ESTs from the same gene and to provide a common annotation. Although the precise approaches taken by the databases vary, they all generally provide high-quality annotation for the cDNAs represented in the public databases. Currently, the most-used databases and their links are:

cDNA DATABASES UniGene | TIGR Gene Indices | STACK | DoTS
IMAGE-PROCESSING SOFTWARE Axon | BioDiscovery | Imaging Research | NHGRI Microarray Project | TIGR software tools | Eisen lab
DATA ANALYSIS TOOLS BioDiscovery | European Bioinformatics Institute (EBI) Expression Profiler | Eisen lab | Silicon Genetics | Spotfire | X Cluster | TIGR software tools | J-express
META-LISTS OF OTHER AVAILABLE SOFTWARE EBI | National Center for Genome Resources | Rockefeller | École Normale Supérieure | Stanford

Adapted from Quackenbush, J. Toxicol Appl Pharmacol 207(2 Suppl.): 195–199, September 1, 2005. With permission.

pixels covering the array. Typically, a probe is composed of up to 300 pixels. Each observed pixel signal is the ensemble consequence of its mixed population of correct and defective sequences. As a consequence of nonspecific and/or somewhat specific levels of hybridization (cross-hybridization between nonperfectly base-paired sequences), defective sequences in the pixel population are likely to partially bond to an assortment of target sequences, creating a weak signal. Figure 5.5 reviews some of the basic principles of image detection.

Several programs are available to extract spot intensity for two-dye arrays after the slides have been scanned (A relatively complete list can be found at http://www.statsci.org/micrarra/image.html). TIGR Spotfinder is a free software program that allows for the rapid, reproducible, and computer-

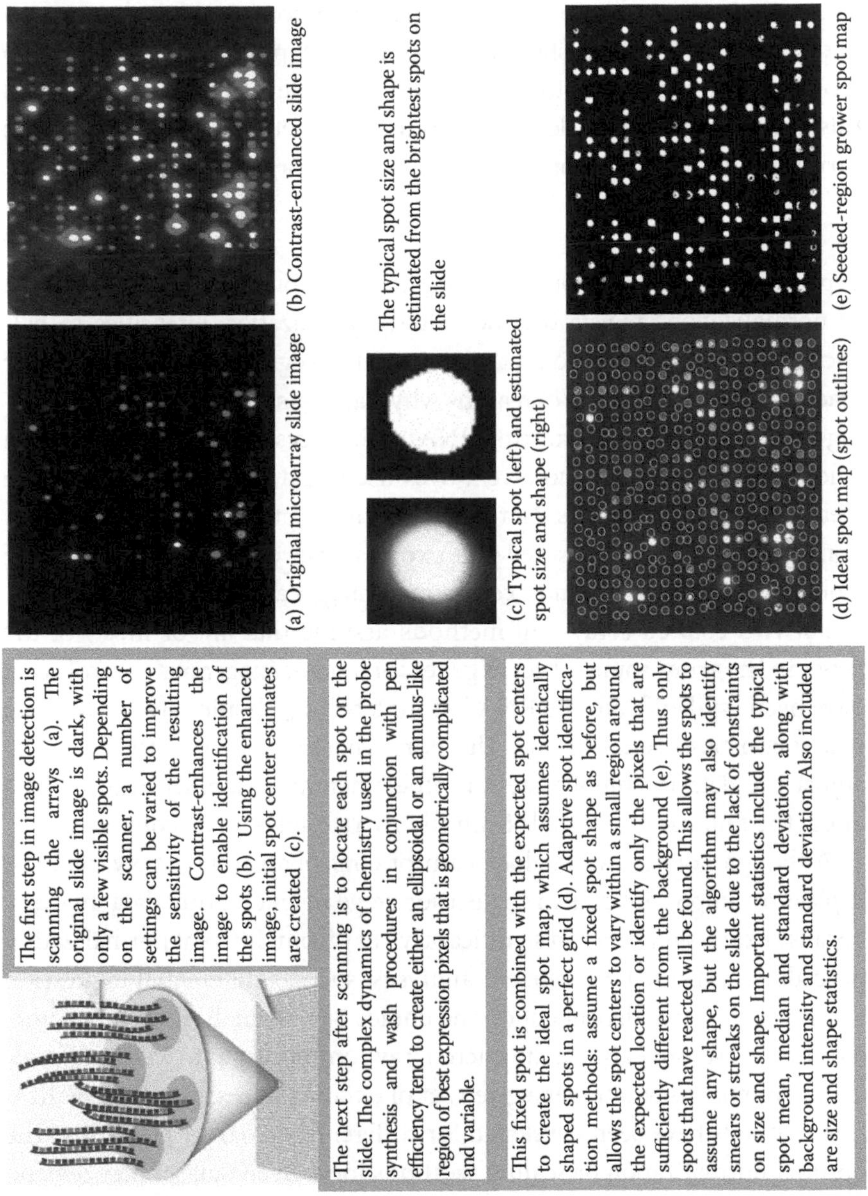
The first step in image detection is scanning the arrays (a). The original slide image is dark, with only a few visible spots. Depending on the scanner, a number of settings can be varied to improve the sensitivity of the resulting image. Contrast-enhances the image to enable identification of the spots (b). Using the enhanced image, initial spot center estimates are created (c).
(a) Original microarray slide image
(b) Contrast-enhanced slide image
The next step after scanning is to locate each spot on the slide. The complex dynamics of chemistry used in the probe synthesis and wash procedures in conjunction with pen efficiency tend to create either an ellipsoidal or an annulus-like region of best expression pixels that is geometrically complicated and variable.
(c) Typical spot (left) and estimated spot size and shape (right)
The typical spot size and shape is estimated from the brightest spots on the slide
This fixed spot is combined with the expected spot centers to create the ideal spot map, which assumes identically shaped spots in a perfect grid (d). Adaptive spot identification methods: assume a fixed spot shape as before, but allows the spot centers to vary within a small region around the expected location or identify only the pixels that are sufficiently different from the background (e). Thus only spots that have reacted will be found. This allows the spots to assume any shape, but the algorithm may also identify smears or streaks on the slide due to the lack of constraints on size and shape. Important statistics include the typical spot mean, median and standard deviation, along with background intensity and standard deviation. Also included are size and shape statistics.
(d) Ideal spot map (spot outlines)
(e) Seeded-region grower spot map

FIGURE 5.5 Image analysis. (Adapted from Automated Microarray Image Analysis (AMIA) Toolbox for MATLAB®, http://www.pnl.gov/statistics/AMIA/. With permission.)

aided analysis of microarray images and the quantification of gene expression. TIGR Spotfinder reads paired 16-bit or 8-bit TIFF image files generated by most microarray scanners and uses a semiautomatic grid construction to define the areas of the slide where spots are expected. This is a desktop program that allows you to explore two available segmentation methods (histogram and Otsu) to define the boundaries between each spot and the surrounding local background (http://www.tm4.org/spotfinder.html).

2. Normalization

Typically, the first transformation applied to expression data, referred to as normalization, adjusts the individual hybridization intensities to balance them appropriately, so that meaningful biological comparisons can be made. There are a number of reasons why data must be normalized, including unequal quantities of starting RNA, differences in labeling or detection efficiencies between the fluorescent dyes used, and systematic biases in the measured expression levels. Normalization attempts to remove such variation, which affects the measured gene expression levels; it does not correct for biological variations. The most common strategies are presented in Box 5.2.

For two-colored arrays, all methods assume that all (or most) of the genes in the array, some subset of genes, or a set of exogenous controls that have been "spiked" into the RNA before labeling, should have an average expression ratio equal to one. The normalization factor is then used to adjust the data to compensate for experimental variability and to "balance" the fluorescence signals from the two samples being compared.

Although normalization alone cannot control all systematic variations, it plays an important role in the earlier stage of microarray data analysis, because expression data can significantly vary from different normalization procedures (MAQC). Subsequent analyses, such as differential expression testing, is therefore dependent on the choice of a normalization procedure. No clear consensus exists in the microarray community as to which normalization method is best under a given set of circumstances. Many normalization methods have been developed for different microarray platforms. The optimal normalization or scaling methods for a given dataset may depend both on the experiment and on many attributes of that microarray data set.

3. Image-Processing Algorithms for Oligo Arrays

Several image-processing methods have been developed for Affymetrix arrays, which are the most commonly used oligonucleotide microar-

BOX 5.2 | COMMON NORMALIZATION STRATEGIES

Total Intensity Normalization

Total intensity normalization data relies on the assumption that the quantity of initial mRNA is the same for both labeled samples. Furthermore, one assumes that some genes are upregulated in the query sample, relative to the control and that others are downregulated. For the hundreds or thousands of genes in the array, these changes should balance out so that the total quantity of RNA hybridizing to the array from each sample, is the same. Consequently, the total integrated intensity computed for all the elements in the array should be the same in both the Cy3 and Cy5 channels. Under this assumption, a normalization factor can be calculated and used to rescale the intensity for each gene in the array.

Normalization Using Regression Techniques

For mRNA derived from closely related samples, a significant fraction of the assayed genes would be expected to be expressed at similar levels. In a scatter plot of Cy5 vs. Cy3 intensities (or their logarithms), these genes would cluster along a straight line, the slope of which would be one if the labeling and detection efficiencies were the same for both samples. Normalization of these data is equivalent to calculating the best-fit slope using regression techniques and adjusting the intensities so that the calculated slope is one. In many experiments, the intensities are nonlinear, and local regression techniques are more suitable, such as LOcally WEighted Scatterplot Smoothing (LOWESS) regression.

Normalization Using Ratio Statistics

To normalize data using ratio statistics, the assumption is that although individual genes might be up- or downregulated, in closely related cells, the total quantity of RNA produced is approximately the same for essential genes, such as "housekeeping genes." Using this assumption, it is possible to estimate an approximate probability density for the ratio Tk = Rk/Gk (where Rk and Gk are, respectively, the measured red and green intensities for the kth array element). This can then be used in an iterative process that normalizes the mean expression ratio to one and calculates confidence limits that can be used to identify differentially expressed genes.

Reproduced from Quackenbush, J. Nat Genet 32 Suppl.: 496–501, December 2002. With permission.

rays. These methods estimate the amount of RNA from fluorescent array images, while trying to minimize the extraneous variation that occurs owing to technical artifacts. Plasmode data sets are real (not computer-simulated) data sets for which the true structure is known. These can be used as a way of testing a proposed analytical method and have been used to evaluate different image-processing normalization methods — one of these data sets has been selected for analysis in the tutorial presented in Section 4. The three core strategies are the model-based expression index, the MAS 5.0 statistical algorithm, and the robust multichip average.

In the Affymetrix system, the mismatched probe (MM) probe contains oligonucleotide sequences identical to the perfect match (PM) probe, except for a single nucleotide at the center of the sequence, which is different and is intended to serve as an internal control of hybridization specificity. The PM and MM intensities for each probe set are combined together to produce biologically meaningful expression values. Ideally, expression indices should be both precise (low variance) and accurate (low bias). In 2001, Affymetrix developed a new summary measure based on Tukey's biweight function, called MAS 5.0 (http://www.wi.mit.edu/CMT/protocols/statisticalalgorithms.pdfalgorithm). This algorithm is implemented on a chip-by-chip basis (http://www.affymetrix.com/products/software/specific/mas.affx).

Model-based expression index uses the invariant set normalization method, which chooses a subset of PM probes with small within-subset rank difference in the two arrays, to serve as the basis for fitting a normalization curve. The fitted curve is the running median curve in the scatter plot of probe intensities of the two arrays. The dCHIP software (http://biosun1.harvard.edu/complab/dchip/) computes the probe set intensity signal using a multiplicative model. Fitting the model, "dCHIP expression measures" are obtained for each probe set. Using this approach, dCHIP allows for a standard error (SE) for each probe set intensity to be measured, which is an indicator of the hybridization quality to the probe set. SEs are useful for discarding probe sets with low hybridization quality.

The robust multiarray analysis (RMA) method models PM intensity as a sum of exponential and Gaussian distributions for signal and background, respectively. It uses quantile normalization and a log-scale expression effect plus probe effect model that is fit robustly (median polish) to define the expression estimate for each gene (conducts a multichip analysis). In the tutorial presented in Section 4, we use this strategy to analyze a plasmode data set available from Affymetrix. The GC-RMA method describes an algorithm similar to RMA, but incorporating the MM using a model based on GC content definition (GC-RMA). GC-RMA substantially refines the RMA algorithm by replacing the model for background correction with a more sophisticated computation that uses each probe's sequence information to adjust the measured intensity for the effects of nonspecific binding, according to the different bond strengths of the two types of base pairs. It also takes into account the optical noise present in data acquisition (Affymetrix; http://www.affymetrix.com/products/software/specific/arrayassist_lite.affx).

A more recent option for oligo array normalization is the application of the PLIER (probe logarithmic intensity error) algorithm. This produces a summary value for a probe set by accounting for experimentally observed patterns in feature behavior and handling error appropriately at low and high abundance. PLIER accounts for the systematic differences between features by means of parameters termed feature responses, using one such parameter per feature (or pair of features, when using MM probes, to estimate cross-hybridization signal intensities for background). Feature responses represent the relative differences in intensity between features hybridizing to a common target. PLIER produces a probe-set signal by using these feature responses to interpret intensity data, applying dynamic weighting by empirical feature performance, and handling error appropriately across low and high abundances. Feature responses are calculated using experimental data across multiple arrays. PLIER also uses an error model that assumes error is proportional to the observed intensity rather than to the background-subtracted intensity. This ensures that the error model can adjust appropriately for relatively low and high abundances of target nucleic acids.

Recently, an online tool, Affycomp II, has been developed to facilitate further research in the field of microarray data processing. This program allows users to benchmark their normalization methods using "known" data sets from Affymetrix GeneChip experiments (http://affycomp.biostat.jhsph.edu/), and makes those benchmark results publicly available.

4. Quality Control

Digitized spot luminosities are not mRNA concentrations. Based on the preceding discussion, tissue contamination, RNA degradation, labeling efficiency, hybridization efficiency, unspecified hybridization, wrong clone, PCR yield, contamination of PCR product, spotting efficiency, DNA-support binding, image segmentation (e.g., overshining), spot quantitation, background correction, and others, all contribute to the numeric value assigned as signal intensity. Quality control (QC) measures have been developed to deal individually with the potential sources of error. Various data-visualizing tools are used to asses the quality of microarray data. Most common are graphic checks: histograms, box plots, images, and residual images (Figure 5.6). Detailed examination of the raw data is a fundamental step in QC, the information gathered is invaluable in determining quality filters during preprocessing of the data. Specific strategies for microarray QC were used extensively as part of the MAQC project to assess different platform performance.

Individual measurements from a single microarray platform do not share the same precision, sensitivity, or specificity. Consider that, for a microarray with 99% accuracy ($P < 0.01$), readouts of 10,000 data points would still yield 100 false-positive signals based solely on random chance. This makes comparisons between different microarray experiments unreliable. Moreover, the annotations across different platforms are not represented by exactly the same gene sequence regions. Competing sequence targets vary from tissue to tissue and from sample to sample, thereby adding to variability in the hybridization-based measurements for any given probe. It has been proposed that the only way to efficiently deal with variability is to create standards for each step of the microarray process. Three technical elements of the system have been highlighted as critical control points where standards may be applied to make results more universally sharable: measurement traceability (www.measurementuncertainty.org/), method validation (http://jbt.abrf.org/cgi/reprint/12/1/11.pdf), and uncer-

FIGURE 5.6 Assessment of data quality. Various data visualizing tools are used to assess the quality of microarray data. Most common are graphic checks: (a) Spot shape QC score (spot area/spot perimeter, for an ideal circle: $\pi R^2/2\pi R = R/2$). Probe intensity (integral: add all pixels in a spot area; median: take median intensities in spot), repeat for background. A "good spot" is above 2*median (background). (b) Image files can reveal substantial artifacts and systematic patterns. (c) MA plots are used observe the distribution of intensity values and log ratios. This is a plot of log-ratio of two expression intensities vs. the mean log-expression of the two. Nondifferentially expressed genes are on a horizontal line at $M = 0$. Ideally, this should be symmetric. (d) Box plot and smoothed histogram of Log2 data for five separate arrays. Look for differences in shape and location. In the example, array 3 deviates from the others significantly. (Panel a was cordially provided by Dr. Vasily Sharov from TIGR. Panel c was cordially provided by Dr. François Papin of McGill University, Canada. Panels b and d are from the authors' laboratory.)

tainty quantification (http://www.springerlink.com/content/xd84qg-dm0yl2agan/). An example of how to improve measurement traceability is the "reference design," the introduction of a universal reference RNA is expected to greatly improve the ability to share data. The MAQC project has set the stage for future protocolization of microarray analysis. Precision-only estimates of uncertainty are typically used to select differentially expressed genes; as the technique evolves and quantitative assess-

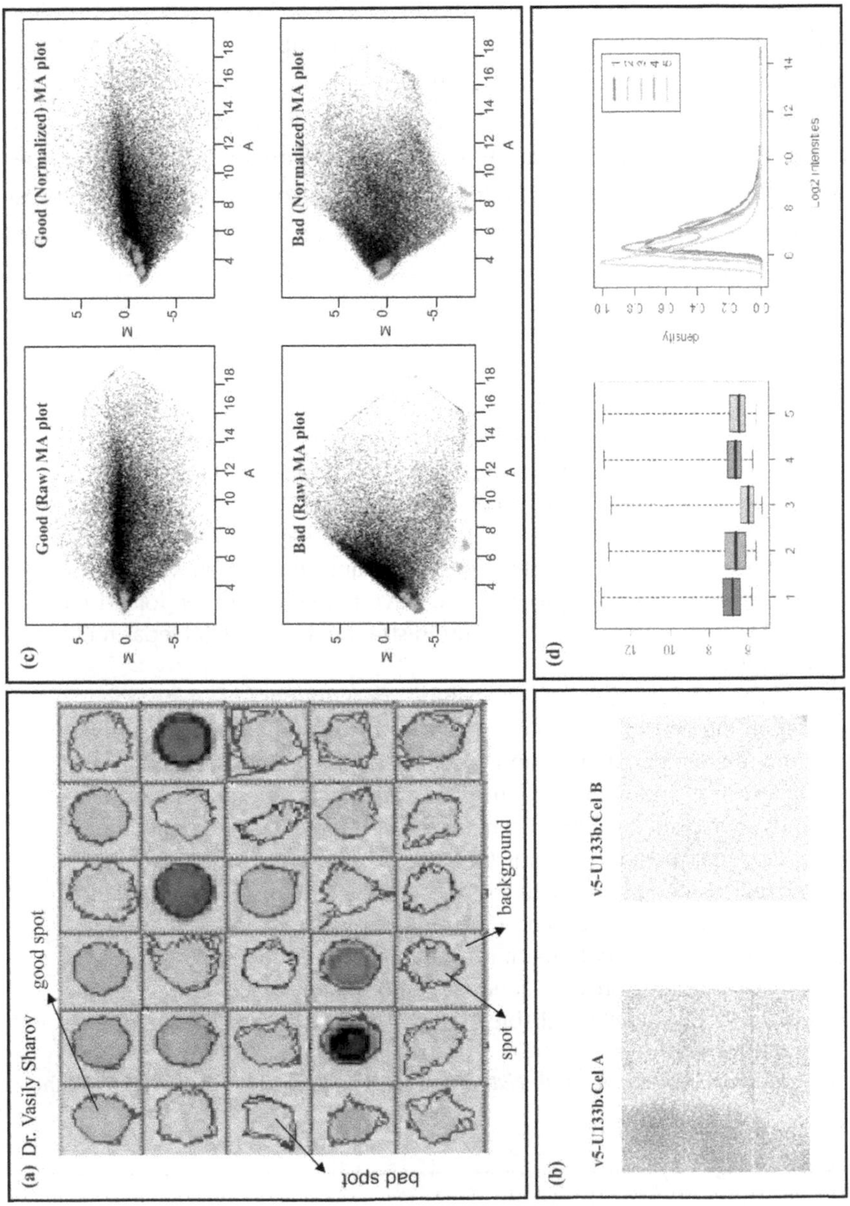
(a) Dr. Vasily Sharov
good spot
bad spot
spot
background
(b)
v5-U133b.Cel A
v5-U133b.Cel B
(c)
Good (Raw) MA plot
Good (Normalized) MA plot
Bad (Raw) MA plot
Bad (Normalized) MA plot
M
A
(d)
density
Log2 intensities

ments become more robust and validation strategies easier to perform on a larger number of genes, this will significantly improve.

Another important component of maintaining quality is generating data that conform to certain standards and, consequently, can potentially be shared. Guidelines for reporting and annotation of microarray data from the Microarray Gene Expression Data (MGED) Society (http://www.mged.org/) — using MIAME (Minimum Information About a Microarray Experiment) standards (Box 5.3) and the Microarray Gene Expression Markup Language (MAGE-ML) represent an important step toward this goal. The efforts of this multinational academic–industry partnership has made it possible to develop databases that can house the many types of microarray data within the same data structure, enabling some data queries between experiments and experimental platforms. The ArrayExpress microarray database (http://www.ebi.ac.uk/arrayexpress/?) is the first

BOX 5.3 | THE MIAME GUIDELINES FOR DATA REPORTING

The Microarray Gene Expression Data Society (MGED) is an international discussion group of microarray experts, with the primary goal of developing methods for data sharing between experimental platforms. The main output of this group has been the Minimum Information About a Microarray Experiment (MIAME) guidelines for microarray data annotation and reporting. The guidelines have been adopted by a number of scientific journals and have recently been endorsed for use by the US Food and Drug Administration and the US Department of Agriculture.

The MIAME guidelines include descriptions of experimental design (number of replicates, nature of biological variables), samples used, extract preparation and labeling, hybridization procedures and parameters, and measurement data and specifications. These guidelines have been most important for the spotted cDNA and oligonucleotide experimental platforms (see Box 5.1), in which the flexibility in microarray design and utilization also leads to considerable variation in array data generation and reporting between different laboratories. The guidelines do not attempt to dictate how experiments should be done, but rather provide adequate information associated with any published or publicly available experiment so that the experiment can be reproduced.

Adapted from Brazma, A. et al. Nat Genet 29(4): 365–371, December 2001. With permission.

major publicly accessible database that adheres to this universal data-presentation platform.

SECTION 3 DATA ANALYSIS

Expression-profiling experiments can be broadly characterized into four primary groups: class discovery, class comparison, class prediction, and mechanistic studies. Briefly, for class comparison, the goal is to identify genes that are differentially expressed between two or more groups. The groups can represent different biological states such as disease state, histological type, or treatment group. For class discovery studies, the aim is to identify groups within the samples examined or among the genes examined. Class prediction studies involve trying to predict group membership for a sample, based on gene expression profiles. For mechanistic studies, the process of discovery is often hypothesis driven, and the goal is to shed light on the mechanisms underlying particular responses. There are multiple ways of achieving the task of extracting biological information from microarray analysis. Irrespective of the specific tool exploited, the process itself can be broken down into three distinct tasks: (1) identification of significantly regulated genes, (2) identification of global patterns of gene expression, and (3) the determination of the biological meaning of both individual genes and group of genes.

1. Differential Expression

The fundamental goal of microarray expression profiling is to identify differential gene expression in the condition of interest. In addition, the identification of discrete patterns of gene expression and subsequent partitioning of the data set based on these expression patterns is commonly done. The process of identifying genes that are differentially expressed is divided into two main parts: (1) applying filtering criteria to identify differential expression, including fold change and statistical significance determined by comparison statistics (inferential analyses) and (2) separation of those differentially expressed messages into discrete groups or clusters, based on expression pattern (classification). Depending on the experimental design, this may be as simple as generating a list of upregulated and downregulated genes or it may involve the use of sophisticated analysis to identify more complex patterns of gene expression.

2. Inferential Analysis

Inference involves making conclusions on the basis of circumstantial evidence rather than on direct observation. In the context of gene expression analysis, it is about making a logical judgment of the truth of a hypothesis that involves unobserved parameters about the whole populations, based on statistical analysis of the population. The methods aimed at minimizing inferential errors — type 1 and type 2 errors — as well as estimating the long-range error rate.

For the most part, inferential analysis involves applying comparison statistics to the data. The criteria applied will influence the content of the list and determine how confident we are that the genes identified are truly differentially regulated. Although it has been commonly used, the concept that fold change alone can give insight into differential gene expression may not be useful in practice, because it does not address the reproducibility of the observed difference, and therefore is not useful in determining the statistical significance of changes in expression. Comparison tests require replicates and exploit the variability within the replicates to assign a confidence level or p-value as to whether specific genes are expressed in differential fashion. The commonly used tests can be grouped based on the number of groups being compared, as well as the number of factors being examined. For two-group comparison, the *t*-test, Welch's *t*-test, and the Wilcoxon rank sum can be used. When three or more groups are being compared, this requires the use of the one-way ANOVA or the Kruskal–Wallis test, provided a single factor is being examined. For multiple groups and multiple factors (more than two), the two-way ANOVA must be used.

Comparison tests are also grouped according to how they deal with differences between groups and the variability within groups. Parametric tests, similar to *t*-tests, are used when the data are normally distributed, and identify genes where there are large differences between the groups but a small amount of variability within the groups; where the values for replicates in each group are similar. Nonparametric tests, such as the Wilcox rank sum test, assume the distribution is not normal, and identify genes where all or most of the values for the replicates in one group are higher than those in the other group. These tests do not assess the variability between replicates in a group; they only determine whether the values in one group are higher than those in another group. A parametric test is especially useful for experiments using model systems, such as animal models or cell lines, in which the researcher expects good agreement between replicates, and where excessive variability indicates either a technical artifact or nonuniform biological response across replicates. In contrast, nonparametric tests can be useful when data from patients are being examined and there may be a large amount of variability between replicates. In addition to standard statistical tests, there are also multiple variations of these tests that have been developed specifically to address important limitations pertaining to microarray data, such as statistical power for studies with limited replicates by calculating variance from a pool of genes with similar expression levels rather than using only the values for each gene being measured.

Inferential analysis carried out in the context of microarray gene expression profiling experiments exploits the concept of "shrinkage" to overcome statistical limitations. Considering each gene separately when conducting statistical tests is terribly inefficient. By using all the data simultaneously, better estimates of variance can be obtained, resulting in more powerful testing. Capitalizing on the parallel nature of microarrays, information can be "borrowed" across genes to improve variance estimates and thereby increase statistical power.

The primary limitation of parametric statistical analysis is that it makes assumptions about the distribution of the data. By contrast, resampling-based inference (RBI) methods rely on resampling the data. Compared with the standard parametric statistics, RBI has the advantage of being robust and flexible to accommodate any new statistics without the need to mathematically derive a statistical distribution. Examples of RBI methods would be bootstrapping and permutation testing. Problems with this strategy are that there are multiple ways to permute the data, and only a few will yield valid inferences. Another is the sampling unit; most methods permute the genes rather than the sampling unit; this strategy has been criticized for ignoring both sample size and nonindependence across genes. Also, because the sample sizes are small, RBI p-value distribution can be coarse or granular, and it will often be algebraically impossible to obtain p-values below some specified level. To overcome this problem, some false discovery rate corrections (FDR) procedures, such as the significance analysis of microarray (SAM) algorithm, combine all resampled test statistics across all genes to obtain very small p-values. This is based on two assumptions: that the null distribution for the test statistics is the same for all transcripts; and that all transcripts are independent — this may indeed not be the case.

Finally, multiple issues remain surrounding the testing of multiple hypotheses in microarray analysis. Intersection-union testing (IUT) is useful when asking "and" or "all" questions, such as which genes are differentially expressed or correlated with each other in all conditions analyzed.

3. Correction for Multiple Comparisons

The p-value calculated for an individual gene represents the chance of a similar difference occurring because of chance, and this risk is cumulative for all tests being performed. In other words, if 10,000 comparisons are performed and a cutoff value of 0.05 for the raw p-value is selected, then by chance alone, 500 genes will be selected as differentially expressed. Consequently, the p-value needs to be adjusted, based on the number of comparisons performed; this is called a multiple comparisons correction. The point of these adjustments is to reduce the number of false-positive changes identified in an experiment. Tests that correct for multiple comparisons can be grouped into two main categories: family-wise error rate (FWER), which adjust the p-value so that it reflects the chance of at least one false-positive being found in the list; that is, if we identified 500 genes with an adjusted FWER p-value of 0.05, then there is a 5% chance of having one false-positive in the list of 500. Examples of FWER methods include Bonferonni and Holm. FDRs adjust the p-value so that it reflects the frequency of false-positives in the list. Therefore, if we identified 500 genes with an adjusted FDR p-value of 0.05, then there are an estimated 25 false-positives. Examples of FDR methods include Benjamini and Hochberg and SAM. For the most part, the FWER methods are much more conservative, and the FDR methods are accepted in "discovery" experiments, in which a small number of false-positives may be acceptable.

Recently, mixture models have become more prevalent. Mixture models treat gene expression arrays as being composed of two or more populations of genes: one represents those genes that are differentially expressed and the other, those genes that are not differentially expressed. Many related mixture-model methods (MMMs) have been devised. MMMs estimate FDRs for genes that are declared differentially expressed, whereas the original Benjamini and Hochberg approach controls the FDR at or below a certain level. Consequently, MMMs have been postulated to be more powerful methods of dealing with multiple corrections. Although there are subtle differences between different methods proposed, they all estimate a "gene-specific" FDR that is interpreted as the Bayesian probability that a gene that is declared to be differentially expressed is a false-positive. Despite significant advances in the understanding of how different statistical tools can be applied to correct multiple comparisons performed during microarray analysis, questions remain about accommodating dependence among genes in FDR estimations.

4. Classification and Clustering

The process of classification entails assigning objects to classes (groups) on the basis of measurements made on these objects. There are two main options for class assignment: (1) supervised classification (class assignment, prediction, or discrimination), in which objects (genes) are placed into preexisting categories that are predetermined (training or learning); this is when a set of labeled objects is used to form a classifier for classification of future observations; or (2) unsupervised clustering, in which no *a priori* knowledge of object classification is used, and objects are allowed to cluster so that novel classes can be discovered from the data. Clustering methods are descriptive or explanatory tools that can be used to identify patterns of gene expression. Although each clustering approach will work with any data set, in practice, they often do not work well for large data sets in which most of the expression levels do not vary. Consequently, it has become common to see clustering strategies being used after the data have already been manipulated using inferential statistics.

5. Unsupervised Clustering

Cluster-analysis algorithms group objects on the basis of a similarity (or dissimilarity) metric that is computed for one or more features or variables. For example, genes can be grouped into classes on the basis of the similarity in their expression profiles across tissues, cases, or conditions. Cluster-analysis approaches entail making several choices regarding which metric to use to quantify the distance or similarity among the objects in question, what criteria to optimize in determining the cluster solution, and how many clusters to include in a solution. No consensus or clear guidelines exist to direct these decisions. Cluster analysis will always produce clustering, but whether a pattern observed in the sample data characterizes a true pattern present in the population remains to be determined. Resampling methods can be exploited to address this issue, but results indicate that most clustering in microarray data sets is unlikely to reflect reproducible patterns in the overall population.

It is useful to consider the values that make up a microarray data set as a matrix, with each row being data for a single gene and each column being data for a single array experiment. Data for a gene in the matrix defines a gene expression vector, which has as many dimensions as there are data points within the vector (Figure 5.7a). Using standard mathematical metrics, the similarity (or dissimilarity) between different vectors can then be measured in conjunction with certain rules (an algorithm); these metrics can then be used to organize data (Figure 5.7b). There are two main metrics that can be used: correlation and distance metrics.

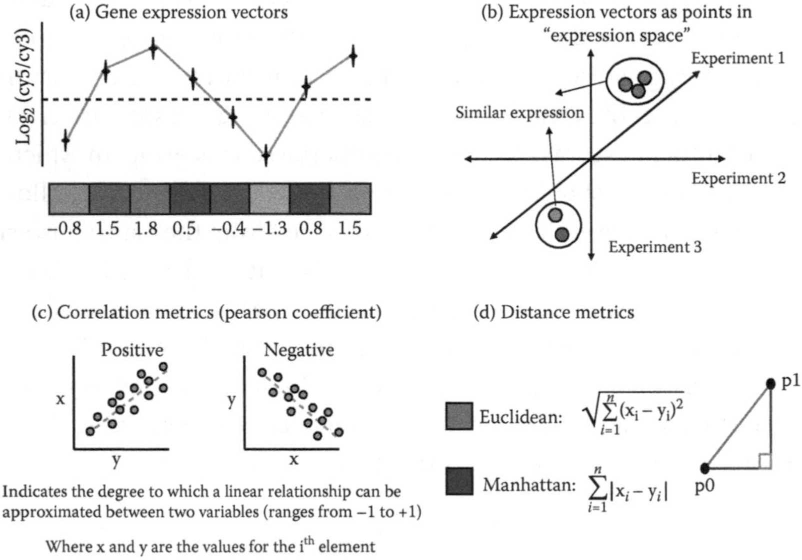
(a) Gene expression vectors
Log_2 (cy5/cy3)
−0.8 1.5 1.8 0.5 −0.4 −1.3 0.8 1.5
(b) Expression vectors as points in "expression space"
Experiment 1
Similar expression
Experiment 2
Experiment 3
(c) Correlation metrics (pearson coefficient)
Positive
Negative
x
y
y
x
Indicates the degree to which a linear relationship can be approximated between two variables (ranges from −1 to +1)
Where x and y are the values for the ith element
(d) Distance metrics
Euclidean: $\sqrt{\sum_{i=1}^{n}(x_i - y_i)^2}$
Manhattan: $\sum_{i=1}^{n}|x_i - y_i|$
p1
p0

FIGURE 5.7 Gene Expression vectors and distance metrics. (a) Expression values for each gene in the matrix defines the gene expression vector. Above +1, this denotes the positive direction of the vector, values less than 1 identify the negative direction of the vector. (b) Expression vectors thus denote expression points in space and can be plotted and grouped accordingly. In (b), this is done according to the expression in specific experiments. Expression vectors can be organized using either correlation of partitioning metrics. (c) Correlation matrices, e.g., Pearson coefficient search for the degree to which a linear relationship can be approximated between two variables (this usually ranges between −1 to 1). This is performed for the expression values x and y for each gene i in the array. (d) Two main distance metrics are used to partition gene expression vector: Euclidean and Manhattan city blocks. Euclidean distance between two vectors is simply the distance in space between the two end points defined by those vectors (p0 to p1, light gray), whereas in Manhattan city blocks, the distance between two vectors is the path one would have to follow between two addresses in an urban downtown, making only right-angle turns (p0 to p1, black). (Adapted from MeV4 Tutorial by John Quackenbush, Dana-Farber Cancer Institute, with permission.) (http://www.tm4.org/)

A. Correlation Metrics

Correlation metrics explore the degree to which a linear relationship can be approximated between two variables (Figure 5.7c). There are several correlation metrics that can be used, and these can be divided into parametric and nonparametric metrics. Parametric metrics make underlying assumptions about the distribution of the data, whereas nonparametric measures of correlation use ranks within the data instead. For the most part, a metric that makes assumptions is more powerful than one that does not make assumptions (if those assumptions are correct). A nonparametric measure is preferred when such assumptions cannot be made safely. In the case of gene expression microarray data, the log-ratio measurements do form a roughly normal distribution, and using the Pearson correlation is reasonable. When these assumptions cannot be made, the Spearman rank correlation, or Kendall's, is more appropriate. There are several statistical tests to determine the "goodness of fit" for data distribution to normal distribution, such as the Shapiro–Wilk test or the D'Agostino–Pearson omnibus test, which provides a p-value for the hypothesis that data were drawn from a normal distribution. Such tests are implemented in various statistical packages. In addition, visual inspection of a frequency histogram of data can also be used to determine whether the distribution deviates grossly from a normal distribution.

B. Distance Metrics

Euclidean distance: The Euclidean distance between two vectors is simply the distance in space between the two end points defined by those vectors (Figure 5.7d). It corresponds to the geometric distance into the multidimensional space. Thus, it is sensitive to the direction of the vectors, like the Pearson correlation, and also to their magnitude. Unlike the Pearson correlation, the Euclidean distance is sensitive to a change in magnitude, but relatively less sensitive to a small change in phase. Euclidean distance may be a more useful metric than the Pearson correlation, when the magnitude of change is an important element of the analysis.

Manhattan or city-block distance: The Manhattan distance between two points can be thought of in terms of the path one would have to follow between two addresses in an urban downtown, making only right-angle turns (Figure 5.7d). The distance is calculated as the sum of absolute values of these orthogonal legs of the journey rather than as the sum of squares of Euclidean distance. This makes the Manhattan distance less sensitive to outlier values, as each element of the vector is weighted linearly rather than quadratically.

6. Criteria to Classify Clustering Algorithms

Briefly, there are several criteria that need to be considered when classifying clustering algorithms: (1) agglomerative vs. divisive (agglomerative adds objects to clusters, whereas divisive splits clusters); (2) monothetic vs. polythetic (calculates distance on the basis of one feature at a time or calculates distance based on all features simultaneously); (3) hard vs. fuzzy (each object belongs to a single cluster vs. each object may belong to several clusters); and (4) incremental vs. nonincremental (describes constraints of execution time and memory space effects). In this paper, we will only spend time in agglomerative vs. divisive clustering.

An agglomerative cluster places each object in its own cluster and gradually merges these atomic clusters into larger and larger clusters until all objects are in a single cluster. A divisive cluster reverses the process by starting with all objects in one cluster and subdividing into smaller pieces. Agglomerative hierarchical clustering is a simple and effective method for exploratory analysis of gene expression microarray data. An exploratory method does not specifically test any particular hypotheses, but instead simply allows the user to explore data. Genes with similar patterns are grouped, so exploring data is much easier than if they were disorganized. Gene expression vectors are organized in a tree structure, with the goal that each vector is closest in the tree to the vectors most similar to it according to the distance metric and linkage rule chosen. Each node in the tree represents a group of similar genes, and the height of the node in the tree indicates the degree of similarity. The data matrix is then reordered according to the tree structure so that again each vector is next to similar vectors. Clustering in both the gene and experiment dimensions may be carried out sequentially on the same matrix. The largest correlation/smallest distance in the matrix defines the two most similar vectors, which are then joined to form a node. This node is then compared to each other's expression vector or node (using some linkage rule), and these results are added to the correlation matrix. Again, the most similar vectors/nodes are joined, and the process is repeated. Thus, single-expression profiles are joined successively to form nodes, which in turn are joined further. The process continues until all individual profiles and nodes have been joined to form a single hierarchical tree. Divisive clustering essentially reverses this process.

7. Rules for Comparing Nodes

A clustering algorithm needs a rule to determine how to compare a node to either a single expression vector or another node. Among the various ways in which this might be done, four are commonly implemented (Figure 5.8):

Single linkage (nearest neighbor): The similarity of two nodes is taken as the best (highest correlation, or shortest distance) of all pairwise comparisons of the members of one node to the other. This produces very loose clusters.

Complete linkage (furthest neighbor): The similarity between two nodes is recorded as the lowest similarity of all pairwise comparisons between the members of one node to the other. Complete linkage tends to produce tight clusters. It is as computationally efficient as single linkage.

Average linkage: The similarity between two nodes is recorded as the average correlation from all pairwise comparisons between the members of one node to the other. Average linkage tends to produce

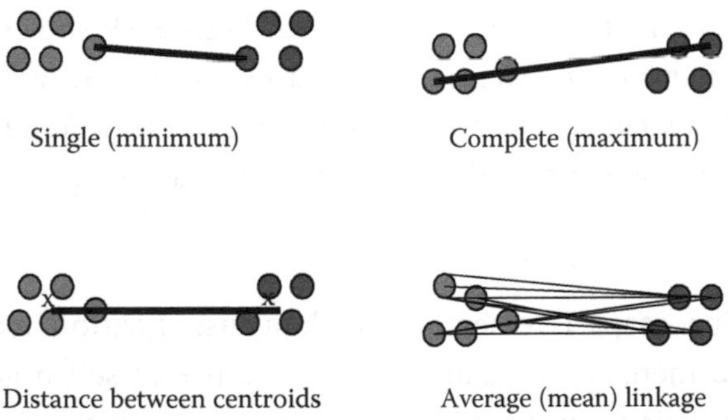
Single (minimum)
Complete (maximum)
Distance between centroids
Average (mean) linkage

FIGURE 5.8 Rules for comparing nodes. Single linkage (nearest neighbor): The similarity of two nodes is taken as the best (highest correlation, or shortest distance) of all pairwise comparisons of the members of one node to the other. Complete linkage (farthest neighbor): The similarity between two nodes is recorded as the lowest similarity of all pairwise comparisons between the members of one node to the other. Average linkage: The similarity between two nodes is recorded as the average correlation from all pairwise comparisons between the members of one node to the other. Centroid linkage: The similarity between two nodes is the similarity between the centroids of those nodes. The centroid of a node is simply calculated by averaging its constituent expression vectors. (From Jean Yee Hwa Yang, http://www.biostat.ucsf.edu/jean/. With permission.)

clusters intermediate between single and complete linkage in terms of internal consistency.

Centroid linkage: The similarity between two nodes is the similarity between the centroids of those nodes. The centroid of a node is simply calculated by averaging its constituent expression vectors.

There are some key limitations to clustering. For example, in centroid linkage, the average vector that is calculated may not represent accurately any of the contained vectors. Irrespective of the method that is used, vectors within a node will become less similar as they approach the root of the tree; that is, nodes become more heterogeneous. Also, any suboptimal joint made early on, cannot be corrected. In addition, when clustering genes by experiment (columns), the similarity between vectors is calculated over the total number of genes within the data set. Therefore, if a group of genes in sample A is most similar to sample B — and this similarity is biologically important — this may be completely obscured by the fact that sample C is overall more similar to sample A. This important detail would be lost in this analysis. Finally, it may be that a hierarchical structure does not apply to the data. An alternative to clustering is partitioning data into more or less homogeneous groups instead. Figure 5.9 shows the conceptual difference between partitioning and clustering. Several such partitioning methods exist; the most commonly used in microarray analysis are self-organizing maps (SOM) and K-means clustering.

8. Self-Organizing Maps

Self-organizing maps is essentially a data visualization technique that reduces the dimensions of data through the use of self-organizing neural networks. This is an unsupervised strategy that is used for partitioning data into a two-dimensional matrix of cells or partitions. Each gene and/or array is assigned to a single partition. The vectors in each partition are most similar to each other; each partition, overall, is more similar to adjacent partitions than to those farther away in the matrix. Prior to initiating the analysis, the user defines a geometric configuration for the partitions, typically a two-dimensional rectangular or hexagonal grid. Random vectors are generated for each partition, but before genes are assigned to the partitions, the vectors are first "trained" using an iterative process that continues until convergence, so that the data are most efficiently separated. SOMs have been applied to gene expression data in a

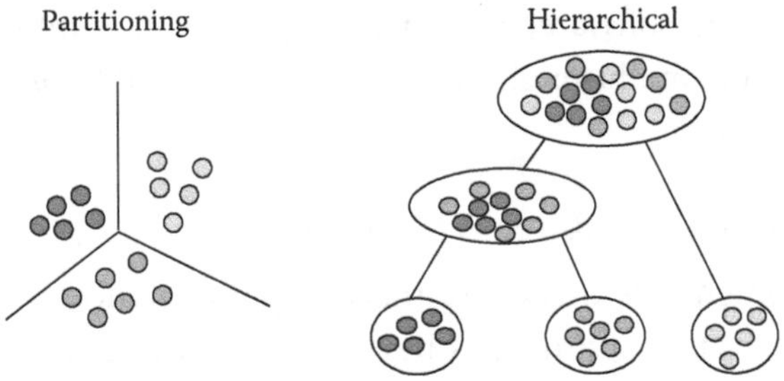
Partitioning
Hierarchical

FIGURE 5.9 Hierarchical vs. partitioning clustering. Schematic representation of conceptual differences between partitioning and hierarchical clustering strategies. (From Jean Yee Hwa Yang, http://www.biostat.ucsf.edu/jean/. With permission.)

number of studies; we will be describing how to perform SOM clustering in the tutorial.

9. K-Means

K-means clustering partitions data in a manner similar to self-organizing maps, with the key difference being that one partition does not influence another directly. In this strategy, objects are partitioned into a fixed number (k) of clusters that are internally similar but externally dissimilar. The K-means algorithm assigns each point to the cluster center (centroid) that is nearest. The center is the average of all the points in the cluster; that is, its coordinates are the arithmetic mean for each dimension separately over all the points in the cluster. Using an iterative model, objects are moved between clusters and intra- and intercluster distances are measured with each move. Objects are allowed to stay in the new cluster only if they are closer to it than the original cluster. The shuffling process continues until moving objects is only making the clusters more variable. K-means clustering has been used successfully to analyze microarray data (see Tutorial in Section A).

10. Limitations of Partitioning Methods

One of the main drawbacks of partitioning methods is the uncertainty in choosing an optimal number and arrangement of partitions. Several methods for determining the correct number of partitions to make have been suggested, including the Gap statistic (http://www.genetics.ucla.edu/horvathlab/Biostat278/GapStatistics.ppt), which was designed with gene expression data in mind. The main goal when partitioning expression data is to reduce the within-cluster dispersion, such that each cluster is reasonably homogeneous while at the same time, the between-cluster dispersion is large.

For references on software, see http://engr.smu.edu/~mfonten/research/MAsoftware.html and http://ihome.cuhk.edu.hk/~b400559/arraysoft_mining_specific.html.

11. Performance of Clustering Methods

Formal assessments of clustering metrics performance are still lacking. Recently, at least two approaches have been used to examine the data to predict which clustering methodology may perform better under individual circumstances. The first approach is the figure of merit (FOM). This is a "leave one out" approach, in which data from all but one array are clustered and then assessed to see how well clustered data predict data in the excluded array. This is repeated for all arrays. The more robust the clustering method, the more predictive the clustering should be of data in the left-out experiment. For real data, single-linkage clustering frequently performed almost identically to random assignment of genes to clusters, whereas the other linkage methods, clustering affinity search technique (CAST) and K-means clustering did significantly better, performing similarly to each other. The second approach explored how coherent the biological annotation for genes within subclusters was, using different metrics and different algorithms. Again, with this approach, single linkage performed poorly. In addition, average linkage performed worse than random as a cluster was cut into more subclusters. The consensus, however, using either a "leave-out approach or exploiting biological annotation to determine cluster performance is that, in terms of measures of dissimilarity, no method outperformed Euclidean distance for ratio-based measurements or Pearson distance for non-ratio-based measurements at the optimal choice of cluster number. Moreover, SOMs were the best approach for both measurement types at higher numbers of clusters. Clearly, unless there is a compelling reason to do so, single linkage should not be used for clustering microarray data, despite being an available option in many software packages. Table 5.3 highlights top clustering software packages. In the tutorial outlined in Section 4, the reader is encouraged to explore different clustering algorithms, using the Affymetrix data set provided.

Unsupervised methods for classification are extremely popular. This probably reflects the fact that they are relatively easy to use and require almost no hypothesis and almost no data assumptions. Moreover, the investigator is guaranteed to obtain clustering, irrespective of the sample size, data quality, or experimental design, or indeed, any biological

TABLE 5.3 Well known and high-performing clustering software

Name	Source/author	Web site
Acuity	Molecular Devices	http://www.moleculardevices.com/pages/software/gnacuity.html
Cluster	Michael Eisen	http://rana.lbl.gov/EisenSoftware.htm http://bonsai.fms.u-tokyo.ac.jp/
Cluster 3.0	Michiel de Hoon	http://bonsai.ims.u-tokyo.ac.jp/~mdehoon/software/cluster/software.htm#ctv
GeneSpring	Agilent	http://www.genespring.com
Hierarchical	Jinwook Seo	http://www.cs.umd.edu/hcil/hce/
J-Express	Molmine	http://www.molmine.com/
MeV	John Quackenbush	http://tm4.org
XCluster	Gavin Sherlock	http://genetics.standford.edu/~sherlock/cluster.html

validity that is associated with the cluster. These are significant limitations. Moreover, little information is available about the absolute validity or relative merits of clustering procedures. The evidence indicates that the clusters that are produced with typical sample sizes (<50) are generally not reproducible. More importantly, the reproducibility of unsupervised classification rarely seems to address the questions that are asked by biologists.

12. Supervised Classification (Class Prediction)

Another powerful way to analyze microarray data is to consider the situation where the goal of the experiment is not to identify function, but rather is to identify genes that can be used to group samples into biologically or clinically relevant classes. In supervised clustering, the experiment typically begins with *a priori* knowledge of the groups represented in the data — although any hypothesis along these lines can be further explored using clustering approaches. Taking into account these initial groups, the next step is to ask whether there are any genes that can be used to separate the relevant classes. For the most part, a computer algorithm finds the rule that best classifies a set of available cases for which the correct type is known. This approach has shown great promise in exploring the use of microarray technology for the purpose of molecular clinical diagnosis. Examples of class prediction analysis strategies include vector support machines (VSM), artificial neural networks (ANN), diagonal linear discriminant analysis (DLDA), K-nearest neighbor (KNN), and various other discrimination methods (http://stat-www.berkeley.edu/users/terry/zarray/Html/discr.html). The ultimate goal is to generalize the trained classifier as a routine diagnostic tool for differentiating between the samples that are difficult or even impossible to classify using available methods. This is also an area of extremely active research, where the disciplines of statistics and machine learning have contributed much (http://cbio.uct.ac.za/arrayportal/data_anls_details.htm).

13. Limitations of Classification Strategies

First, for the user, it quickly becomes evident that many methods yield nonunique solutions or, in other words, can return different solutions of very similar quality (e.g., prediction error rate), which itself leads to the question of how to choose among solutions. A direct way of approaching this problem is via model combination and model averaging. Model averaging is well known, and theory shows that a (weighted) average of predictions from several models should perform better. Other model-averaging strategies have recently been developed. Regardless of which models are used, two general problems can affect all models/algorithms. First, most of the available methods assume additive effects of genes. Nonadditive relationships or interactions, also called synergistic (or antagonistic) effects, are present when the outcome depends not just on the sum of the independent contributions of X and Y, but on their combined effects. Random forests implicitly incorporate interactions as they are an ensemble of classification trees, but the actual interactions are not easy to see. Second, the predictive capacity of many models can be hampered by unrecognized heterogeneity within classes that are regarded as homogeneous. Not much work has been done in this area. A final set of problems involves the biological interpretation of class prediction models. Most methods for building predictors tend not to return models that allow for easy biological interpretation of why and how those predictors are used and how the genes in the predictors affect and relate to the class prediction. These problems have been detailed, and examples are methods that use dimension reduction via principle component analysis (PCA) or proportional hazard regression model for survival analysis (PLS), in which all genes have loadings on all the components, making it virtually impossible to interpret the biological meaning, if any, of the components. In addition, "molecular signatures" or "gene expression signatures" are key features in many studies in cancer research and seem to imply the idea of coordinate expression of subsets of genes, so that some of these sets of coordinate expression would be related to some criterion of interest (e.g., cancer type, or survival). In spite of their apparent relevance, however, there seems to be no approach for identifying molecular signatures; that is, sets of genes that are tightly coexpressed and that can be used as successful predictors. Many unresolved issues remain and, in truth, good classification performance, per se, does not shed any light into the underlying biological or clinical phenomena.

To evaluate the performance of a predictor, it is common to provide the error rate of the predictions. However, many papers report error rates that are biased, leading to overoptimistic claims about the performance of different methods. One possible reason for overinflated estimates may arise from reporting the "resubstitution rate," the error rate computed from the very same observations that were used to build the classifier. This is a problem, because the resubstitution rate is severely biased due to overfitting — this is when the classifier "adapts" to some peculiarity of the data and does great with that particular data set, but poorly as a predictor. To solve this problem either cross validation or bootstrap has been used; both methods build the predictor using a subset of the data, and then predict the values for the remaining data, thus ensuring that the predictions are from data not used for the training. A second common problem is to carry out the cross-validation after the gene selection: all samples are used for gene selection, and the cross-validation process does not include gene selection. The solution is to perform cross-validation or bootstrap so that all steps of the analysis (including gene selection, but also other potential steps such as imputation) are included in the cross-validation. Another potential difficulty in the field is the lack of validation of the methodology; many new methods that are published are not evaluated against standard competing methods.

It is also important to recognize that microarray experiments are observational studies. Observational studies present several potential problems, primarily, background differences between groups and the presence of potential confounding variables. A related problem is interaction, such as when the degree of association between an exposure factor (e.g., expression of gene A) and the disease is different for different levels of the confounding variable, such as sex; there is evidence that this might be the case in lung cancer. Confounding and interaction can be addressed, at least partially, by appropriately using relevant covariates in the statistical models. These factors are of critical importance.

A lingering issue is regarding the sharing of software for microarray data analysis. The Open Bioinformatics Foundation (http://www.open-bio.org/) is "focused on supporting open source programming in bioinformatics." The Free Software Foundation (http://www.fsf.org) and the Open Source Initiative (http://www.opensource.org/) explain free and open-source software.

SECTION 4 TUTORIAL

In the following section, a tutorial based on different microarray analysis tools will be used to allow the reader to analyze a trial set of microarray data, understand the analysis sequence, and become familiar with common software tools available.

1. TM4 Software Overview

For the purpose of this tutorial, all the data will be analyzed in TM4 (http://www.tm4.org/scgi-bin/getprogram.cgi?program=expcnvt). This is free software that can be downloaded into your personal computer. The TM4 suite of tools consist of four major applications, Microarray Data Manager (MADAM), TIGR_Spotfinder, Microarray Data Analysis System (MIDAS), and Multiexperiment Viewer (MeV), as well as a MIAME-compliant MySQL database, all of which are freely available to the scientific research community at TIGR's software download site. Although these software tools were developed for spotted two-color arrays, many of the components can be easily adapted to work with single-color formats such as filter arrays and GeneChips™(Affymetrix). This software is OSI certified (Open Source Initiative [OSI] is a non-profit corporation dedicated to managing and promoting the Open Source Definition for the good of the community, specifically through the OSI Certified Open Source Software certification mark and program [http://www.opensource.org/]). In this tutorial, we will not be using MADAM or TIGR_Spotfire. MADAM is essentially a data management tool used to upload, download, and display various microarray data to and in a database management system (MySQL). This software interfaces with MySQL and allows data to be electronically recorded, captured, administrated, and annotated, enabling data to be shared and used by others within the scientific community. TIGR Spotfinder is an image-processing software created for analysis of image files (TIFF files) generated in microarray expression studies. This software enables spot quantification.

2. Experiment and Data

For this tutorial, we will be using a plasmode data set provided by Affymetrix. The initial step is to download a test data set. Affymetrix has made available on their Web site (Permission obtained to use data at http://www.affymetrix.com/support/technical/sample_data/datasets.affx) two very useful data sets: the human genome U133 data set and human genome U95 data set. Both are arrayed in a Latin square format (a Latin square of order *n* is an *n* by *n* array of *n* symbols, in which every symbol occurs exactly once in each row and column of the array). We will be using the U133A data set in the example. The human genome U133 data set consists of 3 technical replicates of 14 separate hybridizations of 42 spiked transcripts in a complex human background, at concentrations ranging from 0.125 p*M* to 512 p*M*. Thirty of the spikes are isolated from a human cell line, four spikes are bacterial controls, and eight spikes are artificially engineered sequences believed to be unique in the human genome. A total of five files need to be downloaded. The data are available from Affymetrix in a .Cel file format (U133 Data, HG-U133A_tag_Latin_Square.zip, 141 MB); this will allow the data to be processed at the probe level (we will not be reviewing the image file). This data set requires a special, alternate chip description file (CDF, HG-U133A_tag_CDF.zip, 6.9 Kb), available from the same site, containing information about the eight artificial clones. The exact spiked sequences are found in the Excel file describing the experimental design (U133 Description, HG-U133A_Tag_description.zip, 19 KB). For further analysis, the probe sequences (HG-U133A_tag_ProbeSequence.zip, 4.2 MB) as well as the complete library files (HG-U133A_tag_library files, 14 MB) are included. These files are particularly useful in illustrating the use of replicates and spiked controls. They will allow the reader to focus on important technical issues discussed in the text.

3. Preprocessing

MIDAS-RMA can be used to analyze .Cel files generated by Affymetrix. MIDAS-RMA can be downloaded from https://sourceforge.net/projects/midas-tm4/. Make sure you download the 2.0 version; the earlier version

(1.9) cannot perform RMA analysis. Before running the software, make sure you have ready:

1. CEL files (.Cel) you downloaded from Affymetrix — the Latin Square experiment.
2. CDF files (Channel file) in text format (the file should be in a format that you can open and read with Notepad) — the way the tutorial has been set up, the file formats you have downloaded from Affymetrix are appropriate.

Once you have both files in the correct format, download MIDAS package and extract the files to the directory you want. Please make sure you also download the .pdf file containing RMA instructions — this may be very useful and essentially walks you through the operations required for preprocessing. For the most part, unless specified, we will be using all default settings in the interest of simplicity.

Using MIDAS_RMA, we will detect expression signals using RMA as our background correction method (see Figure 5.5). The data will be normalized using quantile normalization (see Box 5.2). The default method for expression summary of RMA data will be RMA Median Polish. The software will also perform virtual trimming — eliminating poor-quality probe signals as per default settings. Figure 5.10 shows MIDAS_RMA window outlining default settings for data preprocessing.

Click on MIDAS.bat to start the application.

Click on "Read All Data Files in a Folder" (this is the third "glasses" icon from the left of the menu bar with caption).

Now click on the blank "Value" column of "Multiple Data Files Names" — in the parameters window. A small pop-up will appear. Make sure to choose "Affymetrix Probe Intensity files (.Cel)" from the top Load expression files of type. The "please specify raw data files…" window that appears will allow you to navigate and select the .Cel files that you want to include in the analysis. To make this example more efficient, load the three replicates for experiments 1, 2, 3, and 4 as follows:

12_13_02_U133A_Mer_Latin_Square_Expt1_R1.Cel

12_13_02_U133A_Mer_Latin_Square_Expt1_R2.Cel

12_13_02_U133A_Mer_Latin_Square_Expt1_R3.Cel

12_13_02_U133A_Mer_Latin_Square_Expt2_R1.Cel

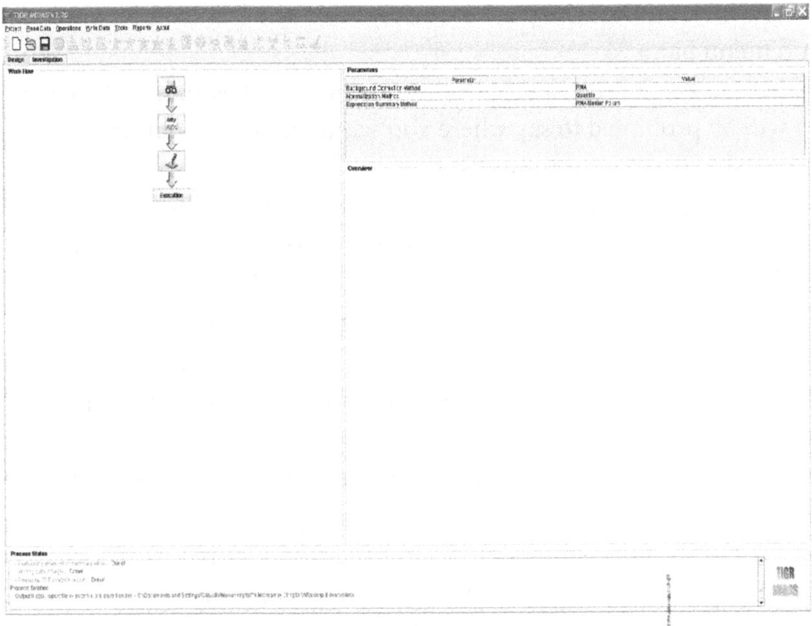

FIGURE 5.10 MIDAS default settings for data preprocessing.

12_13_02_U133A_Mer_Latin_Square_Expt2_R2.Cel

12_13_02_U133A_Mer_Latin_Square_Expt2_R3.Cel

12_13_02_U133A_Mer_Latin_Square_Expt3_R1.Cel

12_13_02_U133A_Mer_Latin_Square_Expt3_R2.Cel

12_13_02_U133A_Mer_Latin_Square_Expt3_R3.Cel

12_13_02_U133A_Mer_Latin_Square_Expt4_R1.Cel

12_13_02_U133A_Mer_Latin_Square_Expt4_R2.Cel

12_13_02_U133A_Mer_Latin_Square_Expt4_R3.Cel

Once the .Cel files you want are highlighted, press "OK" to close the window. At this point the, .Cel file names should be displayed in the "Value" entry for "Multiple Data files Names."

Click on the second icon from the right on the menu bar with "Affy" on it.

In the parameters window, make sure that for "Background Correction Method," you click on "RMA," which should be the default value on this first line.

Now click on the "Output Processed File," the last icon on your right-hand menu bar.

Click on the "execution" button on the Work Flow panel; at this point, you will be prompted to say where you want all of your output files saved. Choose the directory you want and press "Save."

At this stage, the software will request the .CDF file. In the pop-up window, just press "…" and direct the computer to where you saved the .CDF file. Press "Apply."

The application will then start running. You will be able to follow the completion of the analysis in the "Process status" window. Once the process is completed, you will find the process log (.rpt) and result (.txt) under the project directory you have chosen previously.

The Latin Square experiment we initially downloaded in .Cel format has gone through the process of preprocessing.

Your output file should look something like Figure 5.11, in which the first column is the probe ID, and subsequent columns contain the RMA normalized gene expression values. Column names refer to each individual sample.

4. Data Processing

Multiexperiment Viewer (MeV) will be used to process RMA files. MeV can only process normalized and filtered expression files. This is a versatile microarray data analysis tool, incorporating sophisticated algorithms for clustering, visualization, classification, statistical analysis, and biological theme discovery. MeV can handle several input file formats. These include the ".mev" and ".tav" files generated by TIGR Spotfinder and TIGR MIDAS, and also Affymetrix® (".txt") and Genepix® (".gpr") files. MeV generates informative and interrelated displays of expression and annotation data from single or multiple experiments (http://www.tm4.org/mev.html).

Click on TMEV.bat (TMEV batch file) to start the program. Also download the MeV_Manual_4_0.pdf for further detailed instructions.

The main menu bar will appear with four menus: *File*, *Display*, *Window*, and *References*.

Go to *File*, and click on "Load Data."

Select the Tab Delimited, Multiple Sample Files (TDMD) (*,*) options from the drop-down menu, and load TDMS format files. A pop-up window will show you the folder containing the files — this will be displayed in tabular format in the file-loader-preview table. Click the cell in the table

Microsoft Excel - Midas Processed_txt

I46 8.70047322

	A	B	C	D	E	F	G
1	Probe set	Expt1_R1.CEL	Expt1_R2.CEL	Expt1_R3.CEL	Expt2_R1.CEL	Expt2_R2.CEL	Expt2_R3.CEL
2	AFFX-BioB-5_at	6.976390068	6.980343097	7.051200908	6.886644603	7.060962859	7.054205235
3	AFFX-BioB-M_at	7.78781396	7.773149007	7.7820066	7.657352333	7.823090551	7.865910652
4	AFFX-BioB-3_at	6.946027714	7.001656044	6.937830544	6.883148631	7.014980636	6.987203568
5	AFFX-BioC-5_at	8.505889033	8.420137696	8.465973784	8.463086016	8.55740575	8.610151334
6	AFFX-BioC-3_at	7.756114114	7.719882609	7.737692165	7.720865972	7.854981923	7.926294196
7	AFFX-BioDn-5_at	8.74610707	8.74971649	8.838120583	8.578151177	8.854065253	8.873878088
8	AFFX-BioDn-3_at	11.04139842	11.03092834	11.04883825	11.08192876	11.09041716	11.12215142
9	AFFX-CreX-5_at	11.69886955	11.6610895	11.72200612	11.64694472	11.79039964	11.80752069
10	AFFX-CreX-3_at	12.19811649	12.1457598	12.22985207	12.20047197	12.31846255	12.30134037
11	AFFX-DapX-5_at	10.86622497	10.75675706	10.91595027	11.37195397	11.44098003	11.44503145
12	AFFX-DapX-M_at	11.75136944	11.60562547	11.75224814	12.1708651	12.29386849	12.30127987
13	AFFX-DapX-3_at	11.98577566	11.97032535	11.89194065	12.41621164	12.53133794	12.53121824
14	AFFX-LysX-5_at	12.35663217	12.33551567	12.32837112	3.685382098	3.64462173	3.71665486
15	AFFX-LysX-M_at	12.54272968	12.53124697	12.47861255	4.451792555	4.289486873	4.368020539
16	AFFX-LysX-3_at	12.41482935	12.39446838	12.43586981	4.455774217	4.490633253	4.499533445
17	AFFX-PheX-5_at	12.53690204	12.41196271	12.45552814	3.937608855	3.884369563	3.856060694
18	AFFX-PheX-M_at	12.15272208	12.16739456	12.22843048	3.9736913	3.902835796	3.900348466
19	AFFX-PheX-3_at	12.37767633	12.4911577	12.38381887	5.312808132	5.293176842	5.278184056
20	AFFX-ThrX-5_at	12.66692156	12.50183287	12.59448386	4.798291762	4.909175368	4.832326381
21	AFFX-ThrX-M_at	12.56304438	12.45623449	12.53301551	3.874981608	3.887362936	3.83487981
22	AFFX-ThrX-3_at	13.22982444	13.16039102	13.18040613	5.253600772	5.264136114	5.216600011
23	AFFX-TrpnX-5_at	3.611821734	3.654105208	3.6335207	3.670713862	3.647964077	3.638660196
24	AFFX-TrpnX-M_at	3.596549451	3.645139502	3.659361901	3.622705763	3.613217175	3.582255958
25	AFFX-TrpnX-3_at	3.205689389	3.206566796	3.193673547	3.211075346	3.212319994	3.206827897
26	AFFX-HUMISGF3A/M97935_5_at	5.446682788	5.436854502	5.344832167	5.340473415	5.419471879	5.372482617
27	AFFX-HUMISGF3A/M97935_MA_at	7.51681516	7.510774967	7.477231878	7.484107567	7.541710974	7.540296372
28	AFFX-HUMISGF3A/M97935_MB_at	6.896257488	6.858471751	6.899189376	6.756296171	6.920956486	6.939950748
29	AFFX-HUMISGF3A/M97935_3_at	7.989667684	7.87607807	7.909745709	7.721902925	7.953907141	8.053533484
30	AFFX-HUMRGE/M10098_5_at	6.377347118	6.185445975	6.443899528	6.352137585	6.478675511	6.560232153
31	AFFX-HUMRGE/M10098_M_at	7.438735396	7.388947953	7.431117197	7.377805088	7.503973202	7.528657722
32	AFFX-HUMRGE/M10098_3_at	7.882087968	7.832635655	7.87467416	7.979930221	8.130242612	8.152336688
33	AFFX-HUMGAPDH/M33197_5_at	12.86488418	12.82737452	12.85550007	12.84346916	12.92691815	12.9012852
34	AFFX-HUMGAPDH/M33197_M_at	12.97628313	12.95671764	12.96475024	13.01695746	13.00766487	13.01468582
35	AFFX-HUMGAPDH/M33197_3_at	12.75280976	12.63164427	12.76369971	12.80979491	12.80955548	12.81237209
36	AFFX-HSAC07/X00351_5_at	12.89510749	12.89679434	12.88647838	12.92819448	12.92501821	12.91485926
37	AFFX-HSAC07/X00351_M_at	13.02410217	13.04226442	12.98744176	13.01844471	13.04887675	13.06850087
38	AFFX-HSAC07/X00351_3_at	12.64606022	12.57791813	12.65806355	12.63027829	12.70505949	12.68909491

FIGURE 5.11 Excel file showing processed RMA data. The first column contains Probe ID information. All subsequent columns are labeled according to their specific samples and contain information pertinent to expression values for each gene (rows) in each different sample (columns).

that contains the upper-leftmost expression value in the files (not probe ID). The header labels for the annotation fields will be displayed at the bottom of the dialog. Check that the correct fields are listed before clicking on "Load."

Once the loading process is complete, the software generates a Main Expression Image; this is the heat map seen in Figure 5.12.

Note that for each set of expression values loaded, a column is added to the main display (Figure 5.12). This display is an expression Image Viewer. Here, each column represents a single sample, and each row, a single gene. By convention, red is upregulated, and black is downregulated. The color intensity is usually related to the degree of up- or down-regulation, the more intense color indicating extremes of hybridization intensity.

We have downloaded a total of 12 samples (12 chips) each containing expression values for 22,300 genes.

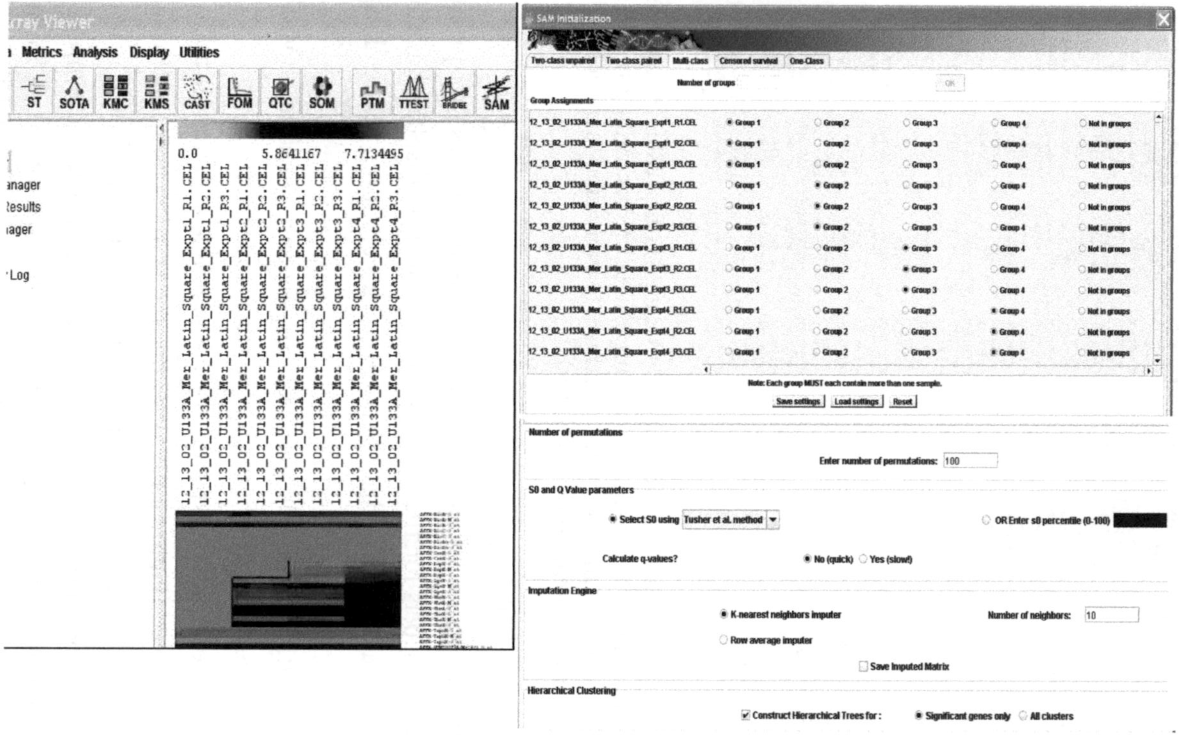
Metrics
Analysis
Display
Utilities
ST
SOTA
KMC
KMS
CAST
FOM
QTC
SOM
PTM
TTEST
BRIDGE
SAM
0.0
5.8641167
7.7134495
SAM Initialization
Two-class unpaired
Two-class paired
Multi-class
Censored survival
One-Class
Number of groups
Group Assignments
Note: Each group MUST each contain more than one sample.
Save settings
Load settings
Reset
Number of permutations
Enter number of permutations: 100
S0 and Q Value parameters
Select S0 using
Tusher et al. method
OR Enter s0 percentile (0-100)
Calculate q-values?
No (quick)
Yes (slow!)
Imputation Engine
K-nearest neighbors imputer
Number of neighbors: 10
Row average imputer
Save Imputed Matrix
Hierarchical Clustering
Construct Hierarchical Trees for :
Significant genes only
All clusters

FIGURE 5.12 MeV heat map of RMA data and SAM initiation window. Left-hand side shows navigation icons highlighting individual processing steps applied to the data. In the middle is the MeV heat map (Main Image) of RMA-processed .Cel files, and in the right-hand side is the SAM initiation window, opened to show selections for multiclass analysis protocol.

5. Significant Analysis of Microarray

The next step is to treat the data to an inferential analysis step — statistical analysis to identify those genes whose expression value is significantly different.

Click on SAM in the horizontal Menu bar. Background information about SAM is discussed briefly in Section 3.

In SAM, the data for each gene are permuted, and a test statistic *d* is computed for both the original and the permuted data for each gene. SAM generates an interactive plot (Figure 5.12) of the observed vs. expected (based on the permuted data) *d*-values. You can change the value of the tuning parameter *delta* using either the slide bar or the text input file below the plot. *Delta* is a vertical distance (in graph units) from the solid line of slope 1 (i.e., where observed = expected). The two dotted lines flanking this solid line, represents the region within +/- delta units from the "observed = expected" line. In the two-class analysis, the genes whose plot values are represented by black dots are considered nonsignificant, those gray dots are considered to be significantly upregulated, and the dark gray ones are thought to be significantly downregulated. You also have the option of applying a fold change criterion to selection of genes. In the multiclass analysis, the direction of change is not provided. The SAM plot shows all significantly altered genes as gray dots.

In the pop-up menu, you will be offered a choice of types of analysis: Two-class unpaired, Two-class paired, Multi-class, Censored survivals, or One-Class.

Click on Multi-class, and the number of groups is "4." Press "OK."

A pop-up window will appear requesting assignment of samples to specific analysis groups. Assign all three replicates from Experiment 1 to group A, all replicates from Experiment 2 to group B, all replicates from Experiment 3 to group C, and all three replicates from Experiment 4 to group D.

Select "100" as the number of permutations.

Select S0 using Tusher et al. method.

To expedite the analysis for "Calculate q-values," say "No" (quick).

Click "K-nearest neighbor" for imputation engine, and "10" for number of neighbors.

Click also on Hierarchical clustering "construct hierarchical Trees" for "significant genes only." When the software prompts you for which clustering method to use, click on "Pearson correlation."

Click "Go."

A pop-up window showing you the SAM plot will appear. This is an interactive window, and it allows you to use the slider to set the value of *delta*. Feel free to adjust *delta* at different values, and look carefully at the two dashed lines flanking the solid line of "expected = observed." As the space between the lines contracts and expands, the color of the dots at the inflection point of the line change from gray dots to black and *vice versa*. Adjust *delta* so that you have the maximum number of significant genes and a minimum number of false significant genes (choose a false discovery rate that is acceptable, usually less or equal to 1%). In this example, we set a *delta* of 0.18785691; there should be 4.44108 median number of false significant genes and 68 (number of) significant genes identified.

Figure 5.13 shows the resulting SAM plot.

Click on SAM in the navigator on the left-hand side. Separate icons should now be available for the SAM graph, Delta Table, Expression Images, Hierarchical Tree (where we used Pearson correlation), Centroid graphs, Expression graphs, Table views, Cluster information, and General information.

If you click on the history icon, a description of all relevant processing steps performed to data on this particular data set should be outlined.

Please review the information in the manual, and use this opportunity to explore SAM.

A. Clustering Strategies

Another analysis option is to use clustering methods to identify gene expression profiles. The gene list contained in the Main View is over 22,000 genes long, and for the most part, most clustering algorithms would have difficulty processing so many genes on a personal computer; usually, computational memory is the limiting factor.

Go to "Adjust data" in the horizontal Menu.

Scroll down to "Data Filters," one-colored arrays and click on "Set Lower Intensity cut-off"; insert 10.0 in the lower intensity cut-off value. Press "Ok."

In the navigational window on the right-hand side, an icon for "Data Filter" should have appeared.

Click on the expression Image; note that a total of 807 genes passed the Low Intensity cut-off filter.

Use this lower number of genes to cluster data using SOMs.

Click on SOM on the horizontal bar.

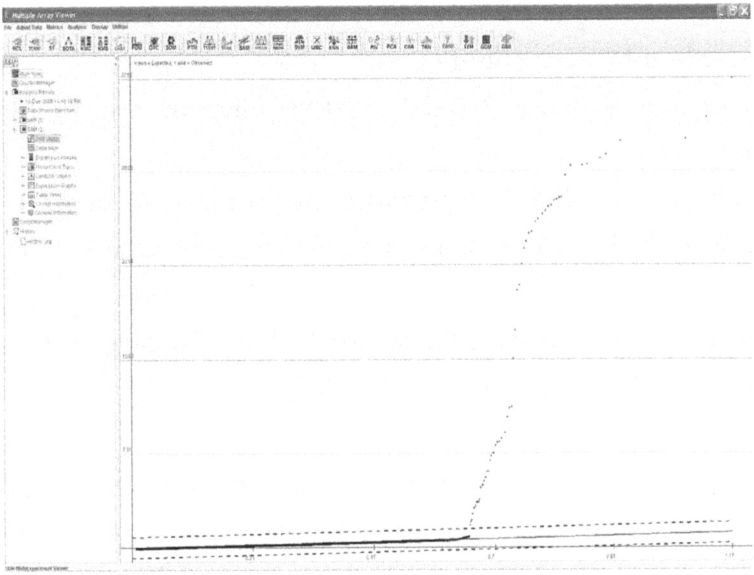

FIGURE 5.13 SAM plot. On the y-axis are the observed values and on the x-axis are the expected values. Solid line is the line that corresponds to gene expression values when expected = observed. Dashed lines are set by delta, which allows you to shrink or expand the area around the solid line, to determine how many genes will be deemed statistically significant (see text). Black dots correspond to the expression values of genes whose expression does not change between expected and observed. Gray dots correspond to expression values for genes whose expression is significantly different in the observed to the expected.

A pop-up window will prompt you to decide whether you want to cluster genes or samples. You can do either; for the purpose of this example, we will cluster genes.

In the distance metric selection, choose Euclidean distance. This is the default metric for SOMs.

For parameters, keep the current default settings: Dimension X and Y = 3; Iterations 2000, alpha 0.05, radius 3.0; Initialization random genes; Neighborhood Gaussian, and Topology Hexagonal.

Do not construct a hierarchical tree.

Press "OK."

An icon should pop up on your left-hand navigator for SOM-genes.

Click on this icon and scroll down to U-matrix color.

Figure 5.14 shows the use of SOM to identify 2 unique clusters (dark gray and black) identified in this data set. Table information identifies individual genes contained in each cluster.

You may vary any of the parameters we mentioned and examine the performance of different strategies.

You may use the same reduced data set to explore the use of K means clustering.

Click on "KMC." The pop-up window that appears will allow you to choose between clustering samples and clustering genes; choose "clustering genes."

Select cluster genes, and use the same default settings suggested.

Figure 5.15 shows KMC plots you will generate going through this operation.

Try running KMC again, and this time instead of clustering genes, cluster samples. Note that all samples in the Affymetrix data set are extremely similar to each other and are grouped in a single cluster. Try repeating the same procedure with the second data set provided.

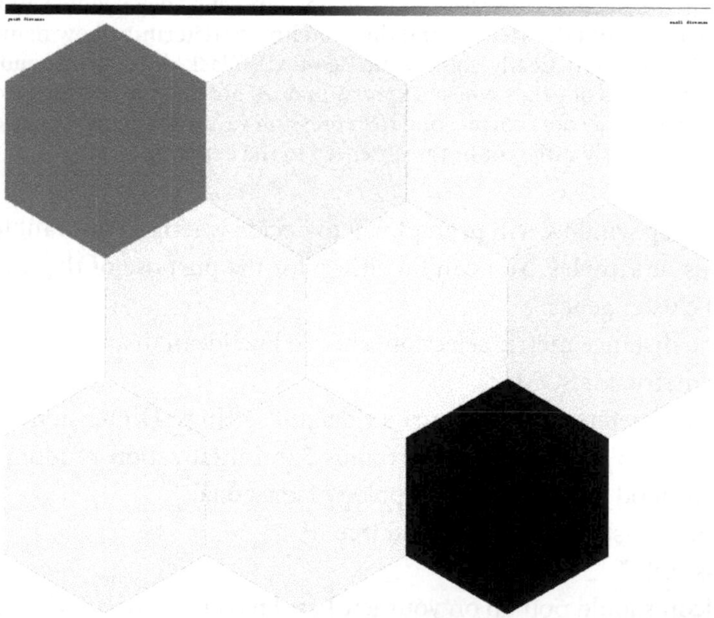

FIGURE 5.14 SOM plots — this is the self-organizing map hexagonal plot showing that the data can be viewed as containing nine clusters, of these two (gray and black) stand out as containing genes whose expression pattern may be perceived as unique.

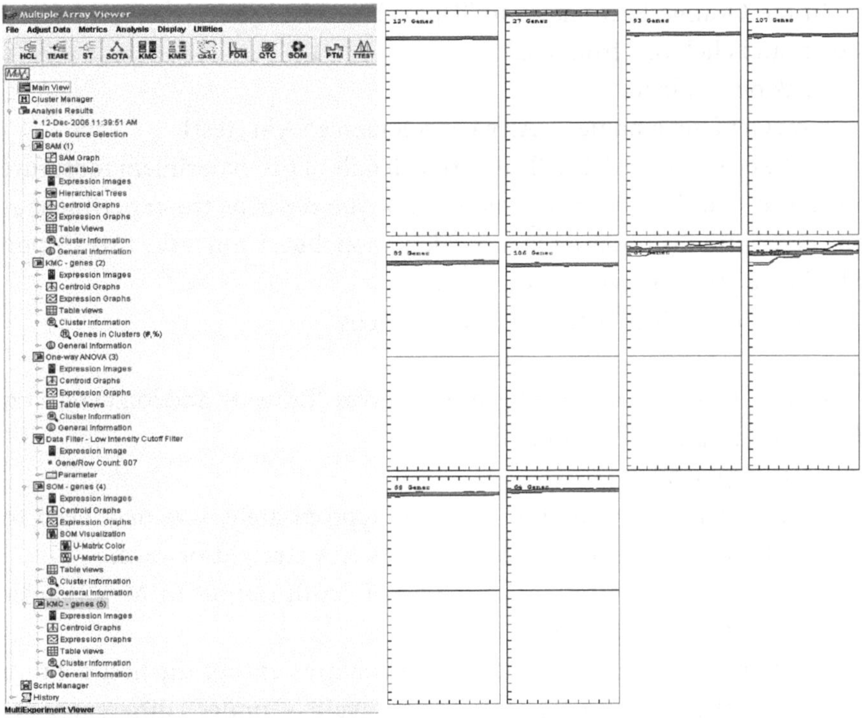
Multiple Array Viewer
File Adjust Data Metrics Analysis Display Utilities
HCL TEASE ST SOTA KMC KMS CAST FOM QTC SOM PTM TTEST
Main View
Cluster Manager
Analysis Results
12-Dec-2006 11:39:51 AM
Data Source Selection
SAM (1)
SAM Graph
Delta table
Expression Images
Hierarchical Trees
Centroid Graphs
Expression Graphs
Table Views
Cluster Information
General Information
KMC - genes (2)
Expression Images
Centroid Graphs
Expression Graphs
Table views
Cluster Information
Genes in Clusters (#,%)
General Information
One-way ANOVA (3)
Expression Images
Centroid Graphs
Expression Graphs
Table Views
Cluster Information
General Information
Data Filter - Low Intensity Cutoff Filter
Expression Image
Gene/Row Count: 807
Parameter
SOM - genes (4)
Expression Images
Centroid Graphs
Expression Graphs
SOM Visualization
U-Matrix Color
U-Matrix Distance
Table views
Cluster Information
General Information
KMC - genes (5)
Expression Images
Centroid Graphs
Expression Graphs
Table views
Cluster Information
General Information
Script Manager
History
MultiExperiment Viewer

FIGURE 5.15 KMC plots — on the left-hand side the navigational bar shows how to keep track of the data processing tools that have been already applied to the data set. If you highlight the icon for Expression graphs and scroll down to all graphs, the expression graphs for all genes clustered using K-means comes up on the right-hand side, showing how genes with different gene expression values are clustered across the data set.

B. Gene Ontology Analysis

Any list generated using the aforementioned strategies, and other strategies that the reader may want to explore, can be further analyzed using Functional Ontologies. This may be particularly useful at any stage of postprocessing.

In this particular example, we shall use the small group of genes selected as statistically significant using SAM.

Make sure you can open your SAM significant gene file in either Notepad or Excel, and that it is in .txt format.

Go to Gene Ontology Tree Machine (http://bioinfo.vanderbilt.edu/gotm/).

You must register to be able to use the software package, but this should be fairly fast.

Once you have registered and you are logged in, click on "WebGestalt."

In the Orange Section, go to "Upload" in the "select method," scroll down, and click on "from file."

Click on "Upload."

Give Gene set a name — Affy Latin Squares SAM_test1.

In the box provided, briefly describe details of the experiment for future reference — it does not really matter how you describe the experiment as the description is not used in any calculation, but if you leave this section blank, the program will not compute.

Under "organism," choose "Homo sapiens."

Under "ID type," choose "HG_U133A."

Under "Upload your file of gene set," press "Browse" and load your .txt file containing the ID column only.

Press "Upload."

Note that for this program to work appropriately, you may need to reformat the file generated by MeV. This is very straightforward.

Go to the Tabular format of the SAM result output in MeV. Under "File," save this as a .txt file.

Open the file in Excel and delete all columns except the first column containing the probe ID data. Save the file again — under a different name — e.g., Affy Latin Squares SAM_test1. This is the file that you will load into GOTM.

Once the file is loaded, a new window will appear requesting that you choose details for further analysis.

Under "Gene Information retrieval tool," you may leave this section blank; this way, the data will be computed by default.

Under "Gene set analysis tool," go to "Go Tree."

Under "select a reference" set scroll down, and click on "wengestalt_HG_U133A."

Under "statistical method," scroll down and select "hypergeometric test."

Under "select a significance value," select "0.05."

Under "select a minimum number of genes," select "4."

Click on "Make Go tree."

The computer will perform the operation; once it is completed, you will be prompted. Press on "check Go Tree."

In the new window you will be able to see how the list of genes selected is significantly enriched for genes overrepresented by specific GO ontologies. Specific functional enrichment for biological process, molecular function, and cellular component will be highlighted.

Functions represented as gray dots are overrepresented in the data set. The p-value of 0.05 indicates the likelihood that this enrichment occurred by chance alone.

In the orange section, in the far right, click on "enriched DAG."

Figure 5.16 shows the DAG (directed acyclic graph) of the enriched GO categories ($p < 0.05$ and at least genes, which are represented as thin boxes) and their nonenriched parents (which are colored black, represented as thick boxes). Click on particular nodes to view the list of gene members. You may adjust the font name and size and redraw the graph.

This software allows you to explore the data from a functional perspective.

This concludes the tutorial. It is far from complete, but presumably has served the purpose of introducing the reader to a variety of software analysis tools freely available and relatively easy to use. For your convenience, a second data set is provided.

An alternative data set is available from GEO (http://www.ncbi.nlm.nih.gov/entrez/query.fcgi?CMD=search&DB=gds). The GSE1318 Web page has links for the authors involved in generating the data, for a PubMed citation (15105423) describing this data, the e-mail address of the submitter (ness@unm.edu), and other information. The files come from an Affymetrix microarray analysis of gene expression changes induced by using a recombinant adenovirus to express the c-Myb transcription factor in human MCF-7 cells. To download the data, go to the GEO Web site: http://www.ncbi.nlm.nih.gov/projects/geo/. In the "Query" section in the middle of the page, type "GSE1318" (without the quotes) into the box labeled "GEO accession," then push "GO."

GSE1318 is a data series, a collection of 25 microarray data sets (GSM21610–GSM21638). To simplify things and make them go faster, we will only use the first six for this tutorial (GSM21610–GSM21615). At the bottom of the page is part of a table showing the data structure. This table lists normalized and raw values for each type of sample. However, the data in the table already has the replicate samples averaged together. We will start with the original data for three sets of replicates instead.

Download the data sets GSM21610–GSM21615. There are links for each set just above the partial table. Follow the following steps to download each data set:

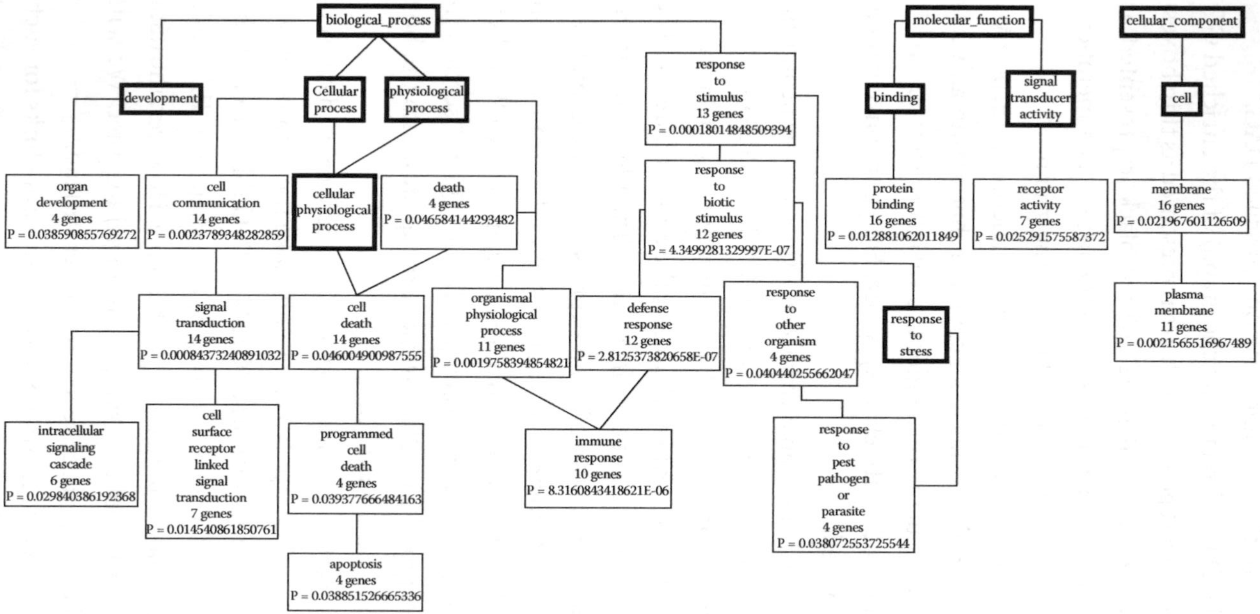
biological_process
molecular_function
cellular_component
development
Cellular process
physiological process
binding
signal transducer activity
cell
organ development 4 genes P = 0.038590855769272
cell communication 14 genes P = 0.0023789348282859
cellular physiological process
death 4 genes P = 0.046584144293482
response to stimulus 13 genes P = 0.00018014848509394
response to biotic stimulus 12 genes P = 4.3499281329997E-07
protein binding 16 genes P = 0.012881062011849
receptor activity 7 genes P = 0.025291575587372
membrane 16 genes P = 0.021967601126509
signal transduction 14 genes P = 0.00084373240891032
cell death 14 genes P = 0.046004900987555
organismal physiological process 11 genes P = 0.0019758394854821
defense response 12 genes P = 2.8125373820658E-07
response to other organism 4 genes P = 0.040440255662047
response to stress
plasma membrane 11 genes P = 0.0021565516967489
intracellular signaling cascade 6 genes P = 0.029840386192368
cell surface receptor linked signal transduction 7 genes P = 0.014540861850761
programmed cell death 4 genes P = 0.039377666484163
immune response 10 genes P = 8.3160843418621E-06
response to pest pathogen or parasite 4 genes P = 0.038072553725544
apoptosis 4 genes P = 0.038851526665336

FIGURE 5.16 Directed acyclic graph (DAG) showing the enriched GO categories generated from the list of significantly altered genes selected using multiclass analysis in SAM. Each box represents a GO category: biological process, molecular function, or cellular component. Each horizontal level represents a level of the GO ontology, so that parent ontologies give rise to subsequent ontology generations. All thin boxes identify statistically significant functional enrichment, whereas thick boxes represents the nonenriched parent ontologies.

1. Click on the link for the appropriate data set (e.g., GSM21610).
2. From the new page, use the pull-down menu to set the "Scope" to "Samples."
3. Set the "Format" to "SOFT" and the "Amount" to "Full."
4. Push "GO" to download the data on your computer.
5. In the "GEO accession" box, change the accession number to the next one (e.g., GSM21611), push "GO" to download, then repeat for the rest of the samples.

Put all the newly downloaded files in one folder.
The samples are the following:

GSM21610 and GSM21611: Control cells, uninfected MCF-7

GSM21612 and GSM21613: Control cells, infected with a control (GFM-only) adenovirus

GSM21614 and GSM21615: Experimental cells, infected with an adenovirus expressing c-Myb and GFP

CONCLUSIONS

In this chapter, we have reviewed some of the important challenges associated with gene array analysis. These arise in part from the inherent variability of expression microarrays at the individual slide and spot level, from the large-scale nature of the data, from the novel structure of microarray data, and in part because the full use of expression profiles for inferring gene function is still only partly explored. The key areas in microarray data analysis include experimental design, the assessment of significance for differential expression, discriminant analysis (supervised learning), and clustering (unsupervised learning). These are supported by equally important but lower-level techniques for data acquisition, storage, linkage to gene databases, normalization, and visualization. As novel and more mature approaches to handling microarray data become available, one critical limitation remains: how to make efficient and maximal use of the information generated from heterogeneous genomic metadata sources such as transcriptomics, proteomics, genotyping, as well as the variety of platforms contributing to the fields. Complicating factors include the presence of unavoidable high level of noise, the high levels of interactions with a large number of traits, and the heterogeneities in time and space. This provides a unique opportunity for computational and systems scientists to develop tools to integrate this vastly complex data. The research in this field is prolific and exciting. Although much effort will have to be spent on optimizing and validating the technology, the promise is undeniable.

The author would like to thank Drs. Pepin, Quackenbush, Sharov, and Yang, for cordially allowing us to reproduce or adapt their figures for this tutorial. We would also like to thank Dr. Quackenbush for corrections of part of this manuscript.

REFERENCES

1. Quackenbush, J. Extracting meaning from functional genomics experiments. *Toxicol Appl Pharmacol* 207(2 Suppl.): 195–199, September 1, 2005.
2. Brazma, A., Hingamp, P., Quackenbush, J. et al. Minimum information about a microarray experiment (MIAME)-toward standards for microarray data. *Nat Genet* 29(4): 365–371, December 2001.
3. Hoheisel, J.D. Microarray technology: beyond transcript profiling and genotype analysis. *Nat Rev Genet* 7(3): 200–210, March 2006.
4. Ness, S.A. Basic microarray analysis: strategies for successful experiments. *Methods Mol Biol* 316: 13–33, 2006.
5. Shippy, R., Fulmer-Smentek, S., Jensen, R.V. et al. Using RNA sample titrations to assess microarray platform performance and normalization techniques. *Nat Biotechnol* 24(9): 1123–1131, September 2006.
6. Canales, R.D., Luo, Y., Willey, J.C. et al. Evaluation of DNA microarray results with quantitative gene expression platforms. *Nat Biotechnol* 24(9): 1115–1122, September 2006.
7. Ayroles, J.F. and Gibson, G. Analysis of variance of microarray data. *Methods Enzymol* 411: 214–233. 2006.
8. Churchill, G.A. Fundamentals of experimental design for cDNA microarrays. *Nat Genet* 32 Suppl.: 490–495, December 2002.
9. Bretz, F., Landgrebe, J., and Brunner, E. Design and analysis of two-color microarray experiments using linear models. *Methods Inf Med* 44(3): 423–430, 2005.
10. Baker, S.C., Bauer, S.R., Beyer, R.P. et al. The External RNA Controls Consortium: a progress report. *Nat Methods* 2(10): 731–734, October 2005.
11. Eisensten, M. Microarrays: quality control. *Nature* 442(7106): 1067–1070, August 31, 2006.
12. Minor, J.M. Microarray quality control. *Methods Enzymol* 411: 233–255, 2006.
13. Timlin, J.A. Scanning microarrays: current methods and future directions. *Methods Enzymol* 411: 79–98, 2006.
14. Rensink, W.A. and Hazen, S.P. Statistical issues in microarray data analysis. *Methods Mol Biol* 323: 359–366, 2006.

15. Allison, D.B., Cui, X., Page, G.P., and Sabripour, M. Microarray data analysis: from disarray to consolidation and consensus. *Nat Rev Genet* 7(1): 55–65, January 2006.
16. Eads, B., Cash, A., Bogart, K., Costello, J., and Andrews, J. Troubleshooting microarray hybridizations. *Methods Enzymol* 411: 34–49, 2006.
17. Brazma, A., Krestyaninova, M., and Sarkans, U. Standards for systems biology. *Nat Rev Genet* 7(8): 593–605, August 2006.
18. Quackenbush, J. Standardizing the standards. *Mol Syst Biol* 2: 2006, 2006.
19. Brooksbank, C. and Quackenbush, J. Data standards: a call to action. *OMICS* 10(2): 94–99, 2006.
20. Salit, M. Standards in gene expression microarray experiments. *Methods Enzymol* 411: 63–78, 2006.
21. Shi, L., Reid, L.H., Jones, W.D. et al. The MicroArray Quality Control (MAQC) project shows inter- and intraplatform reproducibility of gene expression measurements. *Nat Biotechnol* 24(9): 1151–1161, September 2006.
22. Brazma, A., Sarkans, U., Robinson, A. et al. Microarray data representation, annotation and storage. *Adv Biochem Eng Biotechnol* 77: 113–139, 2002.
23. Quackenbush, J. Computational approaches to analysis of DNA microarray data. *Methods Inf Med* 45 Suppl. 1: 91–103, 2006.
24. Quackenbush, J. Microarray analysis and tumor classification. *N Engl J Med* 354(23): 2463–2472, June 8 2006.
25. Breitling, R. Biological microarray interpretation: the rules of engagement. *Biochim Biophys Acta* 1759(7): 319–327, July 2006.
26. Datta, S. and Datta, S. Empirical Bayes screening of many p-values with applications to microarray studies. *Bioinformatics* 21(9): 1987–1994, May 1, 2005.
27. Naiman, D.Q. Random data set generation to support microarray analysis. *Methods Enzymol* 411: 312–325, 2006.
28. Olson, N.E. The microarray data analysis process: from raw data to biological significance. *NeuroRx* 3(3): 373–383, July 2006.
29. Díaz-Uriarte, R. Supervised methods with genomic data: a review and cautionary view. Azuaje, F. and Dopazo, J., eds. *Data Analysis and Visualisation in Genomics and Proteomics*. John Wiley & Sons, 2004.
30. Dudoit, S., Gentleman, R.C., and Quackenbush, J. Open source software for the analysis of microarray data. *Biotechniques* Suppl.: 45–51, March 2003.
31. Quackenbush, J. Microarray data normalization and transformation. *Nat Genet* 32 Suppl.: 496–501, December 2002.
32. Pritchard, C.C., Hsu, L., Delrow, J., and Nelson, P.S. Project normal: Defining normal variance in mouse gene expression. PNAS 98(23), 13266–13271, November 6, 2001.

CHAPTER 6

Gene Expression Profiling by SAGE

Renu Tuteja

CONTENTS

ABSTRACT

SAGE is a powerful technology for measuring global gene expression, through rapid generation of large numbers of transcript tags. It rivals microarray analysis, with the advantage that it is affordable for standard laboratories. It provides a platform to define complete metabolic pathways and has been applied to study responses to drug treatment. The SAGE technique's high sensitivity and its global assessment of the transcriptome suggest that it is a perfect tool for analytical studies on representative samples to find candidate genes, which can be assessed in larger clinical populations. Although technically quite distinct, SAGE has some of the same limitations associated with microarray studies. As with microarrays, comparisons between SAGE studies of hormone-regulated gene expression in different tissues or cell types will likely reveal relatively limited overlaps in gene expression profiles. Comparisons between different SAGE studies may be complicated by the number of different statistical methods used to analyze expression patterns in SAGE libraries, which, like microarrays, must deal with "noise" associated with stochastic variations in gene expression. Nevertheless, SAGE will continue to be significant in building a reference database for gene expression analysis. This chapter describes the SAGE method, differences between SAGE and microarrays, applications of this technology to medical research, steps for SAGE data analysis, and methods to retrieve data from SAGE databases.

SECTION 1 SAGE EXPERIMENTATION

1. What Is SAGE?

Serial analysis of gene expression (SAGE) is a novel method used to give scientists an overview of a cell's complete gene activity. SAGE is a method of large-scale gene expression analyses and was developed in the mid-1990s. SAGE detects transcripts by extracting short tags from them. It is a technique that works by capturing RNAs, and it allows the speedy and quantitative study of a number of transcripts. It is a very powerful technique that has the potential to generate the complete catalog of mRNAs present within a cell population at a given time, along with their prevalence. SAGE is one of the few methods capable of uniformly probing gene expression at a genome level irrespective of mRNA abundance and without a previous knowledge of the transcripts present. It is intriguing to note that owing to limited length, many SAGE tags are shared by the transcripts from different genes, which can complicate gene identification. It is well established now that SAGE has advantages over other hybridization-based methods, for instance, subtractive hybridization and differential display. It is also better than the expressed sequence tag (EST) approach because SAGE can identify the genes, which are expressed at low levels and which correspond to a majority of genes in the human genome. By comparing different types of cells, the researchers expect to generate profiles that will facilitate an understanding of healthy cells and what goes wrong during diseases.

2. Brief Introduction to SAGE Procedure

In simple terms, SAGE works by capturing the RNAs, followed by "rewriting" them into DNA, and cutting a small, fourteen-letter tag from each one. Because it would take a long time to load tens of thousands of single tags into a sequencing machine, the method joins a lot of tags together into long molecules, which are called *concatemers*. The sequencer reads these molecules, counts and analyzes them, and computer programs generate a list of the genes that these tags belong to.

The technique is mainly based on two principles:

1. The short oligonucleotide sequence tags (10–11 base pairs), which contain sufficient information to uniquely identify transcripts, are used to identify genes and the relative abundance of their transcripts within mRNA.
2. The concatenation of short sequence tags allows an effective analysis of transcripts in a serial manner because SAGE uses serial processing such that 25-50 SAGE tags are analyzed on each lane of DNA sequencer.

These sequence data are analyzed to discover each gene expressed in the cell and the levels at which each gene is expressed. This information creates a library that can be used to analyze the differences in gene expression among cells. The occurrence of each SAGE tag in the cloned multimers precisely reflects the abundance of the transcript. Therefore, SAGE provides a precise picture of gene expression at both the quantitative and the qualitative levels. Briefly, some of the SAGE steps are as follows: trap the RNAs with beads and convert the RNA into cDNA. These cDNAs are digested so that there is an end sticking out, and a "docking unit" is ligated to this end. A short tag is cut off, and the two tags are combined together to form a ditag. These ditags are copied using polymerase chain reaction (PCR) and specific primers and are ligated to form concatemers, which are then sequenced. Using software, these cDNAs are identified and characterized, and the sequence of each tag is matched to the gene that produced the RNA.

A. Generation of the Library

The step-by-step procedure for library construction is shown in Figure 6.1. Double-stranded cDNA is synthesized from mRNA by using oligo(dT) primer. To enable recovery of 3'-cDNA fragments, the oligo (dT) primer contains a 5'-biotin moiety. The resulting double-stranded cDNA is then cleaved with restriction enzyme (also known as *anchoring enzyme*) *Nla*III, which recognizes and cleaves to leave a 3' CATG extension CATG (Figure 6.1). This cleavage step results in the formation of a specific position within each transcript for a consequent excision of the adjoining SAGE tag. By using streptavidin-coated magnetic beads, these biotinylated 3'-cDNAs are affinity-purified. These captured cDNAs are then ligated to linkers (or adapters) (Figure 6.1). The linkers used for this purpose are oligonucleotide duplexes, which contain a *Nla*III 4-nucleotide cohesive overhang, a type IIS recognition sequence, and a PCR primer sequence (primer A or primer B). Type IIS restriction enzymes, also known as tagging enzyme, cleave the DNA at a defined distance 14–15 nucleotides 3' of its nonpalindromic recognition sequence, releasing the linker-adapted SAGE tag from each cDNA. The most common enzyme used for this cleavage is *Bsm*FI. The majority of the cleavage sites used for *Bsm*FI are 15 nucleotides downstream of its recognition sequence but ~20% of the sites are 14 nucleotides downstream of its recognition sequence. Hence, for repairing the ends of linker-adapted SAGE tags, DNA polymerase (Klenow) enzyme is used (Figure 6.1). These repaired cDNAs from each pool are mixed together and then ligated using T4 DNA ligase. The ligated tags now serve as the templates for PCR amplification, with the primers specific to each linker. These linker-adapted ditags are amplified by primers A and B, digested with *Nla*III to release the primer-adapters, and the SAGE ditags are purified (Figure 6.1). The cleavage with anchoring enzyme *Nla*III allows the isolation of ditags, which are then polymerized using T4 DNA ligase and cloned into a high-copy plasmid vector. Each cloned insert is organized as a concat-

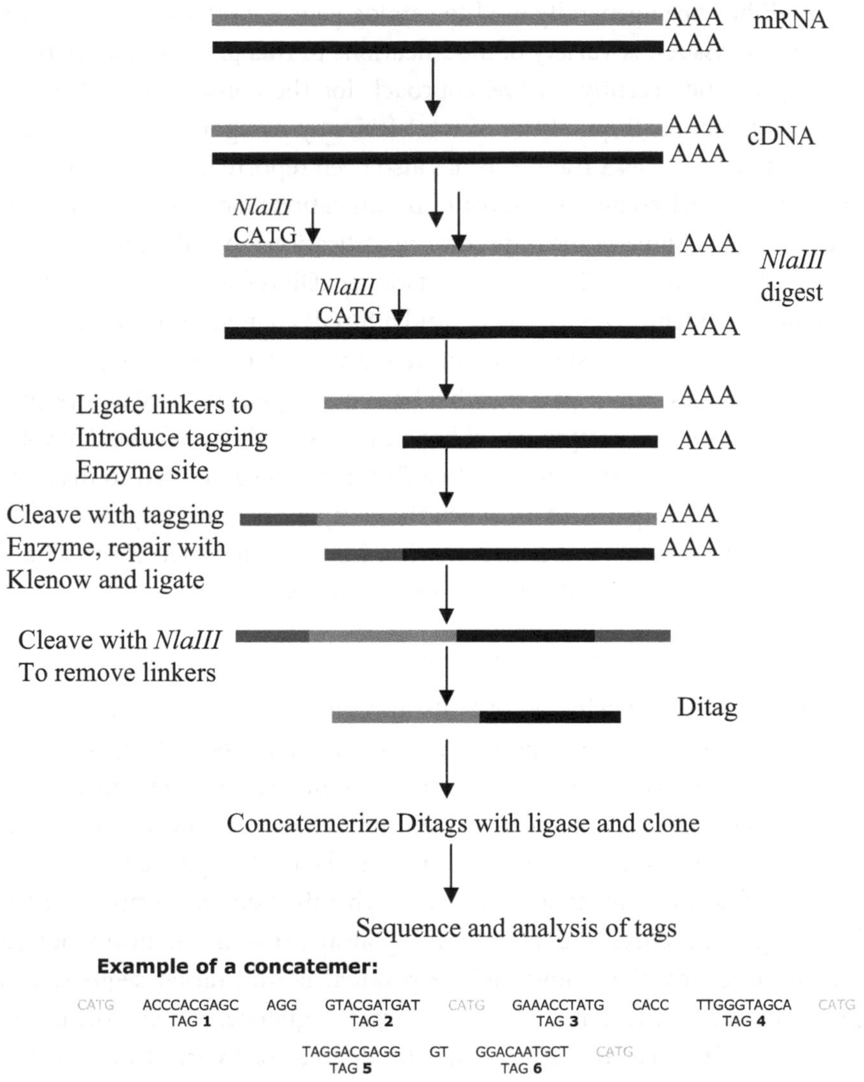
AAA
AAA
mRNA
AAA
AAA
cDNA
NlaIII
CATG
AAA
NlaIII
digest
NlaIII
CATG
AAA
Ligate linkers to
Introduce tagging
Enzyme site
AAA
AAA
Cleave with tagging
Enzyme, repair with
Klenow and ligate
AAA
AAA
Cleave with NlaIII
To remove linkers
Ditag
Concatemerize Ditags with ligase and clone
Sequence and analysis of tags
Example of a concatemer:
CATG ACCCACGAGC AGG GTACGATGAT CATG GAAACCTATG CACC TTGGGTAGCA CATG
TAG 1 TAG 2 TAG 3 TAG 4
TAGGACGAGG GT GGACAATGCT CATG
TAG 5 TAG 6

FIGURE 6.1 Step-by-step procedure for SAGE library construction.

enated series of ditags of 20–22 nucleotides in length, separated by the (4-nucleotide recognition sequence) for the anchoring enzyme *Nla*III. Automated sequencing of concatenated SAGE tags allows the routine identification of ~1000 tags per sequencing gel. This rapid sequencing allows quantitation and identification of cellular transcripts. SAGE can generate an exhaustive profile of the gene expression because each tag individually identifies a transcript.

SAGE has been generally used to catalog gene expression patterns in a number of tissues. A variety of modifications to this procedure have been developed, and recently a novel approach for the construction of SAGE libraries from small quantities of total RNA by using Y linkers to selectively amplify 3' cDNA fragments has also been reported. Since the discovery of SAGE technology, a number of modifications have been reported in the technique. Some of the important modifications include micro-SAGE, "SAGE-lite," mini-SAGE, generation of longer cDNA fragments from SAGE tags for gene identification (GLGI), and a SAGE adaptation for downsized extracts (SADE). A substantially improved version of SAGE, SuperSAGE (which generates 26 base pair tags), has been developed and used to interpret the "interaction transcriptome" of both the host and one of its eukaryotic pathogen. It is interesting to note that SAGE can also be used to discover new genes in the organisms for which the complete genome sequence is not available. If a sequence does not match a known gene, then it must have appeared from a gene that has not been discovered earlier.

3. What is the Difference between SAGE and Microarray?

Microarrays use the principle of Watson–Crick base pairing and are used to identify the profiles of expressed genes in a given tissue at a given time. In DNA microarrays, a number of known sequences (either short oligonucleotides or cDNA sequences) are immobilized on a solid support commonly known as a *chip*; usually fluorescent-labeled target sequences are added and allowed to hybridize. After hybridization the similar probes bind target sequences and, after washing away the nonspecifically bound target sequences, the amount of the residual bound target sequences is proportional to the amount of the target sequence, which forms the basis of nucleic acid quantification. The degrees of hybridization in the control and the test preparation are compared. Microarray analysis has enabled the measurement of thousands of genes in a single RNA sample. A comparison between SAGE and microarray analysis has been shown in Table 6.1.

Two widely used methods for the production of microarrays are (a) *in situ* synthesis of oligonucleotides and (b) robotic deposition of nucleic acids (oligonucleotides or PCR products) onto a glass slide. There are many methods to label the target sequences, and the labeling method depends on the amount of RNA. For a sufficient amount of RNA, i.e., about 20 μg of total RNA, it can be labeled with biotin directly. If the RNA quantity

TABLE 6.1 Difference between Microarray Analysis and SAGE Technology

Parameters	Microarray	SAGE
Principle	Hybridization of labeled cDNA	10–17 base pair tag generation and sequencing
Novel gene discovery	No	Yes
Polymorphism discovery	Yes	No
Data format	Relative intensity	Count of amplified SAGE tags
Low-abundance transcript detection	No	Yes
Setup cost	High	Low
Operation procedure	Easy	Difficult
Multiple sample Handling	Yes	No

is limited, it is first converted to a double-stranded cDNA (for 20–200-fold amplification), and then it is subsequently transcribed into a labeled antisense RNA by *in vitro* transcription. Usually, there are two basic techniques used for the detection of hybridization. The control and test preparations could be placed together on a single chip or separately on two chips. As opposed to SAGE analysis, which is usually performed between single libraries, microarray studies can be performed in triplicate or more. Microarray technology makes possible a more complete and comprehensive experimental approach in which variation in the transcript level of the whole genomes can be simultaneously assayed in response to a variety of stimuli. This genomewide approach to transcriptional analysis or "transcriptional profiling" makes available the comparative data on the relative expression level of various transcripts within an organism and relates this to modifications that occur as a consequence of a defined cellular stimulus.

Microarrays bear a resemblance to the yeast two-hybrid technique because they provide a screening platform to restrict the potential gene target. The major advantage in using microarrays is that they can handle a great number of known genes to establish the profiles of their expression. But further validation for microarray results is absolutely necessary because the false-positive rate for microarray experiments is very high. The microarray techniques generally need commercially produced chips in addition to specialized equipment and sophisticated computing skills and facilities. A number of issues must be considered before establishing a microarray platform and starting the expression profiling studies, in particular, the overall cost. For a cDNA microarray platform, one must purchase a clone set, a robot, printing pins, and the reagents needed for DNA amplification and purification. The cost of these materials can differ considerably, though once the process of printing and hybridizing microarrays has been optimized, the cost per experiment will fall dramatically. Therefore, one must decide if the number of planned experiments is adequate to justify the time and cost of establishing a microarray platform. If not, it may be more sensible to seek the services of an academic microarray core facility or a commercial entity.

4. Application of SAGE to Medical Research

Since the discovery of SAGE in 1995, it has been used to provide a comprehensive analysis of a variety of different tissue samples, each usually consisting of millions of cells. The SAGE method is also used to study global gene expression in cells or tissues in various experimental conditions. The approach has been extended recently to permit analysis of the gene expression in substantially fewer cells, thereby allowing analysis of more homogeneous cell populations or microanatomical structures. SAGE data can also be used to complement studies in cases where other gene expression methods may be inefficient. It is a well-known fact that the gene expression profiles of different types of cells (for example, a muscle cell and a brain cell) are very different. Similarly, the profile of a cancerous cell, or one that has been infected, will also deviate from that of a normal cell. Therefore, by monitoring the complete activity of the genome, SAGE should give researchers strong clues about patterns of gene activity that contribute to a particular disease. It is well established that SAGE can be used in an extensive variety of applications, such as to identify the effect of drugs on tissues, to identify disease-related genes, and to offer insights into the disease pathways (Ye et al. 2002). Another major application of SAGE is in the identification of differentially expressed genes. Modification of the original SAGE method, such as longSAGE for genomic DNA analysis, further enhanced its utility for cancer target identification.

SAGE and its variant longSAGE have been widely used to catalog gene expression patterns in a number of tissues. These techniques have also identified several useful prognostic markers and therapeutic targets in cancer. SAGE profiling has shown that the most intense changes in gene expression coupled with tumorigenesis occur early, and several of these changes influence the expression of secreted proteins. A number of studies have been done that involve SAGE as a tool to study various aspects of cancer. Some of these studies are listed in the following sections.

A. SAGE in Cancer

Gastric carcinoma is the fourth most common cause of cancer death worldwide, but its molecular biology is poorly understood. Gene expression patterns were examined in the gastrointestinal tumors using SAGE to understand the complex differences between normal and cancer cells in humans. The SAGE method has been used to systematically analyze transcripts present in lung and thyroid cancer. Two approaches, SAGE and DNA arrays, have been used to elucidate pathways in breast cancer progression and ovarian cancer by finding genes consistently expressed at different levels in primary breast cancers, metastatic breast cancers, and normal mammary epithelial cells. SAGE profiles of two well-characterized breast tumor cell lines were compared with SAGE profiles of normal breast epithelial cells to identify the differentially expressed genes. The coupling of SAGE and DNA arrays resulted in finding the predicted overexpression of known breast cancer markers HER-2/neu and MUC-1 and also in the identification of genes and potential pathways characteristic of breast cancer.

SAGE was applied to identifying expression of developmental control genes in the neuroblastoma and glioblastoma. The human homologue of the Drosophila Delta gene, Delta like-1 (DLK1), was shown to have an unusually high expression in a SAGE library of the SK-N-FI neuroblastoma cell line. Northern blot analysis confirmed high DLK1 expression in SK-N-FI and several other neuroblastoma cell lines. The data therefore suggest a role for the Delta-Notch pathway in a neuroblast differentiation. The chromosomal position of human genes is rapidly being established. These mapping data were integrated with the genomewide messenger RNA expression profiles as provided by SAGE to study prostate cancer and renal cell carcinoma. Biological effects of androgen on target cells are mediated in part by transcriptional regulation of androgen-regulated genes (ARGs) by androgen receptor. Using SAGE a comprehensive repertoire of ARGs in LNCaP cells (prostate cancer cell line) has been determined. One of the SAGE-derived tags exhibiting homology to an expressed sequence tag was maximally induced in response to synthetic androgen R1881 treatment.

SAGE has been used for the investigation of modulation in gene expression in numerous health conditions and to identify hormone-regulated genes in a variety of experimental models. For example, in a study reported for breast tumors, the gene expression profiles of breast carcinoma cells were compared, and it was observed that the patterns of gene expression in normal tissue were markedly different from those of the tumors. Recently, data regarding the application of SAGE technology to various studies, including various cancers, have been compiled in a few reviews. The SAGE database was also used to identify the differences between solid tumors and the cell lines, and 62 genes were identified to be overexpressed in tumors. It was suggested that the SAGE analysis has offered a molecular window into tumor biology, which in some instances could determine the difference between drug sensitivity and drug resistance. SAGE has also been used for comparative studies of gene expression profiles in hormone-dependent and -independent cancers. In another study using SAGE, it has been shown that histone deacetylases are generally expressed in almost all the tissues, and there were no major differences between the expression patterns of histone deacetylases of normal and the malignant tissues. Therefore, even though derived from only a few tissue libraries, gene expression profiles obtained by using SAGE in all probability represent an impartial yet characteristic molecular signature for most of the human cancers.

B. SAGE in Cardiovascular Diseases

There are only a limited number of studies that have used the SAGE technique to evaluate the cardiovascular system. Some of these studies are analysis of normal human heart transcriptomes, alterations in gene expression induced by hypoxia in human cardiac cells, and differentiation of the pleuripotent cells into cardiomyocytes. SAGE studies in the hematopoietic cells have generated novel insights into the cardiovascular system. SAGE analysis has shown that the genes associated with atherosclerosis and lipid metabolism were induced in cytokine-stimulated human monocytes. With all of these data and the delineation of the cardiovascular transcriptome, the next step will be the application of these SAGE datasets for diagnostic and therapeutic purposes.

C. New Gene Mapping and Karyotyping

The LongSAGE approach has broadened the scope of the application of SAGE. LongSAGE has a complex tag sequence; therefore, it allows for the direct mapping to the genomic sequence database. This advanced technique is very useful for screening candidate genes, and it also allows for the high-resolution mapping of the chromosomal derangements in cancer cells. This technique may possibly be modified in the near future to screen for mitochondrial or nuclear genetic changes taking place in a variety of diseases.

Besides the medical research in human physiology and the diseases, SAGE technology has also been successfully used for the analysis of gene expression in a variety of other species such as the yeast *Saccharomyces cerevisiae*, rice seedlings, the malaria parasite *Plasmodium falciparum*, and *Arabidopsis* roots.

SECTION 2 SAGE DATA ANALYSIS

1. SAGE Data Analysis Steps

The result of SAGE is a long list of nucleotides that have to be investigated by the computer. This analysis will do numerous things: count the tags, determine which ones come from the same RNA molecule, and figure out which ones come from well-known, well-studied genes and which ones are new. SAGE determines the expression level of a gene by measuring the frequency of a sequence tag derived from the corresponding mRNA transcript. However, individual SAGE tags can match many sequences in the reference database, complicating gene identification. The primary data outcome of the SAGE technology is the cloned insert sequence that corresponds to the concatenated tags, the ditags, which are separated by the four base restriction site (*Nla*III site).

The steps for tag extraction as described by Lash et al. (2000) are as follows:

1. The CATG (*Nla*III site) is located within the ditag concatamer.
2. The ditags of 20–26 base pair length, which occur between these sites, are extracted.
3. The repeat occurrences of ditags (including the repeat occurrences in the reverse-complemented orientation also) are removed.
4. Reverse-complementing the right-handed tag, the endmost ten bases of each ditag are defined as "tags."
5. The tags corresponding to linker sequences (such as TCCCCGTACA and TCCCTATTAA) and those with unspecified bases (bases other than A, C, G, or T) are removed.
6. The number of occurrences for each tag is counted.

The outcome of this processing is a catalog of tags with their matching count values and therefore is a digital depiction of cellular gene expression.

An essential step in SAGE library analysis is tag mapping, which refers to the unambiguous determination of the gene represented by a SAGE tag and is called tag-to-gene mapping. UniGene project is an experimental system for automatically separating all the GenBank sequences such as ESTs, proteins, mRNA/cDNA, etc., into a nonredundant set of gene-oriented clusters. Each UniGene cluster contains sequences that correspond to a unique gene and is associated with its corresponding tissue type. These UniGene clusters are created for almost all the organisms for which there are genes in the GenBank. Therefore, the UniGene project provides a single identifier and gene description for each cluster of sequence. These identifiers are utilized in the creation of a SAGE tag-to-gene mapping.

This process of the tag-to-UniGene cluster assignment as described by Lash et al. (2000) is as follows:

1. The individual human sequences from the GenBank submission record that are represented in UniGene are separated.
2. The sequence orientations are allocated by way of combination of identification of polyadenylation tail (a minimum of 8 A's), polyadenylation signal (AATAAA or ATTAAA), and orientation annotation (5' or 3').
3. A 10-base tag that is 3'-adjacent to the most 3'-*Nla*III site is extracted.
4. A UniGene identifier to each human sequence with a SAGE tag is allocated.
5. For every tag–UniGene pair, two frequencies are calculated: the first from the number of times this tag–UniGene pair has been seen divided by, separately, the number of sequences with this tag and, second, the number of sequences with tags in this UniGene cluster.

2. What Online Programs Are Available?

The SAGE300 program is probably the most commonly used application for SAGE analysis. In order to identify SAGE tags, this program compiles a database of tags extracted from human sequences in Genbank. SAGE300 is freely available from http://www.sagenet.org. It has been shown that although in many cases it is probable that a given tag sequence is unique within the genomes, in larger genomes this cannot be safely assumed. Therefore, for a thorough analysis of the SAGE data, the USAGE package has been developed. USAGE is a Web-based application that contains an integrated set of tools, which includes many functions for analyzing and comparing the SAGE data. USAGE is freely accessible for academic institutions at http://www.cmbi.kun.nl/usage/. Additionally, USAGE includes a statistical method for the planning of new SAGE experiments. It is accessible in a multiuser environment offering users the option of sharing data and is interfaced to a relational database to store data and analysis results. USAGE provides the biologist increased functionality and flexibility for analyzing the SAGE data.

For the analysis of gene expression profiles derived from ESTs and SAGE, Expression Profile Viewer (ExProView), a software tool, has been developed. This software visualizes a complete set of classified transcript data in a two-dimensional array of dots, a "virtual chip," in which each dot represents a known gene as characterized in the transcript databases. To evaluate the software, public EST and SAGE gene expression data obtained from the Cancer Genome Anatomy Project at the National Center for Biotechnology Information were analyzed and visualized. In another study, eSAGE, a comprehensive set of software tools for managing and analyzing data generated with SAGE has been described. eSAGE was written in Visual Basic v6.0 (Microsoft Corp.) and is compatible with the Windows 95/98 and NT v4.0 operating systems (Microsoft Corp.).

Recently, using the statistical approach of Audic and Claverie (1997), a useful and flexible tool (WEBSAGE) was developed. WEBSAGE is a software that enables a rapid and thorough analysis of the SAGE data, and this simple tool also performs statistical analysis on SAGE data. It compares a large number of SAGE tags and depicts the comparison of two SAGE libraries in a scatter plot. It is freely available and accessible at http://bioserv.rpbs.jussieu.fr/websage/index.php. The limitation of their method is that more than two libraries cannot be visualized using their tool. However, using WEBSAGE, full SAGE data are comprehensively represented in plots, and WEBSAGE not only gives the identification but also the function of the gene. Therefore, WEBSAGE is a useful Web service, and the user can query with a specific tag and get tag-to-gene assignment and gene function from Kegg, Biocarta, and Gene Ontology databank.

3. Demo: How to Analyze Generated Raw SAGE Data

WEBSAGE is a very useful program for the analysis of SAGE data. The steps to use WEBSAGE are well explained on its Web site and are summarized as follows.

A. Data Input

First click on the input data link, and enter the pathname of your file. The files can be provided in Kinzler's or generic format. Then enter the legends of x-axis and y-axis (for the first library, such as liver; and for the second library, such as kidney). These names are displayed on the graph on the respective places. To compute a scatter plot, press the "submit" button.

B. Scatter Plot

A representative scatter plot is shown in Figure 6.2. In the plot the X and Y axes are normalized and represent the tags of the first (liver) and the second (kidney) library, respectively. The p value is computed according to the formula of Audic and Claverie: $p(y \mid x) = (N2/N1)y(x + y)!/\ x!y!(1+N2/N1)(x+y+1)$, where $N1$ is the size of library 1, $N2$ is the size of library 2, x is the number of tags in library 1, and y is the number of tags in library 2.

The number of analyzed tags is shown under the title. On the legend, the number of tags for each class is written between brackets. It is interesting to note that the color of the plot represents 3 classes of p value: green color represents a significant p value (0.01); yellow color represents p value of medium significance (>0.01 and <0.05), and red color represents a nonsignificant p value (>0.05). The size of the plot is proportional to the number of tags it contains. The scale is logarithmic, and data are normalized.

C. Identification of Tag

By clicking on a plot, a Web page is displayed that contains the tag's information. Mainly, the information consists of the following:

Tag: the tag analyzed. By clicking on the tag link, a new window (NCBI site) gives the chromosome link of the tag.

Number of library legend: the abundance of each tag in each library.

Fold change: the fold increase (Inc) or reduced (Red) gene expression.

GeneBank accession number: accession number in GenBank of the gene corresponding to the tag.

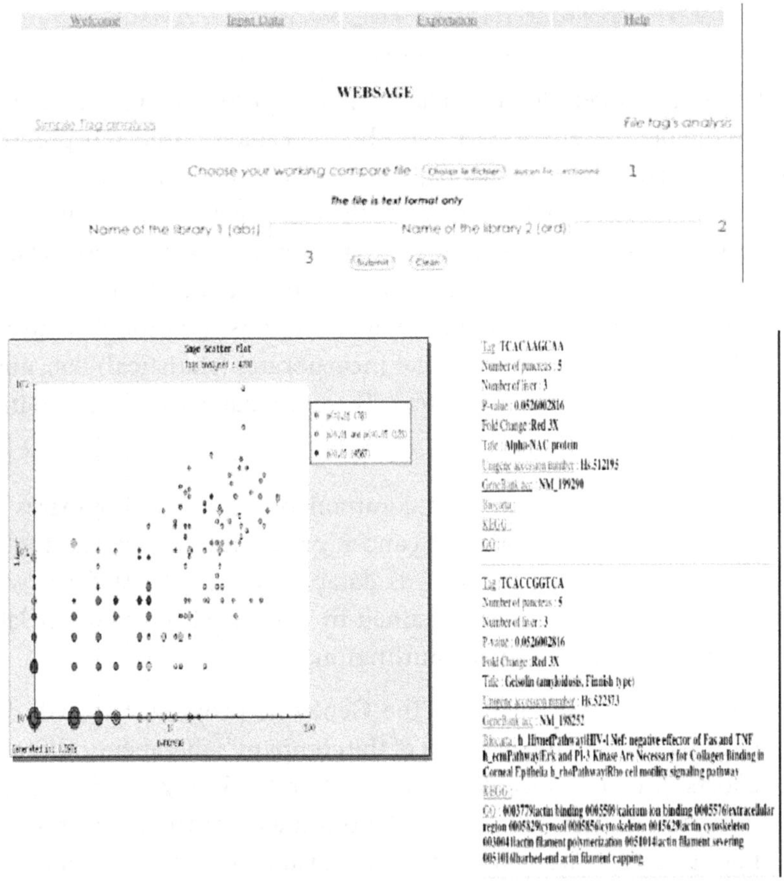
WEBSAGE
File tag's analysis
Choose your working compare file
1
The file is text format only
Name of the library 1 (abs)
Name of the library 2 (ord)
2
3
Submit
Clean
Sage Scatter Plot
Tag: TCACCGGTCA
Number of pancreas : 5
Number of liver : 3
P-value : 0.0526002816
Fold Change: Red 3X
Title: Alpha-NAC protein
Title: Gelsolin (amyloidosis, Finnish type)
KEGG
GO

FIGURE 6.2 WEBSAGE analysis of SAGE data.

Unigene accession number: accession number in Unigene of the gene corresponding to the tag.

Title: the gene name corresponding to the tag.

KEGG, Biocarta, and GO (Gene Ontology): the information corresponding to the gene from these databanks. A new window on the site is displayed by clicking on one of these links.

D. Data Exportation

The export local plot button is clicked to export information corresponding to the plot or clicked to export all the data exportation link.

SECTION 3 RETRIEVE DATA FROM SAGE DATABASES

1. Brief Introduction to Available SAGE Databases

The data analysis steps for a standard SAGE experiment are as follows. To confirm the tag content of the newly obtained SAGE clone, it is processed using the software. Consequently, the main characteristics of the SAGE software are (1) to extract and tabulate the tag sequences and counts from raw sequence files, (2) to correlate the tag abundance among projects, and (3) to match the tag sequences to sequences in other databases.

Therefore, as mentioned in the previous sections, a thorough analysis of SAGE data necessitates software that incorporates (statistical) data analysis techniques with a database system. The main databases are listed in the following text:

1. SAGEnet: Among the most common public SAGE databases is a SAGE tag database for colon cancer, pancreatic cancer, and the corresponding normal tissue. This database is called SAGEnet (www.sagenet.org), and it is maintained by the Vogelstein/Kinzler lab at Johns Hopkins University, Burlington, MA.
2. Genzyme's SAGE database: The Genzyme proprietary SAGE database, which is an integral part of the company's therapeutic discovery efforts, presently includes over 4 million SAGE tags, and is believed to represent over 100,000 genes from major cancers and from normal human tissues. This database is available to commercial licensees for a fee through distributors (Celera Genomics and Compugen). Genzyme also offers a for-fee service whereby it will generate SAGE tag libraries for contracting parties.
3. SAGEmap: SAGEmap is a public gene expression database repository produced by the National Institutes of Health's (NIH's) National Center for Biotechnology Information (NCBI) in conjunction with the NIH's Cancer Genome Anatomy Project (CGAP). To provide the quantitative expression levels on a genomewide scale, the CGAP uses SAGE. Over 7 million SAGE tags from 171 human cell types have been assembled. This database uses SAGE to quantify the transcript levels in both the malignant and normal human tissues. By accessing SAGEmap, the user can compare transcript populations between any of the posted libraries. The WWW and FTP components of this resource SAGEmap are located at http://www.ncbi.nlm.nih.gov/sage and ftp://ncbi.nlm.nih.gov/pub/sage, respectively. In these Web sites

the authors have described SAGE data submission procedures, the construction and characteristics of SAGE tags to gene assignments, the derivation and use of a novel statistical test designed specifically for the differential-type analyses of SAGE data, and the organization and use of this resource. To enhance the utility of this data, the CGAP SAGE project created SAGE Genie, a Web site for the analysis and presentation of SAGE data (http://cgap.nci.nih.gov/SAGE). SAGE Genie provides an automatic link between gene names and the SAGE transcript levels, accounting for alternative transcription and many potential errors. SAGE Genie was created for better analysis and dissemination of the digital gene expression profiles. SAGE Genie automatically identifies SAGE tags from a gene's primary or alternatively polyadenylated transcript. By using the SAGE Genie, a large group of SAGE data can be readily and precisely viewed. Begun originally as a database of gene expression data from brain cancer investigations, SAGEmap can now accept SAGE sequence data from any source. This database presently contains 623 SAGE experiments from 18 organisms. SAGEmap can also construct a user-configurable table of data comparing one group of SAGE libraries with another. By accessing SAGEmap the user can compare transcript populations between any of the posted libraries.

An important feature of SAGEmap is its systematic approach to mapping SAGE tags to UniGene EST clusters. Rather than simply matching SAGE tags across all *Nla*III sites in the sequence clusters, SAGEmap incorporates EST frequency, location, and orientation information as well as a correction for sequencing error to refine the mapping of tags to genes. This mapping represents a "best guess" for matching of observed tags to expressed sequences in the public database.

4. The Mouse SAGE site: This site is maintained on an ongoing basis at the Institute of Molecular Genetics, Academy of Sciences of the Czech Republic, and is accessible at the Internet address http://mouse.biomed.cas.cz/sage/. The database aims to provide mouse geneticists with easy-to-use Web-based tools for utilizing mouse SAGE data. It is a recently developed Web-based database of all the available public libraries generated by SAGE from various mouse tissues and cell lines. The database contains mouse SAGE libraries organized in a uniform way and provides Web-based tools for browsing, compar-

ing, and searching SAGE data with reliable tag-to-gene identification. A modified approach based on the SAGEmap database is used for reliable tag identification.

5. The Gene Expression Omnibus (GEO) at NCBI: This is one of the major fully public repositories for high-throughput molecular abundance data, which are primarily gene expression data. As described, the three central data entities of GEO are platforms, samples, and series.

   Platform: A platform describes a list of probes that define what set of molecules may be detected. A platform record describes the list of elements on the array (e.g., cDNAs, oligonucleotide probe sets, ORFs, and antibodies) or the list of elements that may be detected and quantified in that experiment (e.g., SAGE tags, peptides). Each platform record is assigned a unique and a stable GEO accession number (GPLxxx). A Platform may reference many samples that have been submitted by multiple submitters.

   Sample: A sample explains the set of molecules that are being probed and references a single platform used to generate its molecular abundance data. A sample record also describes the conditions under which an individual sample was handled, the manipulations it underwent, and the abundance measurement of each element derived from it. Each sample record is assigned a unique and stable GEO accession number (GSMxxx). A sample entity must reference only one platform and may be included in the multiple series.

   Series: A series organizes samples into the meaningful data sets, which make up an experiment. A series of record defines a set of related samples considered to be part of a group, how the samples are related, and if and how they are ordered. A series provides a focal point and description of the experiment as a whole. Series records may also contain tables describing extracted data, summary conclusions, or analyses. Each series record is assigned a unique and a stable GEO accession number (GSExxx).

The database is user-friendly and allows the submission, retrieval, and the storage of a number of data types, including SAGE data. The GEO database presently holds more than 30,000 submissions and is publicly accessible through the World Wide Web at http://www.ncbi.nlm.nih.gov/geo.

6. Another SAGE resource comprises the online query tool of the Saccharomyces Genome Database, which allows users to search and present SAGE data obtained for yeast in several ways (http//:genome.www.stanford.edu/cgi-bin/SGD/SAGE/query/SAGE).

2. How to Retrieve Data from GEO

GEO stores a broad collection of high-throughput experimental data that have been generated as a result of processing by multiple means and analyzed by a variety of methods. To address some of these issues, an extra level of curation was added, in which submitted samples were collected into biologically meaningful and statistically comparable GEO DataSets (GDSs). These DataSets (GDSxxx) are curated sets of GEO sample data. These records supply a coherent synopsis about an experiment and offer a basis for downstream data mining and display tools. A GDS record represents a collection of biologically and statistically comparable GEO samples and forms the basis of GEO's suite of data display and analysis tools. Samples within a GDS refer to the same platform; that is, they share a common set of probe elements. Information reflecting the experimental design is provided through GDS subsets.

GEO data can be viewed and downloaded in several formats; several options are available for the retrieval and display of original GEO records. The Scope feature allows display of a single accession number or any (Platform, Sample, or Series) or all (Family) records related to that accession. The amount determines the quantity of data displayed, and the format controls whether records are displayed in HTML or in SOFT (Simple Omnibus Format in Text) format. SOFT is an ASCII text format that was designed to be a machine-readable representation of data retrieved from, or submitted to, GEO.

As suggested by Barrett et al. (2005), there are a number of ways and formats in which GEO data may be retrieved:

1. Using the GEO accession number, the individual platform, sample, series, and GDS records can be accessed on the Web. On the GEO site, these related records are intralinked in such a way that the associated platform, sample, series, and GDS records can be easily navigated. Each GDS record has three options for the download of that dataset. The complete SOFT document contains all information for that dataset, including dataset description, type, organism, subset

allocation, etc., as well as a data table containing identifiers and values. The data-only option allows download of the data table only, whereas the quick view provides dataset descriptive information and the first 20 rows of the data table.

2. Using the title, type, platform, or organism, the GDS records may be browsed at http://www.ncbi.nlm.nih.gov/geo/gds/gds_browse.cgi. The records submitted by the user might also be browsed by category or submitter.

3. Both GDS and GEO data are available for bulk download via file transfer program (FTP). GEO DataSets may be downloaded in complete GDS SOFT format, whereas complete original GEO records, partitioned by GEO Platform, may be downloaded in SOFT format. All of the user-submitted records, raw data, and GDS value matrices with annotation are accessible for bulk download via (FTP). It should be noted that all of the user-submitted records are assembled as compressed series and platform "family" files that combine all related accessions as well. Equivalent files can also be downloaded from each record on the Web.

3. Demo of the Procedure Preceding Section: A Step-By-Step Tutorial

As described on the Web site, GEO data can be retrieved in a number of ways. The Accession Display bar (found at the bottom of the GEO home page and at the top of each GEO record) can be used to look at a particular GEO record for which an accession number is available. This tool has several options for selecting the format and amount of data to view.

For the inquiry of all the GEO submissions in a specific field, or over all fields, the Entrez GEO Datasets or Entrez GEO Profile interfaces can be used. It is to be noted that the Entrez GEO DataSets query all GEO DataSet annotation, allowing identification of experiments of interest. Entrez GEO Profiles query the gene expression/molecular abundance profiles, allowing the identification of genes or sequences or profiles of interest.

For browsing the lists of GEO data and experiments, the GDS browser or the list of current GEO repository contents can be investigated; for example:

NCBI
Entrez GEO DataSets
My NCBI
Welcome renut. [Sign Out]
Search GEO DataSets for sage analysis breast cancer
Go Clear Save Search
Limits Preview/Index History Clipboard Details
The GEO site
GEO FAQ
List GEO Contents
Entrez Help | FAQ
Display Summary Show 20 Send to
All: 2 DataSets: 0 Platforms: 0 Series: 2
Items 1 - 2 of 2
One page.
1: GSE507 record: Novel estrogen and tamoxifen induced genes [Homo sapiens]
Links
Summary: (Submitter supplied) The breast cancer promoting effects of estrogen and the chemopreventive effects of tamoxifen are thought to be mediated by the estrogen receptor, a ligand-dependent transcription factor. Therefore, comprehensive analysis of gene expression profiles following estrogen or tamoxifen treatment may help us better understand the role estrogen plays in tumorigenesis. We utilized SAGE (Serial Analysis of Gene Expression) technology to identify genes regulated by estrogen and tamoxifen in the ZR75-1 estrogen dependent breast cancer cell line. In this manner we have identified several genes that were regulated by estrogen or tamoxifen. Here we report the identification and initial characterization of EIT-6 (Estrogen Induced Tag-6), a novel nuclear protein and a new member of the evolutionarily conserved SM-20 family of growth regulatory immediate-early genes. more...
1 related Platform
Samples: 3
GSM1128: ZR75-1 untreated
GSM1129: ZR75-1 estrogen treated
GSM1130: ZR75-1 tamoxifen
2: GSE14 record: CGAP SAGE [Homo sapiens Mus musculus]
Links
Summary: (Submitter supplied) This series represents the Cancer Genome Anatomy Project SAGE library collection. Libraries contained herein were either produced through CGAP funding, or donated to CGAP. The Cancer Genome Anatomy Project (CGAP: http://cgap.nci.nih.gov) is an interdisciplinary program established and administered by the National Cancer Institute (NCI: http://www.nci.nih.gov) to generate the information and technological tools needed to decipher the molecular anatomy of the cancer cell. Keywords = CGAP Keywords = SAGE
2 related Platforms
Samples: 214 (listing 18)
GSM668: SAGE_293-CTRL
GSM669: SAGE_293-IND
GSM670: SAGE_95-259
GSM671: SAGE_95-260
GSM672: SAGE_95-347
GSM673: SAGE_95-348
GSM674: SAGE_A+
GSM675: SAGE_A2780-9
GSM676: SAGE_BB542_whitematter
GSM677: SAGE_Br_N
GSM678: SAGE_CAPAN1
GSM680: SAGE_CPDR_LNCaP-C
GSM681: SAGE_CPDR_LNCaP-T
GSM682: SAGE_Caco_2
GSM683: SAGE_Chen_LNCaP
GSM684: SAGE_Chen_LNCaP_no-DHT
GSM685: SAGE_Chen_Normal_Pr
GSM686: SAGE_Chen_Tumor_Pr

FIGURE 6.3 Display of GEO Dataset for SAGE analysis in breast cancer.

1. To search for genes analyzed in breast cancer, type the phrase: SAGE analysis breast cancer on the site Entrez GEO Datasets and press "search." (Figure 6.3)

2. A page will appear with the results, showing the results as: all; DataSets; Platform; and Series (Figure 6.4 to Figure 6.7).

3. The results will be displayed as items 1 and 2, etc.

4. By clicking on the record button, the complete record is displayed. This record displays all the relevant and necessary information regarding the data set (Figure 6.8a, Figure 6.8b).

By further clicking on the sample, the details and sequences of the tags and the genes that they belong to can be easily obtained (Figure 6.9, Figure 6.10).

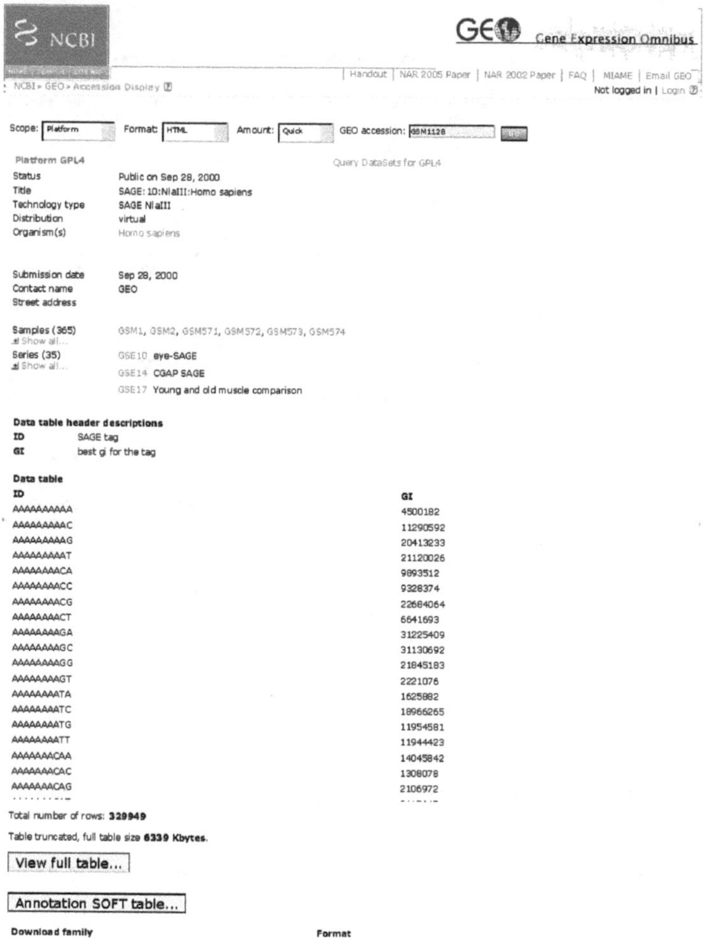
NCBI
GEO Gene Expression Omnibus
Handout | NAR 2005 Paper | NAR 2002 Paper | FAQ | MIAME | Email GEO
NCBI » GEO » Accession Display
Not logged in | Login
Scope: Platform Format: HTML Amount: Quick GEO accession: GSM1128
Platform GPL4
Query DataSets for GPL4
Status Public on Sep 28, 2000
Title SAGE: 10:NlaIII:Homo sapiens
Technology type SAGE NlaIII
Distribution virtual
Organism(s) Homo sapiens
Submission date Sep 28, 2000
Contact name GEO
Street address
Samples (365) GSM1, GSM2, GSM571, GSM572, GSM573, GSM574
Show all...
Series (35)
Show all...
GSE10 eye-SAGE
GSE14 CGAP SAGE
GSE17 Young and old muscle comparison
Data table header descriptions
ID SAGE tag
GI best gi for the tag
Data table
ID GI
AAAAAAAAAA 4500182
AAAAAAAAAC 11290592
AAAAAAAAAG 20413233
AAAAAAAAAT 21120026
AAAAAAAACA 9893512
AAAAAAAACC 9328374
AAAAAAAACG 22684064
AAAAAAAACT 6641693
AAAAAAAAGA 31225409
AAAAAAAAGC 31130692
AAAAAAAAGG 21845183
AAAAAAAAGT 2221076
AAAAAAAATA 1625882
AAAAAAAATC 18966265
AAAAAAAATG 11954581
AAAAAAAATT 11944423
AAAAAAACAA 14045842
AAAAAAACAC 1308078
AAAAAAACAG 2106972
Total number of rows: 329949
Table truncated, full table size 6339 Kbytes.
View full table...
Annotation SOFT table...
Download family
Format

FIGURE 6.4 GEO accession display of platform.

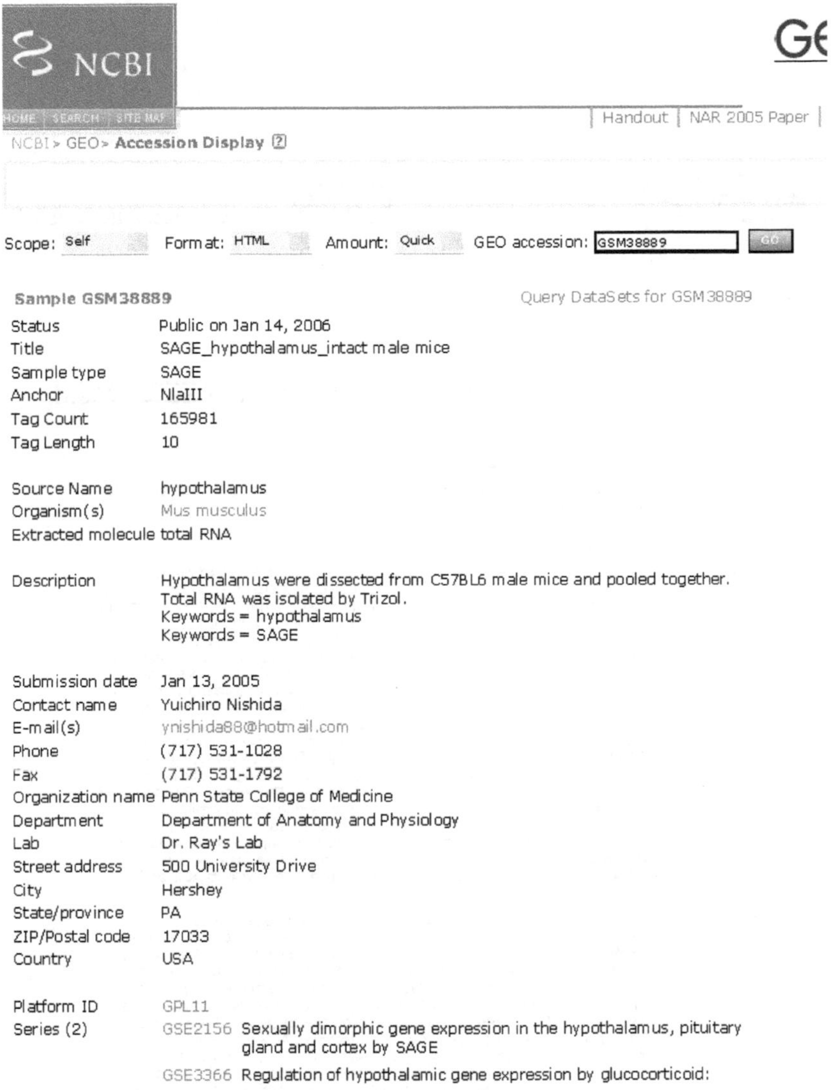

NCBI

HOME | SEARCH | SITE MAP

| Handout | NAR 2005 Paper |

NCBI > GEO > **Accession Display**

Scope: Self   Format: HTML   Amount: Quick   GEO accession: GSM38889   GO

Sample GSM38889   Query DataSets for GSM38889

Status	Public on Jan 14, 2006
Title	SAGE_hypothalamus_intact male mice
Sample type	SAGE
Anchor	NlaIII
Tag Count	165981
Tag Length	10
Source Name	hypothalamus
Organism(s)	Mus musculus
Extracted molecule	total RNA
Description	Hypothalamus were dissected from C57BL6 male mice and pooled together. Total RNA was isolated by Trizol. Keywords = hypothalamus Keywords = SAGE
Submission date	Jan 13, 2005
Contact name	Yuichiro Nishida
E-mail(s)	ynishida88@hotmail.com
Phone	(717) 531-1028
Fax	(717) 531-1792
Organization name	Penn State College of Medicine
Department	Department of Anatomy and Physiology
Lab	Dr. Ray's Lab
Street address	500 University Drive
City	Hershey
State/province	PA
ZIP/Postal code	17033
Country	USA
Platform ID	GPL11
Series (2)	GSE2156 Sexually dimorphic gene expression in the hypothalamus, pituitary gland and cortex by SAGE
	GSE3366 Regulation of hypothalamic gene expression by glucocorticoid:

FIGURE 6.5 GEO accession display of samples.

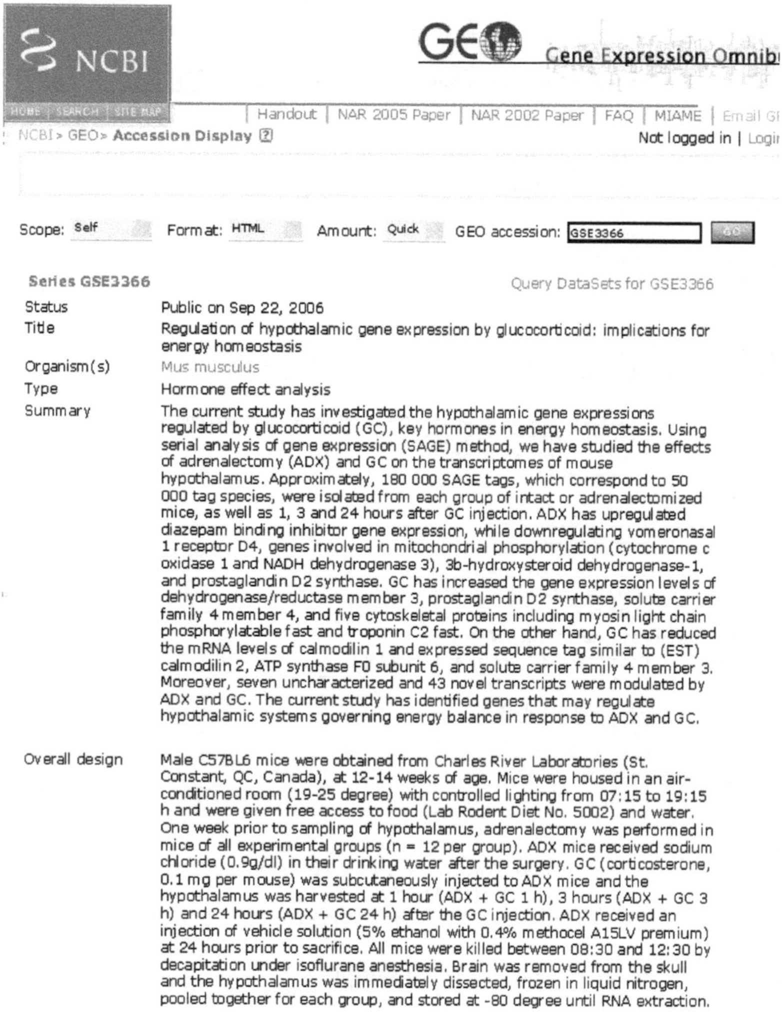
NCBI
HOME | SEARCH | SITE MAP
GEO Gene Expression Omnib
| Handout | NAR 2005 Paper | NAR 2002 Paper | FAQ | MIAME | Email GI
NCBI > GEO > Accession Display
Not logged in | Logi
Scope: Self Format: HTML Amount: Quick GEO accession: GSE3366 GO
Series GSE3366
Query DataSets for GSE3366
Status
Public on Sep 22, 2006
Title
Regulation of hypothalamic gene expression by glucocorticoid: implications for energy homeostasis
Organism(s)
Mus musculus
Type
Hormone effect analysis
Summary
The current study has investigated the hypothalamic gene expressions regulated by glucocorticoid (GC), key hormones in energy homeostasis. Using serial analysis of gene expression (SAGE) method, we have studied the effects of adrenalectomy (ADX) and GC on the transcriptomes of mouse hypothalamus. Approximately, 180 000 SAGE tags, which correspond to 50 000 tag species, were isolated from each group of intact or adrenalectomized mice, as well as 1, 3 and 24 hours after GC injection. ADX has upregulated diazepam binding inhibitor gene expression, while downregulating vomeronasal 1 receptor D4, genes involved in mitochondrial phosphorylation (cytochrome c oxidase 1 and NADH dehydrogenase 3), 3b-hydroxysteroid dehydrogenase-1, and prostaglandin D2 synthase. GC has increased the gene expression levels of dehydrogenase/reductase member 3, prostaglandin D2 synthase, solute carrier family 4 member 4, and five cytoskeletal proteins including myosin light chain phosphorylatable fast and troponin C2 fast. On the other hand, GC has reduced the mRNA levels of calmodilin 1 and expressed sequence tag similar to (EST) calmodilin 2, ATP synthase F0 subunit 6, and solute carrier family 4 member 3. Moreover, seven uncharacterized and 43 novel transcripts were modulated by ADX and GC. The current study has identified genes that may regulate hypothalamic systems governing energy balance in response to ADX and GC.
Overall design
Male C57BL6 mice were obtained from Charles River Laboratories (St. Constant, QC, Canada), at 12-14 weeks of age. Mice were housed in an air-conditioned room (19-25 degree) with controlled lighting from 07:15 to 19:15 h and were given free access to food (Lab Rodent Diet No. 5002) and water. One week prior to sampling of hypothalamus, adrenalectomy was performed in mice of all experimental groups (n = 12 per group). ADX mice received sodium chloride (0.9g/dl) in their drinking water after the surgery. GC (corticosterone, 0.1 mg per mouse) was subcutaneously injected to ADX mice and the hypothalamus was harvested at 1 hour (ADX + GC 1 h), 3 hours (ADX + GC 3 h) and 24 hours (ADX + GC 24 h) after the GC injection. ADX received an injection of vehicle solution (5% ethanol with 0.4% methocel A15LV premium) at 24 hours prior to sacrifice. All mice were killed between 08:30 and 12:30 by decapitation under isoflurane anesthesia. Brain was removed from the skull and the hypothalamus was immediately dissected, frozen in liquid nitrogen, pooled together for each group, and stored at -80 degree until RNA extraction.

FIGURE 6.6 GEO accession display of series.

NCBI
Entrez GEO DataSets
My NCBI
[Sign In] [Reg
Search GEO DataSets for GSE14[RGPL]
Go Clear Save
Limits Preview/Index History Clipboard Details
Display Summary Show 20 Send to
The GEO site
GEO FAQ
List GEO Contents
Entrez
Help | FAQ
All: 2 DataSets: 0 Platforms: 2 Series: 0
Items 1 - 2 of 2
One page.
1: GPL11 record: SAGE:10:NlaIII:Mus musculus [Mus musculus]
Links
Summary: (Submitter supplied)
29 related Series
2: GPL4 record: SAGE:10:NlaIII:Homo sapiens [Homo sapiens]
Links
Summary: (Submitter supplied)
35 related Series
Disclaimer | Write to the Help Desk
NCBI | NLM | NIH

FIGURE 6.7 GEO Dataset records: platform.

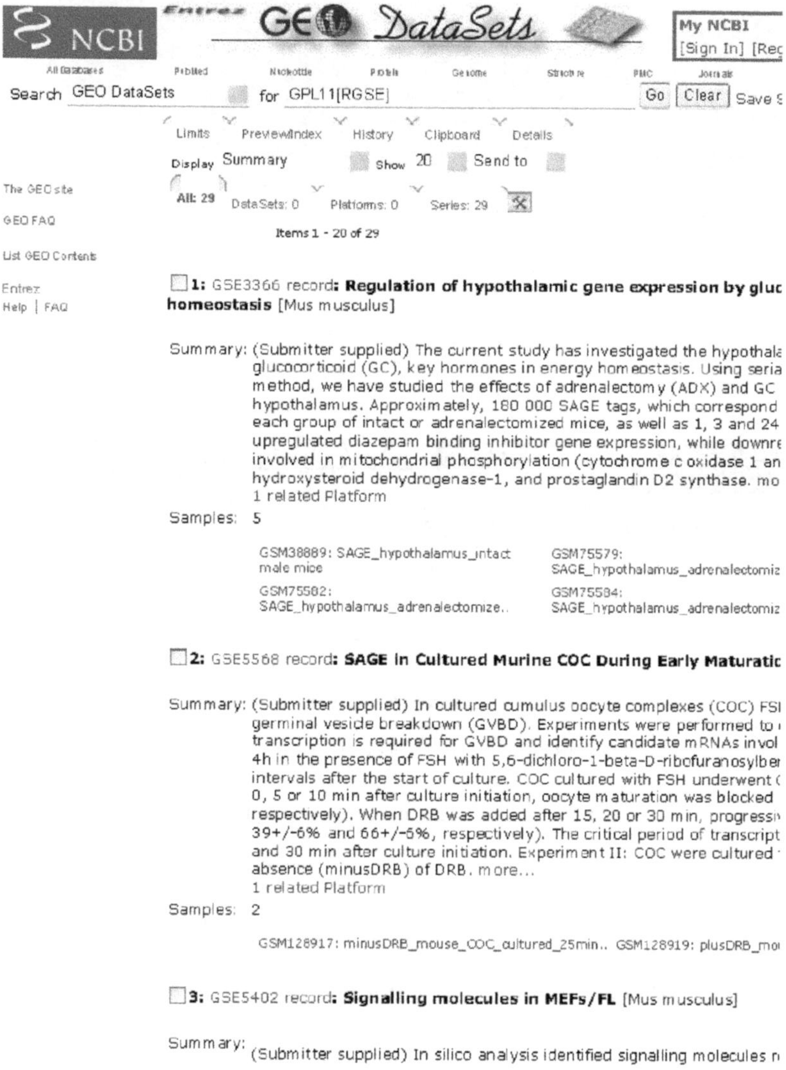
NCBI
Entrez GEO DataSets
My NCBI
[Sign In] [Reg
All Databases
PubMed
Nucleotide
Protein
Genome
Structure
PMC
Journals
Search GEO DataSets for GPL11[RGSE]
Go
Clear
Save S
Limits
Preview/Index
History
Clipboard
Details
Display Summary
Show 20
Send to
The GEO site
GEO FAQ
List GEO Contents
Entrez
Help | FAQ
All: 29
DataSets: 0
Platforms: 0
Series: 29
Items 1 - 20 of 29
1: GSE3366 record: Regulation of hypothalamic gene expression by gluc
homeostasis [Mus musculus]
Summary: (Submitter supplied) The current study has investigated the hypothala
glucocorticoid (GC), key hormones in energy homeostasis. Using seria
method, we have studied the effects of adrenalectomy (ADX) and GC
hypothalamus. Approximately, 180 000 SAGE tags, which correspond
each group of intact or adrenalectomized mice, as well as 1, 3 and 24
upregulated diazepam binding inhibitor gene expression, while downre
involved in mitochondrial phosphorylation (cytochrome c oxidase 1 an
hydroxysteroid dehydrogenase-1, and prostaglandin D2 synthase. mo
1 related Platform
Samples: 5
GSM38889: SAGE_hypothalamus_intact male mice
GSM75579: SAGE_hypothalamus_adrenalectomiz
GSM75582: SAGE_hypothalamus_adrenalectomize..
GSM75584: SAGE_hypothalamus_adrenalectomiz
2: GSE5568 record: SAGE in Cultured Murine COC During Early Maturatic
Summary: (Submitter supplied) In cultured cumulus oocyte complexes (COC) FSI
germinal vesicle breakdown (GVBD). Experiments were performed to
transcription is required for GVBD and identify candidate mRNAs invol
4h in the presence of FSH with 5,6-dichloro-1-beta-D-ribofuranosylbe
intervals after the start of culture. COC cultured with FSH underwent
0, 5 or 10 min after culture initiation, oocyte maturation was blocked
respectively). When DRB was added after 15, 20 or 30 min, progressi
39+/-6% and 66+/-5%, respectively). The critical period of transcript
and 30 min after culture initiation. Experiment II: COC were cultured
absence (minusDRB) of DRB. more...
1 related Platform
Samples: 2
GSM128917: minusDRB_mouse_COC_cultured_25min.. GSM128919: plusDRB_mo
3: GSE5402 record: Signalling molecules in MEFs/FL [Mus musculus]
Summary: (Submitter supplied) In silico analysis identified signalling molecules r

FIGURE 6.8A (a) and (b): GEO Dataset records: detail of series.

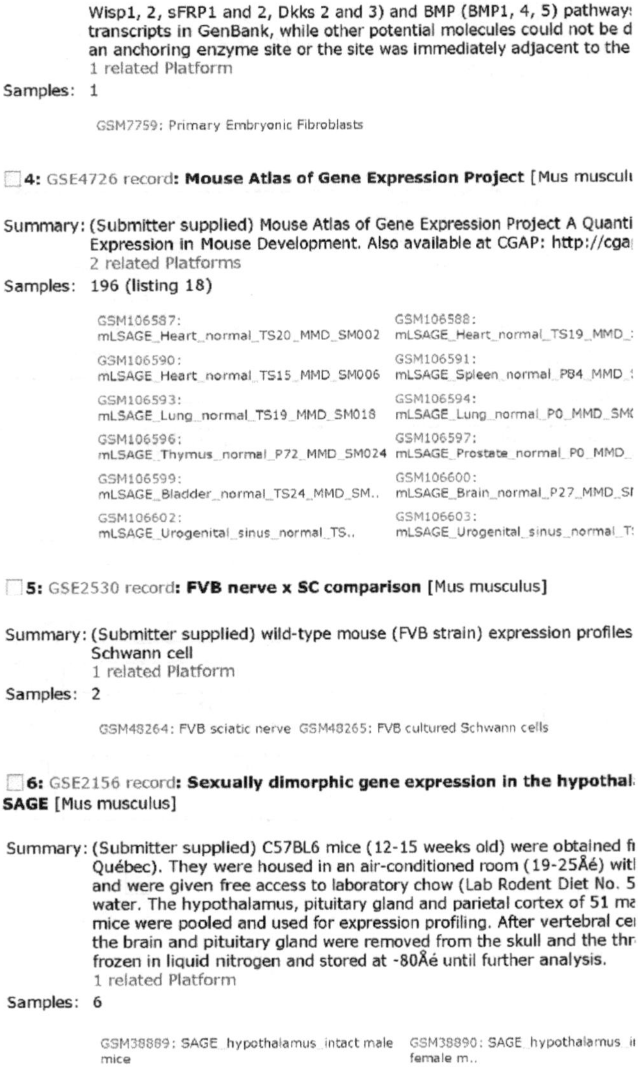
Wisp1, 2, sFRP1 and 2, Dkks 2 and 3) and BMP (BMP1, 4, 5) pathway
transcripts in GenBank, while other potential molecules could not be d
an anchoring enzyme site or the site was immediately adjacent to the
1 related Platform

Samples: 1

GSM7759: Primary Embryonic Fibroblasts

4: GSE4726 record**: Mouse Atlas of Gene Expression Project** [Mus muscul

Summary: (Submitter supplied) Mouse Atlas of Gene Expression Project A Quanti
Expression in Mouse Development. Also available at CGAP: http://cga
2 related Platforms

Samples: 196 (listing 18)

GSM106587: mLSAGE_Heart_normal_TS20_MMD_SM002	GSM106588: mLSAGE_Heart_normal_TS19_MMD_
GSM106590: mLSAGE_Heart_normal_TS15_MMD_SM006	GSM106591: mLSAGE_Spleen_normal_P84_MMD_
GSM106593: mLSAGE_Lung_normal_TS19_MMD_SM018	GSM106594: mLSAGE_Lung_normal_P0_MMD_SM
GSM106596: mLSAGE_Thymus_normal_P72_MMD_SM024	GSM106597: mLSAGE_Prostate_normal_P0_MMD_
GSM106599: mLSAGE_Bladder_normal_TS24_MMD_SM..	GSM106600: mLSAGE_Brain_normal_P27_MMD_S
GSM106602: mLSAGE_Urogenital_sinus_normal_TS..	GSM106603: mLSAGE_Urogenital_sinus_normal_T

5: GSE2530 record**: FVB nerve x SC comparison** [Mus musculus]

Summary: (Submitter supplied) wild-type mouse (FVB strain) expression profiles
Schwann cell
1 related Platform

Samples: 2

GSM48264: FVB sciatic nerve GSM48265: FVB cultured Schwann cells

6: GSE2156 record**: Sexually dimorphic gene expression in the hypothal**
SAGE [Mus musculus]

Summary: (Submitter supplied) C57BL6 mice (12-15 weeks old) were obtained f
Québec). They were housed in an air-conditioned room (19-25Åé) wit
and were given free access to laboratory chow (Lab Rodent Diet No. 5
water. The hypothalamus, pituitary gland and parietal cortex of 51 ma
mice were pooled and used for expression profiling. After vertebral ce
the brain and pituitary gland were removed from the skull and the th
frozen in liquid nitrogen and stored at -80Åé until further analysis.
1 related Platform

Samples: 6

GSM38889: SAGE_hypothalamus_intact male mice GSM38890: SAGE_hypothalamus_i female m..

FIGURE 6.8B

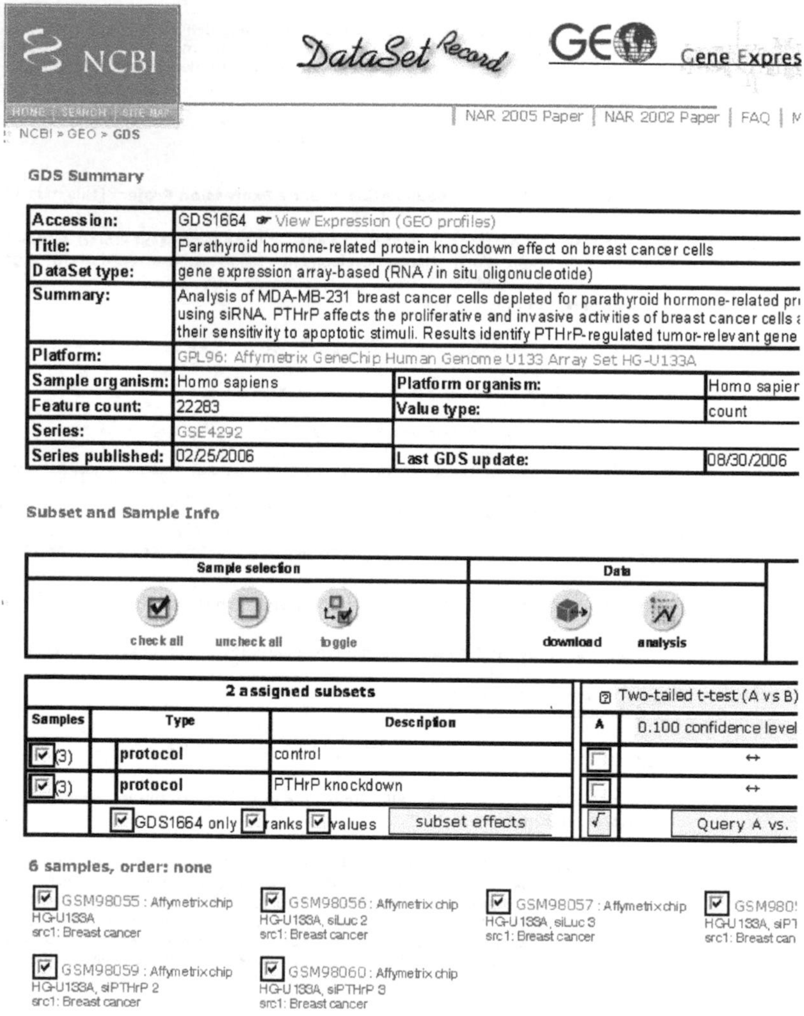
NCBI
HOME | SEARCH | SITE MAP
DataSet Record
GEO
Gene Expres
NAR 2005 Paper | NAR 2002 Paper | FAQ |
NCBI » GEO » GDS
GDS Summary
Accession: GDS1664 View Expression (GEO profiles)
Title: Parathyroid hormone-related protein knockdown effect on breast cancer cells
DataSet type: gene expression array-based (RNA / in situ oligonucleotide)
Summary: Analysis of MDA-MB-231 breast cancer cells depleted for parathyroid hormone-related pr
using siRNA. PTHrP affects the proliferative and invasive activities of breast cancer cells
their sensitivity to apoptotic stimuli. Results identify PTHrP-regulated tumor-relevant gene
Platform: GPL96: Affymetrix GeneChip Human Genome U133 Array Set HG-U133A
Sample organism: Homo sapiens
Platform organism: Homo sapier
Feature count: 22283
Value type: count
Series: GSE4292
Series published: 02/25/2006
Last GDS update: 08/30/2006
Subset and Sample Info
Sample selection
check all
uncheck all
toggle
Data
download
analysis
2 assigned subsets
Samples
Type
Description
(3) protocol control
(3) protocol PTHrP knockdown
GDS1664 only ranks values subset effects
Two-tailed t-test (A vs B)
A
0.100 confidence level
Query A vs.
6 samples, order: none
GSM98055 : Affymetrix chip HG-U133A src1: Breast cancer
GSM98056 : Affymetrix chip HG-U133A, siLuc 2 src1: Breast cancer
GSM98057 : Affymetrix chip HG-U133A, siLuc 3 src1: Breast cancer
GSM980 HG-U133A, siP src1: Breast can
GSM98059 : Affymetrix chip HG-U133A, siPTHrP 2 src1: Breast cancer
GSM98060 : Affymetrix chip HG-U133A, siPTHrP 3 src1: Breast cancer

FIGURE 6.9 Dataset Records.

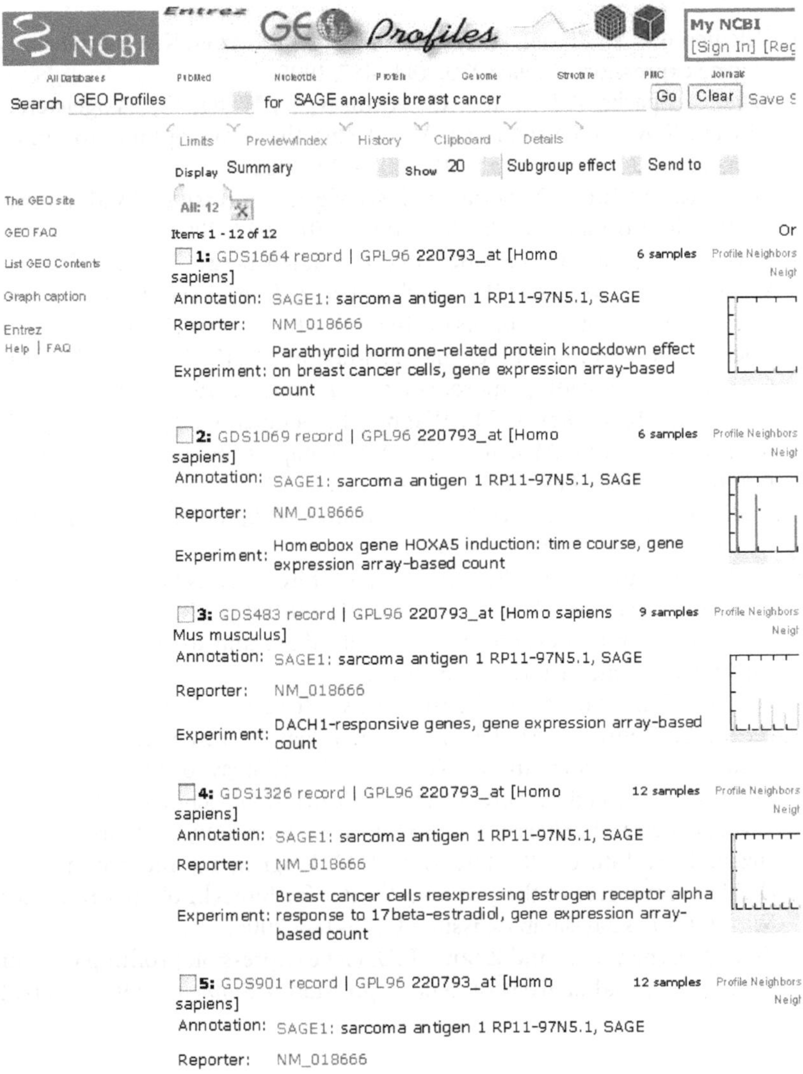
Entrez GEO Profiles
NCBI
My NCBI
[Sign In] [Reg
All Databases PubMed Nucleotide Protein Genome Structure PMC Journals
Search GEO Profiles for SAGE analysis breast cancer Go Clear Save
Limits Preview/Index History Clipboard Details
Display Summary Show 20 Subgroup effect Send to
The GEO site
GEO FAQ
List GEO Contents
Graph caption
Entrez
Help | FAQ
All: 12
Items 1 - 12 of 12
1: GDS1664 record | GPL96 220793_at [Homo sapiens] 6 samples Profile Neighbors
Annotation: SAGE1: sarcoma antigen 1 RP11-97N5.1, SAGE
Reporter: NM_018666
Experiment: Parathyroid hormone-related protein knockdown effect on breast cancer cells, gene expression array-based count
2: GDS1069 record | GPL96 220793_at [Homo sapiens] 6 samples Profile Neighbors
Annotation: SAGE1: sarcoma antigen 1 RP11-97N5.1, SAGE
Reporter: NM_018666
Experiment: Homeobox gene HOXA5 induction: time course, gene expression array-based count
3: GDS483 record | GPL96 220793_at [Homo sapiens Mus musculus] 9 samples Profile Neighbors
Annotation: SAGE1: sarcoma antigen 1 RP11-97N5.1, SAGE
Reporter: NM_018666
Experiment: DACH1-responsive genes, gene expression array-based count
4: GDS1326 record | GPL96 220793_at [Homo sapiens] 12 samples Profile Neighbors
Annotation: SAGE1: sarcoma antigen 1 RP11-97N5.1, SAGE
Reporter: NM_018666
Experiment: Breast cancer cells reexpressing estrogen receptor alpha response to 17beta-estradiol, gene expression array-based count
5: GDS901 record | GPL96 220793_at [Homo sapiens] 12 samples Profile Neighbors
Annotation: SAGE1: sarcoma antigen 1 RP11-97N5.1, SAGE
Reporter: NM_018666

FIGURE 6.10 GEO profile details.

REFERENCES

1. Velculescu, V.E., Zhang, L., Vogelstein, B., and Kinzler, K.W. Serial analysis of gene expression. *Science* 270: 484–487, 1995.
2. Saha, S., Sparks, A.B., Rago, C., Akmaev, V., Wang, C.J., Vogelstein, B., Kinzler, K.W., and Velculescu, V.E. Using the transcriptome to annotate the genome. *Nat Biotechnol* 20(5): 508–512, 2002.
3. Tuteja, R. and Tuteja, N. Serial analysis of gene expression (SAGE): unraveling the bioinformatics tools. *Bioessays* 26: 916–922, 2004.
4. Datson, N.A., van der Perk-de Jong, J., vanden Berg, M.P., de Kloet, E.R., Vreugdenhil, E. MicroSAGE: a modified procedure for serial analysis of gene expression in limited amounts of tissue. *Nucl Acids Res* 27: 1300–1307, 1999.
5. Cheung, V.G., Morley, M., Aguilar, F., Massimi, A., Kucherlapati, R., Childs, G., Making and reading microarrays. *Nat Genet* 21: 15–19, 1999.
6. Lash, A.E., Tolstoshev, C.M., Wagner, L., Schuler, G.D., Strausberg, R.L., Riggins, G.J., and Altschul, S.F., SAGEmap: a public gene expression resource. *Genome Res* 7: 1051–1060, 2000.
7. Audic, S. and Claverie, J.-M. The significance of digital gene expression profiles. *Genome Res* 7: 986–995, 1997.
8. Barrett, T., Suzek, T.O., Troup, D.B., Wilhite, S.E., Ngau, W.C., Ledoux, P., Rudnev, D., Lash, A.E., Fujibuchi, W., and Edgar, R. NCBI GEO: mining millions of expression profiles — database and tools. *Nucl Acids Res* 33(Database issue): D562–D566, 2005.
9. Wheeler, D.L., Barrett, T., Benson, D.A., Bryant, S.H., Canese, K., Chetvernin, V., Church, D.M., DiCuccio, M., Edgar, R., Federhen, S., Geer, L.Y., Helmberg, W., Kapustin, Y., Kenton, D.L., Khovayko, O., Lipman, D.J., Madden, T.L., Maglott, D.R., Ostell, J., Pruitt, K.D., Schuler, G.D., Schriml, L.M., Sequeira, E., Sherry, S.T., Sirotkin, K., Souvorov, A., Starchenko, G., Suzek, T.O., Tatusov, R., Tatusova, T.A., Wagner, L., and Yaschenko, E. Database resources of the National Center for Biotechnology Information. *Nucl Acids Res* 34(Database issue): D173–180, 2006.
10. Ye, S.Q., Usher, D.C., and Zhang, L.Q. Gene expression profiling of human diseases by serial analysis of gene expression. *J Biomed Sci* 9(5): 384–394, 2002.

CHAPTER 7

Regulation of Gene Expression

Xiao-Lian Zhang and Fang Zheng

CONTENTS

The regulation of gene expression is important for any biological process. This chapter introduces several major aspects of regulation on gene transcription and expression, including alternative promoters, alternative splicing, alternative translational initiation, and RNA editing. In each section, a brief theoretical description of the topic will be followed by a step-by-step tutorial on how to retrieve and analyze curated data using the currently available bioinformatics tools.

SECTION 1 ALTERNATIVE PROMOTERS

Part I Introduction

1. What Are Alternative Promoters?

Promoters are modulatory DNA sequences around a transcription initiation site including a complex array of cis-acting regulatory elements required for efficient and accurate initiation of transcription and for controlling expression of a gene. Alternative promoters are special promoters that can lead to significant variation and complexity in the transcription. The transcription initiation levels can be different among alternative promoters and the turnover rate or translation efficiency of mRNA isoforms with different leader exons and promoters can be different. Alternative promoters can have diverse tissue specificity and react differently to signals, and the alternative usage of promoters can lead to the generation of protein isoforms varying at the amino terminus.

2. Why Are Alternative Promoters Important?

Promoters play an important role in regulating transcriptional initiation and efficiency. Emerging evidence suggests that a considerable fraction of human genes probably have alternative promoters. The mammalian genes use alternative promoters, each subjected to different regulatory factors, to regulate and expand their transcriptional and translational potential.

Alternative promoters create elaborate regulations of gene expression in different tissues, cell types, and developmental stages. The product transcripts may encode diverse protein isoforms, or may vary only in their 5' untranslated regions, affecting mRNA stability and the efficiency of translation. Stringency regulation is critical for accurate gene function, and loss of this control may have serious phenotypic effects. There are several examples of diseases associated with alternative promoter usage; for example, in a number of cancers, promoters are specifically activated.

So, a comprehensive description of the transcriptase of cells is the foundation of a complete understanding of the complexities of disease phenotypes, proteome, and the regulation network.

3. What Is the Current Status of Research on Alternative Promoters?

So far no genomewide analysis has been carried out to specifically address the prevalence of alternative promoter usage in mammals. One study analyzed orthologous sequences in both mouse and human genomes, and it was found that 9% of the mouse genes had alternative first exons. On analyzing 152 putative promoter regions in a database of full-length human transcripts generated by the Mammalian Gene Collection, it was noted that 28 of the genes examined contained alternative transcription initiation sites, separated by a sequence more than 500 bp, and these were possible alternative promoters. Later on, 67,000 human transcripts from the NCBI LocusLink database (http://www.ncbi.nlm.nih.gov/LocusLink/) were analyzed, and the evidence suggests that more than 18% of all human genes use alternative promoters.

Typically, genes with alternative promoters contain two or more promoters producing transcripts with identical open reading frames (ORFs). No variation in the product proteins of these genes has been observed.

In these genes, the mRNAs have alternative leader exons with the same downstream exons and identical ORFs. Although these genes produce no protein isoforms, the mRNA variants differ in their transcription patterns and translation efficiencies. Because the resulting protein remains unchanged, variances in different tissue or developmental stages are the main consequences in these cases. In more complex situations, besides transcriptional differences, alternative promoter usage can lead to changed N-termini of proteins or create different ORFs, although the latter possibility is quite rare.

Alternative promoters have different tissue specificity, different developmental activity and expression levels or the variant untranslated regions (UTRs). A common example of a human gene with tissue-specific expression using alternative promoters is the CYP19 gene. The murine Ly49 multigene family, members of which are expressed on the surface of natural killer (NK) cells, is another interesting example of the use of different promoters in different developmental stages. Alternative promoters can also affect transcription procedures by the translation machinery. The p18 (INK4c) gene, a cyclin-dependent kinase inhibitor, illustrates this effect. The human neuronal isoform of the nitric oxide synthase gene (NOS1) is a striking example of multiple promoters, with nine alternative first exons exhibiting variances in tissue specificity and translation efficiency.

The aforementioned examples are of usage of alternative promoters in genes that produce no protein isoforms. Actually, there are other usages of alterative promoters leading to different protein forms with N-termini, some even resulting in a truncated protein such as *p73*. *p73* is a member of the tumor suppressor gene *p53* family.

There are rare cases in which alternative promoters produce different proteins through either alternative reading frames or splicing variation to create new ORFs. The well-documented example is the cyclin-dependent kinase inhibitor 2A gene that influences the activity of *p53*.

The usage of alternative promoters is highly common in mammalian genomes, and these promoters play an important role in organisms. The alternative promoters create diversity at the levels of transcriptome and proteome, and we think this is one of the key mechanisms for organism complexity.

Further research will focus on development of computational tools for annotating experimentally known alternative promoters and exons, conduction of chromatin immunoprecipitation microarray (ChIP-on-chip) and luciferase assays to confirm computationally annotated alternative promoter sequences of human and mouse orthologous genes, and developing computational methods to detect alternative promoters and exons in the human and mouse genomes.

Part II Step-By-Step Tutorial

In this part, we demonstrate how to check whether a human epidermal growth factor receptor (HS_EGFR) utilizes alternative promoters from a Eukaryotic Promoter Database (EPD, http://www.epd.isb-sib.ch).

1. Browsing the List of Alternative Promoters in the EPD

EPD is a compendium of eukaryotic POL II promoters for which the transcription start sites have been examined in experiments. To browse the list of alternative promoters collected in the current release of the EPD, go to the home page by typing the address http://www.epd.isb-sib.ch/ and click the hyperlink "List of alternative promoters" under "Documents." The entire current collection will be displayed. As of December 2, 2006, 220 alternative promoters are present in the database. The first three lines of the list are shown in Figure 7.1. In each collection, there are alternative promoter identification code, independent subset status, accession number, promoter name, and homology status.

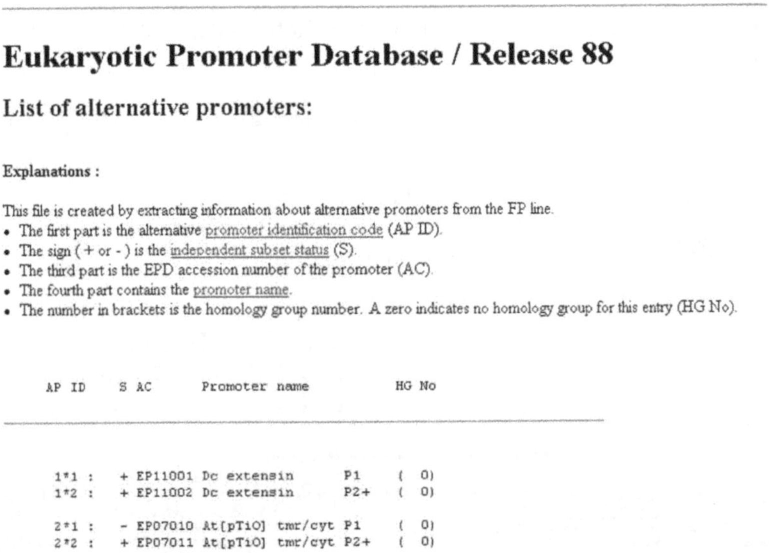

Eukaryotic Promoter Database / Release 88

List of alternative promoters:

Explanations :

This file is created by extracting information about alternative promoters from the FP line.

- The first part is the alternative promoter identification code (AP ID).
- The sign (+ or -) is the independent subset status (S).
- The third part is the EPD accession number of the promoter (AC).
- The fourth part contains the promoter name.
- The number in brackets is the homology group number. A zero indicates no homology group for this entry (HG No).

```
AP ID    S AC      Promoter name          HG No

1*1 :   + EP11001 Dc extensin       P1    (  0)
1*2 :   + EP11002 Dc extensin       P2+   (  0)

2*1 :   - EP07010 At[pTiO] tmr/cyt P1    (  0)
2*2 :   + EP07011 At[pTiO] tmr/cyt P2+   (  0)
```

FIGURE 7.1 Introduction of eukaryotic promoter database.

2. Go to the Search Window of the EPD

Either directly type the following address (http://www.epd.isb-sib.ch) in your browser or simply click the hyperlink EPD home page as displayed in Figure 7.1 to get access to the search window of the EPD as shown in Figure 7.2.

3. Search for Alternative Promoters of HS_EGFR

In the search window by the "Quick Search," type "HS_EGFR." Click "Quick Search" in your browser or press "Enter" on your keyboard. Two alternative promoters, HS_EGFR_1 and HS_EGFR_2, for this particular gene will be displayed as shown in Figure 7.3. Click the hyperlinks to either HS_EGFR_1 or HS_EGFR_2 to obtain the detailed information on each promoter, respectively. This information includes: ID — Identification, AC — Accession number, DT — DaTe, DE — Description, OS — Organism Species, HG — Homology Group, AP — Alternative Promoter, NP — Neighboring Promoter, DR — Database Cross-References, RN — Reference Number, RX — Reference Cross-References, RA — Reference Authors, RT — Reference Title, RL — Reference Location, ME — Methods, SE — Sequence, FL — Full Length, IF — Initiation Frequency, TX — TaXonomy, KW — KeyWords, FP — Functional Position, DO — Documentation, RF — Literature Reference, and // — Termination Line.

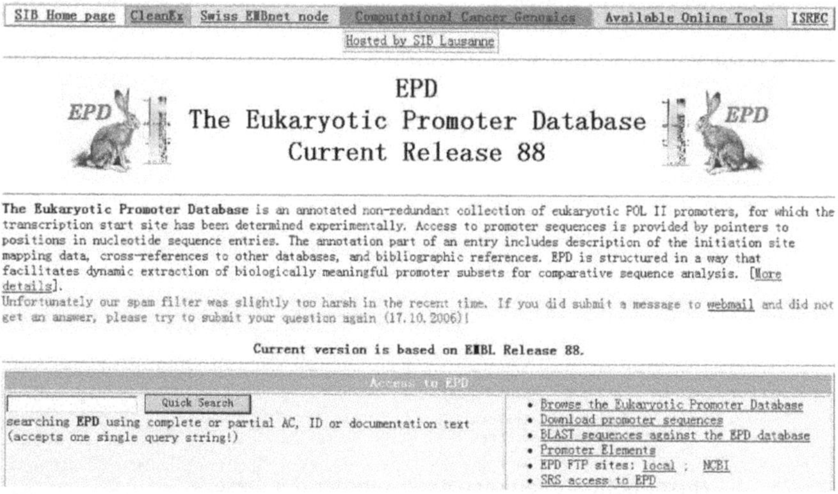
SIB Home page
CleanEx
Swiss EMBnet node
Computational Cancer Genomics
Available Online Tools
ISREC
Hosted by SIB Lausanne
EPD
EPD
EPD
The Eukaryotic Promoter Database
Current Release 88
The Eukaryotic Promoter Database is an annotated non-redundant collection of eukaryotic POL II promoters, for which the transcription start site has been determined experimentally. Access to promoter sequences is provided by pointers to positions in nucleotide sequence entries. The annotation part of an entry includes description of the initiation site mapping data, cross-references to other databases, and bibliographic references. EPD is structured in a way that facilitates dynamic extraction of biologically meaningful promoter subsets for comparative sequence analysis. [More details].
Unfortunately our spam filter was slightly too harsh in the recent time. If you did submit a message to webmail and did not get an answer, please try to submit your question again (17.10.2006)!
Current version is based on EMBL Release 88.
Access to EPD
Quick Search
searching EPD using complete or partial AC, ID or documentation text (accepts one single query string!)
Browse the Eukaryotic Promoter Database
Download promoter sequences
BLAST sequences against the EPD database
Promoter Elements
EPD FTP sites: local ; NCBI
SRS access to EPD

FIGURE 7.2 Home page of eukaryotic promoter database.

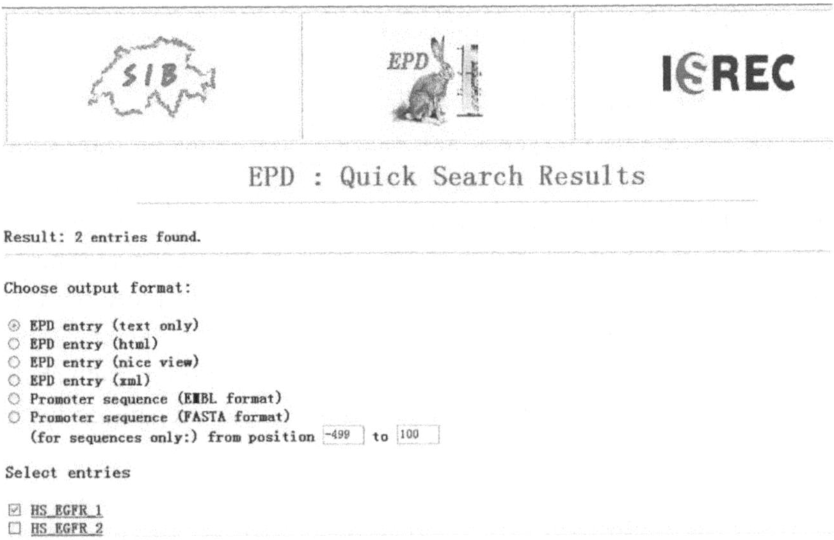
SIB
EPD
ISREC
EPD : Quick Search Results
Result: 2 entries found.
Choose output format:
EPD entry (text only)
EPD entry (html)
EPD entry (nice view)
EPD entry (xml)
Promoter sequence (EMBL format)
Promoter sequence (FASTA format)
(for sequences only:) from position -499 to 100
Select entries
HS_EGFR_1
HS_EGFR_2

FIGURE 7.3 Search result view of eukaryotic promoter database.

These results indicate that human epidermal growth factor receptor gene utilizes two alternative promoters. The consequences of these two different promoter usages in the EGFR gene are different protein isoforms. EGFR 1 protein is the product of alternative promoter #1, and the EGFR 2 protein is the product of alternative promoter #2. Further investigation of alternative promoter usage in the EGFR gene in different tissues, different development stages, and different conditions will help shed light on the functional roles of the EGFR gene in human health and disease.

For the search in the EPD database, it has to be pointed out that *ENTRY_NAME* is a unique entry identifier "HS_EGFR," which obeys rigorous naming principles. It contains two or three fields. The first is the species identification, and the second field is the protein code of SWISS-PROT ID used for gene identification. For human EPD entries, the official gene symbol approved by the HUGO nomenclature committee could be used instead of the SwissProt ID. The third field is optional. The "_" sign serves as a separator. Promoter entries are presented in a similar format as EMBL and SWISS-PROT sequence entries.

Part III Sample Data

The human epidermal growth factor receptor (HS_EGFR).

SECTION 2 ALTERNATIVE SPLICING

Part I Introduction

1. What Is Alternative Splicing?

Most of the eukaryotic genes are mosaic, consisting of intervening sequences separating the coding sequences. The coding sequences are exons, and the intervening sequences are introns. The process by which introns are removed from the pre-mRNA is called *RNA splicing*. Alternative splicing means some pre-mRNAs can be spliced in more than one way, generating alternative mRNAs. A cell can splice the "primary transcript" in different ways and thereby make different polypeptide chains from the same gene. About 60% of the human genes are spliced in this manner.

2. Why Is Alternative Splicing Important?

RNA splicing is very important for the maintenance of gene functions and regulation of the expression of certain genes. The mutations that affect RNA splicing can cause approximately 15% of all genetic diseases. The same pre-mRNA can be spliced variously in different tissues, cell types, and at different developmental stages, and react to various biological signals. So far only about 30,000 genes have been identified, compared with previous estimates of 100,000 or more. This indicates that one gene encodes more than one distinct mRNA; hence, more than one protein may play a critical role in expanding the function of our genomes. The usage of alternative splicing can switch on and off the expression of a certain gene. In this case, one functional protein is produced by a splicing event, and the nonfunctional proteins resulted from other splicing events. A single gene can produce several different kinds of proteins, when different splicing possibilities exist. In one extreme case, the Drosophila DSCAM gene can be spliced in 38,000 alternative ways. In another common case, alternative splicing of the fibronectin gene during its translation is a well-documented phenomenon and accounts at least in part for the known different forms of this protein. The regulation of RNA splicing can also generate protein isoforms in different tissues. The creatine kinase is a good example of a gene that generates specialized forms in different tissues.

Research on alternative splicing is very helpful for understanding the regulations of transcription and genome complexity, and for revealing the mechanisms of some genetic diseases related to alternative splicing.

3. What Is the Current Status of Research in Alternative Splicing?
Basically, sequences within the RNA determine where splicing occurs. The borders between introns and selected exons are marked by specific nucleotide sequences within the pre-mRNAs. The intron is removed in a form called a lariat as the flanking exons are joined. A process named *transsplicing*, in which two exons of different RNA molecules are spliced together, can fuse exons from different RNA molecules.

A large complex called *spliceosome* mediates the splicing. The complexes of snRNA and proteins are called *small nuclear ribonuclear proteins* (snRNPs). The spliceosome is the largest snRNP, and the exact makeup differs at different stages of the splicing reaction.

The splicing pathway includes assembly, rearrangement, and catalysis within the spliceosome. The spliceosome is composed of about 150 proteins and 5 small nuclear RNAs (snRNAs). Many functions of the spliceosome are executed by its RNA components. A small group of introns is spliced by a minor spliceosome. This spliceosome works on a minority of exons, and those have distinct splice-site sequences. The chemical pathway is the same as for the major spliceosome.

The observation of self-splicing introns suggests the RNA can catalyze RNA splicing. The self-splicing intron folds by itself into a specific conformation within the precursor RNA and catalyzes the chemistry of its own release and the exon ligation.

There are two ways for a spliceosome to find the splice sites reliably. In one way, the C-terminal tail of the RNA polymerase II carries genes of various splicing proteins, and cotranscriptional loading of these proteins to the newly synthesized RNA ensures all the splice sites emerging from RNA polymerase II are readily recognized, thus preventing splice site skipping. The other way is dependent on SR proteins, which are composed of splicing factors and regulators and can bind to the ESEs (exonic splicing enhancers) present in the exons and promote the use of the nearby splice sites by recruiting the splicing machinery to those sites.

Alternative splicing can be either constitutive or regulated. RNA splicing can be regulated either negatively, by repressors that prevent the splicing machinery from gaining access to a particular splice site on the RNA, or positively, by activators that help direct the splicing machinery to an otherwise overlooked splice site. There are two kind of regulating sequences, exonic (or intronic) splicing enhancers (ESE or ISE) or silencers (ESS and ISS). The former enhance and the latter repress splicing. Proteins that regulate splicing bind to these specific sites for their action. As we mentioned previously, SR proteins bind to enhancers and act as activators, and hnRNPs bind with RNA and act as repressors.

Thus, there are two outcomes of alternative splicing: (1) producing multiple isoform proteins and (2) regulating the expression level of a given gene with functional and nonfunctional proteins.

Part II Step-By-Step Tutorial

1. Databases of Alternative Splicing

EASED is an online available outline of alternative splice forms in certain organisms (*Arabidopsis thaliana*, *Bos taurus*, *Caenorhabditis elegans*, *Drosophila melanogaster*, *Danio rerio*, *Homo sapiens*, *Mus musculus*, *Rattus norvegicus*, and *Xenopus laevis*). Alternative splice forms are identified by comparison of high-scoring ESTs to mRNA sequences (both from GenBank) with known exon-intron information (from ENSEMBL database) using BLAST. The ends of each aligned sequence pair for deletions or insertions in the EST sequence are compared using filtering programs with defined parameters to reveal the existence of alternative splice usages.

2. Searching the Alternative Splicing Sites of a Given Gene

The following are the search steps:

1. Type the following address (http://eased.bioinf.mdc-berlin.de/) in your browser (Internet Explore or others) as shown in Figure 7.4.
2. You may use either a GenBank or EnsEMBL accession number or keywords to request the database. For instance, if you want to search

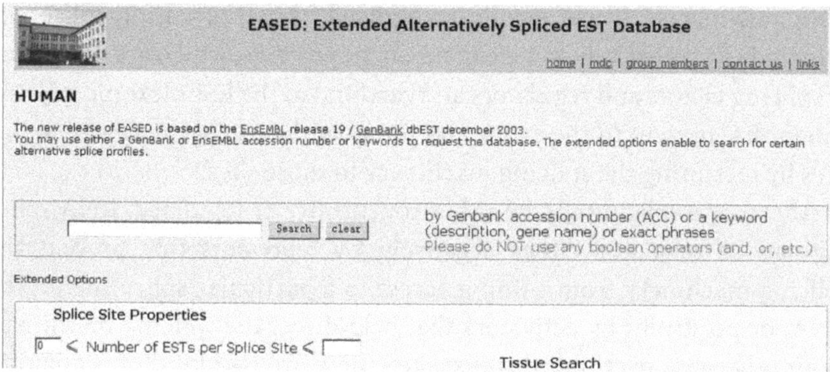
EASED: Extended Alternatively Spliced EST Database
home | mdc | group members | contact us | links
HUMAN
The new release of EASED is based on the EnsEMBL release 19 / GenBank dbEST december 2003.
You may use either a GenBank or EnsEMBL accession number or keywords to request the database. The extended options enable to search for certain alternative splice profiles.
Search
clear
by Genbank accession number (ACC) or a keyword
(description, gene name) or exact phrases
Please do NOT use any boolean operators (and, or, etc.)
Extended Options
Splice Site Properties
0
Number of ESTs per Splice Site
Tissue Search

FIGURE 7.4 Home page of extended alternatively spliced EST database.

the gene "Tranmembrane 4 Superfamily," then type "Tranmembrane 4 Superfamily" in the blank. Click "search" in your browser. The entry of alternative splicing related to Transmembrane 4 Superfamily gene will be displayed (Figure 7.5).

3. Click each entry in blue in your browser, for example "ENST00000003603," to display the information of this entry.

The result page is divided into four major parts. The first part (General Information about the Entry) summarizes the most important information as database id's, organism, and description (Figure 7.6).

The alternative splice profile (ASP) of each human sequence is demonstrated in the second part (Alternative Splice Frequency). ASP comprises the number of alternatively spliced ESTs (NAE), the number of constitutively spliced ESTs (NCE), the number of alternative splice sites (NSS) per mRNA, and the number of ESTs from cancerous tissues. Furthermore, the tissue type and developmental stages are shown in different colors to help the user understand the origins of the matching ESTs.

The third part, the so-called Splice Site View, illustrates all alternative splice sites for the whole transcript.

The fourth part was displayed in Figure 7.7 and named as the Splice Site Profile. The Splice Site Profile shows the parameters mentioned earlier, such as number of alternative spliced ESTs, number of constitutively spliced ESTs, and the number of cancer ESTs as well as the histological sources and developmental stages for each splice site. The accession numbers of the matching ESTs are also shown here. The classification of alternative splice events is categorized by the location of the HSP (high-scoring

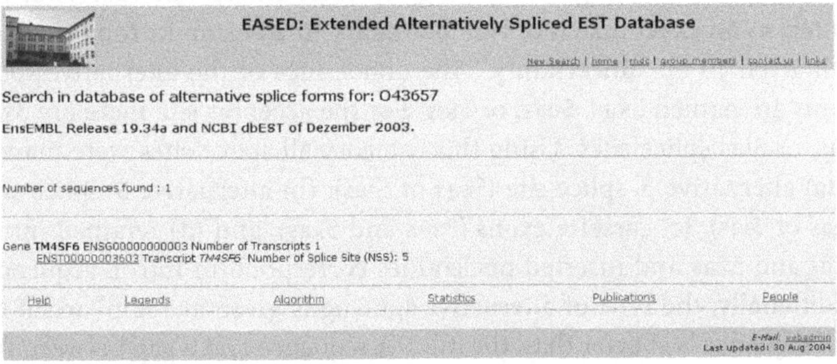
EASED: Extended Alternatively Spliced EST Database
Search in database of alternative splice forms for: O43657
EnsEMBL Release 19.34a and NCBI dbEST of Dezember 2003.
Number of sequences found : 1
Gene TM4SF6 ENSG00000000003 Number of Transcripts 1
ENST00000003603 Transcript TM4SF6 Number of Splice Site (NSS): 5
Help
Legends
Algorithm
Statistics
Publications
People
Last updated: 30 Aug 2004

FIGURE 7.5 Search result view of EASED.

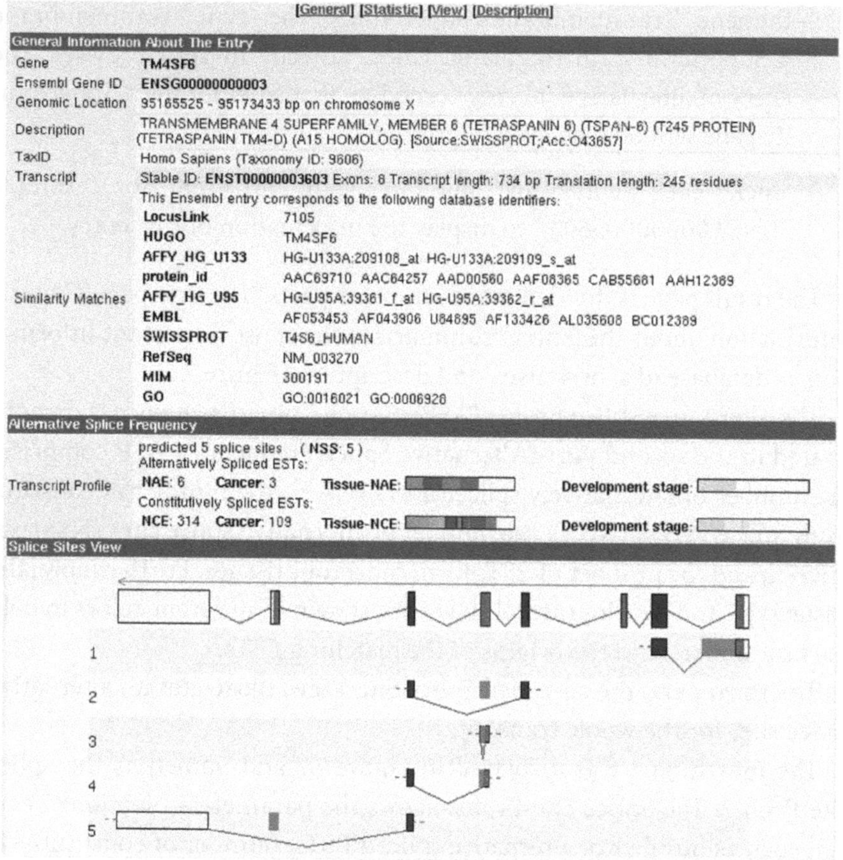
[General] [Statistic] [View] [Description]
General Information About The Entry
Gene TM4SF6
Ensembl Gene ID ENSG00000000003
Genomic Location 95165525 - 95173433 bp on chromosome X
Description TRANSMEMBRANE 4 SUPERFAMILY, MEMBER 6 (TETRASPANIN 6) (TSPAN-6) (T245 PROTEIN) (TETRASPANIN TM4-D) (A15 HOMOLOG). [Source:SWISSPROT;Acc:O43657]
TaxID Homo Sapiens (Taxonomy ID: 9606)
Transcript Stable ID: ENST00000003603 Exons: 8 Transcript length: 734 bp Translation length: 245 residues
This Ensembl entry corresponds to the following database identifiers:
LocusLink 7105
HUGO TM4SF6
AFFY_HG_U133 HG-U133A:209108_at HG-U133A:209109_s_at
protein_id AAC69710 AAC64257 AAD00560 AAF08365 CAB55661 AAH12389
Similarity Matches AFFY_HG_U95 HG-U95A:39361_f_at HG-U95A:39362_r_at
EMBL AF053453 AF043906 U84895 AF133426 AL035608 BC012389
SWISSPROT T4S6_HUMAN
RefSeq NM_003270
MIM 300191
GO GO:0016021 GO:0006926
Alternative Splice Frequency
predicted 5 splice sites (NSS: 5)
Alternatively Spliced ESTs:
Transcript Profile NAE: 6 Cancer: 3 Tissue-NAE: Development stage:
Constitutively Spliced ESTs:
NCE: 314 Cancer: 109 Tissue-NCE: Development stage:
Splice Sites View
1
2
3
4
5

FIGURE 7.6 The first part of the result page: general information about the entry.

pairs) boundaries compared to the given exon–intron boundaries. It was defined as an exact match of HSP boundary to an exon–intron boundary with given 10 bp "uncertainty." The donor sites of the alternative splice events are named 5xas, 5eas, or 5ias. For the acceptor site there are 3xas, 3eas, or 3ias splice sites. Using this category all splice sites were marked as (a) alternative 5' splice site (5eas or 5ias), (b) alternative 3' splice sites (3eas or 3ias), (c) cassette exons (3xas and 5xas), and (d) retained intron (3xas and 5xas and inserted nucleotides correspond to intron sequence). Additionally, the type of alternative splicing is given as "skip" when the EST sequence is shorter than the mRNA sequence and a gap between two HSPs is found on the mRNA. The type of alternative splicing is given as "insert" *vice versa*.

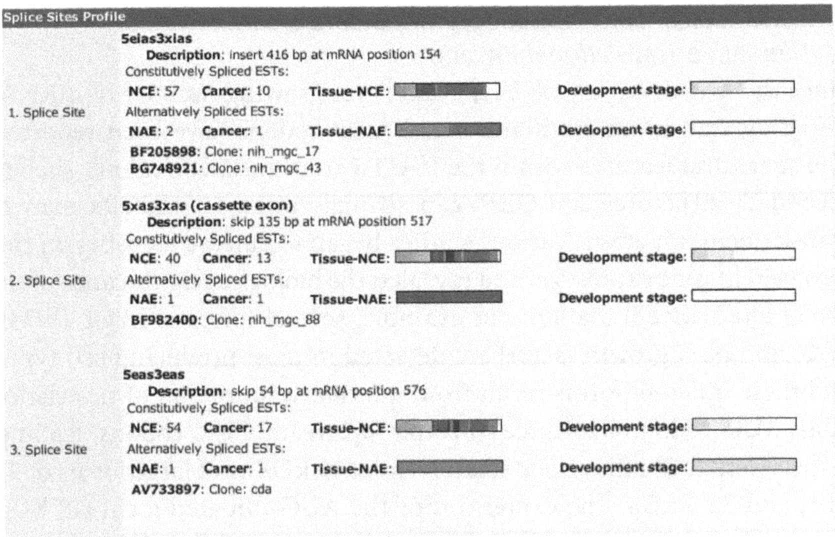
Splice Sites Profile
1. Splice Site
5eias3xias
Description: insert 416 bp at mRNA position 154
Constitutively Spliced ESTs:
NCE: 57 Cancer: 10 Tissue-NCE:
Development stage:
Alternatively Spliced ESTs:
NAE: 2 Cancer: 1 Tissue-NAE:
Development stage:
BF205898: Clone: nih_mgc_17
BG748921: Clone: nih_mgc_43
2. Splice Site
5xas3xas (cassette exon)
Description: skip 135 bp at mRNA position 517
Constitutively Spliced ESTs:
NCE: 40 Cancer: 13 Tissue-NCE:
Development stage:
Alternatively Spliced ESTs:
NAE: 1 Cancer: 1 Tissue-NAE:
Development stage:
BF982400: Clone: nih_mgc_88
3. Splice Site
5eas3eas
Description: skip 54 bp at mRNA position 576
Constitutively Spliced ESTs:
NCE: 54 Cancer: 17 Tissue-NCE:
Development stage:
Alternatively Spliced ESTs:
NAE: 1 Cancer: 1 Tissue-NAE:
Development stage:
AV733897: Clone: cda

FIGURE 7.7 The alternative splice profile, splice site view, and splice site profile of the entry.

Part III Sample Data

Transmembrane 4 Superfamily and its gene bank accession number (AC: O43657)

SECTION 3 ALTERNATIVE TRANSLATION INITIATION

Part I Introduction

1. What Is Alternative Translational Initiation?

In eukaryotic cells, translation is usually initiated according to the ribosome scanning model; that is, the 40S ribosomal subunit and translation initiation factors bind to the 5' end of mRNA (messenger RNA) and scan the RNA molecule in the 3' direction until they reach an AUG codon; then the 80S ribosome assembles and begins protein synthesis.

ATI (alternative translational initiation) is one of the mechanisms that increases the complexity level of an organism by alternative gene expression pathways. The use of ATI codons in a singe mRNA contributes to the generation of protein diversity. The genes produce two or more versions of the encoded proteins, and the shorter version, initiated from a downstream in-frame start codon, lacks the N-terminal amino acids fragment of the full-length isoform version.

2. Physiological and Pathological Implications of Alternative Translational Initiation

Since the first discovery of ATI, a small yet growing number of mRNAs initiating translation from alternative start codons have been reported. The structural features within the 5' UTR (untranslated region), such as secondary structures and ORFs can strongly influence the efficiency of translational initiation. Various studies began to emerge focusing on this new field in gene expression and revealed the biological significance of the use of alternative initiation. For example, several forms of bFGF (bFGF, basic fibroblast growth factor) are detected in most producing cell types, and these different forms result from alternative initiation of translation at an AUG codon or at three in-frame upstream CUG codons, leading, respectively, to synthesis of a small form of 19 kDa or of large forms of 21, 21.5, and 22.5 kDa. The expression of the AUG-initiated form (18 kDa) leads to cell transformation, whereas expression of the CUG-initiated forms leads to cell immortalization. In the translation of bFGF mRNA, five cis-acting elements located in the 5' UTR of bFGF mRNA are able to modulate the global or alternative use of the four initiation codons. As the cis-acting elements in mRNA could be the targets for specific trans-acting factors involved in cell growth and differentiation, control of the alternative initiation of translation of bFGF mRNA will have an important impact on cell behavior.

The glucocorticoid receptor gene, its products, and their actions represent a paradigm that the expression of different isoforms of any protein resulting from ATI could have physiological and pathological implications. Glucocorticoids interact with GRs (GRs, glucocorticoid receptors), through which they exert their effects. It was found that expression of about 20% of the expressed human leukocyte genome was positively or negatively affected by glucocorticoids. There is the report (Chrousos & Kino, 2005) that the GRα regulates expression of bL-Selection and CD11/CD18 on human neutrophils. Variant mRNA was translated from at least eight initiation sites into multiple GRα isoforms termed GRα.-A through GRα-D (A, B, C1 to C3, and D1 to D3). Recently, a convincing association was made between the ER22/23EK polymorphism of the human GR gene and increased human longevity secondary to a healthier metabolic profile. These polymorphisms were previously found to be associated with subtle glucocorticoid resistance. It was found that when the ER22/23EK polymorphism was present, about 15% more GRα-A protein was expressed than when it was absent, whereas total GR levels (GRα-A plus GRα-B) were not affected. These results suggested that transcriptional activity in GRα (ER22/23EK) carriers was decreased because more of the less transcriptionally active GRα-A isoform was formed. The underlying mechanism may be due to an altered secondary mRNA structure. Beyond the production of N-terminal isoforms conferring an additional important mechanism for regulation of GR actions, the mineralocorticoid, ERα (estrogen), and PRs (progesterone receptors) also contain potential alternative translation initiation sites in their N-terminal domains. Therefore, tissue-specific and regulated variable N-terminal isoform production may be a general mechanism that defines target tissue sensitivity to steroid hormones, further adding to the complexity of their own signal transduction systems.

Several other examples can be described here. LP-BM5 MuLV (MuLV, murine leukemia virus)-infected C57BL/6 mice develop profound immunodeficiency and B-cell lymphomas. Recent study suggests the existence of a novel ORF 2 products, with ATI downstream from normal ORF 1 product, that is required for LP-BM5-induced pathogenesis and has potentially broad implications for other retroviral diseases. An ATI at an in-frame internal AUG located three codons downstream of the stop codon mutation R37X, which caused the ATR-X (ATR-X, alpha-thalassemia/mental retardation syndrome), could rescue the phenotype and is associated with levels of ATRX protein that are up to 20% of those seen in the wild type. The synthesis of a foot-and-mouth disease virus initiates at two start codons located 84 nucleotides apart in the same reading frame, which leads to the synthesis of two alternative N-terminal processing products of the viral polyprotein, the leader protein L and L'.

3. Databases on Alternative Translational Initiation

ATID (Alternative Translational Initiation Database, http://166.111.201.26/atie/) is an online resource that collects gene information, alternative products of genes, and domain structures of isoforms. The ATID database is available for public use at http://bioinfo.au.tsinghua.edu.cn/atie/. Supplementary instructions about this database and statistical analyses can also be found on the Web page. The records of alternative translational events are converted and stored in a MySQL database program (http://www.mysql.com). The topological structure of the FGF2 (fiber growth factor) gene in *Homo sapiens* as shown in Figure 7.8. There are four shorter isoforms lacking the N-terminal amino acids fragments from the original 30.8-kD full-length isoform (Figure 7.8).

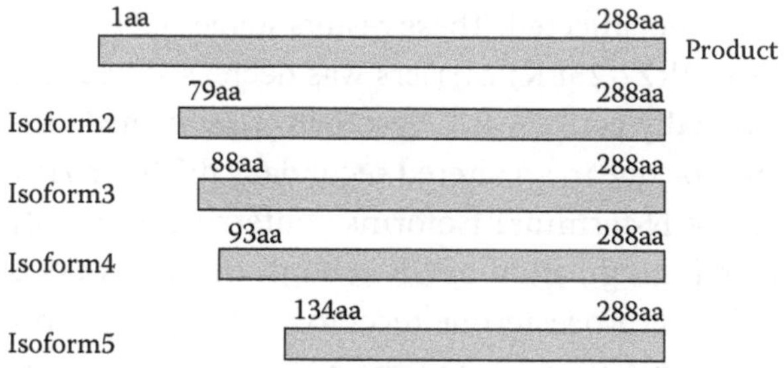
1aa
288aa
Product
79aa
288aa
Isoform2
88aa
288aa
Isoform3
93aa
288aa
Isoform4
134aa
288aa
Isoform5

FIGURE 7.8 ATID screenshots: Topological structure of five alternative isoforms initiated from the FGF2 gene.

In response to the need for systematic studies on genes involving ATI, ATID is established to provide data of publicly available genes, alternatively translational isoforms, and their detailed annotations. This database contains 650 alternatively translated variants assigned to a total of 300 genes. These database records of alternative ATI have been collected from publicly available protein databases, such as SWISS-PROT database and Entrez protein database protein database on NCBI (NCBI, National Center for Biotechnology Information) (http://www.ncbi.nlm.nih.gov/Entrez/). Additionally, the information of 89 ATI events extracted from the published research literature is also included in the database. The genes, contributing to alternative translational initiation, cover many species including *Homo sapiens*, *Mus musculus*, *Bostaurus*, *Saccharomyces cerevisiae*, *Virus*, etc.

Also, a large number of software packages (http://www.expasy.ch/tools/dna.html, or http://www.biology.utah.edu/jorgensen/wayned/ape/) for translating DNA sequences already exists. However, many of these fine tools do not support translating sequences containing degenerate nucleotides, have no or limited support for alternative translation tables (including alternative initiation codons), and in general have problems handling special situations. Virtual Ribosome is a new DNA translation tool (available at http://www.cbs.dtu.dk/services/VirtualRibosome/.) with two areas of focus: (1) full support for the IUPAC degenerate DNA alphabet and all translation tables defined by the NCBI taxonomy group, including the use of alternative start codons; and (2) support for working with files containing intron/exon structure annotation.

Part II Step-By-Step Tutorial

Browser ATID (Alternative Translational Initiation Database) via Web interface pages: (http://166.111.201.26/atie) is a Web-oriented database that uses a browser-based interface to access data under the SQL framework. The Web page allows interaction between the users and the data application. Indexing key identifiers of the database optimizes batch queries from the HTTP Web page interface. When a HTTP request is triggered, we can import items of an SQL database into the application. Then, the results are sent in HTML format on the Web.

1. Search the Alternative Translational Initiation of a Given Gene

The alternative translation initiation of the CRYBA1 gene has been examined in experiments. We will take CRYBA1 as an example:

1. Type following address (http://166.111.201.26/atie) in your browser, and you will find ATID as displayed in Figure 7.9. Then click "search." Figure 7.10 will be displayed. You can type different keywords (ATIE_id, or Gene Name) or input gene sequence; for example, type "CRYBA1" (GeneName), click "GO", and the CRYBA1 gene of

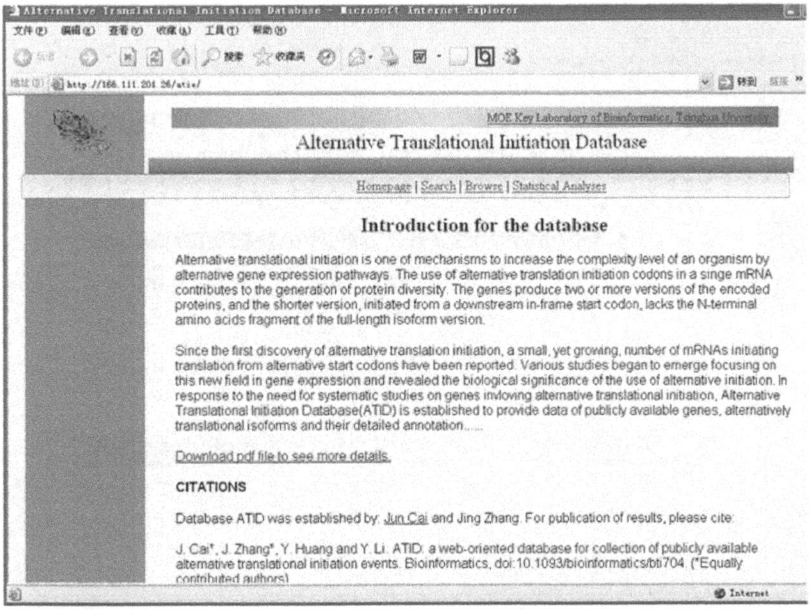
MOE Key Laboratory of Bioinformatics, Tsinghua University
Alternative Translational Initiation Database
Homepage | Search | Browse | Statistical Analyses
Introduction for the database
Alternative translational initiation is one of mechanisms to increase the complexity level of an organism by alternative gene expression pathways. The use of alternative translation initiation codons in a singe mRNA contributes to the generation of protein diversity. The genes produce two or more versions of the encoded proteins, and the shorter version, initiated from a downstream in-frame start codon, lacks the N-terminal amino acids fragment of the full-length isoform version.
Since the first discovery of alternative translation initiation, a small, yet growing, number of mRNAs initiating translation from alternative start codons have been reported. Various studies began to emerge focusing on this new field in gene expression and revealed the biological significance of the use of alternative initiation. In response to the need for systematic studies on genes invloving alternative translational initiation, Alternative Translational Initiation Database(ATID) is established to provide data of publicly available genes, alternatively translational isoforms and their detailed annotation......
Download pdf file to see more details.
CITATIONS
Database ATID was established by: Jun Cai and Jing Zhang. For publication of results, please cite:
J. Cai*, J. Zhang*, Y. Huang and Y. Li. ATID: a web-oriented database for collection of publicly available alternative translational initiation events. Bioinformatics, doi:10.1093/bioinformatics/bti704. (*Equally contributed authors)
Internet

FIGURE 7.9 Search the alternative translation initiation of CRYBA1 gene from the database in the Web site.

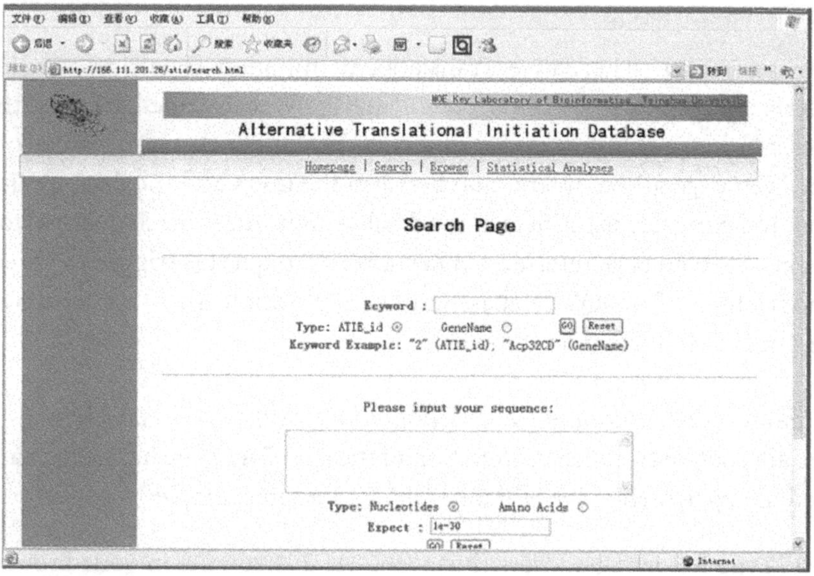
Alternative Translational Initiation Database
Homepage | Search | Browse | Statistical Analyses
Search Page
Keyword :
Type: ATIE_id GeneName
GO Reset
Keyword Example: "2" (ATIE_id); "Acp32CD" (GeneName)
Please input your sequence:
Type: Nucleotides Amino Acids
Expect : 1e-30

FIGURE 7.10 Step-by-step tutorial: keyword queries in alternative translation initiation database (ATID).

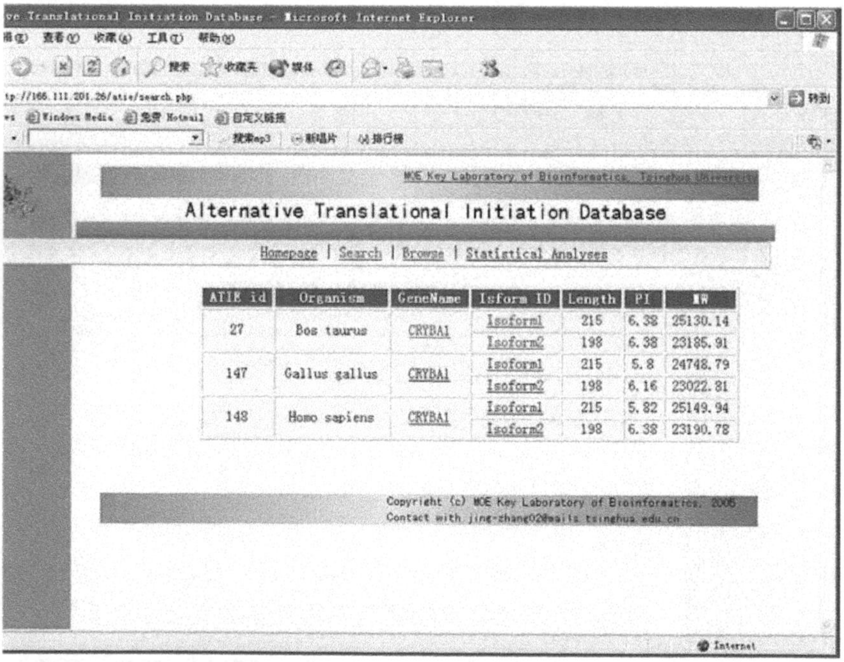
Alternative Translational Initiation Database
Homepage | Search | Browse | Statistical Analyses
ATIE id | Organism | GeneName | Isform ID | Length | PI | MW
27 | Bos taurus | CRYBA1 | Isoform1 | 215 | 6.38 | 25130.14
Isoform2 | 198 | 6.38 | 23185.91
147 | Gallus gallus | CRYBA1 | Isoform1 | 215 | 5.8 | 24748.79
Isoform2 | 198 | 6.16 | 23022.81
148 | Homo sapiens | CRYBA1 | Isoform1 | 215 | 5.82 | 25149.94
Isoform2 | 198 | 6.38 | 23190.78

FIGURE 7.11 Different isoforms (Isoform1 and Isoform2) of the CRYBA1 gene shown in ATID.

different organisms will be displayed. Their isoforms (Isoform1 and Isoform2) are displayed in Figure 7.11.

2. You can also find the alternative translation initiation of CRYBA1 in another way. Click "Browse" in the address (http://166.111.201.26/atie). The result will be displayed as Figure 7.12. Then click "Bos taurus" or" Homo sapiens" under "Mammalian," and you will find the CRYBA1 gene you searched (Figure 7.13). Now, you know how to search the ATI (alternative translation initiation) of different gene; when you find them (Figure 7.13), click the "gene name" and "gene description," references of NCBI database access numbers, and the literature on CRYBA1 gene in *Bostaurus* or *Homo sapiens* will be shown as Figure 7.14A. You can also click a specific region of isoform ID, such as "isoform1," and information on alternative translational products will be displayed. Elements such as point isoelectric (PI) value, molecular weight, domain contexts, and sequence information are designated to annotate the isoform products of genes; the linked Web page for isoform2 of CRYBA1 gene is shown in Figure 7.14B.

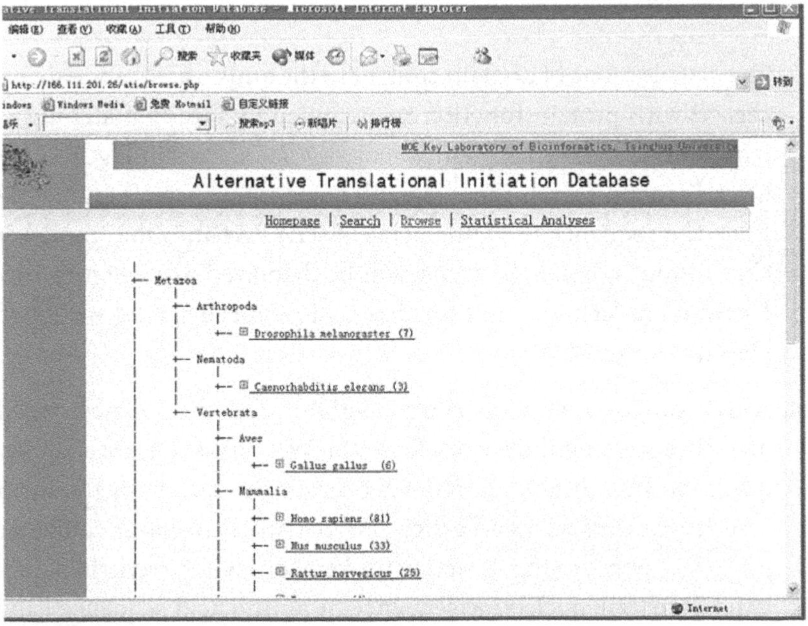
Alternative Translational Initiation Database
Homepage | Search | Browse | Statistical Analyses
Metazoa
Arthropoda
Drosophila melanogaster (7)
Nematoda
Caenorhabditis elegans (3)
Vertebrata
Aves
Gallus gallus (6)
Mammalia
Homo sapiens (81)
Mus musculus (33)
Rattus norvegicus (25)
Internet

FIGURE 7.12 Search the alternative translation initiation of CRYBA1 by browsing from the database in the Web site.

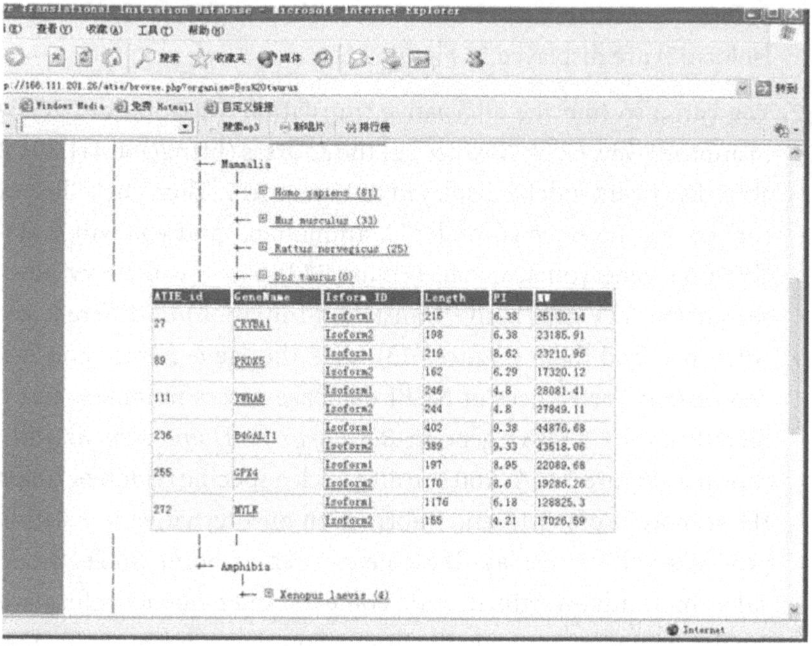
Mammalia
Homo sapiens (81)
Mus musculus (33)
Rattus norvegicus (25)
Bos taurus (6)
ATIE_id
GeneName
Isform ID
Length
PI
27
CRYBA1
Isoform1
215
6.38
25130.14
Isoform2
198
6.38
23185.91
89
PRDX5
Isoform1
219
8.62
23210.96
Isoform2
162
6.29
17320.12
111
YWHAB
Isoform1
246
4.8
28081.41
Isoform2
244
4.8
27849.11
236
B4GALT1
Isoform1
402
9.38
44876.68
Isoform2
389
9.33
43518.06
255
GPX4
Isoform1
197
8.95
22089.68
Isoform2
170
8.6
19286.26
272
MYLK
Isoform1
1176
6.18
128825.3
Isoform2
165
4.21
17026.59
Amphibia
Xenopus laevis (4)
Internet

FIGURE 7.13 Search the CRYBA1 gene from "*Bos taurus*" or "*Homo sapiens*" under "Mammalian."

3. The distributions of domain content in the amino acid sequence concerned with protein function are scanned by family matching system Pfam 21.0 at http://pfam.janelia.org/ (as shown in Figure 7.15A). Click "protein search" and you will enter a new page (Figure 7.15B). Paste the sequence of amino acids of CRYBA1 into the framework, click "inquire," and the result will be displayed as in Figure 7.15C. Lastly, click "crystal," and the description of the crystal will be displayed (Figure 7.15D).

4. ATID supports two querying methods for the entries, besides directly accessing the entries from the Web links. One is a keyword query, and the other is a sequence similarity query. In the former case, keywords such as accession number and gene name can be submitted. The record that contains the keyword will be returned. In the latter case, one nucleotide sequence or amino acid sequence can be submitted in FASTA format through a Web interface. The submitted sequence will be compared with the representative sequences in the database by BLAST (The Basic Local Alignment Search Tool) pro-

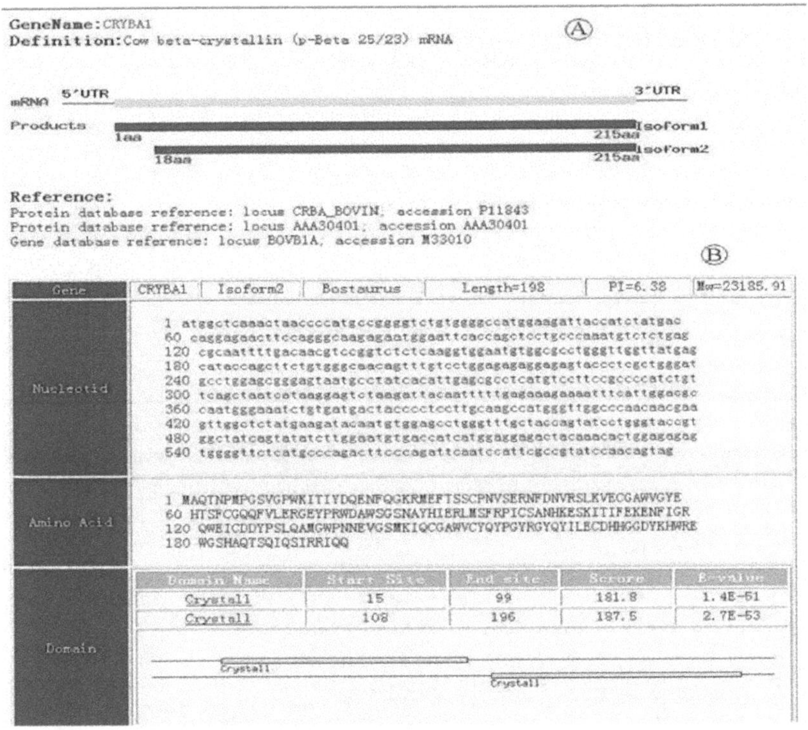

GeneName: CRYBA1
Definition: Cow beta-crystallin (p-Beta 25/23) mRNA
(A)
mRNA
5'UTR
3'UTR
Products
1aa
215aa
Isoform1
18aa
215aa
Isoform2
Reference:
Protein database reference: locus CRBA_BOVIN; accession P11843
Protein database reference: locus AAA30401; accession AAA30401
Gene database reference: locus BOVB1A; accession M33010
(B)
Gene
CRYBA1
Isoform2
Bostaurus
Length=198
PI=6.38
Mw=23185.91
Nucleotid
Amino Acid
Domain
Domain Name
Start Site
End site
Score
E-value
Crystall
15
99
181.8
1.4E-51
Crystall
108
196
187.5
2.7E-53

FIGURE 7.14 Step-by-step tutorial: (A) Gene description and references of NCBI database access numbers and literature of CRYBA1 gene in *Bostaurus* or *Homo sapiens*. (B) A specific region of isoform ID information on alternative translational products. Elements such as point isoelectric (PI) value, molecular weight, domain contexts, and sequence information are designated to annotate isoform2 of the CRYBA1 gene.

gram. The E-value parameter of the BLAST program is adjustable. After the query job is finished, the cluster of database entries that contains the most similar sequences are returned in table format. Detailed information on sequence matches, such as aligned starting sites and aligned ending positions, are given in the display of BLAST result. The Web pages of the database querying system and the returned results of a sequence query are demonstrated on the Web (http://166.111.201.26/atie). Figure 7.16A shows the Web pages of the database querying system, and the returned results of a sequence query are shown in Figure 7.16B.

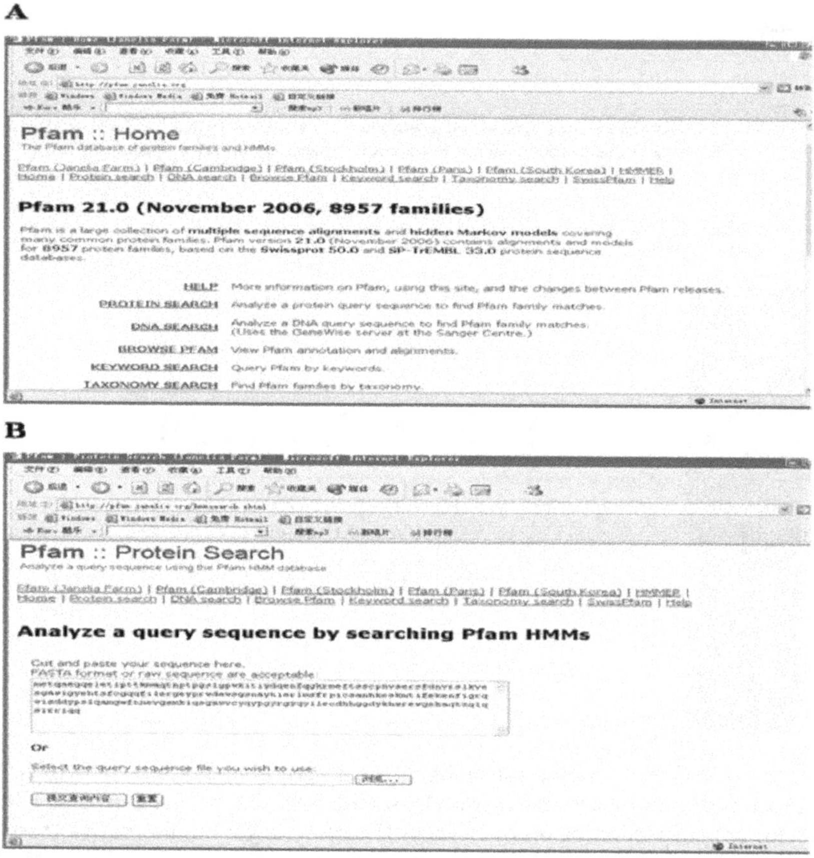
A
Pfam :: Home
Pfam 21.0 (November 2006, 8957 families)
HELP
PROTEIN SEARCH
DNA SEARCH
BROWSE PFAM
KEYWORD SEARCH
TAXONOMY SEARCH
B
Pfam :: Protein Search
Analyze a query sequence by searching Pfam HMMs

FIGURE 7.15 A-B Family matching system Pfam. (A) The distributions of domain content in the amino acid sequence concerned with protein function are scanned by family matching system Pfam 21.0 at http://pfam.janelia.org/. (B) Query sequence by searching Pfam.

Part III Sample Data

CRYBA1 gene: *Homo sapiens* mRNA for crystallin or cow beta-crystallin (p Beta 25/23) mRNA

SECTION 4 RNA EDITING

Part I Introduction

1. What Is RNA Editing?

The term *RNA editing* describes those molecular processes in which the information content is altered in an RNA molecule. RNA editings are posttranscriptional modification machineries that alter the RNA pri-

C

Pfam HMM search results, glocal+local alignments merged (Pfam_ls+Pfam_fs)
Model Seq-from Seq-to HMM-from HMM-to Score E-value Alignment Description
Alignments of top-scoring domains:

D

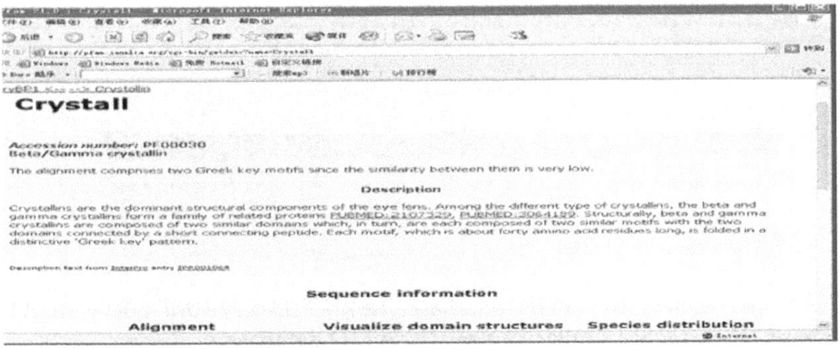
Crystall
Description
Sequence information
Alignment
Visualize domain structures
Species distribution

FIGURE 7.15 C-D Family matching system Pfam. (C) Sequence alignments. (D) Description of crystal.

mary sequence by base modifications, nucleotide insertions or deletions, and nucleotide replacements. After transcription, some RNA molecules are altered to contain bases not encoded in the genome. The diversity of RNA-editing mechanisms includes nucleoside modifications such as C-to-U and A-to-I deaminations, as well as nontemplated nucleotide additions and insertions. Most often this involves the editing or modification of one base to another, but in some organisms can involve the insertion or deletion of a base. RNA editing in mRNAs effectively alters the amino acid sequence of the encoded protein so that it differs from that predicted by the genomic DNA sequence. For example, in mammals, the apo-B(apolipoprotein-B) gene is expressed in liver as a 500-kD protein called Apo-B100, whereas in intestine cells its product is a smaller protein called Apo-B48. Apo-B100 is produced without RNA editing, but Apo-B48 is synthesized from an mRNA whose sequence has been altered by a specific enzyme. This enzyme changes a codon, CAA, in the middle of the original

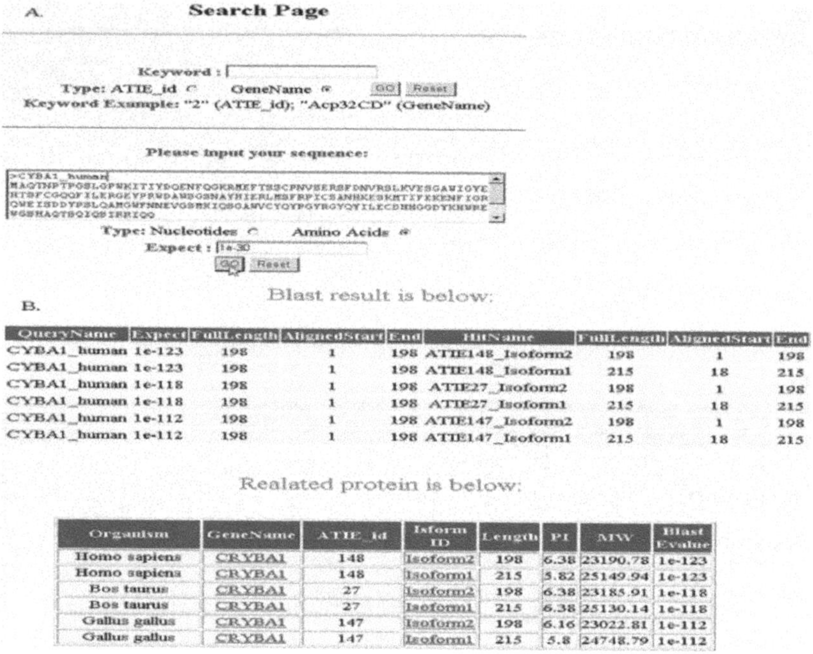
A.
Search Page
Keyword :
Type: ATIE_id GeneName
GO Reset
Keyword Example: "2" (ATIE_id); "Acp32CD" (GeneName)
Please input your sequence:
Type: Nucleotides Amino Acids
Expect :
GO Reset
B.
Blast result is below:
QueryName Expect FullLength AlignedStart End HitName FullLength AlignedStart End
CYBA1_human 1e-123 198 1 198 ATIE148_Isoform2 198 1 198
CYBA1_human 1e-123 198 1 198 ATIE148_Isoform1 215 18 215
CYBA1_human 1e-118 198 1 198 ATIE27_Isoform2 198 1 198
CYBA1_human 1e-118 198 1 198 ATIE27_Isoform1 215 18 215
CYBA1_human 1e-112 198 1 198 ATIE147_Isoform2 198 1 198
CYBA1_human 1e-112 198 1 198 ATIE147_Isoform1 215 18 215
Realated protein is below:
Organism GeneName ATIE_id Isform ID Length PI MW Blast Evalue
Homo sapiens CRYBA1 148 Isoform2 198 6.38 23190.78 1e-123
Homo sapiens CRYBA1 148 Isoform1 215 5.82 25149.94 1e-123
Bos taurus CRYBA1 27 Isoform2 198 6.38 23185.91 1e-118
Bos taurus CRYBA1 27 Isoform1 215 6.38 25130.14 1e-118
Gallus gallus CRYBA1 147 Isoform2 198 6.16 23022.81 1e-112
Gallus gallus CRYBA1 147 Isoform1 215 5.8 24748.79 1e-112

FIGURE 7.16 Query the ATID database. (A) ATID keyword query in ATID database; (B) sequence similarity query in ATID database.

mRNA to the stop codon UAA, thereby causing early termination of the protein synthesis.

Usually there are two classes of RNA editing: the substitution/conversion class and the insertion/deletion class. The insertion/deletion editing occurs in protozoans like *Trypanosoma* and *Leishmania*; in slime molds like *Physarum* spp., and in some viral categories like paramyxoviruses, Ebola virus, etc. To date, the substitution/conversion pathway has been observed in humans along with other mammals, *Drosophila*, and some plants. The RNA-editing processes are known to create diversity in proteins involved in various pathways like lipid transport, metabolism, etc., and may act as potential targets for therapeutic intervention.

To date RNA editing has been observed in mRNA, tRNA (transfer mRNA) and rRNA (ribosomal RNA) molecules of eukaryotes. RNA editings in pre-mRNA that alter protein products are by changing codon or modifying splicing signals, or by altering regulation. tRNA editings are classical RNA modifications, and rRNA and snoRNAs (small nucleolar RNAs) editings are RNA-guided nucleotide modifications.

2. RNA Editing and Human Disease

RNA editing is a physiological mechanism for developmental stages and normal life in both invertebrates and vertebrates. Overexpression or deficiency in RNA editing activities may cause diseases. Hyperediting caused by overexpression of Apobec-1, one of the Apobec enzyme family catalyzing C-to-U RNA editing, leads to carcinomas in model systems, whereas hyperediting of measles virus transcripts has been observed in patients with subacute sclerosing panencephalitis and measles inclusion body encephalitis. ADAR1 (ADAR, adenosine deaminases acting on RNA) knockout mice die embryonically, and ADAR2 null mice are born at full term but die prematurely. Altered RNA-editing activities have also been implicated in the pathogenesis of human malignant gliomas, schizophrenic patients, and suicide victims. RNA editing may be altered in patients with Alzheimer's and Huntington's disease.

3. Types of RNA Editing

RNA editing occurs concomitantly with transcription and splicing processes in the nucleus, as well as in mitochondria and plastids. These modifications have been observed in plants, animals, fungi, and viruses. In mRNA, this editing process can be either in the coding or noncoding region of RNA There are at least 16 different types of RNA editing (C to U, A to G, U to C, A to C, G to C, G to A, A to U, GG to AA, AA to GG, G-, C-, U-, A-, AA-, and C/U insertions, and G deletion). Three selected RNA editing types are briefly described in the following text.

A. C-Insertion and Dinucleotide Insertion Editing The mitochondria of *Physarum polycephalum* and several other members of the phylum Myxomycota display a unique type of RNA editing that is characterized by the insertion of mono- and dinucleotides in RNAs relative to their mtDNA (mitochondrial DNA) template. In addition, four examples of C-to-U base conversion have been identified. The most common mononucleotide insertion is cytidine, although a number of uridine mononucleotides are inserted at specific sites. Adenosine and guanosine have not been observed in mononucleotide insertions. Five different dinucleotide insertions have been observed: GC, GU, CU, AU and AA. Both mono- and dinucleotide insertions create ORFs in mRNA and contribute to highly conserved structural features of rRNAs and tRNAs. If mononucleotide and dinucleotide insertions are considered together, then any one of the four standard ribonucleotides can be inserted. The sites of insertion are distributed relatively uniformly throughout a given mRNA with an average spacing of 25 nucleotides, which varies with a standard deviation of about 10 nucleotides.

The editing site distribution within the rRNAs is also fairly uniform but with an average spacing of 43 nucleotides. The less common dinucleotide insertion sites are intermixed with the mononucleotide insertion sites. The insertion sites are apparently not defined by any consensus sequence in the RNA. Although there is a bias for insertion after purine-pyrimidine dinucleotides, many insertions are after other dinucleotides and numerous purine-pyrimidine dinucleotides are present that do not precede editing sites. To date the insertion of 360 nucleotides at 346 sites in 12 separate RNAs has been identified. With about half of the 60-kb sequence of the mtDNA explored, it is likely that these numbers will double.

A number of significant ORFs have been identified on the mtDNA that apparently do not require insertional RNA editing to create the ORF in the mRNA. The mRNAs that are edited require the insertion of between 9 and 64 nucleotides to create their reading frame. These insertions must both be accurate (the correct nucleotide at the correct location) and efficient to produce functional mRNAs. Inaccurate nucleotide insertion has not been observed for this type of RNA editing. Furthermore, the efficiency of editing (frequency of nucleotide insertion at a given site) is very high, generally greater than 95% in mitochondrial RNA populations.

B. C-To-U Editing C-to-U RNA editing occurs in plant mitochondria and chloroplasts and in apoB (apoprotein B) in mammals. Recently, an APOBEC family of enzymes has been discovered with the ability to deaminate cytidines to uridines on RNA or DNA. The first member of this new family is APOBEC1, which deaminates apoB messenger RNA to generate a premature stop codon. APOBEC1 is evolutionally conserved from bacteria to humans. There is a unique motif containing two phenylalanine residues and an insert of four amino acid residues across the active site motif, which are present in all APOBEC family members, including APOBEC1 AID (activation-induced cytidine deaminase), APOBEC2, and APOBEC3A through APOBEC3G. AID is essential for initiating class-switch recombination, somatic hypermutation, and gene conversion. The APOBEC3 family is unique to primates. They can protect cells from human immunodeficiency virus and other viral infections. Overexpression of enzymes in the APOBEC family can cause cancer, suggesting that the genes for the APOBEC family of proteins are proto-oncogenes.

C. A-To-I Editing RNA editing by adenosine-to-inosine (A-to-I) modification generates RNA and protein diversity in higher eukaryotes selectively altering coding and noncoding sequences in nuclear transcripts. The enzymes responsible for A-to-I editing, ADARs (adenosine deaminases acting on RNA), are ubiquitously expressed in mammals and specifically recognize partially dsRNA (double-stranded RNA) structures where they modify individual adenosines depending on the local structure and sequence environment. Long, extended dsRNAs undergo massive editing, whereas RNA duplex structures with bulges and loops are subject to site-selective editing, as observed in several neurotransmitter receptor mRNAs ensuing single amino acid substitutions. The ADARs (Figure 7.17) specifically target single nucleotides for editing within the partially double-stranded pre-mRNAs of their substrates, such as neuronal glutamate and serotonin receptor transcripts. Because inosine is read as guanosine by the translation machinery, A-to-I editing often leads to codon changes that result in the alteration of protein function. It can also create or destroy pre-mRNA splice signals or lead to alterations in RNA secondary structure. The deficiency or misregulation of A-to-I RNA editing has been implicated in the etiology of neurological diseases such as epilepsy, ALS (amyotrophic lateral sclerosis), and depression in mammals, and it has been shown that a loss of A-to-I editing following the genetic inactivation of ADARs in

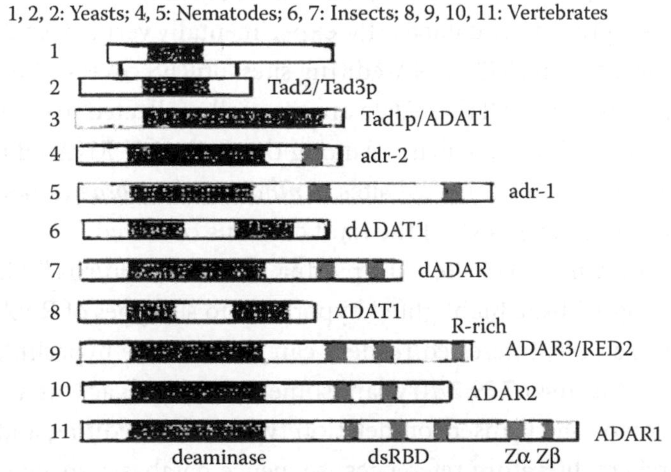
1, 2, 2: Yeasts; 4, 5: Nematodes; 6, 7: Insects; 8, 9, 10, 11: Vertebrates
1
2
Tad2/Tad3p
3
Tad1p/ADAT1
4
adr-2
5
adr-1
6
dADAT1
7
dADAR
8
ADAT1
R-rich
9
ADAR3/RED2
10
ADAR2
11
ADAR1
deaminase
dsRBD
Zα Zβ

FIGURE 7.17 RNA editing by A-to-I (adenosine-to-inosine) modification. Adenosine Deaminase is responsible for A-to-I editing acting on RNA.

mammals, as well as flies and the worms, results in behavioral or neurological dysfunctions or embryonic lethality (ADAR1).

D. RNA-Editing Databases Here, three representative RNA-editing databases are concisely presented. EdRNA (http://edrna.mbc.nctu.edu.tw/) stores putative RNA-editing sites that have been predicted using computational methods. dbRES (http://bioinfo.au.tsinghua.edu.cn/dbRES) collects experimentally verified RNA-editing sites. RNA Editing Website (http://dna.kdna.ucla.edu/rna/index.aspx) gathers information on various type of RNA editing, especially lists of all researchers and literature references on a particular editing field.

EdRNA (http://edrna.mbc.nctu.edu.tw/): EdRNA is a comprehensive RNA-editing database that stores putative RNA-editing sites by computational methodology. In EdRNA, RNA-editing sites are annotated by using some cross-references such as SNP, repeat, UTR, and cross-species-conserved region. EdRNA collects 312,774 RNA-editing sites comprising 6,090 genes, which account for 17.77% of the total number of human genes (34,270) based on the Ensembl database. Computer prediction suggests that the RNA editing occurs mostly in coding regions and thus heavily affects mRNA translations. RNA editing occurs more frequently in 3'UTR regions than in 5'UTR regions.

dbRES (http://bioinfo.au.tsinghua.edu.cn/dbRES): dbRES is a Web-oriented comprehensive database for experimentally verified RNA-editing sites. It now contains 5437 RNA-editing sites. dbRES covers 95 organisms from 251 transcripts. All these data are manually collected from the literature or the Gene Bank database. Among them, C-to-U RNA-editing sites account for 84% of all collected sites. *Anthoceros formosae* ranks number one with 975 sites, i.e., 18% of the total citations collected.

RNA Editing Website (http://dna.kdna.ucla.edu/rna/index.aspx): RNA Editing Website highlighted hyperlinks to six types of RNA-editing sites. For example, interested readers can click on the hyperlinks 1. The Web site (http://164.67.39.27/trypanosome/index.html) acts as a source of information on the U insertion/deletion type of RNA editing and a list of all researchers, literature references, sequence databases specific for this field, upcoming scientific meetings, as well as a section with data taken from published and unpublished research to illustrate research problems and research directions.

Part II Step-By-Step Tutorial

In this part, we show how to search for how many RNA-editing sites exist in the APOB mRNA from the EdRNA Web site (http://edrna.mbc.nctu.edu.tw/):

1. Input the search term: Go to the Web site (http://edrna.mbc.nctu.edu.tw/), and you will find three options for search terms: mRNA ID, Gene symbol, and Gene ID. Select "Gene symbol" and type "APOB" in the search window as displayed in Figure 7.18.
2. Display the search results: After inputting the search term, click "Go"; the result is shown in Figure 7.19. There are 211 predicted RNA-editing sites in the APOB. Click the hyperlinked data for detailed information about positions of each editing sites in its mRNA, gene and chromosome, type, magnitude, and EST (expressed sequence tag) number.

Part III Sample Data

Gene symbol: APOB.

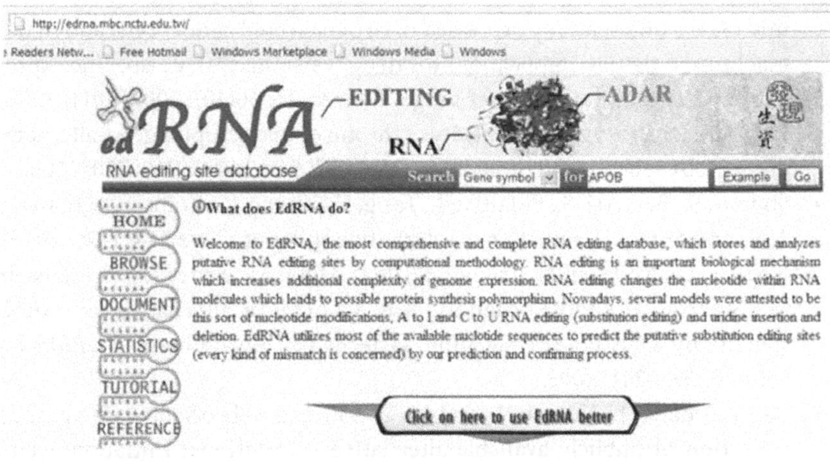
http://edrna.mbc.nctu.edu.tw/
Readers Netw... Free Hotmail Windows Marketplace Windows Media Windows
edRNA
EDITING
ADAR
RNA
RNA editing site database
Search Gene symbol for APOB Example Go
HOME
BROWSE
DOCUMENT
STATISTICS
TUTORIAL
REFERENCE
What does EdRNA do?
Welcome to EdRNA, the most comprehensive and complete RNA editing database, which stores and analyzes putative RNA editing sites by computational methodology. RNA editing is an important biological mechanism which increases additional complexity of genome expression. RNA editing changes the nucleotide within RNA molecules which leads to possible protein synthesis polymorphism. Nowadays, several models were attested to be this sort of nucleotide modifications, A to I and C to U RNA editing (substitution editing) and uridine insertion and deletion. EdRNA utilizes most of the available nuclotide sequences to predict the putative substitution editing sites (every kind of mismatch is concerned) by our prediction and confirming process.
Click on here to use EdRNA better

FIGURE 7.18 Searching RNA-editing sites in the EdRNA Web site.

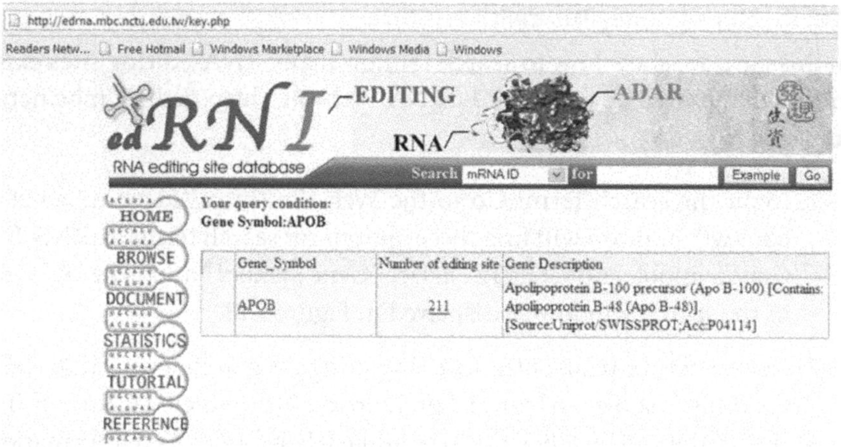
http://edrna.mbc.nctu.edu.tw/key.php
Readers Netw... Free Hotmail Windows Marketplace Windows Media Windows
EDITING
ADAR
RNA
RNA editing site database
Search mRNA ID for Example Go
HOME
BROWSE
DOCUMENT
STATISTICS
TUTORIAL
REFERENCE
Your query condition:
Gene Symbol:APOB
Gene_Symbol
Number of editing site
Gene Description
APOB
211
Apolipoprotein B-100 precursor (Apo B-100) [Contains: Apolipoprotein B-48 (Apo B-48)].
[Source:Uniprot/SWISSPROT;Acc:P04114]

FIGURE 7.19 Query results of RNA-editing sites existing in the APOB mRNA. Two hundred and eleven RNA-editing sites are predicted in the APOB.

REFERENCES

1. Zong, Y., Zhou, S., Fatima, S., and Sorrentino, B.P. Expression of mouse Abcg2 mRNA during hematopoiesis is regulated by alternative use of multiple leader exons and promoters. *J Biol Chem* 281(40): 29625–29632, 2006.
2. Nakanishi, T., Bailey-Dell, K.J., Hassel, B.A., Shiozawa, K., Sullivan, D.M., Turner, J., and Ross, D.D. Novel 5' untranslated region variants of BCRP mRNA are differentially expressed in drug-selected cancer cells and in normal human tissues: implications for drug resistance, tissue-specific expression, and alternative promoter usage. *Cancer Res* 66(10): 5007–5011, 2006.
3. Pan, Y.X. Diversity and complexity of the mu opioid receptor gene: alternative pre-mRNA splicing and promoters. *DNA Cell Biol* 24(11): 736–750, 2005.
4. Stamm, S., Ben-Ari, S., Rafalska, I., Tang, Y., Zhang, Z., Toiber, D., Thanaraj, T.A., and Soreq, H. Function of alternative splicing. *Gene* 344: 1–20, 2005.
5. Nogues, G., Kadener, S., Cramer, P., de la Mata, M., Fededa, J.P., Blaustein, M., Srebrow, A., and Kornblihtt, A.R. Control of alternative pre-mRNA splicing by RNA Pol II elongation: faster is not always better. *IUBMB Life* 55(4–5): 235–241, 2003.
6. Cai, J., Zhang, J., Huang, Y., and Li, Y. ATID: a web-oriented database for collection of publicly available alternative translational initiation events. *Bioinformatics* 21(23): 4312–4314, 2005.
7. Luque, C.M., Perez-Ferreiro, C.M., Perez-Gonzalez, A., Englmeier, L., Koffa, M.D., and Correas, I. An alternative domain containing a Leucine-rich sequence regulates nuclear cytoplasmic localization of protein 4.1R. *J Biol Chem* 278(4): 2686–2691, 2003.

8. Kochetov, A.V., Sarai, A., Rogozin, I.B., Shumny, V.K., and Kolchanov, N.A. The role of alternative translation start sites in the generation of human protein diversity. *Mol Genet Genomics* 273: 491–496, 2005.
9. Chrousos, G.P. and Kino, T. Intracellular glucocorticoid signaling: a formerly simple system turns stochastic. *Sci STKE* 2005(304): pe48, October 4, 2005.
10. Howard, M.T., Malik, N., Anderson, C.B., Voskuil, J.L., Atkins, J.F., and Gibbons, R.J. Attenuation of an amino-terminal premature stop codon mutation in the ATRX gene by an alternative mode of translational initiation. *J Med Genet* 41(12): 951–956, 2004.

CHAPTER 8

MicroRNoma Genomewide Profiling by Microarray

Chang-gong Liu, Xiuping Liu,
and George Adrian Calin

CONTENTS

ABSTRACT

MicroRNAs (miRNAs) are short 19–25 nucleotide RNA molecules that have been shown to regulate the expression of other genes in a variety of eukaryotic systems. MiRNA alterations are involved in the initiation and progression of human cancers. Abnormal expression of miRNAs has been proved to be the main abnormality of the miRNoma in cancer cells. Various methods of microRNA expression profiling of human tumors, mainly by microarray, have identified signatures associated with diagnosis, progression, prognosis, and response to treatment.

SECTION 1 WHAT ARE MIRNAS?

Noncoding RNAs (ncRNAs) range in size from 19 to 25 nt for the large family of microRNAs (miRNAs) that modulate development in several organisms, including mammals, up to more than 10,000 nt for RNAs involved in gene silencing in higher eukaryotes. MiRNAs are typically excised from a 60–110 nt hairpin precursor (fold-back) RNA (named pre-miRNA) structure that is transcribed from a larger primary transcript (named pri-miRNA).

1. MicroRNAs: Strangers in the Genomic Galaxy

First described in *C. elegans* more than a decade ago, over 4000 members of a new class of small ncRNAs, named miRNAs, have been identified in the last four years in vertebrates, flies, worms, and plants, and even in viruses. In humans, the miRNoma (defined as the full spectrum of miRNAs) contains more than 450 experimentally or *in silico* cloned miRNAs, and the total number is expected to pass the 1000 mark. The behavior of miRNA genes differ from the classical paradigms. Genomically, miRNAs represent less than 1% of the size of usual protein coding genes (PCGs), a reason why they "escaped" cloning for such a long time. No open reading frame (ORF) can be identified in the small piece of genome codifying for miRNAs. The splicing reaction requires Dicer RNase III and Argonaute family members. Functionally, it was shown that miRNAs reduce the levels of many of their target transcripts as well as the amount of protein encoded by these transcripts by direct and imperfect miRNA::mRNA interaction. For several miRNAs, participation in essential biological processes for the eukaryotic cell has been proved. For example, the list of proposed functions includes hematopoietic B-cell lineage fate (*miR-181*), B-cell survival (*miR-15a* and *miR-16-1*), cell proliferation control (*miR-125b* and *let-7*), brain patterning (*miR-430*), pancreatic cell insulin secretion (*miR-375*), and adipocyte development (*miR-143*).

2. The "Old" Discovery of miRNAs: Recent Exciting Developments

The discovery of miRNAs began in early 1981 when the heterochronic genes capable of controlling the timing of specific postembryonic developmental events in *C. elegans* were identified. Chalfie et al., in 1981, during a loss-of-function study in *C. elegans*, discovered that mutations in the *lin-4* gene lead to continued synthesis of larval-specific cuticle. Eight years later, Victor Ambros described, in hypodermal cells, an interaction hierarchy of heterochronic regulatory genes *lin-14*, *lin-28*, and *lin-29* to coordinate the "larva-to-adult switch." In 1993 two independent studies, published in the same issue of *Cell* by Ambros and Gary Ruvkun, presented the real nature of the *lin-4* gene and its ability to regulate heterochronic gene expression. These authors, after cloning the *lin-4* gene, demonstrated that the potential ORF does not encode for a protein. They identified two small *lin-4* transcripts of approximately 22 and 61 nt and suggested that the temporal regulation of *lin-14* is guided by *lin-4* RNA via antisense RNA–RNA interaction involving the small RNA *lin-4* and the 3`UTR of *lin-14* whose translation was inhibited. Seven years later, Reinhart et al. showed that *let-7* gene is another heterochronic switch gene coding for a small 21 nt RNA and proposed that the sequential stage-specific expression of *let-7* and *lin-4* RNAs was capable through an RNA–RNA interaction with the 3`UTR of the target genes to trigger the temporal cascade of regulatory heterochronic genes specifying the timing of *C. elegans* developmental events. The miRNA revolution begin in late 2001 when independent researchers in the laboratories of Thomas Tuschl, David Bartel, and Victor Ambros, isolating and cloning RNA from different organisms and cellular systems, by using the same strategy applied to cloned siRNA processed from exogenous dsRNAs in an embryo lysate, were able to isolate a large group of RNAs with the same characteristics of *lin-4* and *let-7*. The data provide evidence for the existence of a large class of small RNAs with potential regulatory roles; because of their small size, the authors referred to these novel RNAs as microRNA (abbreviated miRNA).

SECTION 2 MIRNAS AND HUMAN DISEASES

As a consequence of extensive participation in normal functions, it is quite logical to ask the question, do microRNAs abnormalities play a role in human diseases? The answer to this fundamental question is built on many recent evidences, obtained mainly from the study of human cancers. The present understanding is that miRNAs and proteins involved in the processing of miRNAs are involved in various types of human diseases.

1. MiRNAs and Their Role as Tumor Suppressors and Oncogenes

It was recently shown that miRNA alterations are involved in the initiation and progression of human cancer. Homozygous deletions or the combination mutation + promoter hypermethylation (as is the case for the *miR-15a*/*miR-16a* cluster), or gene amplification (as is the case for *miR-155* or the cluster *miR-17-92*) seem to be the main mechanisms of inactivation or activation, respectively. Because of their small size, the loss-of-function or gain-of-function point mutations represent rare events. miRNAs activity can be influenced either by the repositioning of other genes close to miRNA promoters or regulatory regions (as is the case for *miR-142s* – c-MYC translocation) or by the relocalization of a miRNA near other regulatory elements. The overall effect in the case of miRNA inactivation is the overexpression of target mRNAs, whereas miRNA activation leads to downregulation of target mRNAs involved in apoptosis, cell cycle, invasion, or angiogenesis. To date only few miRNA::mRNA interactions with importance for cancer pathogenesis have been proved. For example, it was elegantly demonstrated that the *let-7* microRNA family regulates RAS oncogenes and that *let-7* expression is lower in lung tumors than in normal lung tissue, whereas RAS protein has an inverse variation. Furthermore, enforced expression of the *miR-17-92* cluster from chromosome 13q32-33 in conjunction with c-myc accelerates tumor development in a mouse B-cell lymphoma model.

2. miRNAs and Other Diseases

Several papers have been published showing a probable link between miRNAs and other human diseases (such as Fragile X syndrome or spinal muscular atrophy), but the precise mechanisms are still not known. Recently, an unexpected mechanism of miRNA involvement in human disease was identified. Sequence variants of a candidate gene on chromosome 13q31.1 named SLITRK1 (Slit and Trk-like 1) were identified in patients with Tourette syndrome, a neurologic disorder manifested particularly by motor and vocal tics and associated with behavioral abnormalities. One variant found in two unrelated patients was located in the 3'UTR binding site for the *miR-189* and might affect SLITRK1 expression. Therefore, it is tempting to propose that germ line mutations or polymorphisms in miRNA genes or interacting sequences in target mRNA might represent a newly described mechanism of predisposition to hereditary disorders.

SECTION 3 SIGNIFICANCE OF MIRNA PROFILING

Proving cancer-specific expression levels for hundreds of miRNA genes is time consuming, requires a high amount of total RNA (at least 10 to 20 ugs for each Northern blot), and uses autoradiographic techniques that require handling of radioactive material. To overcome these limitations and further understand the involvement of miRNAs in human cancer, our group was the first to developed an miRNA microarray and established a novel detection methodology of miRNA expression that overcomes the size limitation of these very small molecules.

1. miRNAs Microarrays as Profiling Tools

The most commonly used high-throughput technique for the assessment of cancer-specific expression levels for hundreds of microRNAs in a large number of samples is represented by oligonucleotide miRNA microarrays (Table 8.1 and Figure 8.1). Several technical variants were independently developed in the last few years, and the main differences between them are included in Table 8.2. Another method to determine miRNA expression levels involves the use of a bead-based flow cytometric technique. Other developments include the quantitative RT-PCR for precursor miRNA or active miRNA or the miRAGE, the genomewide miRNA analysis with serial analysis of gene expression (SAGE). Each of these techniques has its strengths and caveats, and the confirmatory use of a second technique is mandatory at present time.

TABLE 8.1 Examples of High-Throughput Methods for miRNA Expression

Type	Principle	Advantages	Ref.
miRNA microarray microchips	Thousands of oligonucleotide probes used to hybridize with mature/precursor miRNA cDNA probe	Concomitant screening of a large number of miRNAs through extensive sample collections	Liu et al., 2004, review in Calin and Croce, 2006a
Bead-based technology	Single-miRNA oligos coating polystyrene beads hybridized with biotin labeled dsDNA target, followed by flow cytometry signal detection	Higher specificity and accuracy	Lu et al., 2005
Stem-loop qRT-PCR for mature product	Stem-loop RT primer cDNA synthesis followed by quantitative conventional TaqMan PCR	Specific quantification of the mature miRNA	Chen et al., 2005
qRT-PCR for precursor miRNA	Hairpin-specific primers used to amplify cDNA	Specific quantification of the precursor miRNA	Schmittgen et al., 2004
miRAGE (SAGE)	Serial analysis of gene expression (SAGE) adapted for small RNAs.	Mixture of cloning and prexpression profiling: adequate to discover new miRNAs	Cummins et al., 2006

Note: ds DNA — double-stranded DNA; LNA — fluorescent locked nucleic acid.

FIGURE 8.1 Principles of microarray technology used for miRNA profiling. Microarray-based miRNA profiling is presented as described in the majority of profiling studies on primary tumors, initially developed by Liu et al. in 2004. This strategy involves four main steps (presented on the left side): target labeling, hybridization, staining, and signal detection. The different replicates of the spots on the glass slide represent different oligonucleotide sequences corresponding to sequences from the precursor miRNA or active miRNA, molecule. The main advantage of the microarray-based miRNA profiling is the high level of standardization of the procedure, allowing the processing of tens of samples in parallel. Modified with permission from *Nature Reviews Cancer* (Calin, G.A., Croce, C.M. MicroRNA signatures in human cancers. *Nat Rev Cancer.* 6, 857–866 (2006)) Copyright (2006) Macmillan Magazines.

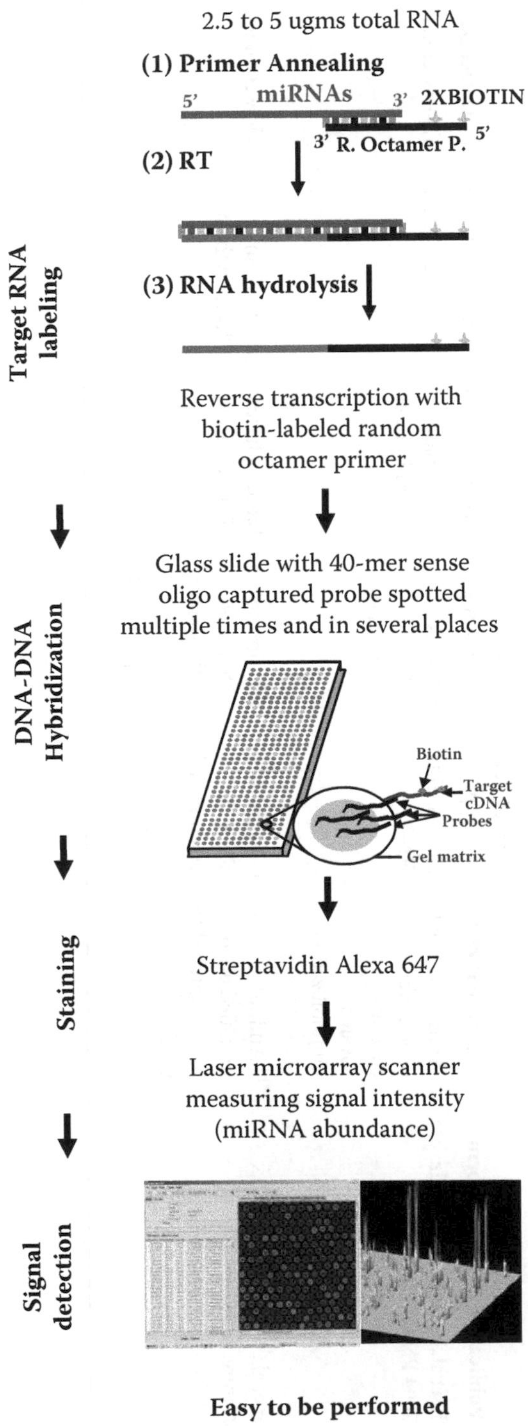
2.5 to 5 ugms total RNA
(1) Primer Annealing
5' miRNAs 3' 2XBIOTIN
3' R. Octamer P. 5'
(2) RT
(3) RNA hydrolysis
Target RNA labeling
Reverse transcription with biotin-labeled random octamer primer
DNA-DNA Hybridization
Glass slide with 40-mer sense oligo captured probe spotted multiple times and in several places
Biotin
Target cDNA
Probes
Gel matrix
Staining
Streptavidin Alexa 647
Laser microarray scanner measuring signal intensity (miRNA abundance)
Signal detection
Easy to be performed

TABLE 8.2 Examples of Differences between Various Platforms of Genomewide miRNA Profiling

Publications	Platform	Capture Probe	Start Material (μg)	Target Labeling	Hybridization	Report Molecules	Detection
Liu et al.,June 2004 PNAS	Glass slide	5'Amine modified oligo probe; chemically covalent immobilization	2.5–5 μg Total RNA	5'Biotins labeled first-strand cDNA	Solid-phase hybridization	Streptavidin Alexa 647	Laser scanner
Lu et al., Nature, June 2005	Beads	5' Amine modified oligo probe captured to beads	1–10 μg Total RNA fractionated	Fractionate RNA ligated with adaptor and PCR amplification	Liquid-phase hybridization	Streptavidin Phycoerythrin	Flow cytometer

2. Biological Significance of miRNA Profiling

After several years of studies employing these technologies and the analyses of more than 1000 primary tumors, three common characteristic themes of miRNA deregulation in human tumors emerged. First, miRNA expression profiles classify every type of human cancer. Second, common miRNA genes are differentially expressed in various types of cancers, suggesting common altered regulatory pathways. Third, onco-miRNAs and suppressor-miRNAs may represent two different versions of the same microRNA gene with respect to the specific mechanism of inactivation or the tissue type where it occurred.

MiRNA alterations are involved in the initiation and development of human cancers. Abnormal expression of miRNAs has been proved to be the main abnormality of the miRNoma in cancer cells. The causes of the widespread differential expression of miRNA genes between malignant and normal cells can be explained by the genomic location of these genes in cancer-associated genomic regions, by epigenetic mechanisms, as well as by alterations of proteins included in processing machinery. miRNA profiling achieved by various methods (Table 8.1) has been exploited to identify miRNAs that are potentially involved in the pathogenesis of human cancers and has allowed the definitions of signatures associated with diagnosis, staging, progression, prognosis, and response to treatment of human tumors.

SECTION 4 SIGNIFICANCE OF MIRNA CGH ASSAY

Several arguments suggest that the highly significant association between the location of miRNAs and chromosomal or molecular genomic aberrations is not without consequences. In order to investigate these effects at the genomewide levels, a specific technology of array comparative genomic hybridization (aCGH) was developed by Zhang et al. in 2006.

1. Genomic Investigation of miRNA Loci by CGH

Most of miRNA genes cloned and verified have been localized onto chromosome loci (Sanger Institute, miRBase http://microrna.sanger.ac.uk/cgi-bin/sequences/browse.pl). The principle of the technology for miRNA DNA copy number detection is as the same as for aCGH for genomic DNA copy number detection. Zhang et al. (2006) first reported the miRNA genomic alteration in human ovarian cancer by using a CGH array with BAC clones. The pair of genomic DNA samples in the microgram was labeled with Cy3 and Cy5 separately by using a Bioprimer random-primed labeling kit (Invitrogen). The labeled control reference and test DNA samples were combined and coprecipitated with human Cot-1 DNA to reduce nonspecificity. Labeled DNA was resuspended in the hybridization buffer and hybridized to CGH array. The hybridized and processed a-CGH array was scanned with both green and red lasers simultaneously. The difference of signal intensities of cyanine dye on the same spot of single BAC clone was considered as a genomic copy number in the specific miRNA locus. The BAC a-CGH has its limitation concerning the resolution of 1 megabase (MB) in regard to the very small miRNA precursor genomic size of 60–110 nt. Fortunately, oligo-based aCGH arrays with higher resolutions in the range of kilobases (Kb) have been developed and are commercially available (http://www.chem.agilent.com/ and http://www.nimblegen.com/).

2. Biological Significance of CGH Investigation

If the location of miRNAs is relevant to tumorigenesis, then structural or functional alterations of miRNAs should be identified in various types of cancers. A growing number of reports are providing such evidence and suggest that abnormal expression of miRNAs is central to cancer pathogeny. The majority of miRNAs causally linked to human tumorigenesis are located in genomic regions altered in cancer, and the genomic abnormality is concordant with expression deregulation (genomic deletion for downregulation and amplification for upregulation, respectively). The combination of nonrandom chromosomal abnormalities and other types of genetic or epigenetic events could contribute to downregulation or overexpression of miRNAs. An extensive study of high-resolution array-based comparative genomic hybridization by Zhang et al. (2006) on 227 human ovarian cancer, breast cancer, and melanoma specimens clearly proved that regions hosting miRNAs exhibit high-frequency genomic alterations in human cancer. Strengthening the importance of these findings is the fact that miRNA copy changes correlate with miRNA expression. The analyses of the same-histotype, breast ductal carcinomas performed by two independent groups using distinct techniques revealed overlapping sets of miRNAs differentially expressed and DNA copy number gains or losses compared to normal breast tissues.

SECTION 5 GENOMEWIDE MIRNA PROFILING BY MICROARRAY

For the technological steps such as target preparation and array hybridization, the detailed protocols can be found in Liu et al. (2004) (online Supplemental Information) (Figure 8.2). Also, the definitions of the main technical terms used in this effort are presented in the glossary (Table 8.3).

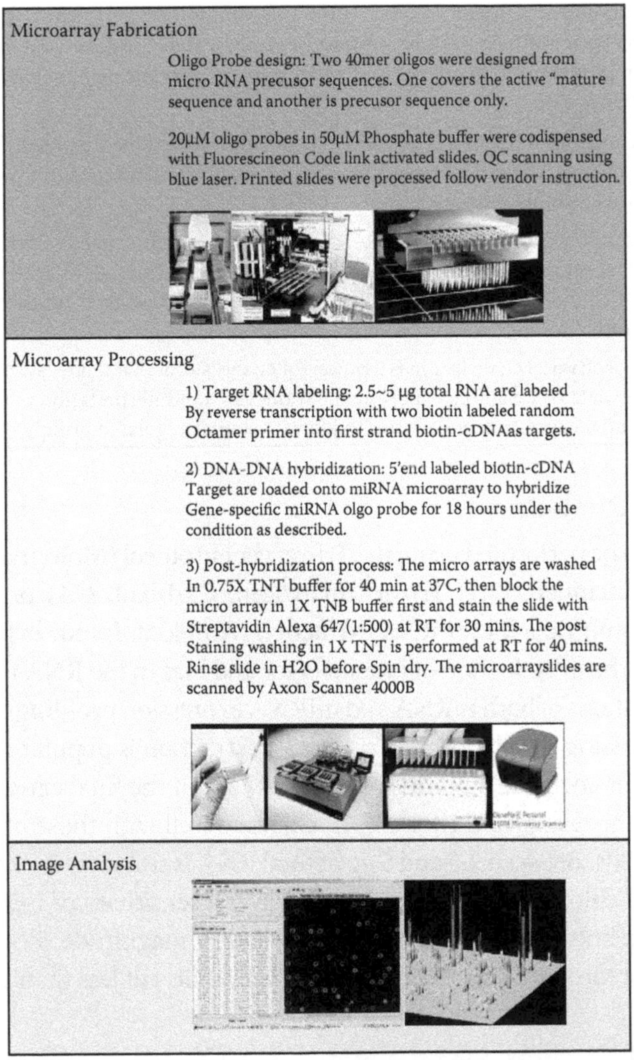
Microarray Fabrication
Oligo Probe design: Two 40mer oligos were designed from micro RNA precusor sequences. One covers the active "mature sequence and another is precusor sequence only.
20μM oligo probes in 50μM Phosphate buffer were codispensed with Fluorescineon Code link activated slides. QC scanning using blue laser. Printed slides were processed follow vendor instruction.
Microarray Processing
1) Target RNA labeling: 2.5~5 μg total RNA are labeled By reverse transcription with two biotin labeled random Octamer primer into first strand biotin-cDNAas targets.
2) DNA-DNA hybridization: 5'end labeled biotin-cDNA Target are loaded onto miRNA microarray to hybridize Gene-specific miRNA olgo probe for 18 hours under the condition as described.
3) Post-hybridization process: The micro arrays are washed In 0.75X TNT buffer for 40 min at 37C, then block the micro array in 1X TNB buffer first and stain the slide with Streptavidin Alexa 647(1:500) at RT for 30 mins. The post Staining washing in 1X TNT is performed at RT for 40 mins. Rinse slide in H2O before Spin dry. The microarrayslides are scanned by Axon Scanner 4000B
Image Analysis

FIGURE 8.2 Step-by-step tutorial for genomewide miRNA profiling by microarray.

TABLE 8.3 Glossary of Terms Used to Describe Microarray Research

miRNoma: The full spectrum of miRNAs expressed in a particular cell type.
Gene-expression profiling: Determination of the level of expression of hundreds or thousand of genes through the use of microarrays. Total RNA extracted from the test tissue or cells and labeled with a fluorescent dye is tested for its ability to hybridize to the spotted nucleic acids.
Hierarchical clustering technique: A computational method that groups genes (or samples) into small clusters and then group these clusters into increasingly higher-level clusters. As a result, a dendrogram (i.e., tree) of connectivity emerges.
Genelist: A group of genes or proteins with some common property, such as putative interaction with miRNAs or same expression profiles. They are generated by target prediction programs or by calculations performed by GeneSpring or some other bioinformatics tool.
Prediction analysis of microarrays (PAM): A statistical technique that identifies a subgroup of genes that best characterizes a predefined class and uses this gene set to predict the class of new samples.
Significance analysis of microarrays (SAM): A statistical method used in microarray analyzes that calculates a score for each gene and thus identifies genes with a statistically significant association with an outcome variable such as transfection with specific miRNAs.
GeneSpring software: GeneSpring is a powerful analysis tool that analyzes the scanned microarray data by assigning experiment parameters and interpretation to filter genes for differential expression and cluster to identify similar regulated groups.

1. Total RNA Isolation

This has to be performed using the Tri-reagent protocol (Molecular Research Center, Cincinnati, OH) Trizol (Invitrogen, Carlsbad, CA) or the newly developed mirVana kit for RNA extraction (Ambion). In our experience, to minimize the introduction of human error and bias in the RNA isolation for the applications of both mRNA and miRNA expression profiling of the same sample crossing the platforms, Tri-reagent extraction is popular and simple. It maintains adequate quantities of small RNAs in the final extraction solution, and the results of miRNACHIP correlate well with those of confirmation methods. Between 2.5 and 5 μg of total RNA is sufficient for each sample expression chip, and the detection assay gives a sensitivity of 1–3 copies per cell, with a linear dynamic range of 2.5 orders of magnitude, 90–94% specificity, and reproducibility CV (coefficient variability) of less than 10%.

2. MicroRNACHIP Production and Description

The current OSU miRNA microarray (version 3.0, miRNACHIPv3) contains probes against 578 miRNA precursor sequences (329 *Homo sapiens*, 249 *Mus musculus*, and 3 *Arabidopsis thaliana* as negative controls). These correspond to human and mouse miRNAs found in the miRNA Registry at www.sanger.ac.uk/Software/Rfam/mirna/ (October 2005) or collected from published papers. All the sequences were confirmed by BLAST alignments with the corresponding genome at www.ncbi.nlm.nih.gov and the hairpin structures were analyzed at www.bioinfo.rpi.edu/applications/mfold/old/rna. All 40-mer oligos were screened for their cross-homology to all genes of the relevant organism, number of bases in alignment to a repetitive element, amount of low-complexity sequence, maximum homopolymeric stretch, global and local G + C content, and potential hairpins (self 5-mers). The best oligo that contained each active site of each miRNA was selected. Next, we attempted to design an oligo that did not contain the active site for each cluster, when it was possible to choose such an oligo that did not overlap the selected oligos by more than 10 nt. To design each of these additional oligos, we required <75% global cross-homology and <20 bases in any 100% alignment to the relevant organism, <16 bases in alignments to repetitive elements, <16 bases of low-complexity, homopolymeric stretches of no more than 6 bases, G + C content between 30 and 70%, and no more than 11 windows of size 10 with G + C content outside 30–70%, and no self 5-mers. In addition, we designed oligos for seven mouse tRNAs and eight human tRNAs, using similar design criteria. We selected a single oligo for each, with the exception of the human and mouse initiators Met-tRNA-i, for which we selected two oligos.

3. Target Preparation

An amount of 2.5–5 μg of total RNA is separately added to a reaction mix in a final volume of 12 μl, containing 1 μg of [3'(*N*)8-(A)12-biotin-(A)12-biotin 5'] oligonucleotide primer. The total RNA and oligo primer mixture is incubated for 10 min at 70°C for specific annealing first and chilled on ice. With the mixture remaining on ice, 4 μl of 5X first-strand buffer, 2 μl of 0.1 *M* DTT, 1 μl of 10 m*M* dNTP mix, and 1 μl Superscript™ II RNaseH$^-$ reverse transcriptase (Invitrogen) (200 U/μl) is added to a final volume of 20 μl and the mixture incubated for 90 min in a 37°C water bath. After incubation for first-strand cDNA synthesis, 3.5 μl of 0.5 *M* NaOH/50 m*M* EDTA is added to 20 μl of first-strand reaction mix and incubated at 65°C for 15 min to denature the RNA/DNA hybrids and degrade RNA templates. Then, 5 μl of 1 *M* Tris-HCI, pH 7.6 (Sigma), is added to neutralize the reaction mix, and first-strand-labeled cDNA targets are stored in 28.5 μl at −80°C until chip hybridization.

4. Array Hybridization

Labeled first-strand cDNA targets from 5 μg of total RNA is used for hybridization on each OSU-CCC miRNA microarray containing probes specific for 578 miRNA precursor sequences (329 *Homo sapiens*, 249 *Mus musculus*, and 3 *Arabidopsis thaliana*). All probes on these microarrays are 5' amine-modified 40-mer oligonucleotides spotted on Codelink-activated slide (GE Healthcare, Piscataway, NJ) by contacting technologies and covalently attached to a polymeric matrix. The microarrays are hybridized in 6X SSPE/30% formamide at 25°C for 18 h, washed in 0.75X TNT at 37°C for 40 min, and processed using direct detection of the biotin-containing transcripts by Streptavidin-Alexa647 conjugate. Processed slides are scanned using an Axon 4000B Scanner (Molecular Devices Corp., Sunnyvale, CA) with the red laser set to 635 nm, at power 100% and PMT 800 setting, and a scan resolution of 10 μm.

5. Raw Data Analysis

This is an important aspect of miRNA profiling, because these arrays are low-density and no clear strategy for data normalization was available. With the exception of the U6 small nuclear RNA, no other small RNA was used in Northern blotting experiments for normalization. In spite of this, we were able to prove that our approach to data analysis gives data that is easily confirmed. During this step we are assisted by experienced bioinformaticians at CCC-OSU and University of Ferarra, Italy. Raw data are normalized and analyzed in GeneSpring® software (Silicon Genetics, Redwood City, CA). Expression data are median-centered using both GeneSpring normalization option or Global Median normalization of the Bioconductor package (www.bioconductor.org). We did not find any substantial difference. Statistical comparisons are done using both the GeneSpring ANOVA tool and the SAM software (Significance Analysis of Microarray, http://www-stat.stanford.edu/~tibs/SAM/index.html). miRNA predictors are calculated by using PAM software (Prediction Analysis of Microarrays, http://www-stat.stanford.edu/~tibs/PAM/index.html); the Support Vector Machine tool of GeneSpring is used for the cross-validation and test-set prediction. In this way, the miRNAs able to best separate the groups are identified and confirmed by two different methods of raw data normalization. All data are submitted using MIAMExpress to the Array Express database at http://www.ebi.ac.uk/arrayexpress/.

6. Validation of miRNAs Results

Confirmation of microarray data is done by Northern blots and quantitative real-time RT-PCR analysis. For Northern blot analysis, 10 to 20 μg of total RNA is used for each sample to run on a 15% polyacrylamide denaturing (urea) Criterion precast gel (Bio-Rad), and then transferred onto Hybond-N+ membrane (Amersham Pharmacia Biotech). The blots are performed as described. Quantitative RT-PCR for miRNA precursors and active molecules are performed as described for precursor miRNAs or for active molecules of miRNA.

7. Sample Data of miRNA Profiling by Microarray

The output of raw data from the microarray facility is shown in the following text:

1. The .gpr raw data extracted by GenePix 6.0 software in .txt file (Figure 8.3).
2. The raw data txt file converted into Microsoft Excel file for further statistical data analysis (Figure 8.4).

The end-result data will mainly consist of (1) dendrograms showing the clustering of multiple samples according to the miRNAs expression (Figure 8.5) and (2) a gene list containing genes differentially expressed at high statistical significance ($P < 0.01$) and biologically significant folds between the different categories of samples. It is important to note that the filter on fold-change is set to 1.2 because this threshold, already used for miRNAs analyzed with the same chip in published papers, was demonstrated to reflect a real biological difference. An example of SAM output is presented in Table 8.4.

ACKNOWLEDGMENTS

Work in Dr. Calin laboratory is supported by a Kimmel Foundation Scholar award and by the CLL Global Research Foundation. Dr. Liu is the Director of Microarray Facility at Comprehensive Cancer Center at Ohio State University. We acknowledge Dr. Carlo Croce for continuing support, and Drs. Massimo Negrini, Manuela Ferracin, Cistian Taccioli, and Stefano Volinia for the statistical analyses of miRNA microarray data. We apologize to many colleagues whose work was not cited because of space limitations.

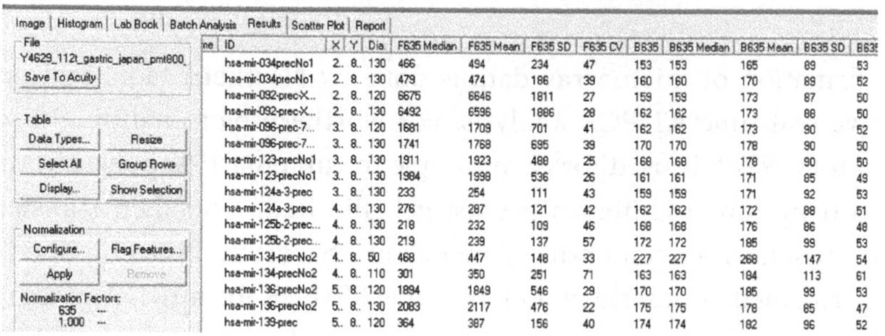
Image | Histogram | Lab Book | Batch Analysis | Results | Scatter Plot | Report
File
Y4629_112t_gastric_japan_pmt800_
Save To Acuity
Table
Data Types...
Resize
Select All
Group Rows
Display...
Show Selection
Normalization
Configure...
Flag Features...
Apply
Remove
Normalization Factors:
635 ---
1.000 ---

ID	X	Y	Dia.	F635 Median	F635 Mean	F635 SD	F635 CV	B635	B635 Median	B635 Mean	B635 SD	B63
hsa-mir-034precNo1	2..	8..	130	466	494	234	47	153	153	165	89	53
hsa-mir-034precNo1	2..	8..	130	479	474	186	39	160	160	170	89	52
hsa-mir-092-prec-X...	2..	8..	120	6675	6646	1811	27	159	159	173	87	50
hsa-mir-092-prec-X...	2..	8..	130	6492	6596	1886	28	162	162	173	88	50
hsa-mir-096-prec-7...	3..	8..	120	1681	1709	701	41	162	162	173	90	52
hsa-mir-096-prec-7...	3..	8..	130	1741	1768	695	39	170	170	178	90	50
hsa-mir-123-precNo1	3..	8..	130	1911	1923	488	25	168	168	178	90	50
hsa-mir-123-precNo1	3..	8..	130	1984	1998	536	26	161	161	171	85	49
hsa-mir-124a-3-prec	3..	8..	130	233	254	111	43	159	159	171	92	53
hsa-mir-124a-3-prec	4..	8..	130	276	287	121	42	162	162	172	88	51
hsa-mir-125b-2-prec...	4..	8..	130	218	232	109	46	168	168	176	86	48
hsa-mir-125b-2-prec...	4..	8..	130	219	239	137	57	172	172	185	99	53
hsa-mir-134-precNo2	4..	8..	50	468	447	148	33	227	227	268	147	54
hsa-mir-134-precNo2	4..	8..	110	301	350	251	71	163	163	184	113	61
hsa-mir-136-precNo2	5..	8..	120	1894	1849	546	29	170	170	185	99	53
hsa-mir-136-precNo2	5..	8..	130	2083	2117	476	22	175	175	178	85	47
hsa-mir-139-prec	5..	8..	120	364	387	156	40	174	174	182	96	52

FIGURE 8.3 The GenePix Results (GPR) .gpr raw data extracted by GenePix 6.0 software in .txt file. This is a text file format developed by Axon Instruments that is used to save GenePix Results data. ID = probe ID; X and Y are the coordinations of individual oligo spot on the microarray slide; Dia. = spot diameter based on the signal intensity of the spot detected on the assayed image; F635 median = median signal pixel intensity value of spot detected at laser 635 nm; F635 Mean = mean (average) signal pixel intensity value of spot detected at laser 635 nm; F635 SD = Standard Deviation of pixel intensity value of the spot detected at laser 635 nm; F635CV = Coefficient Variability of pixel intensity value of the spot at laser 635 nm; B635 = pixel intensity of local Background in surrounding area of the spot detected at laser 635. B635 Median = median pixel intensity of local background detected at laser 635; B635 mean = mean (average) pixel intensity of local background detected at laser 635, and so on. For the details of Genepix Pro 6.0 software, please visit http://www.moleculardevices.com/pages/software/gn_genepix_pro.html.

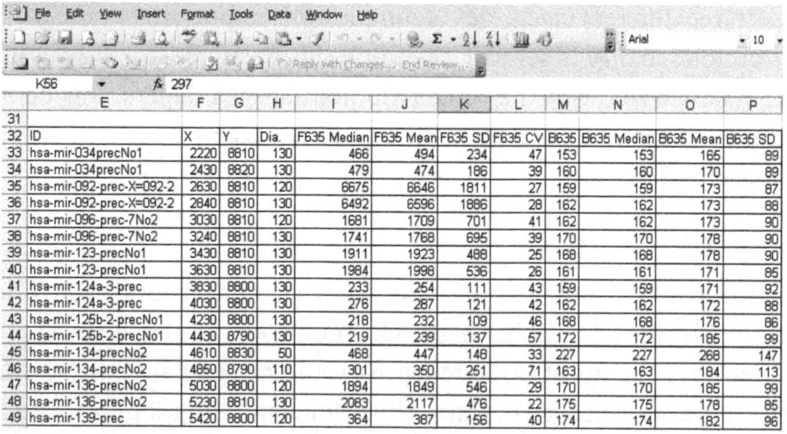
File Edit View Insert Format Tools Data Window Help
Arial
10
Reply with Changes... End Review...
K56
297

	E	F	G	H	I	J	K	L	M	N	O	P
31												
32	ID	X	Y	Dia.	F635 Median	F635 Mean	F635 SD	F635 CV	B635	B635 Median	B635 Mean	B635 SD
33	hsa-mir-034precNo1	2220	8810	130	466	494	234	47	153	153	165	89
34	hsa-mir-034precNo1	2430	8820	130	479	474	186	39	160	160	170	89
35	hsa-mir-092-prec-X=092-2	2630	8810	120	6675	6646	1811	27	159	159	173	87
36	hsa-mir-092-prec-X=092-2	2840	8810	130	6492	6596	1886	28	162	162	173	88
37	hsa-mir-096-prec-7No2	3030	8810	120	1681	1709	701	41	162	162	173	90
38	hsa-mir-096-prec-7No2	3240	8810	130	1741	1768	695	39	170	170	178	90
39	hsa-mir-123-precNo1	3430	8810	130	1911	1923	488	25	168	168	178	90
40	hsa-mir-123-precNo1	3630	8810	130	1984	1998	536	26	161	161	171	85
41	hsa-mir-124a-3-prec	3830	8800	130	233	254	111	43	159	159	171	92
42	hsa-mir-124a-3-prec	4030	8800	130	276	287	121	42	162	162	172	88
43	hsa-mir-125b-2-precNo1	4230	8800	130	218	232	109	46	168	168	176	86
44	hsa-mir-125b-2-precNo1	4430	8790	130	219	239	137	57	172	172	185	99
45	hsa-mir-134-precNo2	4610	8830	50	468	447	148	33	227	227	268	147
46	hsa-mir-134-precNo2	4850	8790	110	301	350	251	71	163	163	184	113
47	hsa-mir-136-precNo2	5030	8800	120	1894	1849	546	29	170	170	185	99
48	hsa-mir-136-precNo2	5230	8810	130	2083	2117	476	22	175	175	178	85
49	hsa-mir-139-prec	5420	8800	120	364	387	156	40	174	174	182	96

FIGURE 8.4 The raw data txt file converted into a Microsoft Excel file for further statistical data analysis. All the terms at row 32 are described in the legend of .gpr files in Figure 8.3.

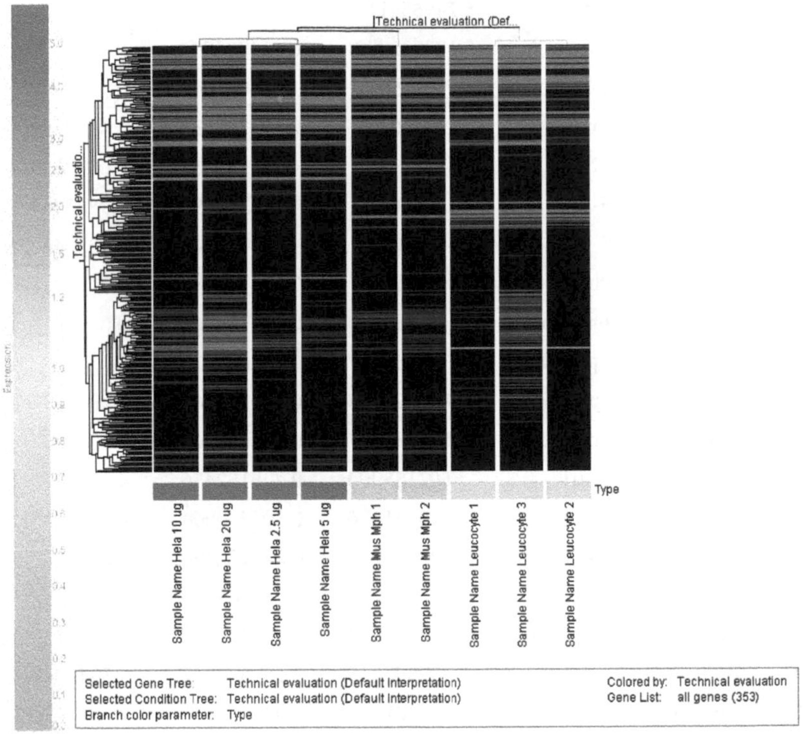
Technical evaluation (Def...
Technical evaluatio...
Type
Sample Name Hela 10 ug
Sample Name Hela 20 ug
Sample Name Hela 2.5 ug
Sample Name Hela 5 ug
Sample Name Mus Mph 1
Sample Name Mus Mph 2
Sample Name Leucocyte 1
Sample Name Leucocyte 3
Sample Name Leucocyte 2
Selected Gene Tree: Technical evaluation (Default Interpretation)
Selected Condition Tree: Technical evaluation (Default Interpretation)
Branch color parameter: Type
Colored by: Technical evaluation
Gene List: all genes (353)

FIGURE 8.5 Hierarchical clustering of samples according to microRNA expression. Unsupervised cluster of duplicate samples from human and mouse tissues (right side) or same sample with different starting amounts of total RNA (left side). Samples are in column, miRNAs in rows. A green-colored gene is downregulated compared to its median expression in all samples, red is upregulated, and yellow means no variation.

TABLE 8.4 The SAM Output of Microarray Data (positive scores denote overexpressed genes, whereas negative scores denote downregulated genes in one state — for example, cancer — relative to controls)

Row	Gene ID	Score (d)	Fold Change	q-value (%)
441	mmu-mir-345No1	10.37102596	5.867706418	0
383	mmu-mir-25No2	5.749654115	4.635803765	0
207	hsa-mir-345No1	5.178522772	6.624705736	0
103	hsa-mir-138-2-prec	3.740915915	13.78013905	0
153	hsa-mir-204-precNo2	3.490530828	7.492364538	4.050939964
451	mmu-mir-375No1	3.278119476	4.714569587	4.050939964
333	mmu-mir-185-prec	2.985666337	4.828383793	5.988346033
17	hsa-mir-007-1-prec	2.902430956	4.369463547	5.988346033
355	mmu-mir-200bNo1	-6.05691285	0.055498783	0
79	hsa-mir-123-precNo1	-4.45235779	0.191715062	0
87	hsa-mir-128b-precNo2	-4.386959727	0.058091275	0
85	hsa-mir-128a-precNo2	-3.856824003	0.103230162	4.304123711
86	hsa-mir-128b-precNo1	-3.817435507	0.069642266	4.304123711
465	mmu-mir-467No1	-3.801185995	0.206152456	4.304123711
281	mmu-mir-128aNo1	-3.673031526	0.106300064	4.304123711
24	hsa-mir-010a-precNo1	-3.523466759	0.024127996	4.304123711
283	mmu-mir-128bNo1	-3.498762752	0.083489905	4.304123711
285	mmu-mir-128-precNo1	-3.482973272	0.107292352	4.304123711
331	mmu-mir-182-prec	-3.437858835	0.159670902	4.304123711
280	mmu-mir-127-prec	-3.162901964	0.096361137	7.651775487
324	mmu-mir-181aNo1	-2.964061959	0.145973668	7.651775487
369	mmu-mir-218-2-precNo1	-2.958819217	0.281340441	7.651775487
458	mmu-mir-425No2	-2.901648286	0.22693507	7.651775487

Note: Gene ID represents the ID of the oligonucleotide from the chip.

Score (d) represents the value of t-test.

Fold change: If not otherwise specified, it represents ratio between Tumor against Normal (T/N), Treated against Control (T/C), and so on.

Q-value is the lowest false discovery rate (FDR) at which the gene is called significant. The FDR is the expected percentage of false predictions.

REFERENCES

1. Ambros, V. MicroRNA pathways in flies and worms: growth, death, fat, stress, and timing. *Cell* 113, 673–676, 2003.
2. Bartel, D.P. MicroRNAs: genomics, biogenesis, mechanism, and function. *Cell* 116, 281–297, 2004.
3. Calin, G.A. and Croce, C.M. MicroRNA signatures in human cancers. *Nat Rev Cancer* 6, 857–866, 2006a.
4. Calin, G.A. and Croce, C.M. MicroRNA-cancer connection: the beginning of a new tale. *Cancer Res* 66, 7390–7394, 2006b.
5. Calin, G.A. and Croce, C.M. MicroRNAs and chromosomal abnormalities in cancer cells. *Oncogene* 25, 6202–6210.
6. Calin, G.A., Dumitru, C.D., Shimizu, M., Bichi, R., Zupo, S., Noch, E., Aldler, H., Rattan, S., Keating, M., Rai, K. et al. Frequent deletions and down-regulation of micro- RNA genes miR15 and miR16 at 13q14 in chronic lymphocytic leukemia. *Proc Natl Acad Sci USA* 99, 15524–15529, 2002.
7. Calin, G.A., Ferracin, M., Cimmino, A., Di Leva, G., Shimizu, M., Wojcik, S., Iorio, M.V., Visone, R., Sever, N.I., Fabbri, M. et al. A Unique MicroRNA signature associated with prognostic factors and disease progression in B cell chronic lymphocytic leukemia. *N Engl J Med* 352, 1667–1676, 2005a.
8. Calin, G.A., Sevignani, C., Dumitru, C.D., Hyslop, T., Noch, E., Yendamuri, S., Shimizu, M., Rattan, S., Bullrich, F., Negrini, M., and Croce, C.M. Human microRNA genes are frequently located at fragile sites and genomic regions involved in cancers. *Proc Natl Acad Sci USA* 101, 2999–3004, 2004b.
9. Chen, C., Ridzon, D.A., Broomer, A.J., Zhou, Z., Lee, D.H., Nguyen, J.T., Barbisin, M., Xu, N.L., Mahuvakar, V.R., Andersen, M.R. et al. Real-time quantification of microRNAs by stem-loop RT-PCR. *Nucl Acids Res 33*, e179, 2005.
10. Croce, C.M. and Calin, G.A. miRNAs, cancer, and stem cell division. *Cell* 122, 6–7, 2005.
11. Cummins, J.M., He, Y., Leary, R.J., Pagliarini, R., Diaz, L.A.J., Sjoblom, T., Barad, O., Bentwich, Z., Szafranska, A.E., Labourier, E. et al. The colorectal microRNAome. *Proc Natl Acad Sci USA* 103, 3687–3692, 2006.
12. Lagos-Quintana, M., Rauhut, R., Lendeckel, W., and Tuschl, T. Identification of novel genes coding for small expressed RNAs. *Science* 294, 853–858, 2001.
13. Lau, N.C., Lim, L.P., Weinstein, E.G., and Bartel, D.P. An abundant class of tiny RNAs with probable regulatory roles in Caenorhabditis elegans. *Science* 294, 858–862, 2001.
14. Lee, R.C. and Ambros, V. An extensive class of small RNAs in Caenorhabditis elegans. *Science* 294, 862–864, 2001.
15. Lee, R.C., Feinbaum, R.L., and Ambros, V. A short history of a short RNA. *Cell* S116, S89–S92, 2004.
16. Liu, C.G., Calin, G.A., Meloon, B., Gamliel, N., Sevignani, C., Ferracin, M., Dumitru, C.D., Shimizu, M., Zupo, S., Dono, M. et al. An oligonucleotide microchip for genome-wide microRNA profiling in human and mouse tissues. *Proc Natl Acad Sci USA* 101, 9740–9744, 2004.

17. Lu, J., Getz, G., Miska, E.A., Alvarez-Saavedra, E., Lamb, J., Peck, D., Sweet-Cordero, A., Ebert, B.L., Mak, R.H., Ferrando, A.A. et al. MicroRNA expression profiles classify human cancers. *Nature* 435, 834–838, 2005.
18. Schmittgen, T.D., Jiang, J., Liu, Q., and Yang, L. A high-throughput method to monitor the expression of microRNA precursor. *Nucl Acids Res* 32, 43–53, 2004.
19. Sevignani, C., Calin, G.A., Siracusa, L.D., and Croce, C.M. Mammalian microRNAs: a small world for fine-tuning gene expression. *Mamm Genome* 17, 189–202, 2006.
20. Zhang, L., Huang, J., Yang, N., Greshock, J., Megraw, M.S., Giannakakia, A. et al. MicroRNAs exhibit high frequency genomic alterations in human cancer. *Proc Natl Acad Sci USA* 103, 9136–9141, 2006.

CHAPTER 9

RNAi

Li Qin Zhang

CONTENTS

Since it burst into epigenetic biology a few years ago, RNA interference (RNAi) as a gene-silencing strategy has transformed biological research and become one of most popular technologies to control gene expression. RNAi, occurs in plants, animals, and humans. RNAi via its tool small interfering RNAs (siRNAs), opens up exciting new avenues for use in gene technology. It is widely applied in many disciplines. Nearly every biological or biomedical lab is using this tool to turn off genes to determine gene functions and their roles in cell division, organogenesis, and pathogenesis of plant, animal, and human diseases as well as therapeutic utilities. RNAi has revolutionized genetics and heralded the start of a new research field. In this chapter, we will first concisely expound the concept, historical discovery, design guideline, and applications of siRNA, and then briefly introduce some online siRNA resources in a step-by-step tutorial format.

SECTION 1 INTRODUCTION TO RNAI AND SIRNA

1. What Are RNAi and siRNA?

RNAi stands for RNA interference. It is a mechanism in eukaryotic cells by which short fragments of double-stranded ribonucleic acid (dsRNA) interfere with the expression of a particular gene whose sequence is complementary to the dsRNA. The RNAi process is initiated by the ribonuclease protein Dicer, which binds and cleaves exogenous double-stranded RNA molecules to produce double-stranded fragments of 20–25 base pairs with a few unpaired overhang bases on each end. The short double-stranded fragments produced by Dicer, called siRNAs, are also known as short interfering RNA or silencing RNA. siRNAs have a well-defined structure: a short (usually 21-nt) double strand of RNA (dsRNA) with 2-nt 3' overhangs on either end. Each strand has a 5' phosphate group and a 3' hydroxyl (-OH) group. *In vivo* siRNAs are derived from either long dsRNAs or hairpin RNAs via the conversion of Dicer, whereas *in vitro* they can be artificially synthesized before being transfected into target cells to bring about the specific knockdown of a gene of interest. Essentially any gene could be a target of an siRNA based on its sequence complementarity to the appropriately tailored siRNA. This has made siRNAs an important tool for elucidating any gene function in the postgenomic era.

2. Historical Discovery of siRNA

RNAi is one of the most important historical discoveries in modern scientific history. In the early 1980s it was first noted that small RNA molecules (about 100 nucleotides in length) in *Escherichia coli (E. coli)* could inhibit protein translation. Today, about 25 cases of regulatory trans-acting antisense RNAs have been found in *E. coli.* In the early 1990s, similar phenomena were observed in the worm *Caenorhabditis elegans. In 2001, a*n extensive class of small RNAs called microRNA (miRNA) was identified in *Caenorhabditis elegans.* The miRNAs can regulate gene expression by base-pairing to mRNA, which results in either degradation of the mRNA or suppression of translation. Today, it is estimated that there are about 500 miRNAs in mammalian cells, and that about 30% of all genes are regulated by miRNAs. The RNAi phenomenon was also observed in plants. In the early and middle 1990s, researchers such as David Baulcombe of the Sainsbury Laboratory in Norwich, U.K., determined that adding genes to plants sometimes turned off the endogenous counterparts, a phenomenon then called *cosuppression.* A few years later plant virologists observed that plants carrying only short regions of viral RNA sequences not coding for any viral protein showed enhanced tolerance, to or even resistance against, virus infection. They called this phenomenon *virus-induced gene silencing,* or simply *VIGS.* These phenomena are collectively called *posttranscriptional gene silencing* (PTGS). A PTGS-like process called *quelling* was also established in the fungus *Neurospora crassa.* The real breakthrough of RNAi was made by Andrew Fire and Craig Mello on the regulation of gene expression in the nematode worm *Caenorhabditis elegans.* When they injected either a "sense" or "antisense" mRNA molecules encoding a muscle protein into a worm separately, no changes in the behavior of the worms were found. But when Fire and Mello injected sense and antisense RNA together, they observed that the worms displayed peculiar twitching movements. Similar movements were seen in worms that completely lacked a functioning gene for the muscle protein. Fire and Mello deduced that double-stranded RNA can silence genes, and that this RNA interference is specific for the gene whose code matches that of the injected RNA molecule; thus, Fire and Mello proposed that this is RNA interference, now commonly abbreviated to RNAi. Fire and Mello published their landmark findings in the journal *Nature* on February 19, 1998. Their discovery clarified many confusing and contradictory experimental observations and revealed a natural mechanism for controlling the flow of genetic information. Because of this immensely significant work, The Nobel Assembly at Karolinska Institutet bestowed the Nobel Prize in Physiology or Medicine for 2006 jointly upon Andrew Z. Fire and Craig C. Mello for their discovery of "RNA interference — gene silencing by double-stranded RNA."

3. The RNAi Mechanism

RNAi is an RNA-dependent gene silencing process that is mediated by the RNA-induced silencing complex (RISC). The process is initiated by the ribonuclease protein Dicer, which binds and cleaves exogenous or endogenous double-stranded RNA molecules to produce double-stranded fragments of 20–25 base pairs called siRNAs, with a few unpaired overhang bases on each end. The siRNA are separated and integrated into the active RISC complex. It is the catalytically active components of the RISC complex, known as *argonaut proteins* (endonucleases) in animals, that mediate the siRNA-induced cleavage of the target mRNA. Argonaute proteins have been identified as localized to specific regions in the cytoplasm called cytoplasmic bodies, which are also local regions of high mRNA decay rates. Because the fragments produced by Dicer are double stranded, they could each in theory produce a functional siRNA; however, only one of the two strands — known as the *guide strand* — binds the argonaute protein and leads to gene silencing. The other *anti-guide strand* or *passenger strand* is degraded as an RISC substrate during the process of RISC activation. The strand selected as the guide tends to be the strand whose 5' end is more stable, but strand selection is not dependent on the direction in which Dicer cleaves the dsRNA before RISC incorporation.

The natural occurrence of the RNA interference machinery is not fully understood, but it is known to be involved in miRNA processing and the resulting translational repression. miRNAs, which are encoded in the genome and have a role in gene regulation, typically have incomplete base pairing and only inhibit the translation of the target mRNA; in contrast, RNA interference as used in the laboratory typically involves perfectly base-paired dsRNA molecules that induce mRNA cleavage. After integration into the RISC, siRNAs base-pair to their target mRNA and induce the RISC component protein argonaute to cleave the mRNA, thereby preventing it from being used as a translation template.

Organisms vary in their cells' ability to take up foreign dsRNA and use it in the RNAi pathway. The effects of RNA interference are both systemic and heritable in plants and in *C. elegans*, although not in *Drosophila* or mammals, due to the absence of RNA replicase in these organisms. In plants, RNAi is thought to propagate through cells via the transfer of siRNAs through plasmodesmata.

4. siRNA Applications

siRNAs have been widely applied to downregulate gene expression. This application has two clear objectives. The first is to investigate the function of genes, and the second is to determine which genes are involved in diseases. Its current applications are summarized in the following text.

A tool for functional genomics. After the completion of human sequencing and a number of animal and plant genome sequencings, the challenge has shifted from the discovery of new genes to the elucidation of gene functions. Gene functions can usually be inferred from their expression patterns in different time, space, and conditions. The examination of siRNA's effect on its sequence-specific inhibition of target gene make siRNA a valuable tool in linking genes to their cellular function. Previously, gene knockdown in mammalian organisms was mainly restricted to mouse species. Gene knockout mouse strategy requires dedicated technical training and facilities such that it limits its broad applicability into the determination of gene functions. With the advent of a convenient and economical siRNA tool, gene knockdown technology is accessible to all researchers and will accelerate the progress of functional characterization of all genes. A significant number of researchers have utilized siRNA technology in their endeavors to comprehend the functional roles of individual genes in physiology and pathology. siRNA can also be a powerful tool to dissect any signal transduction pathway. After treating a cell with a gene-specific siRNA, profiling the differentially expressed genes can reveal how many pathways and how many genes in each pathway are affected by the gene of interest. The order of each gene in a particular pathway can be assigned by sequentially knocking down them with their cognate siRNAs and examining their effects.

Identification of disease-associated genes and drug targets. Since its initial discovery, siRNA has been rapidly employed for the identification of disease-associated genes. By knocking down genes in cell-based studies and animal models for the characterization of their roles in various cellular processes including endocytosis, apoptosis, and the cell cycle, siRNA serves as a tool not only for functional genomics but also for the Identification of disease-associated genes. Several examples are summarized in the following text. The implication of the chemokine receptor chemokine (C–X–C motif) receptor 4 (CXCR4) in the proliferation of breast cancer was revealed when its expression was inhibited by its siRNA in the breast cancer cell line MDA-MB-231 and the rate of cell expansion slowed dramatically. The application of siRNA also showed $\alpha_6\beta_4$ integrin and EpCAM receptors on the same cells involved in cellular invasion and metastasis. Silencing of viral genes by RNAi has shed light on some viral diseases, such as HIV, hepatitis, and severe acute respiratory syndrome (SARS)-associated coronavirus. siRNA has also been increasingly used in identification and validation of drug targets for the treatment of diseases. Reducing the expression of a potential therapeutic target by siRNA and evaluating the desired phenotype results should provide a reference on whether an inhibitor of the same target gene would have any therapeutic value or not. This approach has been met with varying degrees of success to improve target therapeutics for cancer, metabolic, inflammatory, infectious, neurological, and other types of disease. Given the initial success and promising potential, RNAi has drawn increasing attention for its potential clinical applications. Currently, there are a number of biotechnology companies developing clinical applications of siRNA in various human diseases. In addition to the ease of synthesis and low production costs relative to protein or antibody therapies, data indicate that siRNA has favorable pharmacokinetic properties and can be delivered to a wide range of organs. However, blood stability, delivery, poor intracellular uptake, and nonspecific immune stimulation still present significant challenges for the development of RNAi reagents for clinical use.

It is anticipated that further understanding of RNAi biology coupled with rigorous performance evaluation will yield reliable and powerful tools for biological inquiry as well as of disease-associated gene and drug target identifications.

5. siRNA Design

No matter what methods, synthetic (chemical, *in vitro* transcription or PCR expression cassettes) or vector based (plasmid or virus), are used to produce siRNA, the first step in all processing is to design siRNA by choosing the siRNA target site from a given cDNA sequence. Although a number of online siRNA design softwares (free or commercial sources) are available and each may have its own design algorithm, siRNAs can be designed in-house by following some generally accepted rules on how to choose an ideal siRNA target site: (1) Targeted regions on the cDNA sequence of a targeted gene should be located 50–100 nt downstream of the start codon (ATG) on the target mRNA sequence; this will enhance the chance of success for the intended siRNA as the region surrounding ATG is usually occupied by translational and regulational proteins. (2) Search for sequence motif $AA(N_{19})TT$ or $NA(N_{21})$, or $NAR(N_{17})YNN$, where N is any nucleotide, R is purine (A, G), and Y is pyrimidine (C, U). (3) Avoid sequences with > 50% G + C content. (4) Avoid stretches of four or more nucleotide repeats. (5) Avoid sequences that share more than 16 or 17 contiguous base pairs of homology with other related or unrelated genes. Homology search is essential to minimize off-target effects of any siRNA. Here are some BLAST tools for homology search: NCBI Blast tool — http://www.ncbi.nlm.nih.gov/BLAST/ (blastn); Blat tool on UCSC Genome Web site — http://genome.ucsc.edu/cgi-bin/hgBlat; and Ensembl Blast — http://www.ensembl.org/Multi/blastview. Normally 3 to 4 siRNAs per gene should be tested for specificity and efficiency of gene silencing. At the same time, a negative control siRNA should be included, which is a scramble nucleotide sequence of the gene-specific siRNA lacking homology to any other genes.

6. siRNA Resources

Rich siRNA resources such as siRNA design software, large collaborative siRNA projects, and siRNA suppliers are available.

siRNA design software. Numerous online siRNA design software is a click away. Three selected free siRNA design software packages are briefly introduced here:

1). siRNA at Whitehead (Whitehead Institute for Biomedical Research): siRNA at Whitehead (http://jura.wi.mit.edu/bioc/siRNAext/) helps select oligos to knock down a gene of interest based on its position within the sequence, the snps, and other criteria. The results are made available directly on the Web and will be emailed to the user when ready.

2). Gene-specific siRNA selector (Bioinformatics Facility, The Wistar Institute): Gene-specific siRNA selectors (http://hydra1.wistar.upenn.edu/Projects/siRNA/siRNAindex.htm) scan a target gene for candidate siRNA sequences that satisfy user-adjustable rules. Selected candidates are then screened to identify those siRNA sequences that are specific to the gene of interest.

3). siDirect: siDirect (http://design.RNAi.jp/) is a Web-based online software system for computing highly effective siRNA sequences with maximum target specificity for mammalian RNAi. Most commercial suppliers of siRNA products also provide free online siRNA design software such as Block-iT RNAi Designer (Invitrogen), siRNA Target Finder (Ambion), and siRNA Design (Integrated DNA Technologies).

Large collaborative siRNA projects. Several representative large collaborative siRNA projects are briefly described in the following text:

1). The RNAi Consortium (TRC): TRC is a public-private consortium based at the Broad. It consists of Broad Institute, Harvard Medical School, the Massachusetts Institute of Technology, Dana-Farber Cancer Institute, the Whitehead Institute for Biomedical

Research, Novartis, Eli Lilly, Bristol-Myers Squibb, Sigma-Aldrich, and research institute Academia Sinica in Taiwan. Their 3-year, $18 million initiative, would create a library of materials to conduct RNAi experiments on 15,000 human genes and 15,000 mouse genes. A total of 150,000 custom-designed plasmids that express short and unique pieces of RNA (known as short hairpin RNAs, or shRNAs) which target specific genes would be created and validated. This fundamental resource has been made available to scientists worldwide through Sigma-Aldrich (http://www.sigmaaldrich.com/) and Open.Biosystem (http://www.openbiosystems.com).

2). CGAP RNAi at NCI: The NCI is part of the consortium supporting the preparation of human and mouse libraries containing RNAi constructs that target cancer-relevant and other genes. The clones, prepared in the laboratory of Greg Hannon of Cold Spring Harbor, produce small RNA molecules, shRNAs, that are available to the public from Open.Biosystem (http://www.openbiosystems.com). As of October 14, 2006, there are 82,306 human and 50,875 mouse siRNA shRNA clones prepared. siRNA searchers can also find genes containing RNAi constructs from http://cgap.nci.nih.gov/RNAi.

3). The Arabidopsis Small RNA Project: The Arabidopsis Small RNA Project is supported by a 2010 Project grant from the National Science Foundation. The project uses facilities and resources provided by the Department of Botany and Plant Pathology, the Center for Gene Research and Biotechnology, Oregon State University. This project seeks to characterize and functionally analyze the two major classes of endogenous small RNAs: microRNAs (miRNAs) and short-interfering RNAs (siRNAs). The specific aims of the project include (1) discovery of new miRNAs and siRNAs, (2) functional analysis of miRNAs and siRNAs, and (3) Functional analysis of multidomain RNaseIII-like genes. Readers can get further information from their Web site (http://asrp.cgrb.oregonstate.edu/).

siRNA Databases. A number of siRNA databases exist. Here, three selective siRNA databases are briefly described:

1. RNAi resources at the NCBI: RNAi resources at the NCBI (http://www.ncbi.nlm.nih.gov/projects/genome/RNAi/) provide a portal to the information on the knowledgebase of RNAi, several NIH- or NSF-funded siRNA consortiums, hyperlinks to a number of siRNA design software and most commercial suppliers of siRNA products as well as stored sequences of RNAi reagents, and experimental results generated using those reagents via The Probe database.

2. HuSiDa: HuSiDa (http://www.hnman-siRNA-database.net), the human siRNA database was established by Matthias Truss and colleagues at Universitatsmedizin and Humboldt-University, Berlin, Germany. This database is a collection of the published siRNA data in PubMed. The database provides sequences of published functional siRNA molecules targeting human genes and important technical details of the corresponding gene silencing experiments, including the mode of siRNA generation, recipient cell lines, transfection reagents and procedures, and direct links to published references. The database also includes information on the quality of the siRNA, such as the silencing activity and the homology to its target gene mRNA sequence. To estimate the siRNA sequence off-target effects, a value given to each entry that has been blasted against the RefSeq database search to identify the length of the longest contiguous part of the siRNA that, in addition to the target mRNA, also matches other mRNA sequences.

3. Protein Lounge siRNA database. Protein Lounge siRNA database (http://www.proteinlounge.com/sirna), is a commercially available Web site. This siRNA database contains siRNA targets against all known mRNA sequences through a variety of organisms including *Homo sapiens*, *Mus musculus*, *Rattus norvegicus*, *Bos Taurus*, *Danio rerio*, *Drosophila melanogaster*, *Anopheles gambiae*, *Caenorhabditis elegans*, *Arabidopsis thaliana*, and *Saccharomyces cerevisiae* (ten different species). You also can select siRNA according to the different protein function subtypes such as kinases, phosphatases, transcription factors, and disease genes. The entire siRNA targets in the database were designed by following the algorithm created by Dr. Thomas Tuschl (Max-Planck Institute). Many of the targets have been tested through Western blot to see if these targets knocked down gene production. The remaining targets in the database have been validated through statistical comparisons. All siRNA targets have been

screened to remove any siRNA that shares homology with other sequences, thus producing targets which are specific to the gene of interest. Also, they claim that in many cases the siRNA was able to produce a knockdown of about 99%. To retrieve data from this Web site, just search the gene name or the GenBank accession number.

SECTION 2 SEARCH FOR SIRNA INFORMATION

In this part, *Homo sapiens* epidermal growth factor receptor (EGFR) gene will be used to demonstrate how to search for its siRNA information from siRecords and display its siRNA data in a record:

1. Search Demo

1). Go to the siRecords home page (http://sirecords.umn.edu/siRecords/index.php). You can start search either by Genbank accession or GI number of the gene of interest. You also can go to Advanced Search, using other information like Host Species, Cell Type, Method, and Sequence to begin the search. If you are looking for particular paper, the search can be started by entering the Pubmed ID, Authors, Title of Article, and Abstract of Article. Here, the Genbank accession number is demonstrated. Type the EGFR Genbank accession number NM_005228 in the text box at the top of the page as displayed in Figure 9.1.

2). Click "search Gene," and the next screen shows the gene name under the accession number (Figure 9.2).

3). Click "NM_005228.3/41327737," and the gene information table will be shown; underneath are the siRNA records of the gene table, which include siRNA sequences (Figure 9.3).

4). Click one of the sequences; in the record fold will be displayed all the detailed information about this particular siRNA, targeting mRNA, and resource article (Figure 9.4). To select a different siRNA sequence, just click "back" to the previous page and choose another one.

2. Sample Data for Section 2

Homo sapiens epidermal growth factor receptor (EGFR) cDNA (NM_005228).

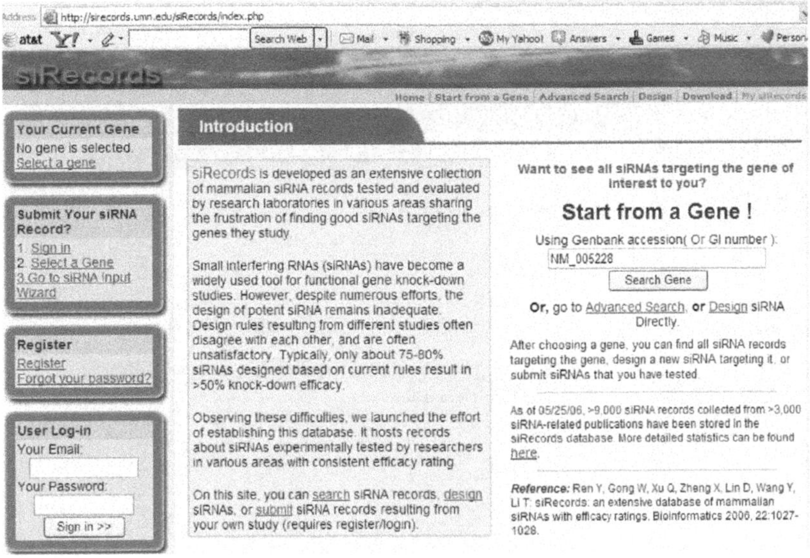
http://sirecords.umn.edu/siRecords/index.php
siRecords
Home | Start from a Gene | Advanced Search | Design | Download
Your Current Gene
No gene is selected.
Select a gene
Submit Your siRNA Record?
1. Sign in
2. Select a Gene
3. Go to siRNA Input Wizard
Register
Register
Forgot your password?
User Log-in
Your Email:
Your Password:
Sign in >>
Introduction
siRecords is developed as an extensive collection of mammalian siRNA records tested and evaluated by research laboratories in various areas sharing the frustration of finding good siRNAs targeting the genes they study.
Small interfering RNAs (siRNAs) have become a widely used tool for functional gene knock-down studies. However, despite numerous efforts, the design of potent siRNA remains inadequate. Design rules resulting from different studies often disagree with each other, and are often unsatisfactory. Typically, only about 75-80% siRNAs designed based on current rules result in >50% knock-down efficacy.
Observing these difficulties, we launched the effort of establishing this database. It hosts records about siRNAs experimentally tested by researchers in various areas with consistent efficacy rating.
On this site, you can search siRNA records, design siRNAs, or submit siRNA records resulting from your own study (requires register/login).
Want to see all siRNAs targeting the gene of interest to you?
Start from a Gene !
Using Genbank accession(Or GI number):
NM_005228
Search Gene
Or, go to Advanced Search, or Design siRNA Directly.
After choosing a gene, you can find all siRNA records targeting the gene, design a new siRNA targeting it, or submit siRNAs that you have tested.
As of 05/25/06, >9,000 siRNA records collected from >3,000 siRNA-related publications have been stored in the siRecords database. More detailed statistics can be found here.
Reference: Ren Y, Gong W, Xu Q, Zheng X, Lin D, Wang Y, Li T. siRecords: an extensive database of mammalian siRNAs with efficacy ratings. Bioinformatics 2006, 22:1027-1028.

FIGURE 9.1 Search for EGFR siRNA information from siRecords. This screenshot displays the initiation of the search for the EGFR siRNA information using the Genbank accession number NM_005228.

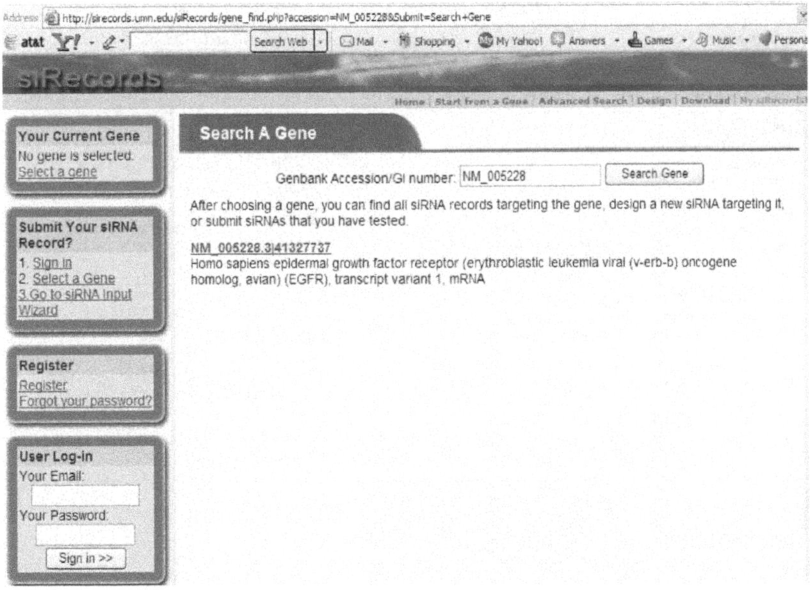
http://sirecords.umn.edu/siRecords/gene_find.php?accession=NM_005228&Submit=Search+Gene
siRecords
Home | Start from a Gene | Advanced Search | Design | Download
Your Current Gene
No gene is selected.
Select a gene
Submit Your siRNA Record?
1. Sign in
2. Select a Gene
3. Go to siRNA Input Wizard
Register
Register
Forgot your password?
User Log-in
Your Email:
Your Password:
Sign in >>
Search A Gene
Genbank Accession/GI number: NM_005228
Search Gene
After choosing a gene, you can find all siRNA records targeting the gene, design a new siRNA targeting it, or submit siRNAs that you have tested.
NM_005228.3|41327737
Homo sapiens epidermal growth factor receptor (erythroblastic leukemia viral (v-erb-b) oncogene homolog, avian) (EGFR), transcript variant 1, mRNA

FIGURE 9.2 The result matching the query.

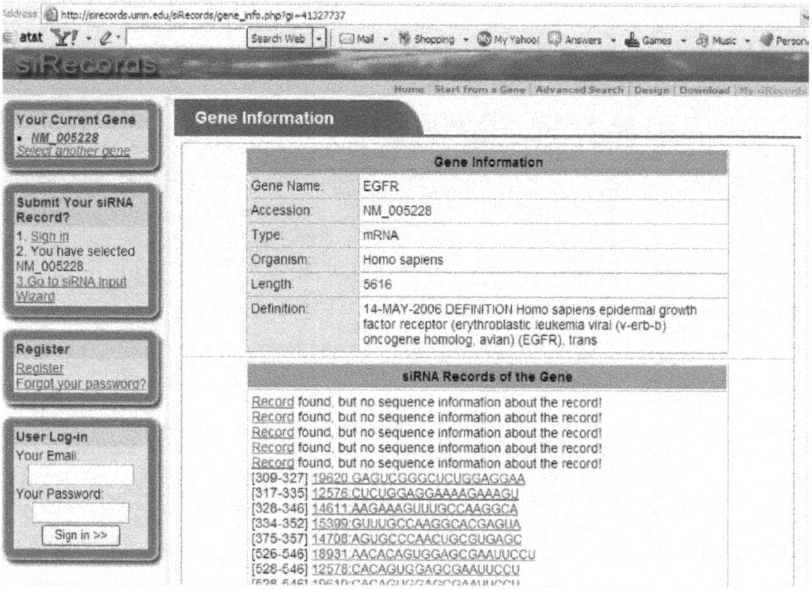
Your Current Gene
• NM_005228
Select another gene
Submit Your siRNA Record?
1. Sign in
2. You have selected NM_005228.
3. Go to siRNA Input Wizard
Register
Register
Forgot your password?
User Log-in
Your Email:
Your Password:
Sign in >>
Gene Information
Gene Information
Gene Name: EGFR
Accession: NM_005228
Type: mRNA
Organism: Homo sapiens
Length: 5616
Definition: 14-MAY-2006 DEFINITION Homo sapiens epidermal growth factor receptor (erythroblastic leukemia viral (v-erb-b) oncogene homolog, avian) (EGFR), trans
siRNA Records of the Gene
Record found, but no sequence information about the record!
Record found, but no sequence information about the record!
Record found, but no sequence information about the record!
Record found, but no sequence information about the record!
Record found, but no sequence information about the record!
[309-327] 19620:GAGUCGGGCUCUGGAGGAA
[317-335] 12576:CUCUGGAGGAAAAGAAAGU
[328-346] 14611:AAGAAAGUUUGCCAAGGCA
[334-352] 15399:GUUUGCCAAGGCACGAGUA
[375-357] 14708:AGUGCCCAACUGCGUGAGC
[526-546] 18931:AACACAGUGGAGCGAAUUCCU
[528-546] 12578:CACAGUGGAGCGAAUUCCU

FIGURE 9.3 Output of the gene information (upper panel) and siRNA record of the gene (lower panel).

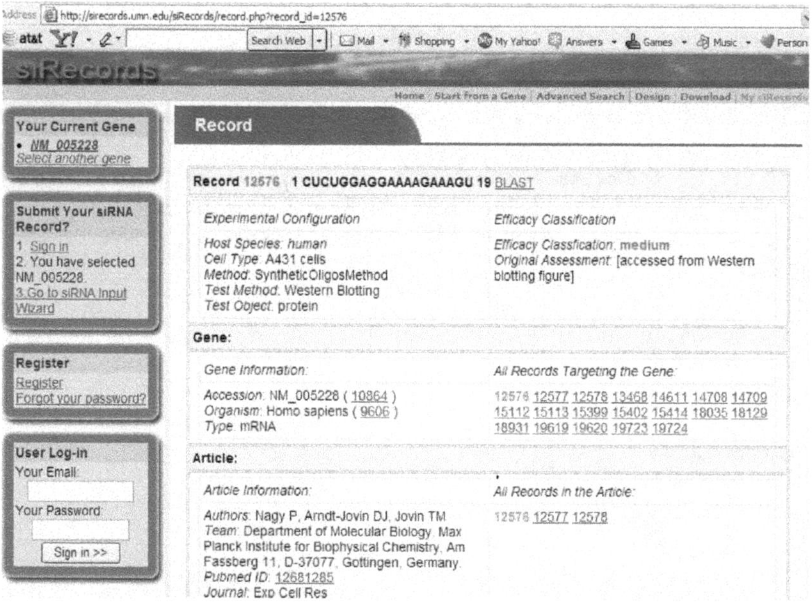
Your Current Gene
• NM_005228
Select another gene
Submit Your siRNA Record?
1. Sign in
2. You have selected NM_005228.
3. Go to siRNA Input Wizard
Register
Register
Forgot your password?
User Log-in
Your Email:
Your Password:
Sign in >>
Record
Record 12576 1 CUCUGGAGGAAAAGAAAGU 19 BLAST
Experimental Configuration
Host Species: human
Cell Type: A431 cells
Method: SyntheticOligosMethod
Test Method: Western Blotting
Test Object: protein
Efficacy Classification
Efficacy Classification: medium
Original Assessment: [accessed from Western blotting figure]
Gene:
Gene Information:
Accession: NM_005228 (10864)
Organism: Homo sapiens (9606)
Type: mRNA
All Records Targeting the Gene:
12576 12577 12578 13468 14611 14708 14709 15112 15113 15399 15402 15414 18035 18129 18931 19619 19620 19723 19724
Article:
Article Information:
Authors: Nagy P, Arndt-Jovin DJ, Jovin TM
Team: Department of Molecular Biology, Max Planck Institute for Biophysical Chemistry, Am Fassberg 11, D-37077, Gottingen, Germany.
Pubmed ID: 12681285
Journal: Exp Cell Res
All Records in the Article:
12576 12577 12578

FIGURE 9.4 Display the record of a particular EGFR siRNA. The top panel shows the siRNA record, the middle panel gene information, and the lower panel reference article information.

REFERENCES

1. Fire, A., Xu, S.Q., Montgomery, M.K., Kostas, S.A., Driver, S.E., and Mello, C.C. Potent and specific genetic interference by double-stranded RNA in *Caenorhabditis elegans*. *Nature* 391: 806–811, 1998.
2. McManus, M.T. and Sharp, PA. Gene silencing in mammals by small interfering RNAs. *Nat Rev Genet* 3(10): 737–747, 2002.
3. Advanced Information on The Nobel Prize in Physiology or Medicine 2006 RNA INTERFERENCE (http://nobelprize.org/nobel_prizes/medicine/laureates/2006/adv.html), October 2, 2006.
4. Zhou, D., He, Q.S., Wang, C., Zhang, J., and Wong-Staal, F. RNA interference and potential applications. *Curr Top Med Chem* 6(9): 901–911, 2006.
5. Lee, R.C. and Ambros, V. An extensive class of small RNAs in Caenorhabditis elegans. *Science* 294, 862–864, 2001.
6. Lau, N.C., Lim, L.P., Weinstein, E.G., and Bartel, D.P. An abundant class of tiny RNAs with probable regulatory roles in Caenorhabditis elegans. *Science* 294, 858–862, 2001.
7. Lagos-Quintana, M., Rauhut, R., Lendeckel, W., and Tuschl, T. Identification of novel genes coding for small expressed RNAs. *Science* 294, 853–858, 2001.
8. Leung, R.K. and Whittaker, P.A. RNA interference: from gene silencing to gene-specific therapeutics. *Pharmacol Ther* 107(2): 222–239, 2005.
9. http://www.ncbi.nlm.nih.gov/projects/genome/RNAi/.
10. http://www.rnaiweb.com.

CHAPTER 10

Proteomic Data Analysis

Yurong Guo, Rodney Lui, Steven T. Elliott, and Jennifer E. Van Eyk

CONTENTS

Proteomics is an emerging scientific field that involves the identification, characterization, and quantification of proteins in cells, tissues, or body fluids. Protein characterization, in an ideal situation, includes amino acid sequence analysis, determination of specific splice variant, polymorphism, and posttranslational modifications (PTMs), identification of protein binding partners, cellular localization, and its potential function. Two-dimensional gel electrophoresis (2DE), mass spectrometry, and protein microarrays are important technologies in proteomics, and the related data analysis tools allow us to interpret the data and get meaningful information. In this chapter, proteomic data analysis, specifically, analysis of 2DE-derived data, mass-spectrometry-derived data, and data derived from protein microarrays will be reviewed. In each section, the basic steps and commonly used data analysis programs will be reviewed, and a step-by-step tutorial of how to use the most popular program will be given.

SECTION 1 ANALYZING 2DE-DERIVED DATA

Part I Introduction

1. Two-Dimensional Gel Electrophoresis: What Is It and Why Is It Useful?

Two-dimensional gel electrophoresis (2DE) is a powerful protein separation tool and one of the cornerstones of proteomic analyses. It separates intact proteins in the first dimension based on intrinsic pI (isoelectric focusing, IEF), and in the second dimension by molecular weight (MW or mass). The combined result of these separations is a two-dimensional spot pattern that can be compared to the spot pattern of a sample that has undergone a different treatment to examine differences in protein regulation.

The power of 2DE is displayed in its ability to resolve hundreds to thousands of proteins into distinct spots that can be visualized with several varieties of protein-staining methods. When combined with the identification potential of mass spectrometry, 2DE can be used to generate gel map protein databases, which are useful for subsequent differential analyses and can provide novel posttranslational modification information. Posttranslational modifications often shift the mass or isoelectric point of a protein, and these shifts can be apparent on 2DE gels. As a result, if a protein is posttranslationally modified, it can be identified from multiple spots, or a spot significantly different from its theoretical position on a gel.

2DE can provide a researcher with an enormous quantity of new data, but the analysis process can be quite involved. To quantify the differences between two or more samples separated by 2DE, the gels must be scanned and the images analyzed by 2DE-specific software. When the gels have been uploaded into the software, all protein spots must be detected, circled (to determine spot boundaries for quantification), and matched. The spot volumes should be corrected for aberrations in background (if necessary) and normalized to a specific gel. Any detected differences should be validated using multiple gels and statistical analyses. Before any analysis can take place, the gels should be qualitatively evaluated for consistency, and the staining and scanning of these gels should be of a high standard of consistency.

2. What Is Involved in Performing Analyses of 2DE Gels?

This section is intended to give the reader insight into 2DE analysis. It will briefly cover preparation and scanning of gels for objective image analysis and explain in detail the use of Progenesis software for quantitative image analysis.

A. Gel Preparation 2DE gels should be prepared and run as consistently as possible to ensure that any changes between gels can be attributed to biological significance and not technical error. Generally, protein loads should be regulated between gels using protein assays to ensure that the appearance of the gel image can be quantitatively linked to the loaded sample. Protein labeling or staining should be consistent, and technical replicates assessed for consistency of the methods. Generally, it is easier to obtain consistent staining results with end-point stains such as Coomassie-brilliant blue.

Prior to any comparisons, the total spot numbers and volumes (summation for all spot volumes per gel) should be compared between all samples for consistency. Measuring both total spot volume and spot number per gel can provide information on whether it is valid to use gels for comparison. If two gels show very minor variation in spot number and the majority of the spots in the sample fall in a linear quantification range and have roughly the same number and size of saturated spots, the researcher will be able to normalize the samples to a single experimental gel. However, if a gel with consistently larger spot volumes has many more than other gels in the set, some spots may be below detection thresholds on other gels, preventing optimal volume correction through normalization. This is more of an issue with stains that have a small linear range. Cy-labeling methods offer a linear quantification over several orders of magnitude (well beneath human visual thresholds), allowing normalization to be more readily applied between gels. Similar total spot volume with dramatically different spot numbers between gels may be the result of poor focusing or resolving issues. The consistency of the analyzed gels may be very important for determining the types of biological changes that can be determined from the gel set. As the technical variation is increased, changes between disease and control will have to be larger to appear statistically significant.

B. Scanning Parameters As with protein loading and staining, digital image capturing is a key factor in determining the quality of a 2DE analysis. Gel images should be captured at an appropriate size and image depth to allow enough contrast between samples in order to accurately mine the data for points of interest. Gels should be scanned at a size that allows the capture of all spot details and can allow the highest reasonable level of accurate spot detection. The term *reasonable level* is used because increasing image size for improving gel analysis results in diminishing returns: the larger image captures more of the gel's characteristics, but as the image's size grows, it approaches the resolution limits of 2DE. Once the image reaches this size, further increases will not improve analysis, but they will continue to increase the processing time required for analysis. Generally, images with resolutions between 200 and 350 dpi are ideal. For directly visual stains (silver, Coomassie, deep purple, etc.), a calibrated, linear, transmissive light scanner will yield the best results for analysis. This scanner should have 16-bit grayscale capability to allow the generation of 65,536 shades of gray instead of the 256 shades that an 8-bit scan allows. This feature allows for a more accurate quantification of the detected spots as the spot volume is more accurately reflected in the more subtle gradient provided by the 16-bit scan. Another scanner property of interest is optical density (OD) (Figure 10.1). The scanner utilized should be able to ensure that all (or as many as possible) of the spots detected are within a linear quantification range. Scanners for these gels should be able to measure at least 3.6 OD units.

If fluorescence is used as a means of quantification, several pieces of equipment can generate a useful gel image. However, few are able to generate an image that utilizes the entire linear range of the stain or dye. Ideally, a laser densitometer with the appropriate excitation wavelengths is best for quantitative analysis.

3. 2DE Gel Analysis Software

There are several varieties of software for the analysis of 2DE gels, ranging from simple viewers to massive display/analysis/statistical validation all-in-one proteomics packages. These software include Progenesis, DeCyder,

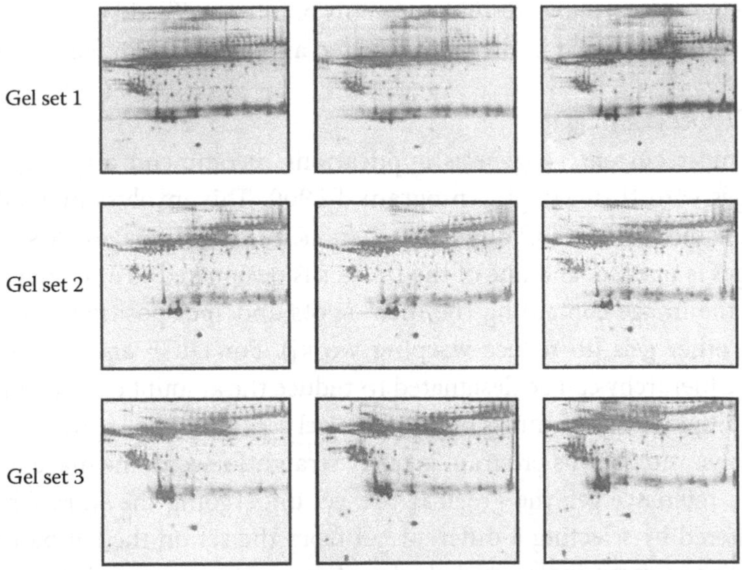
Gel set 1
Gel set 2
Gel set 3

FIGURE 10.1 The gel images used for the sample analysis. Silver-stained 24-cm pH 4–7 gels of 200 μg protein loading of albumin and IgG-depleted human serum.

Melanie, Z3, Image Investigator, Dymension, and many others. The components of these programs are similar in that they all have spot detection, matching, and quantification algorithms as well as a means for generation of statistically validated data, but they differ in cost, performance, differential in gel electrophoresis (DIGE) /non-DIGE capability, automation, and display and analysis method. Despite the large number of differences between the software, they all have the same desired end points: all spots detected perfectly with their boundaries accurately trace the edges of the spot (to allow perfect quantification), and all spots between gels match perfectly to their equivalent spot in another gel.

Part II Step-By-Step Tutorial

In this section, Progenesis-based analysis of 2DE gels will be demonstrated. Progenesis-based analysis was selected because it is one of the most common 2DE software packages, it is able to perform analysis on single or cross-stained gels, it offers automated analyses, and it has the option of manual correction. It is not the intention of this section to guide the user through every aspect and feature of the software as the programs have detailed manuals for this. The purpose of this section is to provide a start-to-finish protocol from the array of tools to quantitatively examine a gel set (Figure 10.2) and determine proteins of interest. To aid in clarification, the icons mentioned in the text are listed and displayed in Figure 10.3.

1. TT900 Warping

The most current Progenesis applications recommend aligning all gel images with their warping program TT900. This involves manually (or with some automation) aligning all spots on all gel images. A set of gel images is opened, and one of the images is designated as a reference (ideally, the image containing the most spots and spot positions consistent with other gels [to reduce warping work]). For DIGE analysis, a DIGE image hierarchy can be designated to reduce the amount of warping work as multiple image channels of the same gel can be aligned simultaneously. The operation of this program is fairly straightforward: the magenta layer is the reference gel; the green is the gel undergoing the warp (this can be altered by selecting a different gel from the set on the left panel). All misaligned green spots are clicked and dragged to the matching magenta spots, where the opposite colors overlay to give a grayscale matched image (Figure 10.4). The fade and checkerboard windows on the right side of the screen can be used to aid in the alignment. The image can be updated to

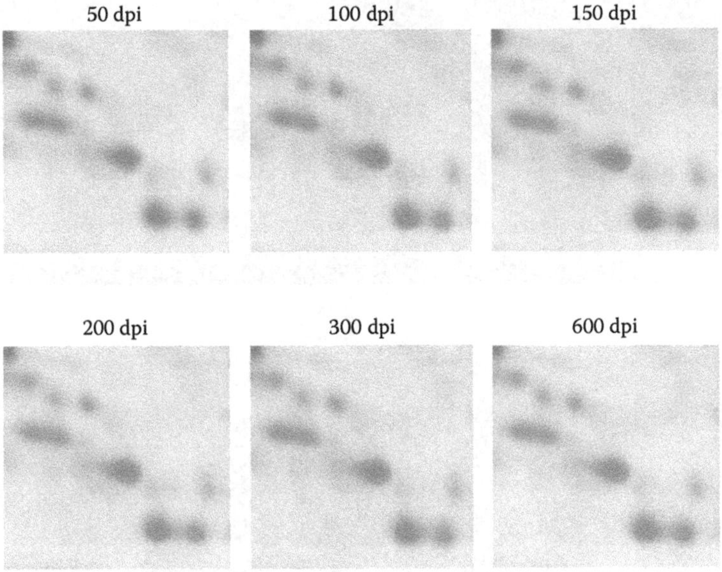
50 dpi
100 dpi
150 dpi
200 dpi
300 dpi
600 dpi

FIGURE 10.2 A comparison of dpi scan settings. At the current magnification, gel images scanned at 200 dots per inch (dpi) and above appear identical, but may subtly influence quantitative results. The quality of analysis decreases rapidly with progressive decrease from 200 dpi.

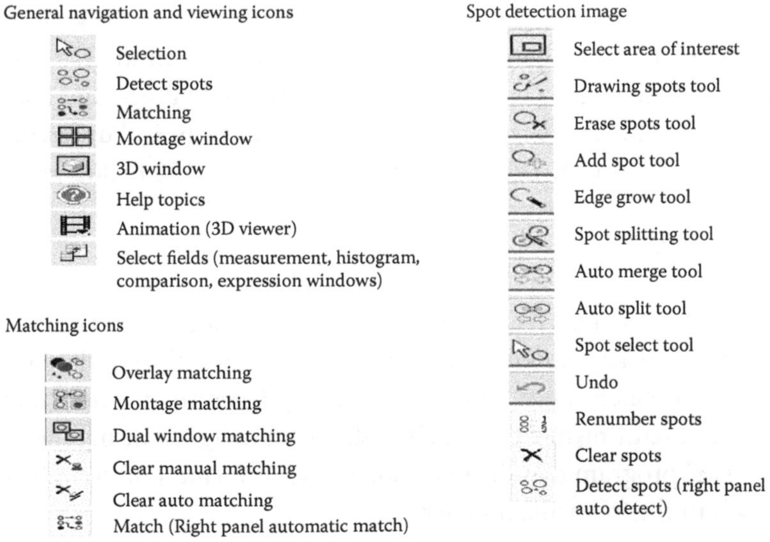
General navigation and viewing icons
Selection
Detect spots
Matching
Montage window
3D window
Help topics
Animation (3D viewer)
Select fields (measurement, histogram, comparison, expression windows)
Matching icons
Overlay matching
Montage matching
Dual window matching
Clear manual matching
Clear auto matching
Match (Right panel automatic match)
Spot detection image
Select area of interest
Drawing spots tool
Erase spots tool
Add spot tool
Edge grow tool
Spot splitting tool
Auto merge tool
Auto split tool
Spot select tool
Undo
Renumber spots
Clear spots
Detect spots (right panel auto detect)

FIGURE 10.3 Progenesis tool icons.

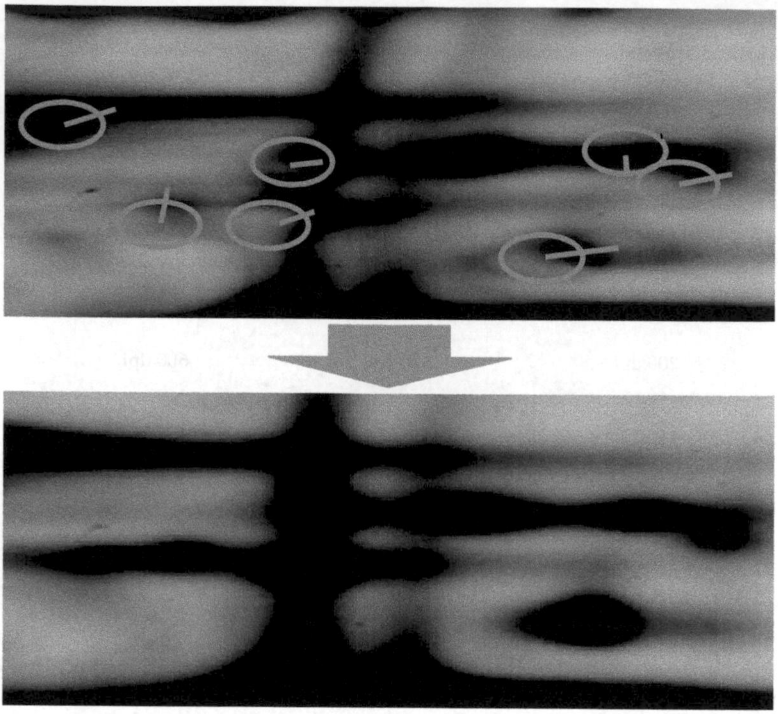

FIGURE 10.4 TT900 warping. Directly align as many spots as possible to master image.

view the overlay progress by occasionally applying warp to the vectors that are generated. The alignments should be performed until all matching spot constellations are completely overlaid—the only green and magenta spots should have come from actual biological or technical differences in the gel, not alignment. This can take anywhere from 100 to 300 vectors.

2. *Progenesis Analysis*

A. Automatic Analysis Gels selected for analysis should first be oriented and cropped in a consistent manner to minimize the effort of the software to align and match spots. This can be accomplished in an external drawing or photo-editing tool that can handle 16-bit images (editing the gels in an external program may, however, alter pixel information) or in Progenesis itself in the following manner:

- Open Progenesis
    - Under the heading marked "Tools" select "Image Manipulation"

- Select the gel to manipulate. The gel will appear in the new window. Flip, rotate, or crop as desired.
  - Select "Save" As to create a new image for the alterations or "Done" to apply the alterations to the file image. Be sure to place all images to be used for analysis in the same folder and store them on the local analysis PC drive. This makes operations faster and smoother than from a CD or network drive.

(It is not advisable to crop DIGE images as many software applications require matched pixel numbers between each channel of the dyes, and cropping these images may affect the ability to overlay the dual channel images in a project.)

When gel images are positioned and cropped as desired, begin the automated analysis as follows:

- Under the "Analysis" heading select "Automatic Analysis" — this will open up the analysis wizard (Figure 10.5a)
  - Name the experiment, and select the analysis type. Cross-stain analysis is for multiple images of the same gel (DIGE experiments), and any other experiments, including single-channel fluorescence, should be single-stain analysis.
    - The next step is selecting the images for analysis. Open the folder of interest and highlight all gels to be included in the analysis (Figure 10.5b).
      - If you are performing a DIGE analysis, you can organize your images at this point. Group images of each gel from all channels and highlight the internal standard by right-clicking on the Cy2 standard.
- The next window (Figure 10.5c) allows regions of interest to be selected for analysis. Spots will only be automatically detected and matched within the green window. Generally, it is best to start with a global image analysis (gel box covering the entire image), and remove incorrect or nonrelevant spot information later.
- The following window (Figure 10.5d) provides the option of combining technical replicates to generate a representative gel for a single sample. This option can reduce technical variation inherent to 2DE, improve the precision of a proteomic analysis and makes navigation

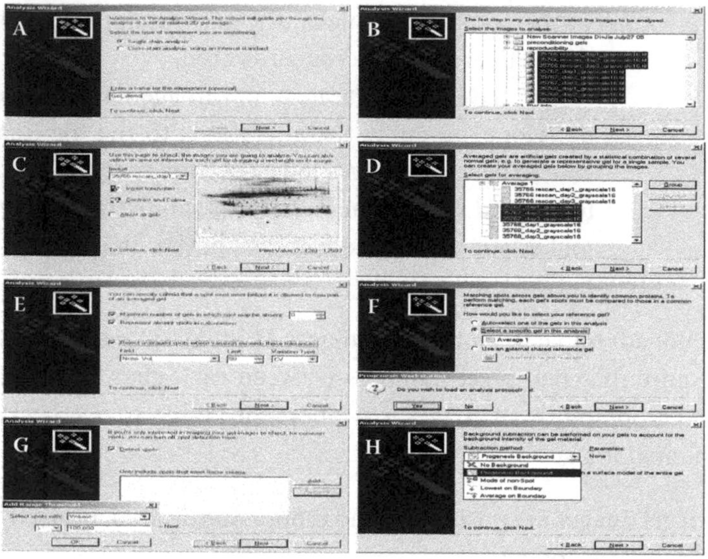
A
B
C
D
E
F
G
H

FIGURE 10.5 a) Automatic analysis. Enter the experiment name and stain type: single or cross-stain. b) Gel selection. Go to the experiment folder and select desired gels. c) Inspect images. View and highlight images that are of interest for analysis. d) Average gels. Group samples or technical replicates together and assign a master image. e) Average gel parameters. Select the conditions for the average gels. f) Select a reference gel. Pick a reference gel template for the experiment. g) Spot detection criteria. Select criteria for inclusion of a spot during detection, or leave it blank and use the default settings. h) Background subtraction. Select the choice algorithm for background correction.

of a DIGE experiment easier. However, it also makes any manual editing of the other images more difficult. It may be advisable to run the automatic analysis both with and without the creation of average gels.

- If the user elects to create average gels, the next window (Figure 10.5e) will allow you to select analysis options to remove outlying spots from the technical replicates. The parameters selected here are dependent on the consistency of the gels analyzed. If most are highly consistent, tight acceptance parameters can be utilized. If gels require manual editing, do not use average gels.
- The next window (Figure 10.5f) prompts you to load an analysis protocol. If you have used Progenesis analysis software before on the sample and optimized the conditions for analysis, you can select a previously generated protocol or you can enter the spot detection

parameters on your own, based on spot volume, circularity, area, etc. As it is difficult to gauge which values to use without visual trial and error from spot analysis, it is recommended to run through the analysis with no analysis protocol, and determine a reasonable protocol for future experiments using the spot-filtering tool that will be discussed in the manual editing section.

- The next window prompts you to select an image from your experiment to serve as a reference gel or import one from a previous experiment. This image is used to perform matching. The gel selected should appear consistent with the other gels in the sample and, if possible, contain the most spots. This is done to make the reference a good representative of as many experimental spots as possible.
- The next window (Figure 10.5g) allows you to toggle the spot detection option on and off and includes detection parameters saved in an analysis protocol. Again, it is beneficial to examine detection without the introduction of spot parameters. However, if the optimal settings are known, implement them.
- The next window (Figure 10.5h) allows selection of background subtraction options. Mode of nonspot is the most common (nonspot pixel grayscale value); lowest on boundary and average on boundary are background subtractions based on the pixel grayscale value of the border of the spot in question. Progenesis Background is a proprietary background correction algorithm. It may be prudent to compare several varieties of background subtraction and determine which works best in your particular analysis.
- The next window (Figure 10.6a) is the gel warping and matching window. If TT900 has been utilized prior to analysis, turn off warping (the images would already be aligned using a superior program), otherwise leave the warp feature on as it aids in generating matches between images. The "Match detected spots to spots in the reference gel" feature can save manual image processing time and is essential to Progenesis analyses. The "combine warping and matching" feature adds to processing time. However, it may yield better results through iterative matching. This feature should not be used if the images are warped with TT900, as the matching process will be very different.

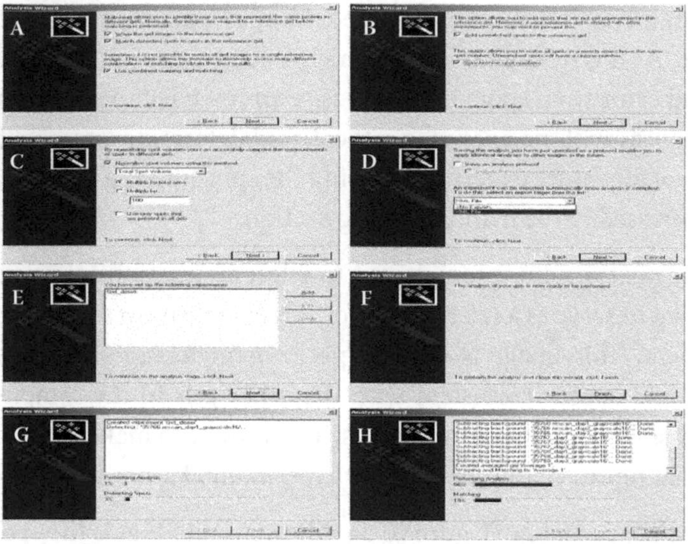
A
B
C
D
E
F
G
H

FIGURE 10.6 a) Matching parameters. Select warping and matching options b) Reference gel options. Select how spot information will be cataloged and incorporated with the reference gel. c) Normalization. Select the type of normalization for the experiment. d) Save and export experiment information. Save analysis protocols and export experiment XML files. e) View experiments to be run. Compile list of experiments to be analyzed. f) Initiation of analysis. Start automated analysis. g) Progression of compilation of analysis. h) Compilation of analyses. Follow experiment progress to completion and view current compilations.

- The next window (Figure 10.6b) deals with spot numbers and unmatched spots. Adding unmatched spots to the reference gel can be toggled on or off. This allows spots not present on the reference gel to be incorporated into the experiments. Synchronizing spot numbers gives all matched identifications the same ID numbers between gels. This is very useful for exporting data for external statistical tests or viewing a spot of interest with the "Go to spot tool" under "Edit."
- The next window (Figure 10.6c) controls normalization. The user can select:
    - Ratiometric — Normalizes images based on ratios of internal standard (only appropriate for DIGE analyses).
    - Total spot volume — Normalizes base on the cumulative volume of all spots in an image (or all matched spots only if this option is selected).

- Total volume ratio — The volume of each spot in each gel is normalized by multiplying the ratio of total spot volume of the image by the base image (image used to generate a reference image).
- Value — Every spot is multiplied by a specific value.
- Match ratio — Every spot present in the image and base image has its volume corrected by multiplying the average of the spot ratios of the base image over the current image and multiplying this by the area of a single pixel in mm^2.
  - The specific conditions of the user's experiments may require specific normalization methods, but generally, total spot volume is appropriate for single-stained gel images with equal protein loading and ratiometric normalization is appropriate for DIGE-format images with a pooled Cy2 internal standard.

- The next window (Figure 10.6d) allows the user to save the analysis protocol for future experiments. If specific, customized detection settings are used, they can be saved at this point.
- The next window (Figure 10.6e) lists the experimental setup to be run. If those are multiple experiments or multiple analysis conditions for the same set of samples, they can all be set up here and will run sequentially. The user can then begin the automated analysis (Figure 10.6f). This can take minutes to hours depending on the size, number, and complexity of the images as well as the analysis options chosen (Figure 10.6g and Figure 10.6h).

B. Evaluating the Automatic Analysis When the analysis is complete, it is a good idea to briefly overview the results (Figure 10.7). Use the arrow icon to select regions of the gel to display in the montage window, or if the data set is too large to effectively view all the images at once, the animation window can be used to quickly view the same region across multiple gels. In addition to the spot detection, turn on match vector from the display options menu on the gel window and check that spot shifts on a gel relative to the reference gel are consistent for regions on the image (vectors in a recognizably different direction or magnitude from their immediate neighbors are signs of an incorrect match). If you are satisfied with the appearance of the analyzed images, data can be exported for statistical validation. Spot statistics can be performed within newer versions of

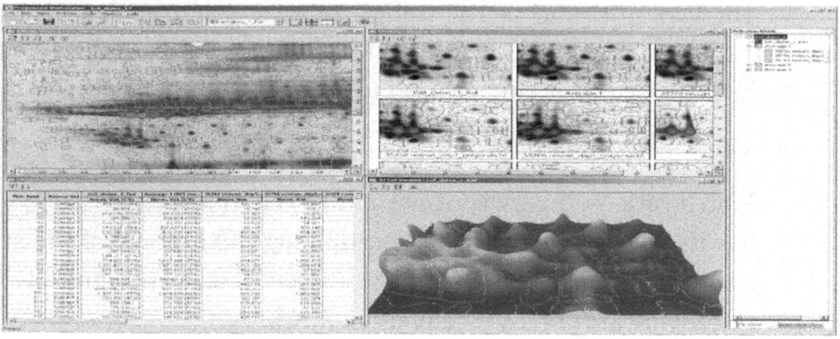

FIGURE 10.7 Experimental navigation. Automatic analyses are complete; examine data using quantitative and visual tools.

Progenesis. However, the newer versions may require the user to set up experiments in a specific manner (average gels), and this may not always be desired.

C. Validation of Spot Information Exporting the data into a spreadsheet program can be easily accomplished, and a greater variety of validation tests can be performed (Figure 10.8). Exporting the data can be performed using the "view" header and selecting the desired data from measurement, comparison, or expression window, then selecting the type of information to be extracted using the field selection icon on the top left of the measurement, comparison, or expression window (usually, normalized volume is desired). The comparison window is usually the most useful, as it organizes the actual spot information in columns by gel, suitable for easy export for statistical validation. The order of the gels in the columns can be altered by selecting the selection tool icon and manipulating the gel order on the right side of the screen by right-clicking on a gel image and selecting "move up" or "move down." If the data are arranged as desired, they can be exported by copying (Ctrl C) and pasting (Ctrl V) the data from the measurement, comparison, and expression windows into an external statistics program, and the desired statistical tests can be performed. Generally, this will be Student's T-tests or analysis of variance. There is debate over the appropriate statistical tests and confidence level to use for analysis, and the majority of this debate is beyond the scope of this section.

Any proteins of interest identified from an external list should be validated by manual examination. Simply put, the person examining the list should check every protein of interest to make sure that the changes are real (actual difference in spot volume, not background aberration) and

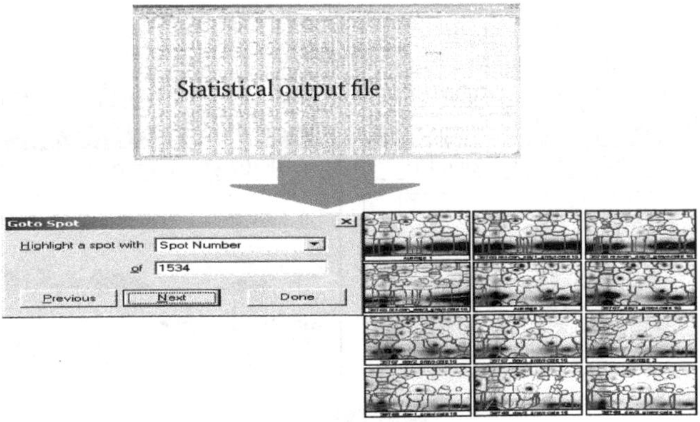
Statistical output file
Goto Spot
Highlight a spot with
Spot Number
of
1534
Previous
Next
Done

FIGURE 10.8 Identify proteins of interest. Use data output to identify potential proteins of interest using statistical software.

both matching and spot boundaries are correct. This can be accomplished by using the "Go to Spot" feature under the "Edit" header. Make sure you have the correct gel of interest in the primary viewer, and enter the spot number in the box. Usually, it is advantageous to have the montage window open to check matching and the 3D viewer to assess the spot boundaries. The images from both the montage and 3D viewer can be exported easily to presentation programs like PowerPoint using copy (Ctrl C) and paste (Ctrl V) tools. If the image set is too large (more than 12 gels), it may be easier to simply use the animation tool on the 3D viewer as follows:

- Open the 3D window and select the icon that looks like a film strip. This will allow the user to select which gels to be included, animation speed and rotation of the 3D image.

D. Manual Spot Detection/Correction Any spots that appear incorrectly matched or detected can be corrected manually. The methods for doing this are discussed here (Figure 10.9).

The images can be navigated and organized using the arrow tool from the menu (Figure 10.3). This tool allows selection of spots by clicking on them, and regions of interest drawing a box around the desired region. This box will also be the area used in montage and 3D views. Gel editing is selected with the "detect spots" icon and performed using the following tools outlined in Figure 10.3.

Manual spot detection is performed primarily with a few of the offered tools: the "draw spots," "erase spots," "auto merge," and "spot splitting"

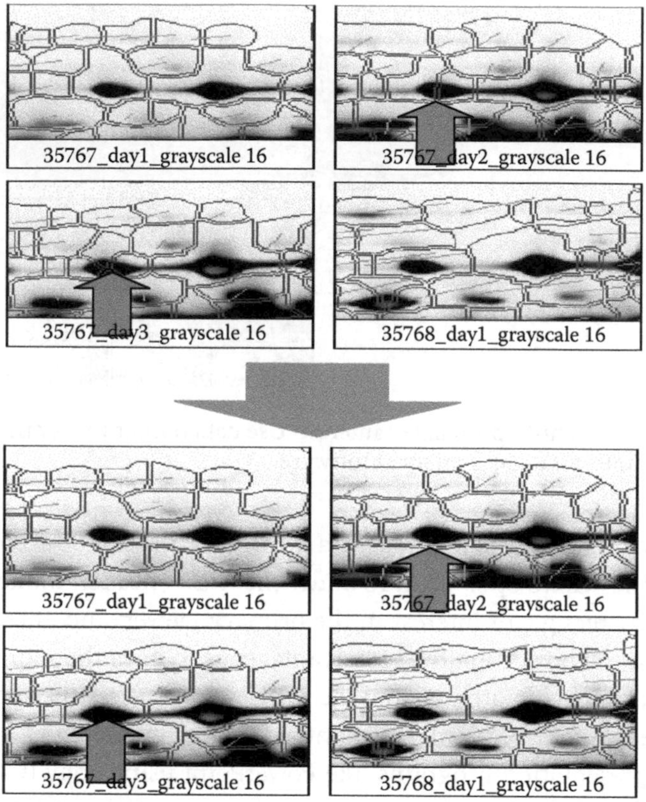
35767_day1_grayscale 16
35767_day2_grayscale 16
35767_day3_grayscale 16
35768_day1_grayscale 16
35767_day1_grayscale 16
35767_day2_grayscale 16
35767_day3_grayscale 16
35768_day1_grayscale 16

FIGURE 10.9 Spot correction. Regions of poor detection are corrected using the "auto-merge" spot tool.

tools. False spots can quickly be removed with the "erase spots" tool. Correcting poorly detected spots is most easily performed by cutting with the spot splitting tool, which cleaves spots in line segments, or expanding the boundaries with the draw spots tool, which can add to the edges of the spot to appropriately engulf the spot. The draw spots and spot splitting tools are both useful for splitting spots as they offer more control than the "auto split" tool. Incorrectly defined spots in crowded gel regions are best resolved by merging the incorrect spots into one spot with the auto merge tool, and then using the spot splitting tool to define the specific spots correctly. The "add spot" tool is not often used as the spots that are effectively identified with this tool are usually correctly detected by the automatic analysis, and the others are easier to identify with the draw spots tool. The "edge grow" tool detects a small central region around the peak of a spot that grows outward with each mouse click. This is generally a slow way to reach the correct boundaries of a spot. Using the auto split tool is faster than the cutting tool, but it frequently makes errors, so the use of this tool should be restricted to occasions where it is the only option, such as on cross-stain analyses.

The version of the software used will also influence how the analysis is performed. If using an older version of the software, the spot corrections should be performed on the actual gel image, whereas warping with TT900 and using SameSpots™ requires the reference image to be the one modified. Any spot identification changes should be checked in the 3D viewer window to ensure the boundaries are correct.

E. Manual Spot Matching Any changes introduced to the spots will result in their match to the corresponding spot in the reference gel to be broken, and the spots must be rematched after the spot editing. This is accomplished by selecting one of the matching options (Figure 10.3). The overlay matching tool is used to align the spots with the reference spots by overlaying the red and blue spot boundaries on the same image. This tool can be used to rapidly correct highly consistent gels. The montage (Figure 10.10) and window matches are performed by first selecting the desired spot for matching in the reference gel, followed by selecting the desired spot in the edited slave gels. When matched, the spots should appear highlighted and have match vectors indicating the direction of shift

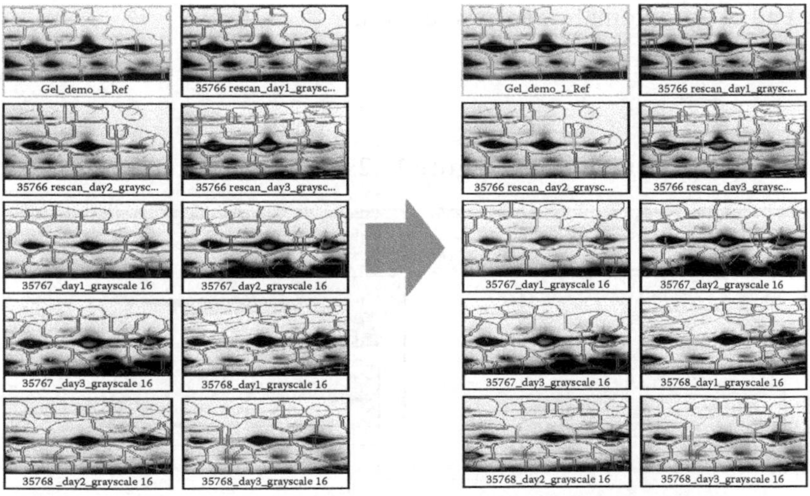
Gel_demo_1_Ref
35766 rescan_day1_graysc...
35766 rescan_day2_graysc...
35766 rescan_day3_graysc...
35767_day1_grayscale 16
35767_day2_grayscale 16
35767_day3_grayscale 16
35768_day1_grayscale 16
35768_day2_grayscale 16
35768_day3_grayscale 16
Gel_demo_1_Ref
35766 rescan_day1_graysc...
35766 rescan_day2_graysc...
35766 rescan_day3_graysc...
35767_day1_grayscale 16
35767_day2_grayscale 16
35767_day3_grayscale 16
35768_day1_grayscale 16
35768_day2_grayscale 16
35768_day3_grayscale 16

FIGURE 10.10 Matching correction. Mismatched or newly edited spots are linked to the reference gel image.

in the match relative to the reference gel. Generally, correcting matches from the montage works best as it provides a comparative view of many or all the gels simultaneously.

If the slave gel used to generate the reference is modified, the matching procedure is more complicated. Any modified spot in this gel must be introduced to the reference gel with the "add to reference" feature under analysis. Before this is done, any spots overlapping the edited spot should be deleted from the reference gel — most quickly accomplished using the spot editing tools. This will remove all of the matches to the deleted reference spots. Adding spots to the reference gel can be accomplished using the "add to reference" feature and selecting the spot to add in the slave gel. The spot should then appear in the reference gel and can be matched to the corresponding spots in other gels as described previously.

If using a version of Progenesis with SameSpots™, matching will not be an issue as the spot boundaries from the reference gel will be applied to all slave gels. Issues arise only if the user wants to introduce a spot or a specific set of spot boundaries. To do this, delete interfering spots from the reference and use the "add to reference" tools as outlined previously, then use SameSpots™ again to apply the boundary information to all slave gels.

At the end of all manual editing, the normalization and background subtraction will be removed and must be reapplied from the analysis menu (Figure 10.11). The data can then be extracted and queried as mentioned at the end of the automatic analysis section.

Part III Sample Data

The sample gels are shown in Figure 10.2.

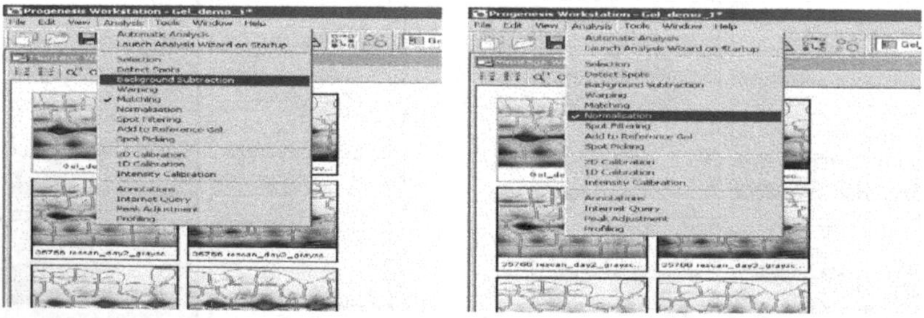
Progenesis Workstation
File Edit View Analysis Tools Window Help
Automatic Analysis
Launch Analysis Wizard on Startup
Background Subtraction
Matching
Normalisation
Add to Reference Gel
Spot Picking
2D Calibration
1D Calibration
Intensity Calibration
Internet Query
Peak Adjustment
Profiling

FIGURE 10.11 Background subtraction and normalization. After manual editing, the gels are background-corrected and normalized for data mining as shown in Figure 10.8 and the "Progenesis Analysis" section.

SECTION 2 ANALYZING MASS-SPECTROMETRY-DERIVED DATA

Part I Introduction

1. The Scientific Utility of Mass Spectrometry and Mass Spectrometry Database Search Programs

Proteomics has been greatly advanced with the development of mass spectrometry methods. Mass spectrometry and the its related database search programs are important tools in proteomics, and they can be used to (1) identify peptides/proteins present in a sample, (2) characterize PTMs, and (3) quantitate proteins expressed in healthy and diseased tissues or body fluids, or cells treated with and without a therapeutic agent or agonist. In this section, the basic steps and commonly used mass spectrometry data analysis programs will be reviewed, and a step-by-step tutorial on how to use the Mascot Daemon program will be given.

2. What Is Involved in Mass-Spectrometry-Derived Data Analysis?

Four basic steps are involved in analyzing mass spectrometry data: understanding the algorithm of the database search program, setting up database search, interpreting the results, and further processing the results. This further processing is necessary so that a minimum set of protein sequences that adequately accounts for the observed peptides is reported.

A. Understanding the Algorithm of the Database Search Program Typically, the input for mass spectrometry database searches can be divided into three categories: accurate peptides molecular weights obtained from the enzymatic digestion of a protein (a peptide mass fingerprint, or PMF), MS/MS data from one or more peptide (an MS/MS ions search), and the combination of mass data with explicit amino acid sequence data or physicochemical data which can be used to infer the amino acid sequence or composition (a sequence query). Several algorithms and computer programs have been described in the literature for protein identification/quantification by searching a sequence database using mass spectrometry data, and the general approach to searching in all cases is similar. In order to identify the "unknown" protein, the experimental data are compared with calculated peptide mass or fragment ion mass values obtained by applying appropriate cleavage rules to the entries in a sequence database. Corresponding mass values are then counted or scored in a way that allows the peptide or protein that best matches the data to be identified. The aim is to pull out the correct entry if the "unknown" protein is in the database.

If the "unknown" protein is not in the database, then the aim is to identify those entries that exhibit the closest homology, most often equivalent proteins from related species, or different proteins isoforms. The alternative is to carry out *de novo* sequencing which extracts the amino acid sequence information without the use of databases.

Four general types of algorithms have been used to match mass spectra to peptide sequences: probability-based matching that calculates a score based on the statistical significance of a match between an observed peptide fragment and those calculated from a protein sequence database or sequence search library; cross-correlation methods, referred to as heuristic algorithms, that correlate experimental spectra with theoretical spectra; the use of unambiguous "peptide sequence tags" derived from spectra that are used to search known amino acid sequences; and *de novo* calling of the sequence directly from the spectrum.

B. Setting Up a Database Search The first step to setting up a database search is to create or download the protein database, generally in FASTA format. FASTA format is a text-based format for representing both nucleic and protein sequences, in which base pairs or proteins are represented using a single-letter code. The databases can be downloaded from a lot of Web-based resources, among which are National Center for Biotechnology Information (ftp://ftp.ncbi.nih.gov/blast/db/FASTA/), European Bioinformatics Institute (ftp://ftp.ebi.ac.uk/pub/databases), and Expert Protein Analysis System (ftp://us.expasy.org/databases/). The second step is to set up the search parameters, typically including the database, enzyme used for digest, number of missed cleavages allowed (typically two), fixed and variable modifications (for example, acetylation, oxidation), peptide mass tolerance (the allowed peptide mass error between experimental and theoretical data; instrument dependant), MS/MS ion mass tolerance (the allowed peptide fragment ion error between experimental and theoretical data; instrument dependent), mass type (monoisotopic or average; instrument dependent), and peptide charge. The third step is to input the experimental mass spectrometry data file, and sometimes the name and designation of the output file.

C. Interpreting the Mass Spectrometry Database Search Results For different database search programs, the output files will differ. Even so, the output file can be manipulated to find the most significant match between the experimental and theoretical mass spectra. For the probability-based matching algorithm, the smaller the probability, the better the match. Typically, matches with $p < 0.05$ are accepted as significant. For cross-correlation-based algorithm, the higher the X-corr (correlation score) and the bigger the ΔCn (normalized correlation score), the better the match. A typically cutoff filter is X-corr 1.9 for +1 ions, 2.2 for +2 ions, and 3.75 for +3 ions with ΔCn > 0.1. In the cases with no significant match, database search parameters would need to be changed, and another round of database search would be required. If a protein is identified with a single peptide (one-hit wonder), manual validation is absolutely necessary before it is reported.

D. Further Processing of the Mass Spectrometry Database Search Results In the experimental strategies based on proteolytic digestion of protein mixtures, proteins are digested to peptides, and the peptide sequences are identified through database search of mass spectrometry data and reassembled into proteins or a protein group. This strategy results in loss of connectivity between peptides and their protein precursors, and two outcomes are possible: distinct peptides that map to only one protein sequence, or shared peptides that map to more than one protein sequence. Although the identification of shared peptides implies that multiple related protein sequence are present, the initial assumption should be that only a single form is detected, and therefore a minimum set of protein sequences, or a nonredundant protein list should be reported. Programs such as ProteinProphet™ (http://proteinprophet.sourceforge.net/, Institute for Systems Biology) and Pro Group™ (Applied Biosystems) are designed for this purpose. Pro Group™ is an integral part of the ProteinPilot™ database search program (Applied Biosystems).

For large-scale experiments, the results of any additional statistical analyses that indicate or establish a measure of identification certainty, or allow a determination of the false-positive rate, need to be provided. For false-positive rate estimation, a reversed database is created by precisely reversing the order of the amino acid sequence for each protein so that the C terminus becomes the N terminus. A composite database is created by adding the reversed database after the forward database. All MS/MS spectra are searched against this indexed database, and the results are filtered using the same criteria for the forward database. The false-positive rate is estimated by doubling the number of peptides found from the reversed database and dividing the result by the total number of identified peptides from both databases according to the following formula: % *fal* =

$2[n_{rev}/(n_{rev} + n_{real})]$, where % *fal* is the estimated false-positive rate, n_{rev} is the number of peptides identified (after filtering) from the reversed database, and n_{real} is the number of peptides identified (after filtering) from the real database.

3. Some Commonly Used Mass Spectrometry Database Search Programs

Mass spectrometry has become the method of choice for protein identification, quantitation, and characterization. Algorithms and computer programs for mass spectrometry database search have improved rapidly in speed, sensitivity, and specificity. In general each instrument company has its own database search programs that are compatible with its instrument data format, and there are also programs that are compatible with multiple data formats. Table 10.1 lists some commonly used mass spectrometry database search programs, and some of their features are briefly discussed in the following text.

Mascot. Mascot is the most commonly used database search program and can be freely accessed at the uniform resource locator (http://www.matrixscience.com). It is a probability-based program, and it is compatible with all three types of input of mass spectrometry database search: PMF data, MS/MS data, and sequence queries. It is also compatible with multi-

TABLE 10.1 Some Commonly Used Mass Spectrometry Database Search Programs

Program	Algorithm	Open-Source or License Required	Associated Companies (license required) or URLs (open-source programs)
Mascot	Probability-based	Open-source, but many more functions with license	http://www.matrixscience.com
Bioworks™	Cross-correlation-based and probability-based	License required	Thermo Electron Corp.
ProteinPilot™	Probability-based	License required	Applied Biosystems/MDS Sciex
OMMSA	Probability-based	Open-source	http://pubchem.ncbi.nlm.nih.gov/omssa/
X!Tandem	Cross-correlation-based	Open-source	http://www.thegpm.org/TANDEM/index.html

ple data formats from different mass spectrometry instrumentation companies, for example, Voyager DAT files from Applied Biosystems, Analyst WIFF data files from Applied Biosystems/MDS Sciex Analyst, Thermo Electron Xcalibur/BioWorks Raw files, and Waters MassLynx sample list. If Macot Distiller (which requires an additional license) is installed, a wide range of native file formats can be processed.

BioWorks™. BioWorks™ from Thermo Electron is used to search data files generated specifically from Thermo Electron Instruments. The old version of BioWorks is based on Sequest, a cross-correlation-based algorithm. The latest version also incorporated a probability-based algorithm, so that both the correlation scores and probabilities of peptides and proteins are reported, and that the results can be compared with those generated from other probability-based search engines. The latest BioWorks contains a tool for easy creation of reverse databases for false-positive rate evaluation. One unique feature of BioWorks is that it can search an indexed database, which contains the mass of each peptide generated by using the appropriate enzyme to cleave each protein sequence entries in the protein FASTA database, and therefore speed up the database searches. PepQuan in BioWorks can be used for peptide/protein quantitation with data generated from isotope-coded affinity tag (ICAT/c-ICAT), isobaric tagging for relative and absolute quantitation (iTRAQ™, Applied Biosystems), and stable isotope labeling with amino acids (SILAC) experiments.

ProteinPilot™. ProteinPilot™ (Applied Biosystems/MDS Sciex) is an integration of two steps: peptide/protein identification and quantitation using the Paragon™ algorithm and further processing of the results to report a minimum group of proteins using the Pro Group™ algorithm. It can be used to analyze data files generated using the quadrupole time-of-flight and hybrid triple quadrupole linear ion trap mass spectrometers from Applied Biosystems. The Paragon™ search algorithm enables over 150 peptide modifications simultaneously, and the Pro Group™ algorithm can be used to distinguish protein isoforms, protein subsets, and suppress false-positives.

Open Mass Spectrometry Search Algorithm (OMMSA). OMMSA is an efficient search engine for identifying MS/MS peptide spectra by searching libraries of known protein sequences. It is compatible with collision-activated dissociation MS/MS (CAD-MS/MS) spectra from ion trap mass spectrometers and electron transfer dissociation MS/MS (ETD-MS/MS) spectra. OMSSA scores significant hits with a probability-based algorithm. OMSSA is free and in the public domain (http://pubchem.ncbi.nlm.nih.gov/omssa/).

X!tandem. X!tandem is an open-source search engine (http://www.thegpm.org/TANDEM/index.html) that has been optimized for speed. It generates theoretical spectra for the peptide sequences using knowledge of intensity patterns associated with particular amino acid residues. These spectra are then correlated with the experimental data and, subsequently, an expectation value is calculated. This algorithm is straightforward to implement using open-source software, and it results in speed improvements by factors exceeding 1000X, compared to the conventional, single-step algorithm.

For large-scale MS/MS data, different database search programs will generate comparable but somewhat different protein identifications. Application of a rescoring algorithm, such as PeptideProphet, could improve the results. PeptideProphet automatically validates peptide assignments to MS/MS spectra made by database search algorithms such as Sequest. Merging and combining database search results using different search algorithms has merit, and the scores could be considered as independent and orthogonal.

Part II Step-By-Step Tutorial

Mascot Daemon, a client application that automates the submission of searches to a Mascot server, is used for demonstration purposes because it is compatible with a wide range of data platforms, and it is by far the most commonly used program. Here, we will demonstrate how to use Mascot Daemon to analyze the MS/MS data acquired on the Thermo Electron linear ion trap (LTQ) mass spectrometer. Data acquired from other instruments can be analyzed when the settings are adjusted accordingly.

1. Setting Up a Database Search

The demo example is a raw file, demo.raw. Briefly, mouse bronchoalveolar lavage fluid (1 μg) was separated on one-dimensional gel electrophoresis, and the resulting gel was silver stained. The band around 50 kD was excised. The protein was reduced with dithiothreitol, alkylated with iodoacetamide, and *in situ* digested with trypsin. The peptides were extracted and analyzed on LTQ interfaced with reversed-phase HPLC (Agilent Technologies, Foster City, CA).

A. Set Up the Path to the Data Import Filter for Data File Search and the Directory for Storing the Peak List Generated from Data Import Filter To do this, open Mascot Daemon, go to Edit→ Preferences → Data import filters, and then use the Browse button to set up these two paths (Full path to LCQ_DTA.EXE has to point directly to extract_msn.exe itself); then save and close the window.

B. Define the Search Parameters Click on the "Parameter Editor" tab and you will see the default setting. Some of the default parameters need to be changed according to experimental conditions. In this case, the final parameters are as following (as shown in Figure 10.12):

Search title: Demo

Taxonomy: Mus musculus

Database: NCBInr

Enzyme: Trypsin

Maximum missed cleavages: 2

Fixed modifications: Carbamidomethyl (sample was alkylated with iodoacetamide)

Variable modifications: Oxidation (M)

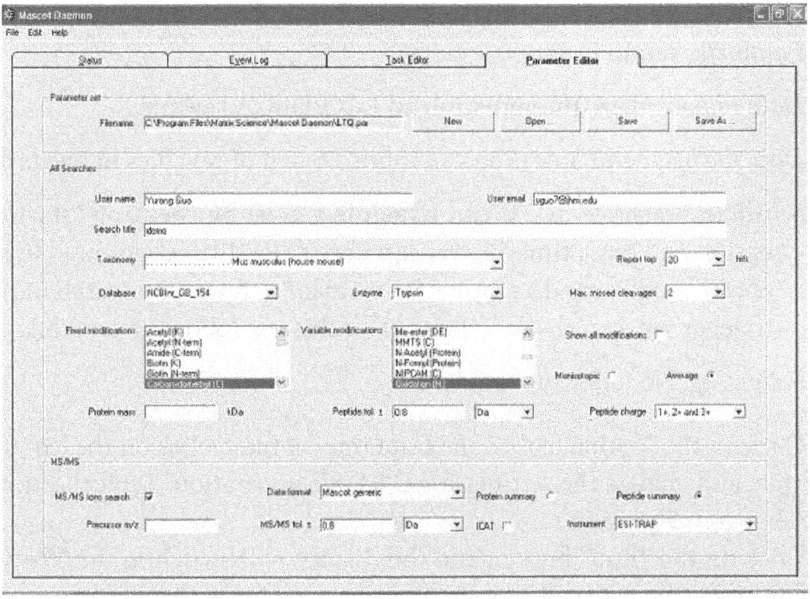
Mascot Daemon
File Edit Help
Status
Event Log
Task Editor
Parameter Editor
Parameter set
Filename
New
Open...
Save
Save As...
All Searches
User name
Yurong Guo
User email
Search title
demo
Taxonomy
Mus musculus (house mouse)
Report top
20
hits
Database
Enzyme
Trypsin
Max missed cleavages
2
Fixed modifications
Acetyl (K)
Acetyl (N-term)
Amide (C-term)
Biotin (K)
Biotin (N-term)
Variable modifications
Me-ester (DE)
MMTS (C)
N-Acetyl (Protein)
N-Formyl (Protein)
NIPCAM (C)
Show all modifications
Monoisotopic
Average
Protein mass
kDa
Peptide tol. ±
0.8
Da
Peptide charge
1+, 2+ and 3+
MS/MS
MS/MS ions search
Data format
Mascot generic
Protein summary
Peptide summary
Precursor m/z
MS/MS tol. ±
0.8
Da
ICAT
Instrument
ESI-TRAP

FIGURE 10.12 Define the search parameters.

Monoisotopic or average: Average (on LTQ, the peptide masses are average masses)

Peptide tolerance ±: 0.8 Da (LTQ has unit resolution)

Peptide charge: 1+, 2+ and 3+

MS/MS ions search: Checked

Data format: Mascot Generic

Protein summary or peptide summary: Peptide summary

MS/MS tolerance ±: 0.8 Da

Instrument: ESI-TRAP

Click on the "Save As" button to save the parameter file as "LTQ.par."

C. Create and Run a Task Click on the "Task Editor" tab and you will see the default screen. Change the settings accordingly and input the data files. The data file for search has to be on the local computer. In this case, the final "Task Editor" tab is shown as in Figure 10.13 with the following settings:

Task: demo.

Parameter set: LTQ.par.

Data import filter: ThermoFinnigan LCQ/DECA raw file.

Data file list: demo.raw (You can input a batch of raw files in one task);

Schedule: Start now. (You can schedule a time by checking "start at" option and put a time; or you can check "Real-time monitor" if you are still acquiring data on the instrument and want to search in real time; or you can check "follow up" if this is a follow-up search.)

Follow-up: No follow-up required.

Click on the "option tab" in the Data import filter, click on the lcq_dta.exe tab, and change the settings for DTA file generation. Typical settings are shown in Figure 10.14.

Click on the "Run" button, and the display will switch to the "Status" tab. Initially, the task will show as an hourglass. Once the task is running and the search is submitted, the icon will change to a clock face. You can switch to the "Event Log" tab at this time to see if there is any problem.

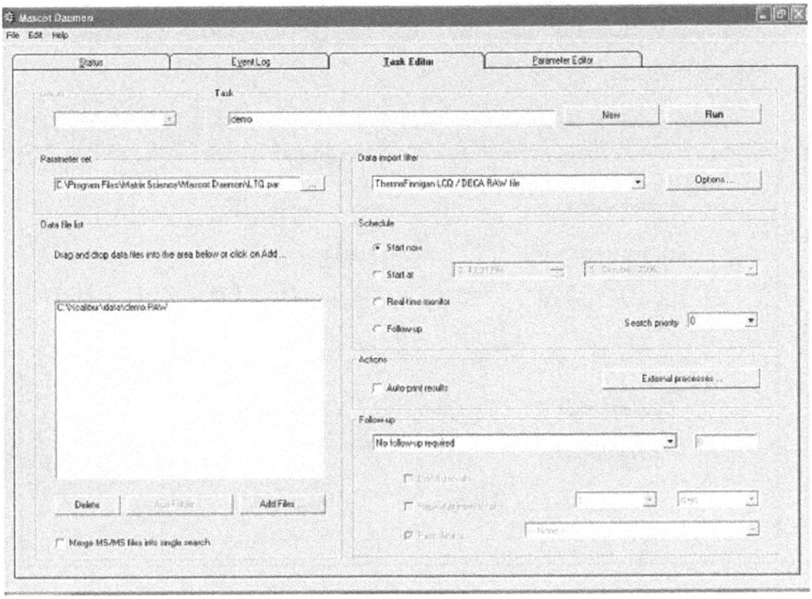
Mascot Daemon
File Edit Help
Status
Event Log
Task Editor
Parameter Editor
Task
New
Run
Parameter set
Data import filter
Options...
Data file list
Drag and drop data files into the area below or click on Add...
Schedule
Start now
Start at
Real-time monitor
Follow-up
Search priority
Actions
Auto-print results
External processes...
Follow-up
No follow-up required
Delete
Add Files...
Merge MS/MS files into single search

FIGURE 10.13 Change the settings and input data files on the “Task Editor” tab.

If there is a problem, e.g., peak list not recognized or no output from the lcq_dta.exe, the details will be recorded in the “Event Log” tab. If everything is normal, as the task is being processed, the Event Log would state “Starting new task,” “Search submitted,” “Search completed,” and finally “Task completed.” On the status tab, a completed task is indicated by a green tick.

2. View and Interpret the Results

Once the search is done, on the “Status tab” you can expand the task node and click on the result node to show the result details on the right. A limited amount of information for each result can be displayed by clicking on the result node. To view the full report, click once on the blue hyperlink, and the report will be loaded to your default Web browser. Figure 10.15 shows the results of this database search.

A. Header and Search Parameters These are displayed on the top of the report, as shown in Figure 10.15A.

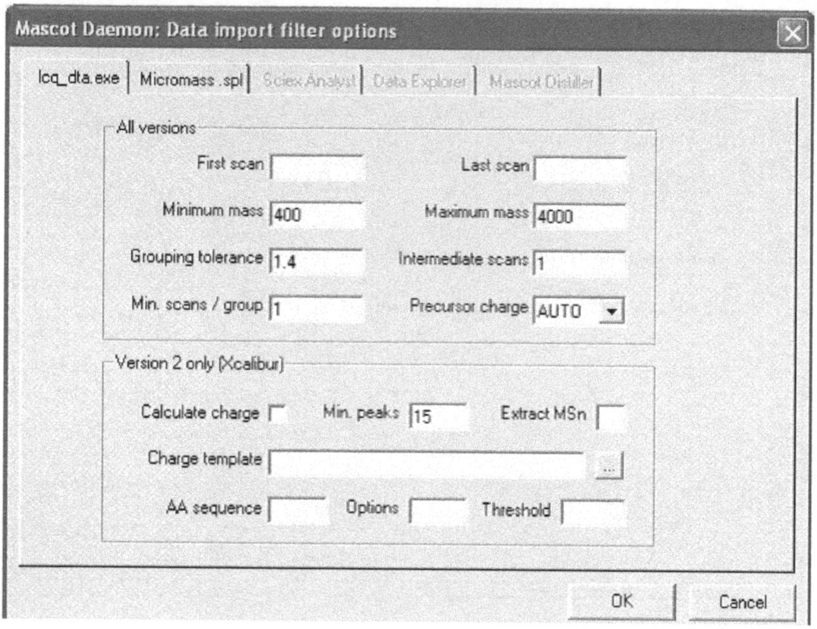
Mascot Daemon: Data import filter options
lcq_dta.exe
Micromass .spl
Sciex Analyst
Data Explorer
Mascot Distiller
All versions
First scan
Last scan
Minimum mass
400
Maximum mass
4000
Grouping tolerance
1.4
Intermediate scans
1
Min. scans / group
1
Precursor charge
AUTO
Version 2 only (Xcalibur)
Calculate charge
Min. peaks
15
Extract MSn
Charge template
AA sequence
Options
Threshold
OK
Cancel

FIGURE 10.14 Typical settings for DTA file generation using lcq_dta.exe.

B. Types of Report and Format Controls You can select the type of report by pulling down the arrow in the "Select Summary Report" session, and then click the "Format As" button. There is a list of variable formats you can choose from. The default format for PMF results is a "Concise Protein Summary," where proteins that match the same set, or a subset, or mass values are grouped into a single hit. For MS/MS searches of less than 1000 spectra, the default report format is "Peptide Summary," which provides the clearest and most complete picture of the results, especially if the sample is a protein mixture. In the Peptide Summary, a minimum set of proteins that completely accounts for the peptide matches found the experimental data is reported. For MS/MS searches of more than 1000 spectra, the best format is "Select Summary," which is similar to Peptide Summary, but provides a very compact view of the results. In this case, we have more than 1000 MS/MS spectra, so the best format is Select Summary (protein hits) as shown in Figure 10.15A. If you switch to Select Summary (unassigned), you will see the list of peptide matches not assigned to protein hits, and if there are no details it means no match.

There are some other format controls in addition to types of report:

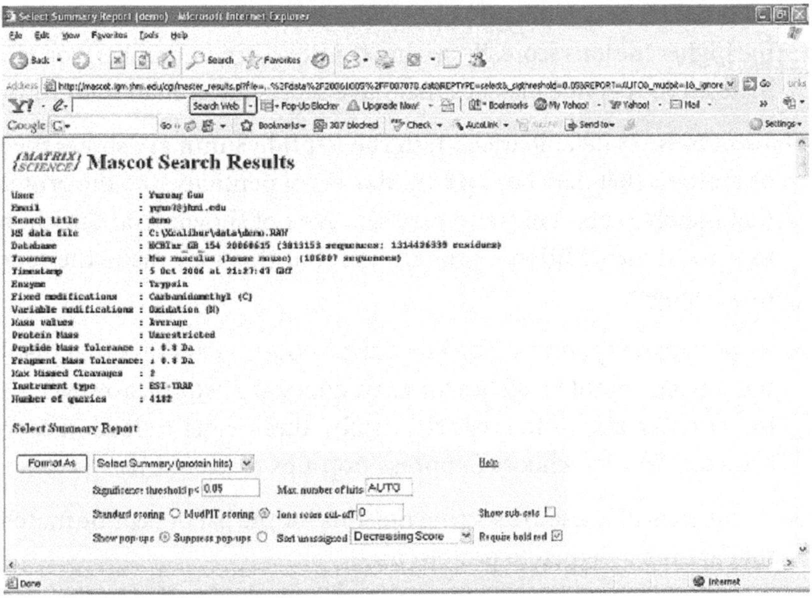
Mascot Search Results
User
Email
Search title : demo
MS data file
Database
Taxonomy : Mus musculus (house mouse)
Timestamp
Enzyme : Trypsin
Fixed modifications : Carbamidomethyl (C)
Variable modifications : Oxidation (M)
Mass values : Average
Protein Mass : Unrestricted
Peptide Mass Tolerance
Fragment Mass Tolerance
Max Missed Cleavages
Instrument type : ESI-TRAP
Number of queries
Select Summary Report
Format As
Select Summary (protein hits)
Help
Significance threshold p< 0.05
Max. number of hits AUTO
Standard scoring
MudPIT scoring
Ions score cut-off 0
Show sub-sets
Show pop-ups
Suppress pop-ups
Sort unassigned
Decreasing Score
Require bold red
Done
Internet

FIGURE 10.15A A typical display of results. (A) Header and search parameters, and types of report and format controls display.

Significance threshold: The default is $p < 0.05$. You can change this to any value in the range 0.1 to 1E–18.

Maximum number of hits: This is the maximum number of protein hits to report, and this can be set when the search is submitted. For protein Summary Report (format for PMF), the maximum number is 50. There is no limit for Peptide Summary or Select Summary for MS/MS results. If you put Auto or 0, it will display all the hits that have a protein score exceeding the significance threshold, plus one extra hit. In our case, we do not know how many proteins are there, so the input is Auto.

Standard or MudPIT scoring: Multidimensional protein identification technology (MudPIT) is a more aggressive protein scoring. This removes many of the junk protein hits, which have high protein score but no high-scoring peptide matches. It is the default for searches that have more than 1000 queries.

Ions score cutoff: In Mascot, the score for an MS/MS match is based on the absolute probability (p) that the observed match between the experimental data and the database sequence is a random event. The reported score is –10Log(P). So, the lower the *P* (the better the match), the higher the ions score. By setting this to a certain number, you eliminate all the peptide matches that have scores lower than this number.

Show subsets: By default, each hit in the Peptide Summary shows the set of proteins that matches a particular set of peptides, and the proteins that match a subset of those peptides are not shown. You can choose to show these additional protein hits, but this may make the report much longer.

Show or suppress pop-ups: The JavaScript pop-up windows that display the top ten peptide matches for each query are helpful; however, they increase the size of the report, which takes longer to load in a Web browser. You can choose Suppress pop-ups to avoid this.

Sort unassigned: These are sorting options for the list of peptide matches that are not assigned to protein hits.

Require bold red: In the results, you can see that peptides could be in bold red, bold dark, red but not bold, or dark plain text. The red means that it is the rank number 1 peptide match for that query, and the bold means that this is the first time this peptide match to a query appears in the report. Requiring a protein hit to include at least one bold red peptide match is a good way to remove duplicate homologous proteins from a report.

C. Body of Report The body of the Peptide Summary report contains a tabular listing of the proteins, sorted by descending protein score, a section of which is shown in Figure 10.15B. The number of proteins reported can be set by putting a number in the "Max. number of hits" in format control.

For each protein, the first line contains the accession string, the protein molecular weight, a nonprobabilistic protein score (which is derived from the ions scores), and the number of peptide matches. The accession string links to the corresponding Protein View of peptide matches; an example is shown is Figure 10.16. The matched peptides are shown in bold red. The second line is a brief descriptive title for the protein, and this is followed by a table summarizing the matched peptides. The contents of the table columns are described briefly in Table 10.2. If any variable modifications are found for that peptide, they will be listed after the peptide sequence. However, the exact sites of modification will not be listed here, and you

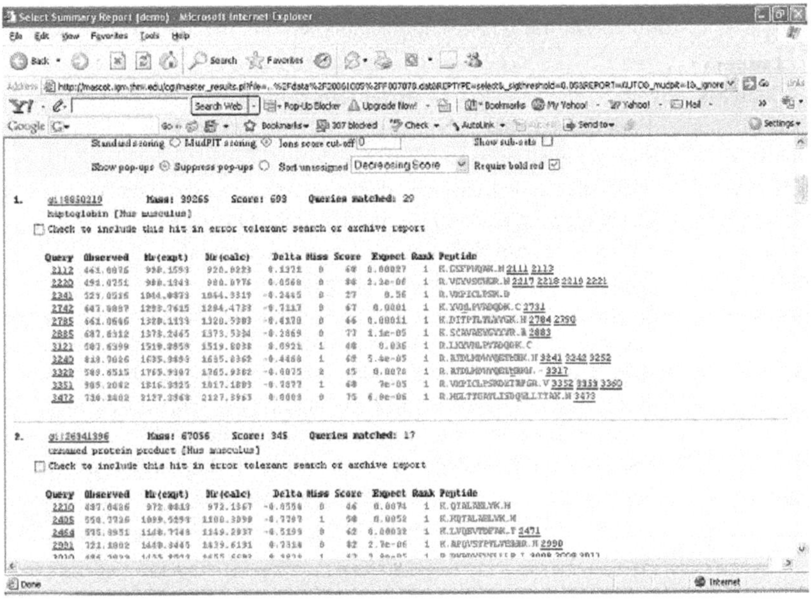

FIGURE 10.15B A typical display of results. (B) body of report display.

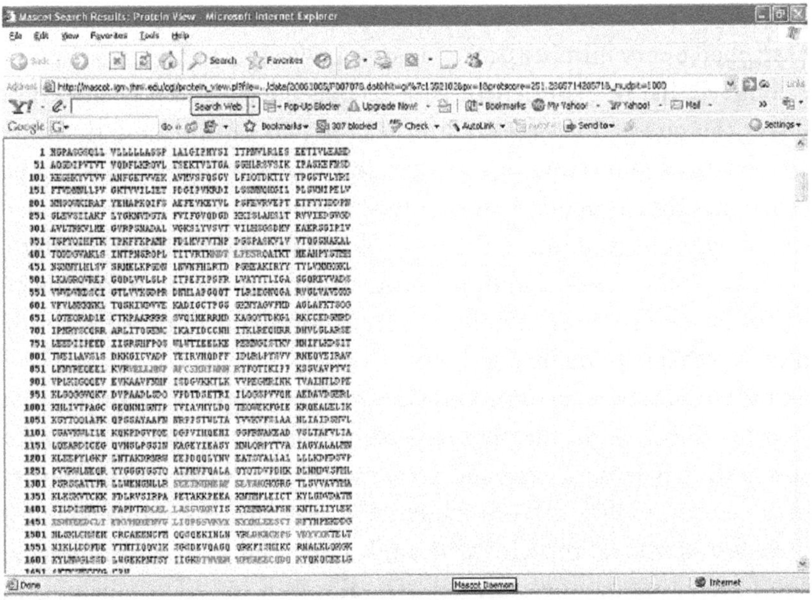

FIGURE 10.16 An example of Protein View.

TABLE 10.2 Contents of the Table Columns Summarizing the Matched Peptides

Column Contents	Brief Description
Query	Hyperlinked query number
Observed	Experimental m/z ratio
Mr (expt)	Molecular mass transformed from experimental m/z ratio
Mr (calc)	Calculated molecular mass of the matched peptide
Delta	Difference (error) between Mr (expt) and Mr (calc)
Miss	Number of missed enzyme cleavage sites
Ions score	Ions score for that match; if there are multiple matches to the same peptide, then the lower-scoring matches are shown in brackets.
Expect	Expectation value for the peptide match, e.g., the number of times we would expect to obtain an equal or higher score purely by chance. The lower this value, the more significant the result.
Rank	Rank of the ions match. On a scale of 1 to 10, 1 is the best match.
Peptide	Peptide sequence in 1-letter code. If any variable modifications are found for this peptide, they will be listed after the sequence string.

have to use Peptide View, which displays the MS/MS spectrum with the matched ions in red, and a list of b and y ion masses with the matched ions highlighted in bold red. Clicking on the hyperlinked query number opens the Peptide View for the match in a new browser window. An example of Peptide View is shown in Figure 10.17. Resting the mouse cursor over the hyperlinked query number causes a pop-up window to appear, displaying the complete list of peptide matches for that query. The pop-up window displays the query title, followed by one or two significance thresholds. Then, there is a table containing information on the highest-scoring peptide matches for the query. In the table, the "Hit" column is the number of the (first) protein containing the peptide match, and a plus sign after the hit number indicates that multiple proteins contain a match for this peptide. The protein column contains the accession string of the (first) protein containing this peptide match.

In most cases, there is prior knowledge of the origin of a sample, so it is only natural to look for matches to proteins that fit. In our case, it is from mouse and it is from a gel band around 40 kD, so the correct proteins should be around 40 kD from mouse. We searched the mouse database, and we found several proteins with high score and reasonable molecular weight, including haptoglobin (39 kD), albumin (67 kD), Fibrinogen, B beta polypeptide (55 kD), and sulfated glycoprotein-2 isoform 2 (44 kD). These proteins all have high-score peptides, and they seem real. However, before we state that they do exist in the sample, we need to open the Pep-

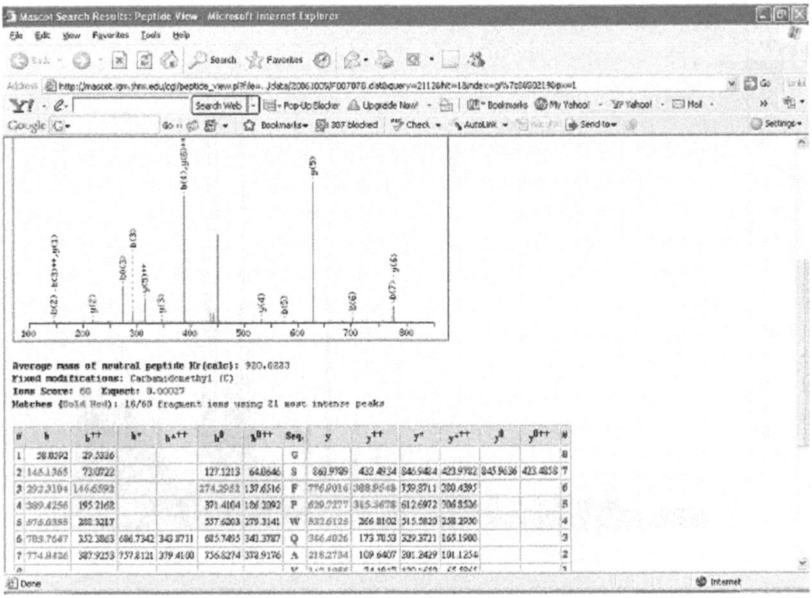
Mascot Search Results: Peptide View - Microsoft Internet Explorer
Done
Internet

FIGURE 10.17 An example of Peptide View.

tide View to confirm at least a couple of peptides for each protein. We also have 7 bold red proteins from Complement C3 precursor, which is a 188-kD protein. Most likely, we are detecting a degradation product from it. This is confirmed by clicking on the accession number of Complement C3 precursor (Figure 10.16). All the peptides, except the first one, are toward the C-terminal of Complement C3 precursor. The first peptide KDTLPESR does not seem real, because it has a rank number of 4 and ions score of 11. From the Peptide View we are able to confirm that it is not real. We also detected keratin family proteins; most likely these are contaminants from sampling and in gel digest.

If the search is restricted to a particular species and the genome of the species is not completely sequenced, there is no guarantee that the sequence of the analyte protein is actually present in the database. In this case, we may need to search a bigger database to find homologous proteins from other species. These searches, however, will take longer to search and have a potential for increased false-positives.

Part III Sample Data

Figure 10.18 shows the total ion current chromatography for the raw file demo.raw.

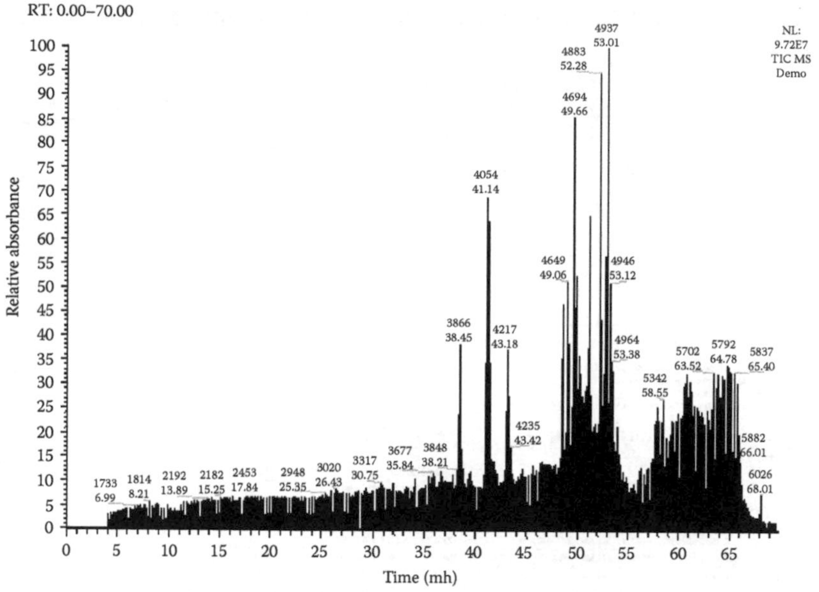
RT: 0.00–70.00
NL: 9.72E7 TIC MS Demo
Relative absorbance
Time (mh)
4937 53.01
4883 52.28
4694 49.66
4054 41.14
4649 49.06
4946 53.12
3866 38.45
4217 43.18
4964 53.38
5702 63.52
5792 64.78
5837 65.40
5342 58.55
4235 43.42
5882 66.01
3677 35.84
3848 38.21
3317 30.75
3020 26.43
2948 25.35
2453 17.84
2182 15.25
2192 13.89
1814 8.21
1733 6.99
6026 68.01

FIGURE 10.18 Total ion current chromatography for demo.raw. The *x*-axis shows time and the y-axis shows the total ion current intensity.

SECTION 3 ANALYZING DATA DERIVED FROM PROTEIN MICROARRAYS

Part I Introduction

1. *Types of Protein Microarrays*

Protein microarrays are being adopted as a critical tool in many fields of biochemical and molecular biological research. Most protein microarray platforms are created by using specialized robots to dispense micro volumes of sample onto slides. Over the last several years, protein microarrays have become more complex, with advances in both the number of samples printed onto each platform and the number of functions that each is able to accomplish.

There are two major classes of protein microarrays that are currently available: analytical and functional protein microarrays. The most common are the analytical microarrays, which include antibody microarrays. A major advantage of antibody microarrays is their capacity to generate high-throughput quantitative and qualitative information from a small volume of a single biological sample. With advances in the production of highly specific and selective antibodies, the increase in the number of binding molecules on each microarray has enabled the generation of a plethora of information in a single experimental run. However, the quality of each antibody, the interaction of the antibodies and analytes in combination (when multiplexed), and matrix affects will determine the accuracy of the multiplex. The functional microarray platforms are also being increasingly applied and so far have focused on protein interactions with molecules, including proteins, lipids, DNA, drugs, and peptides as a means of accessing biological or cellular functionality. This induces the capture of single proteins, protein complexes, organelles, and even cells. Other applications of functional microarrays have been used to analyze other protein functions such as biochemical activity (e.g., kinase activity) and the induction of immune responses. This capture will deal only with protein–protein (antibody) interactions, but similar data processing and analysis occur regardless of type or complexity of the moiety captured on the array.

2. Why Is Protein Microarray Analysis Important?

Like mass spectrometry and mass-spectrometry-related database search programs, protein microarrays and their related data analysis software can be used for protein identification, protein quantitation, and characterization of protein PTMs. A major benefit of using protein microarrays is the increased throughput, which allows researchers to study theoretically thousands of known or candidate proteins at the same time. Similar to gene arrays, each protein array experiment generates large amounts of data, potentially allowing a global perspective. This has pushed biological research to expand into the field of bioinformatics, which use complex techniques of applied mathematics, informatics, statistics, and computational algorithms to analyze biological data. Bioinformatics tools still remain the principal method for protein microarray dataset interpretation of protein abundance, quantitation, and PTMs. More importantly, bioinformatics tools can be applied to compare and profile groups of datasets to determine the level of "relatedness" between samples in multiple dimensions.

3. Procedures in Executing Protein Microarray Data Analysis

To analyze data obtained from protein microarrays and to quantify and identify changes between datasets, the data need to be processed. Steps necessary for data analysis include the following; (1) spots and perimeter identification, (2) importing data files from the microarray scanning source, (3) normalization and quantification, (4) analyzing all datasets statistically, and finally (5) clustering and making multidimensional comparison of samples.

A. Identification of Spots and Perimeter Identification There are numerous ways to scan a protein array slide and obtain the corresponding spot intensities. Of the available commercial scanners, Molecular Devices' GenePix (formally known independently as Axon Instruments) microarray scanners (www.moleculardevices.com) are one of the most popular scanners available for image analysis of protein arrays that have been printed on microscope slide platforms. Other scanning instruments currently available are Agilent Technologies DNA Microarray scanner (www.agilent.com) and GE Healthcare's (formally known as Amersham Biosciences) Typhoon 4910 scanner (www.amershambiosciences.com). Generally, all scanners have application software (e.g., GenePix Pro with GenePix scanners) to control the scanner settings, generate high-resolution tagged image format files (.tiff), and perform image analysis. One of the most crucial steps in evaluating the signal of microarray spots after scanning is to define the limits of the signal by assigning perimeters (borders) around each spot. The printing of irregular spots as seen in most microarray platforms complicates the analysis in many dimensions. Fortunately, the discrepancies could be alleviated by using an automated system that identifies signal from the background. An example of the automated spot perimeter identification in GenePix Pro can be seen in Figure 10.19. One of the benefits of these programs is that it also allows users to personally customize or alter any minor differences that are not corrected by the automated analysis. Once perimeters have been assigned, spot intensity quantifications are exported as spreadsheets, usually in the format of text (.txt) files. The text file format is a widely accepted format for most microarray analysis programs.

B. Importing Data Files from Microarray Scanning Sources Most of the popular software available contain a wide range of analytical tools and algorithms to calculate statistical differences and graphically display the information as graphs and tables. Although some applications in these programs are specifically designed for gene arrays only, the majority of the analysis tools are applicable to proteins arrays. Such programs include Agilent Technologies' GeneSpring GX and, for the more computer-literate individual, Bioconductor of R (www.bioconductor.org). It is very important to have (or convert if necessary) the files in the correct format for the specific software that will be used. The advantage of most spreadsheets is that they can be opened in a number of programs and can be edited and analyzed independently. Each spreadsheet contains the scanning parameters and spot information. An example of a spreadsheet generated from GenePix Pro is illustrated in Figure 10.19. Once the appropriate file has been imported into the analysis program, the user can analyze the information and see the quantification values for each spot on the array. In most programs, parametric values created for each array include block numbers, column and row coordinates, diameter and volumes of spots targeted, and the intensity and background values for individual spots within the array. Datasheets also include basic statistical information such as mean, median, and standard deviation values. Furthermore, normalization is generally conducted prior to import into the analysis programs.

C. Normalization and Quantification of Protein Microarrays Comparing data generated from different protein arrays can be problematic. Differences with the extent or type of protein labeling, affinity and avidity of protein capturing, spot sizes, slide batches, and scanner settings are some factors that strongly influence the experimental results. Therefore, it is necessary that all datasets are normalized in the same manner or with a subset of controls run across all platforms. In almost all cases,

Type=GenePix Pro
DateTime=2006/5/30 08:58:48
PixelSize=10
Wavelengths=635
PMTGain=680
ScanPower=100
LaserPower=3.51
ScanRegion=372,152,1940,6992

Block	Column	Row	Name	ID	X	Y	Dia.	F635 Median	F635 Mean	F635 SD	F635 CV	B635	B635 Median	B635 Mean	B635 SD	B635 CV
1	1	1			4840	2070	250	3957	4087	2620	64	181	181	261	322	123
1	2	1			5380	2070	250	3834	3974	2371	59	181	181	245	280	114
1	3	1			5930	2070	250	3756	3858	2284	59	161	161	203	175	86
1	4	1			6480	2070	250	4292	4174	2805	67	154	154	192	164	85
1	5	1			7040	2070	250	368	420	248	59	156	156	195	197	101
1	6	1			7590	2070	250	337	388	239	61	155	155	203	230	113
1	7	1			8130	2070	250	332	392	248	63	149	149	194	213	109
1	8	1			8680	2070	250	357	428	366	85	148	148	195	185	94
1	1	2			4840	2620	250	214	270	235	87	166	166	221	242	109
1	2	2			5380	2620	250	222	276	247	89	168	168	214	244	114
1	3	2			5930	2620	250	180	228	169	74	149	149	192	196	102
1	4	2			6470	2620	250	182	219	173	78	143	143	182	160	87
1	5	2			7010	2620	250	153	191	135	70	162	162	222	429	193
1	6	2			7560	2620	250	148	175	99	56	158	158	208	227	109
1	7	2			8100	2620	250	149	194	173	89	149	149	189	160	84
1	8	2			8650	2620	250	143	174	111	63	143	143	185	161	87
1	1	3			4840	3160	250	176	211	148	70	162	162	221	299	135
1	2	3			5380	3160	250	168	203	144	70	165	165	213	191	89
1	3	3			5930	3160	250	151	187	136	72	144	144	192	297	154
1	4	3			6470	3160	250	152	182	125	68	144	144	184	258	140
1	5	3			7010	3160	260	226	274	178	64	164	164	218	411	188
1	6	3			7560	3160	250	169	214	165	77	153	153	198	226	114
1	7	3			8100	3160	250	185	212	131	61	154	154	198	200	101
1	8	3			8650	3160	250	173	207	122	58	146	146	193	224	116
1	1	4			4840	3710	250	137	181	218	120	151	151	210	324	154
1	2	4			5380	3710	250	129	174	304	174	149	149	203	310	152
1	3	4			5930	3710	250	143	165	101	61	145	145	183	150	81
1	4	4			6470	3710	250	136	168	128	76	144	144	181	141	77
1	5	4			7010	3710	250	173	210	130	61	153	153	195	234	120
1	6	4			7560	3710	250	185	217	135	62	176	176	233	512	219
1	7	4			8100	3710	250	181	219	144	65	163	163	209	215	102
1	8	4			8650	3710	250	182	222	150	67	152	152	196	177	90

FIGURE 10.19 Example of spreadsheet generated from GenePix Pro software.

it is not possible to compensate for all the possible variations that can potentially alter experimental results. Like most protein quantification assays, normalization in protein arrays is compensated by having known purified samples as standards to compare with the unknowns. This alleviates many of the complications from variations between experiments and batches as seen in gene arrays. Once standard quantifications have been established, normalization across samples can be easily done for the unknown samples. Quantification can be carried out using automated algorithms that are incorporated into microarray software packages or independently using spreadsheet programs such as Microsoft Office Excel (www.microsoft.com). The data from some microarrays could not be normalized if protein standards are not available. An example is cell arrays in which the signal detected is dependent on the intensity of refraction from cells bound to each spot. As with gene arrays, normalization is conducted after image analysis but prior to subsequent data analysis. The most common procedure for normalization in these cases is to do a one-per-spot normalization together with one-per-chip normalization. Normalized datasets are generally compared using the median intensities of each array.

D. Statistical Analysis of Protein Microarray Results Statistics is used to identify significant differences in protein quantity levels between samples and components across experiments. This comparison is performed for each protein within the identifier and across array samples, and the most significant identifications are returned, i.e., proteins identified with the smallest *p*-value. The most common statistical model used for analyzing samples in experimental pools has been using analysis of variance (ANOVA) tests. In practice, there are several types of ANOVA tests, depending on the number of treatments and the way they are applied to the subjects in the experiment. When performing each test, it is important to choose the specific test to compare your samples with and also set the parameters you want the program to apply. When performing an ANOVA test, specific parameters that define your group of tests, parametric and nonparametric tests, *p*-value cutoff, and the multiple testing correction options need to be defined.

E. Clustering and Comparing of Protein Microarray Results A fundamental reason for protein microarray analysis is to understand the biological mechanisms by which experimental systems differ from one another at multiple dimensional perspectives. One of the complications with protein microarray analysis is the complexity of information that is generated from a single experiment and between multiple samples within an experimental pool. Combining statistical analysis with clustering algorithms enables researchers to visually deconvolute large datasets of information to a simplified format (refer to "Clustering Technique in Gene Arrays" section for additional information). The principle of this tool is to analyze and compare protein expression (quantity or ratios) profiles in multiple dimensions. This is a very powerful tool in that it has the capability to cluster components from large sets of data and, in most cases, cluster coregulated proteins that often belong to the same biological pathway or protein complexes into a single group. What clustering does is to extrapolate patterns of protein quantity into subgroups and subclasses. Most clustering algorithms can generally perform two-dimensional clustering. The first dimension usually clusters sample together in relationship or "expression" profiles of samples. An example of this is looking at longitudinal studies (e.g., prognosis of a disease) in which classes of samples should group together based on the groups or types of proteins altered at a particular stage of disease development. Second, dimensional analysis can be carried out to further cluster proteins with expressed at similar levels (or quantities or ratio) across the pool of samples (cross-sectional analysis). The clustering of these proteins generally signifies the groups of proteins responsible for the changes between sample groups.

4. Examples of Protein Microarray Analysis Programs

Aligent Technologies' GeneSpring GX. GeneSpring GX is a powerful visualization and analysis program that has been designed for use with expression data from gene arrays and also quantitation levels of other array platforms such as protein arrays. One of the advantages of this program is it allows researchers to identify targets quickly and reliably. By providing statistically (ANOVA, Student's *T*-test) meaningful results, GeneSpring GX enables prediction of clinical outcomes and characterization of novel "expression" patterns. Furthermore, it is part of an integrated analysis suite that enables a visual and analytical comparison between quantity, genotyping, proteins, metabolites, and other data types to answer complex biological questions.

Bioconductor for R. Bioconductor is an open-source and open-development software project for the analysis and comprehension of microarray data that have been specifically designed for the R programming language. Although the program has been specifically designed to analyze genomic data, users of the software are able to reprogram the algorithms for compatibility with all microarray platforms, including protein arrays. Generally, there are a number of subpackages that can be installed with the program for certain analysis purposes. In addition, the R package system itself provides implementations for a broad range of state-of-the-art statistical and graphical techniques, including linear and nonlinear modeling, cluster analysis, prediction, resampling, survival analysis, and time-series analysis.

Part II Step-By-Step Tutorial

1. Agilent Technologies: GeneSpring GX

GeneSpring GX is used in this demonstration for statistical analysis purposes. It is one of the most user-friendly and comprehensive programs available for microarray analysis. GeneSpring GX is a program that has been designed principally for the analysis of image files exported from scanning devices such as GenePix and Affymetrix (www.affymetrix.com) scanners. Although many of the applications apply only to gene arrays, it has the capability of analyzing all microarray platforms, including protein arrays.

The demonstration here is to identify the differences in protein quantity of a series of known candidate proteins extracted from tissue obtained from patients with ischemic heart disease or idiopathic dilated cardiomyopathy and compared them to those from healthy donors. Normalized datasets (Table 10.3) are obtained from cell protein arrays that capture cluster of differentiation (CD) surface proteins on white blood cells using the image analysis program GenePix Pro.

1. Initiating new experimental analysis in GeneSpring GX
    a. Open the program by double-clicking the shortcut "GeneSpring GX 7.3.1."
    b. Once the program is opened, load the specific array platform that you wish to analyze, e.g., CD antibody array. You will see a list of proteins that will be examined in the data analysis, as shown in Figure 10.20.
    c. Go to File → Import Data and a window should pop up. At this stage, only one file is needed to initiate a new analysis.
    d. Select the array platform that you are using.

TABLE 10.3 Normalized Data Sets of Cell Protein Microarrays

HEALTHY						IDCM						IHD					
HEALTHY1.txt		HEALTHY2.txt		HEALTHY3.txt		IDCM1.txt		IDCM2.txt		IDCM3.txt		IHD1.txt		IHD2.txt		IHD3.txt	
Protein	Signal	Protein	Signal	Protein	Signal	Protein	Signal	Protein	Signal	Protein	Signal	Protein	Signal	Protein	Signal	Protein	Signal
TCR α/β	2.0	TCR α/β	1.2	TCR α/β	1.9	TCR α/β	1.0	TCR α/β	1.0	TCR α/β	1.0	TCR α/β	0.8	TCR α/β	0.2	TCR α/β	0.2
TCR γ/δ	0.1	TCR γ/δ	0.0	TCR γ/δ	0.0	TCR γ/δ	0.0	TCR γ/δ	0.0	TCR γ/δ	0.0	TCR γ/δ	0.0	TCR γ/δ	0.0	TCR γ/δ	0.0
CD1a	0.1	CD1a	0.0	CD1a	0.0	CD1a	0.0	CD1a	0.0	CD1a	0.0	CD1a	0.0	CD1a	0.0	CD1a	0.0
CD2	4.0	CD2	3.6	CD2	3.3	CD2	3.6	CD2	2.7	CD2	1.2	CD2	2.1	CD2	1.0	CD2	1.5
CD3	2.5	CD3	2.5	CD3	2.8	CD3	4.1	CD3	3.6	CD3	3.0	CD3	1.7	CD3	0.8	CD3	0.5
CD4	2.2	CD4	1.3	CD4	2.2	CD4	2.6	CD4	3.4	CD4	3.2	CD4	2.2	CD4	0.3	CD4	0.6
CD5	2.4	CD5	1.6	CD5	1.8	CD5	3.3	CD5	4.2	CD5	3.3	CD5	0.5	CD5	0.0	CD5	0.2
CD7	3.4	CD7	2.2	CD7	2.9	CD7	1.1	CD7	0.9	CD7	1.9	CD7	1.1	CD7	1.2	CD7	1.3
CD8	1.4	CD8	1.8	CD8	1.5	CD8	0.1	CD8	0.7	CD8	0.9	CD8	0.0	CD8	0.7	CD8	0.7
CD9	3.0	CD9	3.3	CD9	2.8	CD9	3.6	CD9	4.0	CD9	3.6	CD9	1.6	CD9	1.6	CD9	1.5
CD10	0.2	CD10	0.4	CD10	0.3	CD10	0.2	CD10	0.0	CD10	0.0	CD10	0.5	CD10	0.1	CD10	0.1
CD11a	3.6	CD11a	3.0	CD11a	2.7	CD11a	4.3	CD11a	3.9	CD11a	3.0	CD11a	3.9	CD11a	2.2	CD11a	1.7
CD11b	1.8	CD11b	2.4	CD11b	2.3	CD11b	2.9	CD11b	3.7	CD11b	2.7	CD11b	2.4	CD11b	2.7	CD11b	2.2
CD11c	0.7	CD11c	0.9	CD11c	1.0	CD11c	3.2	CD11c	2.8	CD11c	2.2	CD11c	2.4	CD11c	2.0	CD11c	1.7
CD13	0.4	CD13	0.4	CD13	0.3	CD13	1.6	CD13	2.9	CD13	1.8	CD13	1.1	CD13	2.6	CD13	1.7
CD14	0.4	CD14	0.5	CD14	0.4	CD14	2.0	CD14	2.9	CD14	1.8	CD14	1.9	CD14	2.5	CD14	2.6
CD15	0.0	CD15	0.0	CD15	0.0	CD15	2.5	CD15	3.5	CD15	2.1	CD15	1.9	CD15	2.2	CD15	1.7
CD16	0.2	CD16	0.5	CD16	0.4	CD16	3.2	CD16	2.3	CD16	1.7	CD16	1.4	CD16	2.1	CD16	1.2
CD19	0.7	CD19	0.7	CD19	0.3	CD19	0.3	CD19	0.1	CD19	0.2	CD19	0.1	CD19	0.1	CD19	0.0
CD20	1.0	CD20	1.0	CD20	1.1	CD20	0.1	CD20	0.1	CD20	0.1	CD20	0.0	CD20	0.0	CD20	0.0
CD21	0.7	CD21	0.8	CD21	0.7	CD21	0.1	CD21	0.0	CD21	0.2	CD21	0.0	CD21	0.0	CD21	0.0
CD22	0.4	CD22	0.3	CD22	0.5	CD22	0.1	CD22	0.0	CD22	0.2	CD22	0.2	CD22	0.4	CD22	0.0
CD23	0.4	CD23	0.3	CD23	0.3	CD23	0.0	CD23	0.0	CD23	0.0	CD23	0.0	CD23	0.0	CD23	0.0
CD24	0.3	CD24	0.2	CD24	0.4	CD24	0.1	CD24	0.0	CD24	0.2	CD24	0.4	CD24	0.0	CD24	0.0
CD25	0.7	CD25	0.7	CD25	0.9	CD25	0.0	CD25	0.0	CD25	0.0	CD25	0.0	CD25	0.1	CD25	0.0
CD28	2.1	CD28	1.6	CD28	2.0	CD28	1.2	CD28	0.8	CD28	1.2	CD28	0.0	CD28	0.2	CD28	0.0
CD29	5.5	CD29	5.1	CD29	5.0	CD29	5.1	CD29	4.6	CD29	3.2	CD29	6.4	CD29	4.4	CD29	2.5
CD31	5.0	CD31	5.1	CD31	5.6	CD31	5.2	CD31	3.4	CD31	3.3	CD31	6.2	CD31	2.1	CD31	2.4
CD32	1.4	CD32	1.0	CD32	1.4	CD32	0.0	CD32	0.1	CD32	0.0	CD32	0.5	CD32	0.0	CD32	0.0
CD33	0.2	CD33	0.2	CD33	0.9	CD33	0.0	CD33	0.0	CD33	1.3	CD33	1.7	CD33	2.3	CD33	1.5
CD34	1.0	CD34	0.6	CD34	1.2	CD34	0.0	CD34	0.0	CD34	0.0	CD34	0.0	CD34	0.0	CD34	0.0
CD36	1.8	CD36	1.4	CD36	2.0	CD36	4.1	CD36	3.8	CD36	3.7	CD36	4.0	CD36	4.0	CD36	3.3
CD37	1.3	CD37	1.5	CD37	1.9	CD37	0.0	CD37	0.3	CD37	0.1	CD37	0.1	CD37	0.0	CD37	0.0
CD38	5.0	CD38	4.2	CD38	4.7	CD38	1.8	CD38	1.6	CD38	0.5	CD38	3.9	CD38	3.7	CD38	3.0
CD40	1.9	CD40	1.8	CD40	1.8	CD40	0.3	CD40	0.2	CD40	0.2	CD40	0.1	CD40	0.0	CD40	0.1
CD41	2.5	CD41	1.9	CD41	2.5	CD41	0.7	CD41	0.3	CD41	1.1	CD41	0.9	CD41	0.9	CD41	0.6
CD42a	3.1	CD42a	2.5	CD42a	3.0	CD42a	0.5	CD42a	0.2	CD42a	2.1	CD42a	0.9	CD42a	0.6	CD42a	0.7
CD43	5.3	CD43	4.4	CD43	5.6	CD43	5.5	CD43	3.7	CD43	4.2	CD43	5.7	CD43	4.9	CD43	4.4
CD44	5.1	CD44	5.0	CD44	5.8	CD44	4.7	CD44	5.2	CD44	3.7	CD44	5.7	CD44	5.1	CD44	4.8
CD45	5.5	CD45	5.3	CD45	5.3	CD45	3.0	CD45	3.4	CD45	3.4	CD45	6.1	CD45	4.6	CD45	4.9
CD45RA	2.9	CD45RA	2.6	CD45RA	3.1	CD45RA	0.6	CD45RA	1.2	CD45RA	0.5	CD45RA	2.1	CD45RA	0.6	CD45RA	0.0
CD45RO	1.4	CD45RO	0.9	CD45RO	1.2	CD45RO	1.9	CD45RO	2.4	CD45RO	2.3	CD45RO	2.7	CD45RO	2.1	CD45RO	1.0
CD49d	4.7	CD49d	3.8	CD49d	4.5	CD49d	1.7	CD49d	2.0	CD49d	1.1	CD49d	3.8	CD49d	2.6	CD49d	3.6
CD49e	4.7	CD49e	4.2	CD49e	5.3	CD49e	4.0	CD49e	2.8	CD49e	4.3	CD49e	4.3	CD49e	2.5	CD49e	2.9
CD52	1.8	CD52	1.8	CD52	1.8	CD52	0.0	CD52	0.0	CD52	0.3	CD52	0.0	CD52	0.0	CD52	0.0
CD54	0.4	CD54	0.3	CD54	0.6	CD54	0.0	CD54	0.0	CD54	0.0	CD54	1.3	CD54	0.0	CD54	0.0
CD56	0.4	CD56	0.3	CD56	0.5	CD56	0.2	CD56	0.4	CD56	0.1	CD56	0.0	CD56	0.4	CD56	0.1
CD57	0.4	CD57	0.3	CD57	0.6	CD57	0.0	CD57	0.0	CD57	0.0	CD57	0.0	CD57	0.0	CD57	0.0
CD60	0.7	CD60	0.7	CD60	0.5	CD60	0.0	CD60	0.0	CD60	0.0	CD60	0.0	CD60	0.0	CD60	0.0
CD61	1.9	CD61	1.9	CD61	2.3	CD61	2.2	CD61	0.0	CD61	1.6	CD61	1.0	CD61	0.9	CD61	0.9
CD62L	0.9	CD62L	0.5	CD62L	1.0	CD62L	0.0	CD62L	0.0	CD62L	0.0	CD62L	0.0	CD62L	0.0	CD62L	0.0
CD62E	2.0	CD62E	2.3	CD62E	2.8	CD62E	0.7	CD62E	0.5	CD62E	1.9	CD62E	0.2	CD62E	1.1	CD62E	1.1
CD62P	2.5	CD62P	2.8	CD62P	3.3	CD62P	1.3	CD62P	0.9	CD62P	2.1	CD62P	0.3	CD62P	1.0	CD62P	1.0
CD64	0.3	CD64	0.2	CD64	0.3	CD64	0.0	CD64	0.1	CD64	0.1	CD64	2.3	CD64	1.2	CD64	0.6
CD65	0.1	CD65	0.1	CD65	0.2	CD65	0.0	CD65	0.0	CD65	0.0	CD65	0.0	CD65	0.0	CD65	0.0
CD66c	0.1	CD66c	0.6	CD66c	0.0	CD66c	0.3	CD66c	0.0	CD66c	0.2	CD66c	1.6	CD66c	0.1	CD66c	1.4
CD71	2.5	CD71	1.3	CD71	1.8	CD71	0.1	CD71	0.5	CD71	1.6	CD71	0.1	CD71	0.2	CD71	0.8
CD77	0.2	CD77	0.1	CD77	0.4	CD77	0.0	CD77	0.0	CD77	0.0	CD77	0.0	CD77	0.0	CD77	0.0
CD79a	0.2	CD79a	0.1	CD79a	0.5	CD79a	0.0	CD79a	0.0	CD79a	0.0	CD79a	0.0	CD79a	0.0	CD79a	0.0
CD79b	0.7	CD79b	0.5	CD79b	0.6	CD79b	0.1	CD79b	0.0	CD79b	0.1	CD79b	0.0	CD79b	0.0	CD79b	0.0
CD80	0.7	CD80	0.5	CD80	0.8	CD80	0.0	CD80	0.0	CD80	0.0	CD80	0.0	CD80	0.0	CD80	0.0
CD86	0.8	CD86	0.9	CD86	0.8	CD86	0.6	CD86	0.4	CD86	0.5	CD86	0.7	CD86	0.1	CD86	0.2
CD88	0.1	CD88	0.0	CD88	0.4	CD88	0.0	CD88	0.0	CD88	0.5	CD88	1.6	CD88	0.0	CD88	0.0
CD95	1.0	CD95	1.3	CD95	2.0	CD95	0.9	CD95	1.0	CD95	0.6	CD95	2.5	CD95	2.6	CD95	0.8
CD102	2.2	CD102	1.3	CD102	1.7	CD102	0.0	CD102	0.0	CD102	0.0	CD102	0.1	CD102	0.0	CD102	0.0
CD103	0.2	CD103	0.2	CD103	0.1	CD103	0.0	CD103	0.0	CD103	0.0	CD103	0.0	CD103	0.0	CD103	0.0
CD117	0.0	CD117	0.0	CD117	0.1	CD117	0.0	CD117	0.0	CD117	0.0	CD117	1.2	CD117	0.0	CD117	0.0
CD120a	0.2	CD120a	0.0	CD120a	0.2	CD120a	0.0	CD120a	0.0	CD120a	0.0	CD120a	0.2	CD120a	0.0	CD120a	0.7
CD122	0.7	CD122	0.3	CD122	0.8	CD122	0.0	CD122	0.2	CD122	0.0	CD122	0.0	CD122	0.0	CD122	0.0
CD126	1.0	CD126	0.4	CD126	0.9	CD126	0.0	CD126	0.0	CD126	0.0	CD126	0.0	CD126	0.1	CD126	0.0
CD128	0.2	CD128	0.2	CD128	0.3	CD128	2.2	CD128	2.1	CD128	1.0	CD128	1.5	CD128	1.6	CD128	0.7
CD130	1.4	CD130	1.0	CD130	1.6	CD130	0.0	CD130	0.0	CD130	0.0	CD130	0.1	CD130	0.0	CD130	0.0
CD134	0.1	CD134	0.1	CD134	0.7	CD134	0.0	CD134	0.0	CD134	0.0	CD134	0.5	CD134	0.0	CD134	0.0
CD135	0.1	CD135	0.1	CD135	0.2	CD135	0.0	CD135	0.0	CD135	0.0	CD135	0.0	CD135	0.0	CD135	0.0
CD138	0.2	CD138	0.0	CD138	0.3	CD138	0.0	CD138	0.1	CD138	0.0	CD138	0.0	CD138	0.0	CD138	0.0
CD154	2.0	CD154	1.5	CD154	1.5	CD154	0.0	CD154	0.0	CD154	0.0	CD154	0.0	CD154	0.0	CD154	0.0
CD235a	0.6	CD235a	0.3	CD235a	0.8	CD235a	0.0	CD235a	0.0	CD235a	0.0	CD235a	0.0	CD235a	0.0	CD235a	0.0
HLA-DR	0.8	HLA-DR	1.1	HLA-DR	0.7	HLA-DR	4.7	HLA-DR	0.8	HLA-DR	1.0	HLA-DR	2.1	HLA-DR	0.0	HLA-DR	1.0
FMC7	0.4	FMC7	0.2	FMC7	0.6	FMC7	0.0	FMC7	0.0	FMC7	0.1	FMC7	0.0	FMC7	0.0	FMC7	0.0
κ	1.3	κ	2.0	κ	1.2	κ	3.0	κ	2.9	κ	2.9	κ	0.7	κ	0.0	κ	0.5
λ	2.1	λ	1.2	λ	1.8	λ	2.8	λ	2.0	λ	2.5	λ	2.5	λ	0.2	λ	0.4
sIg	1.1	sIg	0.8	sIg	0.8	sIg	0.0	sIg	0.2	sIg	0.0	sIg	0.0	sIg	0.0	sIg	0.0

e. At the top of each column, select from the scroll bar at the top of each column the "Gene Identifier" and the "Signal" for the normalized data as seen in Figure 10.21.

f. Import the rest of the files that the user would like to analyze by clicking on each individual file or by clicking "Add All" if all your files are in the same directory.

g. Fill in as many parameters in the sample attributes table.

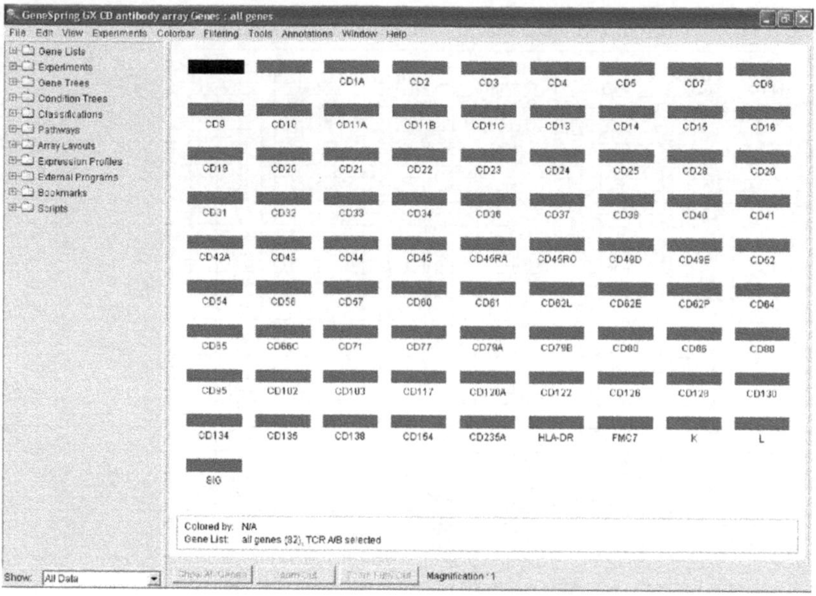
GeneSpring GX CD antibody array Genes : all genes
File Edit View Experiments Colorbar Filtering Tools Annotations Window Help
Gene Lists
Experiments
Gene Trees
Condition Trees
Classifications
Pathways
Array Layouts
Expression Profiles
External Programs
Bookmarks
Scripts
CD1A CD2 CD3 CD4 CD5 CD7 CD8
CD9 CD10 CD11A CD11B CD11C CD13 CD14 CD15 CD16
CD19 CD20 CD21 CD22 CD23 CD24 CD25 CD28 CD29
CD31 CD32 CD33 CD34 CD36 CD37 CD38 CD40 CD41
CD42A CD43 CD44 CD45 CD45RA CD45RO CD49D CD49E CD52
CD54 CD56 CD57 CD60 CD61 CD62L CD62E CD62P CD64
CD65 CD66C CD71 CD77 CD79A CD79B CD80 CD86 CD88
CD95 CD102 CD103 CD117 CD120A CD122 CD126 CD128 CD130
CD134 CD135 CD138 CD164 CD235A HLA-DR FMC7 K L
SIG
Colored by: N/A
Gene List: all genes (82), TCR A/B selected
Show: All Data
Magnification: 1

FIGURE 10.20 Opening window with list of proteins to be analyzed in protein array.

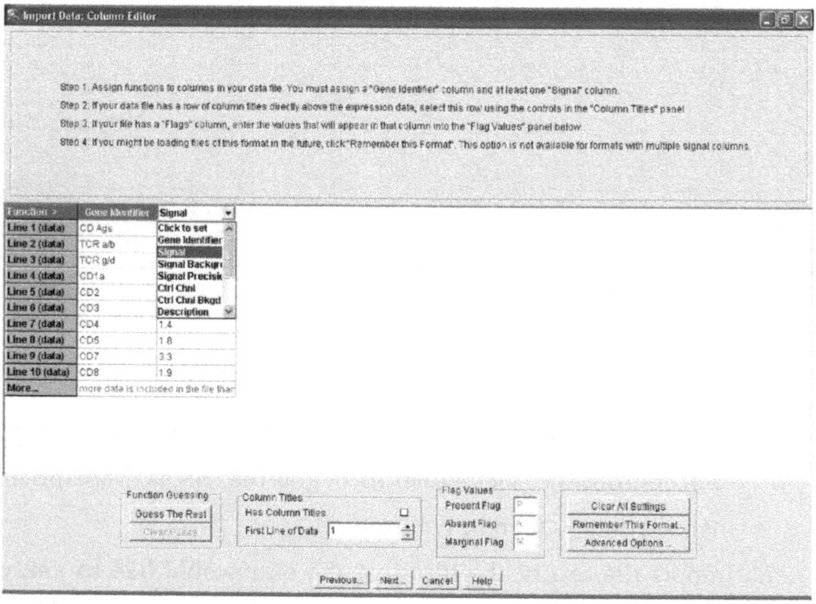
Import Data: Column Editor
Step 1. Assign functions to columns in your data file. You must assign a "Gene Identifier" column and at least one "Signal" column.
Step 2. If your data file has a row of column titles directly above the expression data, select this row using the controls in the "Column Titles" panel.
Step 3. If your file has a "Flags" column, enter the values that will appear in that column into the "Flag Values" panel below.
Step 4. If you might be loading files of this format in the future, click "Remember this Format". This option is not available for formats with multiple signal columns.
Function > Gene Identifier Signal
Line 1 (data) CD Ags
Line 2 (data) TCR a/b
Line 3 (data) TCR g/d
Line 4 (data) CD1a
Line 5 (data) CD2
Line 6 (data) CD3
Line 7 (data) CD4 1.4
Line 8 (data) CD5 1.8
Line 9 (data) CD7 3.3
Line 10 (data) CD8 1.9
More... more data is included in the file than
Click to set
Gene Identifier
Signal
Signal Backgr
Signal Precisi
Ctrl Chnl
Ctrl Chnl Bkgd
Description
Function Guessing
Guess The Rest
Column Titles
Has Column Titles
First Line of Data 1
Flag Values
Present Flag P
Absent Flag A
Marginal Flag M
Clear All Settings
Remember This Format...
Advanced Options...
Previous... Next... Cancel Help

FIGURE 10.21 Assignment of titles to columns in imported text (.txt) files.

h. Click "Ok" to verify the list of samples in the pool for analysis and then finally save the experiment in a subfolder in the "Experiments" folder.

i. Once completed, a "New Experiment Checklist" window will pop up. This window contains options for normalization of certain parameters in the user's imported datasets, as seen in Figure 10.22. If necessary, apply the built-in algorithm ("Per Chip" or "Per Gene" options) to normalize data if you have not done so.

j. The user has completed a new experimental run. The main data analysis window should pop up with a control column on the left, colored bars on in the center, and a color intensity bar scale on the right, as seen in Figure 10.23.

2. Statistical analysis of normalized data
   a. To run statistical analysis, go to Tools → Statistical Analysis (ANOVA).
   b. Assign the proper categories and parameters for your test and also the samples which you want to compare with. This is shown in Figure 10.24. A common practice with this specific run is that the user can do multiple analyses on a number of sample groups

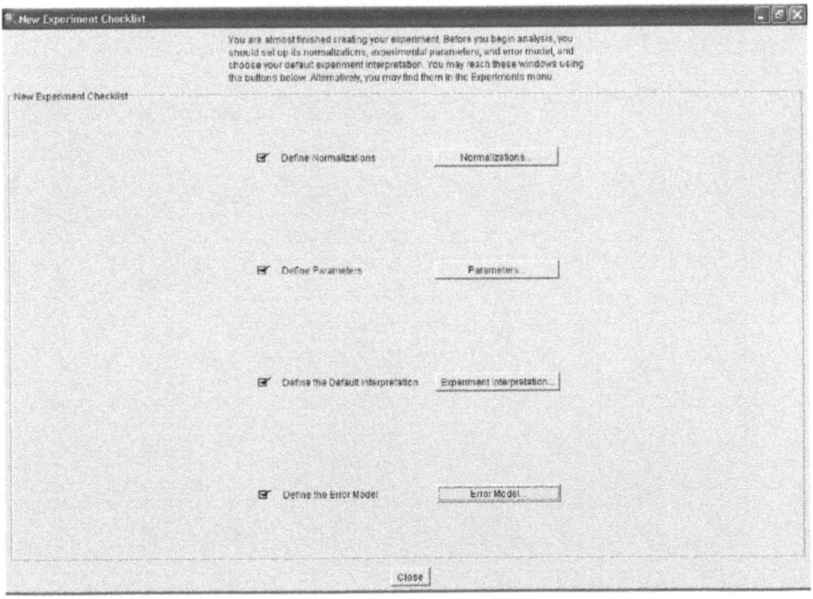
New Experiment Checklist
You are almost finished creating your experiment. Before you begin analysis, you should set up its normalizations, experimental parameters, and error model, and choose your default experiment interpretation. You may reach these windows using the buttons below. Alternatively, you may find them in the Experiments menu.
New Experiment Checklist
Define Normalizations
Normalizations...
Define Parameters
Parameters...
Define the Default interpretation
Experiment interpretation...
Define the Error Model
Error Model...
Close

FIGURE 10.22 Data setup options.

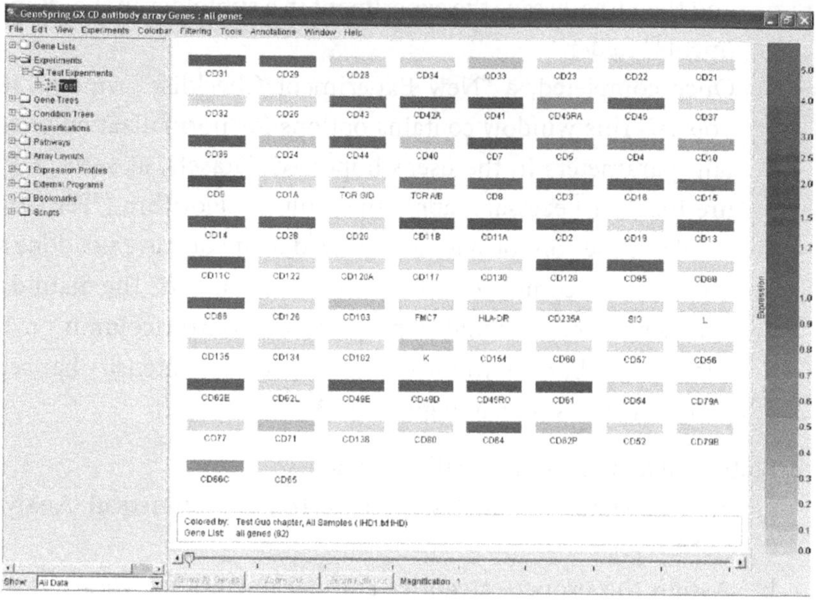
GeneSpring GX CD antibody array Genes : all genes
File Edit View Experiments Colorbar Filtering Tools Annotations Window Help
Gene Lists
Experiments
Test Experiments
Gene Trees
Condition Trees
Classifications
Pathways
Array Layouts
Expression Profiles
External Programs
Bookmarks
Scripts
CD31
CD29
CD28
CD34
CD33
CD23
CD22
CD21
CD32
CD43
CD42A
CD41
CD45RA
CD37
CD44
CD40
CD7
CD5
CD4
CD10
TCR A/B
CD8
CD3
CD15
CD14
CD38
CD11B
CD11A
CD2
CD13
HLA-DR
FMC7
Colored by: Test Guo chapter, All Samples
Gene List: all genes (82)
Magnification
Show
All Data
Expression

FIGURE 10.23 Primary analysis window.

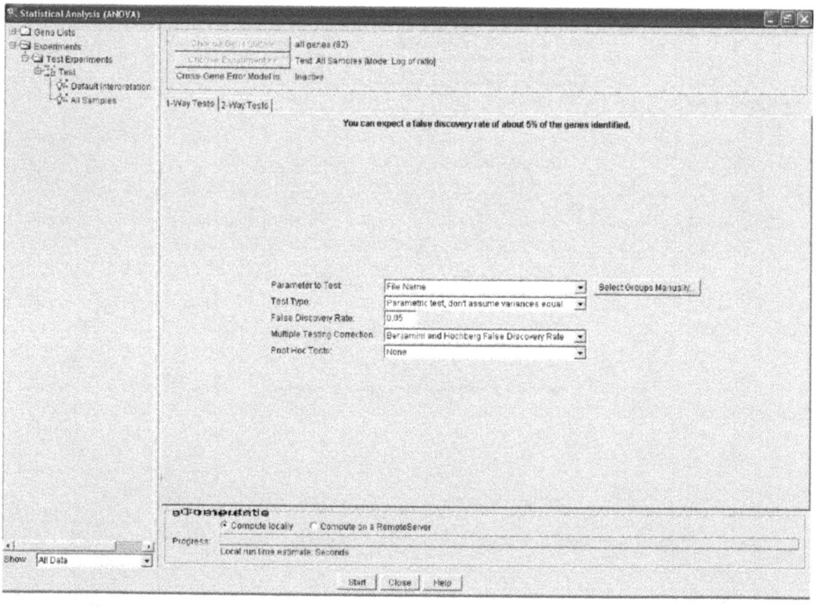
Statistical Analysis (ANOVA)
Gene Lists
Experiments
Test Experiments
Default Interpretation
All Samples
all genes (82)
Cross-Gene Error Model is: Inactive
1-Way Tests
2-Way Tests
You can expect a false discovery rate of about 5% of the genes identified.
Parameter to Test
File Name
Select Groups Manually
Test Type
Parametric test, don't assume variances equal
False Discovery Rate
0.05
Multiple Testing Correction
Benjamini and Hochberg False Discovery Rate
Post Hoc Tests
None
Compute locally
Compute on a RemoteServer
Progress
Start
Close
Help
Show
All Data

FIGURE 10.24 Statistical analysis using ANOVA.

and parameters. You can use the number of parameters to set the number of tests.

c. A table of significant changes will be listed. Save the list to a designated folder, as shown in Figure 10.25. Tables can be copied and saved as spreadsheets.

3. Clustering and sample comparison
   a. To run the clustering analysis, go to Tools → Clustering → Gene Tree. This tool clusters together proteins that belong to the same biological pathway or protein complexes into a single group. In this instance, we could cluster the list of proteins identified from the ANOVA tests. Find the file from the control column and select the file that contains the list of proteins. Again, make sure that you use the correct parameters (mainly similarity measurement test and clustering algorithm) that the user would like the samples to be analyzed with. Once the computational analysis has completed, a block and a Gene Tree are generated. Save the file in the designated folder.
   b. Repeat the same procedures for Tools → Clustering → Condition Tree. Save the file that is generated in the "Condition Tree" folder.

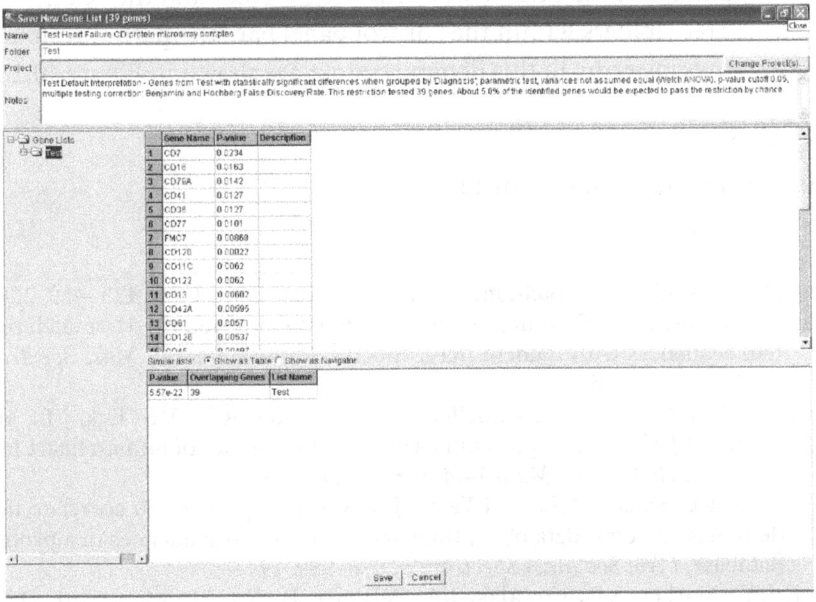
Save New Gene List (39 genes)
Name
Folder
Test
Project
Notes
Gene Lists
Gene Name
P-value
Description
Similar lists
Show as Table
Show as Navigator
P-value
Overlapping Genes
List Name
5.57e-22
39
Test
Save
Cancel

FIGURE 10.25 ANOVA test results.

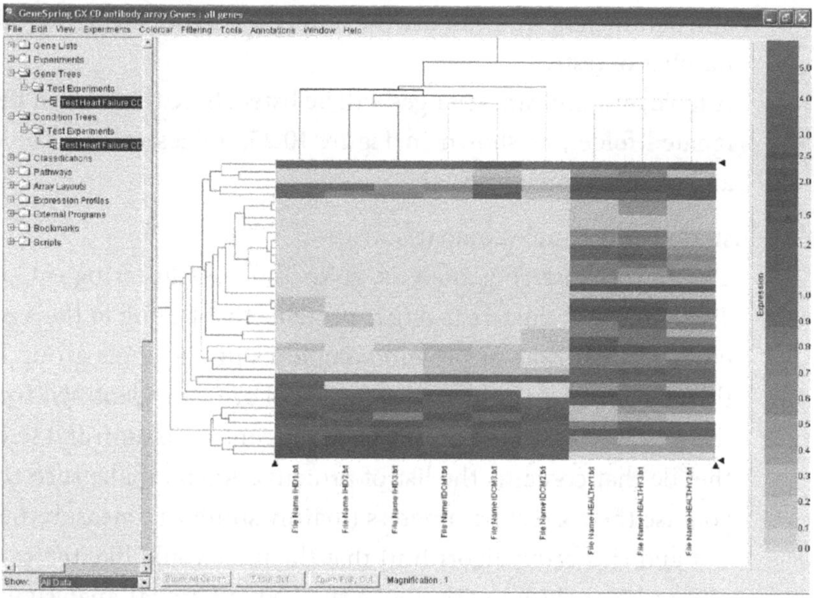
Gene Lists
Experiments
Gene Trees
Test Experiments
Test Heart Failure CC
Condition Trees
Test Experiments
Test Heart Failure CC
Classifications
Pathways
Array Layouts
Expression Profiles
External Programs
Bookmarks
Scripts
Show: All Data
Magnification: 1
Expression
5.0
4.0
3.0
2.5
2.0
1.5
1.2
1.0
0.9
0.8
0.7
0.6
0.5
0.4
0.3
0.2
0.1
0.0

FIGURE 10.26 **Final clustering analysis results.**

c. The clustering analysis is now completed. The user can review the two-dimensional analysis results by going to the main window and select the saved files in the “Gene Tree” and the “Condition Tree” folders within the control panel bar. The partial results of the clustering in this demonstration are shown in Figure 10.26.

Part III Sample Data

The sample data are shown in Table 10.3.

REFERENCES

Chen, C.S. and Zhu, H. Protein microarrays. *Biotechniques* 40(4): 423–429, 2006.

Craig, R. and Beavis, R.C. A method for reducing the time required to match protein sequences with tandem mass spectra. *Rapid Commun Mass Spectrom* 17: 2310–2316, 2003.

dos Remedios, C.G., Liew, C.C., Allen, P.D., Winslow, R.L., Van Eyk, J.E., and Dunn, M.J. Genomics, proteomics and bioinformatics of human heart failure. *J Muscle Res Cell Motil* 24(4–6): 251–260, 2003.

Eng, J.K., McCormack, A.L., and Yates, J.R. 3rd. An approach to correlate tandem mass spectral data of peptides with amino acid sequences in a protein database. *J Am Soc Mass Spectrom* 5: 976–989, 1994.

Geer, L.Y., Markey, S.P., Kowalak, J.A., Wagner, L., Xu, M., Marynard, D.M., Yang, X., Shi, W., and Bryant, S.H. Open mass spectrometry search algorithm. *J Proteome Res* 3: 958–964, 2004.

Hultschig, C., Kreutzberger, J., Seitz, H., Konthur, Z., Bussow, K., and Lehrach, H. Recent advances of protein microarrays. *Curr Opin Chem Biol* 10(1): 4–10, 2006.

Hunt, D.F. Personal commentary on proteomics. *J Proteome Res* 1: 15–19, 2002.

Johnson, R.S. and Taylor, J.A. Searching sequence databases via de novo peptide sequencing by tandem mass spectrometry. *Mol Biotechnol* 22: 301–315, 2002.

Kapp, E.A., Schutz, F., Connolly, L.M., Chakel, J.A., Meza, J.Z., Miller, C.A., Fenyo, D., Eng, J.K., Adkins, J.N., Omenn, G.S., and Simpson, R.J. An evaluation, comparison, and accurate benchmarking of several publicly available MS/MS search algorithms: sensitivity and specificity analysis. *Proteomics* 5: 3475–3490, 2005.

Keller, A., Nesvizhskii, A.I., Kolker, E., and Aebersold, R. Empirical statistical model to estimate the accuracy of peptide identifications made by MS/MS and database search. *Anal Chem* 74: 5383–5392, 2002.

Kozarova, A., Petrinac, S., Ali, A., and Hudson, J.W. Array of informatics: applications in modern research. *J Proteome Res* 5(5): 1051–1059, 2006.

Ma, B., Zhang, K., Hendrie, C., Liang, C., Li, M., Doherty-Kirby, A., and Lajoie, G. PEAKS: powerful software for peptide de novo sequencing by tandem mass spectrometry. *Rapid Commun Mass Spectrom* 17: 2337–2342, 2003.

Peng, J., Elias, J.E., Thoreen, C.C., Licklider, L.J., and Gygi, S.P. Evaluation of multidimensional chromatography coupled with tandem mass spectrometry (LC/LC-MS/MS) for large-scale protein analysis: the yeast proteome. *J Proteome Res* 2: 43–50, 2003.

Perkins, D.N., Pappin, D.J.C., Creasy, D.M., and Cottrell, J.S. Probability-based protein identification by searching sequence database using mass spectrometry data. *Electrophoresis* 20: 3551–3567, 1999.

Pevzner, P.A., Dancik, V., and Tang, C.L. Mutation-tolerant protein identification by mass spectrometry. *J Comput Biol* 7: 777–787, 2000.

Sunyaev, S., Liska, A.J., Golod, A., and Shevchenko, A. MultiTag: multiple error-tolerant sequence tag search for the sequence-similarity identification of proteins by mass spectrometry. *Anal Chem* 75: 1307–1315, 2003.

Tabb, D.L., Saraf, A., and Yates, J.R. 3rd. GutenTag: high-throughput sequence tagging via an empirically derived fragmentation model. *Anal Chem* 75: 6415–6421, 2003.

CHAPTER 11

Protein Sequence Analysis

Jun Wada, Hiroko Tada, and Masaharu Seno

CONTENTS

SECTION 1 PRIMARY STRUCTURE ANALYSIS

Part I Introduction

1. What Is Primary Structure Analysis?

Most natural polypeptides contain between 50 and 2000 amino acid residues and are commonly referred as proteins. The mean molecular weight of an amino acid residue is about 110, and so the molecular weights of most proteins are between 5,500 and 220,000. Each protein has a unique, precisely defined amino acid sequence, and it is often referred as its primary structure. Analyzing the amino acid sequences using a primary structure analysis program is the initial step to predict the functions and three-dimensional structures of proteins. A classical method to determine amino acid sequence is Edman degradation, in which amino acid residues are removed stepwise from the N-terminus by reaction with phenylisothiocyanate. This method was named after Pehr Victor Edman (1916–1977), a Swedish protein chemist, who described the method in 1956. Frederick Sanger, an English biochemist and a two-time Nobel laureate, determined the complete amino acid sequence of insulin in 1955, which earned him his first Nobel Prize in Chemistry in 1958. This section will not describe direct amino acid sequencing but will cover the bioinformatic analysis of primary structure, including (1) computation of theoretical *pI* (isoelectric point) and Mw (molecular weight) of proteins, (2) *de novo* repeat detection in protein sequences, (3) hydropathy plot for proteins, and (4) hydrophobic cluster analysis of proteins.

2. What Is Involved in Primary Structure Analysis?

Two simple steps are involved in the primary structure analysis of proteins: selecting protein sequences and computing them, using specific programs such as Compute pI/Mw, RADAR, PlotScale, and Drawhca.

Selecting protein sequences. For primary sequence analyses, you can submit protein sequences in simple text or FASTA format. The FASTA sequence format is a widely accepted format. It starts with the greater than symbol (>) gene identification number, its reference protein accession number, and its name followed by the sequence.

Computation of pI and Mw of proteins. The tool "Compute pI/Mw" (http://us.expasy.org/tools/pi_tool.html) computes the pI and Mw of proteins. Protein pI is calculated using pK values of amino acids, which were defined by examining polypeptide migration between pH 4.5 and 7.3 in an immobilized pH gradient gel environment with 9.2 *M* and 9.8 *M* urea at 15°C or 25°C. Protein Mw is calculated by the addition of average isotopic masses of amino acids in the protein and the average isotopic mass of one water molecule. This program does not account for the effects of posttranslational modifications; thus modified, proteins on a 2D gel may migrate to a position quite different from that predicted. In addition to pI and Mw, ProtParam (http://us.expasy.org/tools/protparam.html) can compute various physicochemical properties that can be deduced from a protein sequence. The parameters computed by ProtParam include the molecular weight, theoretical pI, amino acid composition, atomic composition, extinction coefficient, estimated half-life, instability index, aliphatic index, and grand average of hydropathicity. Molecular weight and theoretical pI are calculated as in Compute pI/Mw.

A. *De Novo* Repeat Detection in Protein Sequences RADAR stands for Rapid Automatic Detection and Alignment of Repeats in protein sequences (http://www.ebi.ac.uk/Radar/). Many large proteins have evolved by internal duplication, and many internal sequence repeats correspond to functional and structural units. RADAR uses an automatic algorithm to segment query sequences into repeats; it also identifies short-composition-biased as well as gapped approximate repeats and complex repeat architectures involving many different types of repeats in your query sequence.

B. Hydropathy Plot for Proteins PlotScale (http://us.expasy.org/tools/protscale.html) allows computation and representation of the profile produced by any amino acid scale on a selected protein. An amino acid scale is defined by a numerical value assigned to each type of amino acid. ProtScale can be used with fifty predefined scales entered from the literature. You can set several parameters that control the computation of a scale profile, such as the window size, the weight variation model, the window edge relative weight value, and scale normalization.

C. Hydrophobic Cluster Analysis (HCA) of Proteins The HCA method is based on the use of a bidimensional plot, called the HCA plot (http://bioserv.rpbs.jussieu.fr/RPBS/cgi-bin/Ressource.cgi?chzn_lg=an&chzn_rsrc=HCA). The bidimensional plot originates from the drawing of the sequence on α helix (3.6 residues/turn, connectivity distance of 4 residues separating two different clusters), which has been shown to offer the best correspondence between clusters and regular secondary structures. Examination of the HCA plot of a protein sequence allows identification of globular and nonglobular domains. The HCA plot signature is also useful in the comparison of families of highly divergent sequences and allows detection at low levels of relevant similarities.

Part II Step-By-Step Tutorial

The demo example is human vaspin (visceral adipose-tissue-derived serine protease inhibitor), mouse galectin-9, human collectrin, and human α1-antitrypsin.

1. Selecting Protein Sequences

1. Fetch human vaspin protein sequences from the NCBI Web site.
   a. As demonstrated in Figure 11.1, type the following address (http://www.ncbi.nlm.nih.gov/entrez/query.fcgi?db=Protein) in your browser (Internet Explorer or others). In the "Search," Protein category is now the default setting. In the "for," type "human vaspin."
   b. Click "Go" in your browser or press "Enter" on your keyboard. The accession number of human vaspin protein sequence will be displayed as in Figure 11.2.
   c. Click "NP_776249" to display all human vaspin protein sequence information. Then in the Display, select "FASTA" format, and you will get the human vaspin protein sequence in the FASTA format (Figure 11.3).

2. Compute pI/Mw

1. Type the following address (http://us.expasy.org/tools/pi_tool.html) in your browser (Internet Explorer or others).
2. Enter a protein sequence in the box by cutting and pasting the single-letter code as plain text (Figure 11.4). This program does not accept FASTA format sequence; however, it accepts one or more UniProtKB/Swiss-Prot protein identifiers (ID) (e.g., *ALBU_HUMAN*) or UniProt Knowledgebase accession numbers (AC) (e.g., *P04406*), separated by spaces, tabs, or newlines.

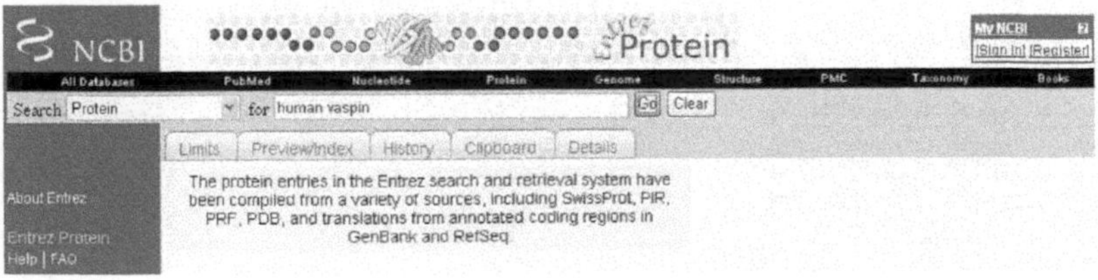
NCBI
Protein
My NCBI
[Sign In] [Register]
All Databases
PubMed
Nucleotide
Protein
Genome
Structure
PMC
Taxonomy
Books
Search Protein
for human vaspin
Go
Clear
Limits
Preview/Index
History
Clipboard
Details
The protein entries in the Entrez search and retrieval system have been compiled from a variety of sources, including SwissProt, PIR, PRF, PDB, and translations from annotated coding regions in GenBank and RefSeq
About Entrez
Entrez Protein
Help | FAQ

FIGURE 11.1 Search for human vaspin protein sequence from NCBI protein database.

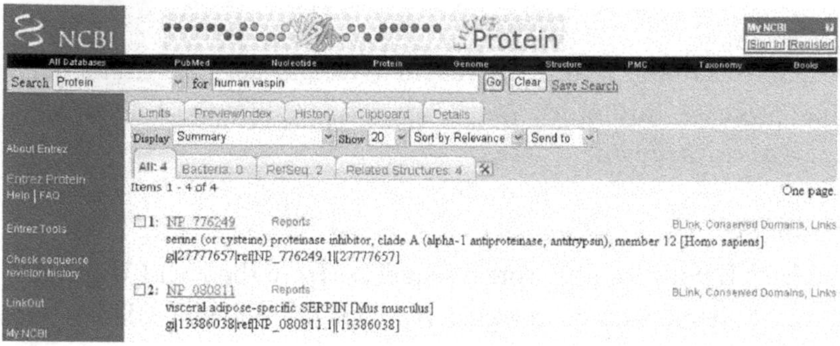
NCBI
Protein
Search Protein for human vaspin
Go Clear Save Search
Display Summary Show 20 Sort by Relevance Send to
Items 1 - 4 of 4
One page.
1: NP 776249
serine (or cysteine) proteinase inhibitor, clade A (alpha-1 antiproteinase, antitrypsin), member 12 [Homo sapiens]
gi|27777657|ref|NP_776249.1|[27777657]
2: NP 080811
visceral adipose-specific SERPIN [Mus musculus]
gi|13386038|ref|NP_080811.1|[13386038]

FIGURE 11.2 Display the accession number of human vaspin (SERPINA12) protein sequence.

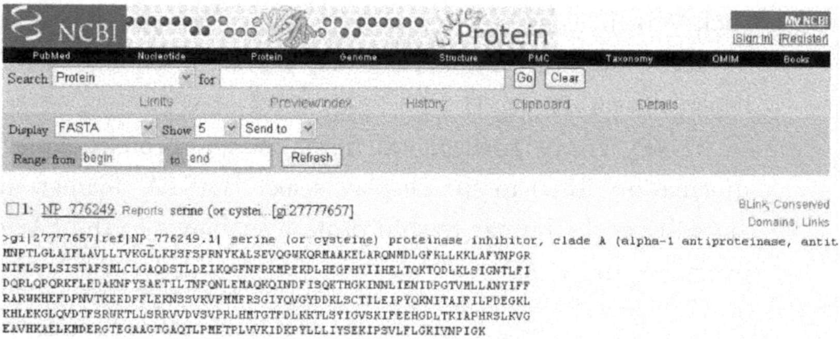
NCBI
Protein
Search Protein for
Go Clear
Display FASTA Show 5 Send to
Range from begin to end Refresh
1: NP 776249 Reports serine (or cystei...[gi:27777657]
>gi|27777657|ref|NP_776249.1| serine (or cysteine) proteinase inhibitor, clade A (alpha-1 antiproteinase, antit

FIGURE 11.3 Human vaspin (SERPINA12) protein sequence in the FASTA format.

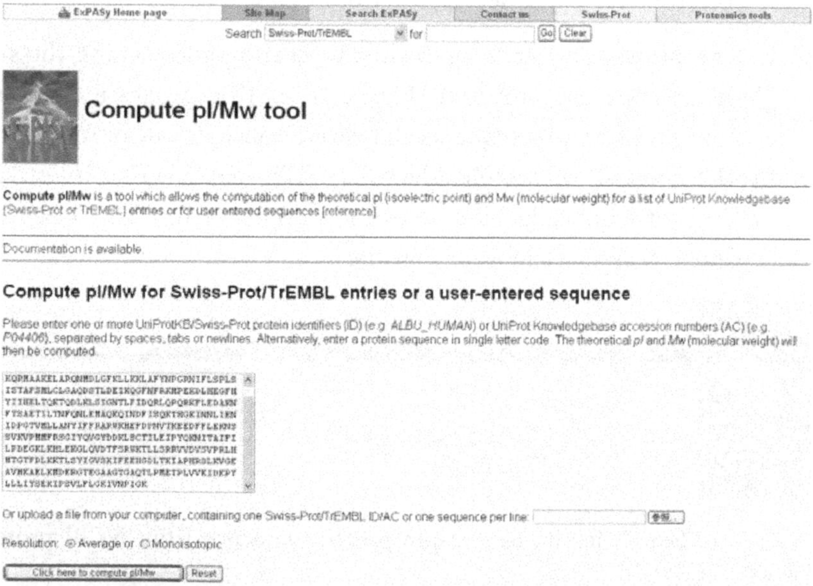
Compute pI/Mw tool
Compute pI/Mw is a tool which allows the computation of the theoretical pI (isoelectric point) and Mw (molecular weight) for a list of UniProt Knowledgebase (Swiss-Prot or TrEMBL) entries or for user entered sequences [reference]
Documentation is available.
Compute pI/Mw for Swiss-Prot/TrEMBL entries or a user-entered sequence
Resolution: Average or Monoisotopic
Click here to compute pI/Mw
Reset

FIGURE 11.4 Submit human vaspin sequence in Compute pI/Mw tool.

3. Click "Click here to compute pI/Mw," and you will have the theoretical pI/Mw (Figure 11.5).

3. *RADAR (Rapid Automatic Detection and Alignment of Repeats in Protein Sequences)*

1. Type the following address (http://www.ebi.ac.uk/Radar/) in your browser (Internet Explorer or others).
2. Enter a protein sequence in the box by cutting and pasting the entire FASTA format sequence (Figure 11.6).

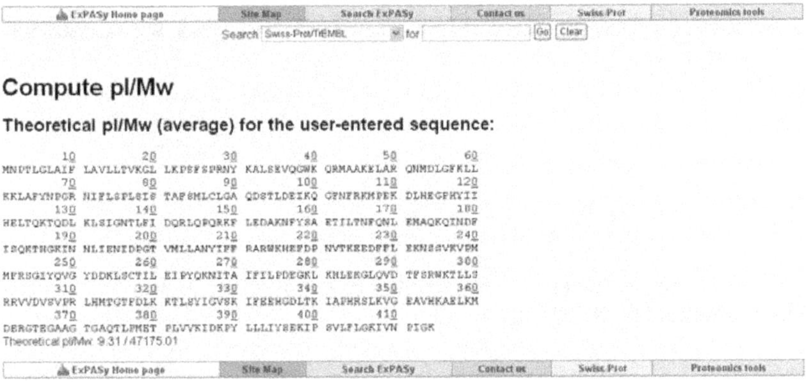

ExPASy Home page
Site Map
Search ExPASy
Contact us
Swiss-Prot
Proteomics tools
Search Swiss-Prot/TrEMBL for
Go
Clear
Compute pI/Mw
Theoretical pI/Mw (average) for the user-entered sequence:
10 20 30 40 50 60
MNPTLGLAIF LAVLLTVKGL LKPSFSPRNY KALSEVQGWK QRMAAKELAR QNMDLGFKLL
70 80 90 100 110 120
KKLAFYNPGR NIFLSPLSIS TAFSMLCLGA QDSTLDEIKQ GFNFRKMPEK DLHEGFHYII
130 140 150 160 170 180
HELTQKTQDL KLSIGNTLFI DQRLQPQRKF LEDAKNFYSA ETILTNFQNL EMAQKQINDF
190 200 210 220 230 240
ISQKTHGKIN NLIENIDPGT VMLLANYIFF RARWKHEFDP NVTKEEDFFL EKNSSVKVPM
250 260 270 280 290 300
MFRSGIYQVG YDDKLSCTIL EIPYQKNITA IFILPDEGKL KHLEKGLQVD TFSRWKTLLS
310 320 330 340 350 360
RRVVDVSVPR LHMTGTFDLK KTLSYIGVSK IFEEHGDLTK IAPHRSLKVG EAVHKAELKM
370 380 390 400 410
DERGTEGAAG TGAQTLPMET PLVVKIDKPY LLLIYSEKIP SVLFLGKIVN PIGK
Theoretical pI/Mw: 9.31 / 47175.01
ExPASy Home page
Site Map
Search ExPASy
Contact us
Swiss-Prot
Proteomics tools

FIGURE 11.5 Report of human vaspin in Compute pI/Mw tool.

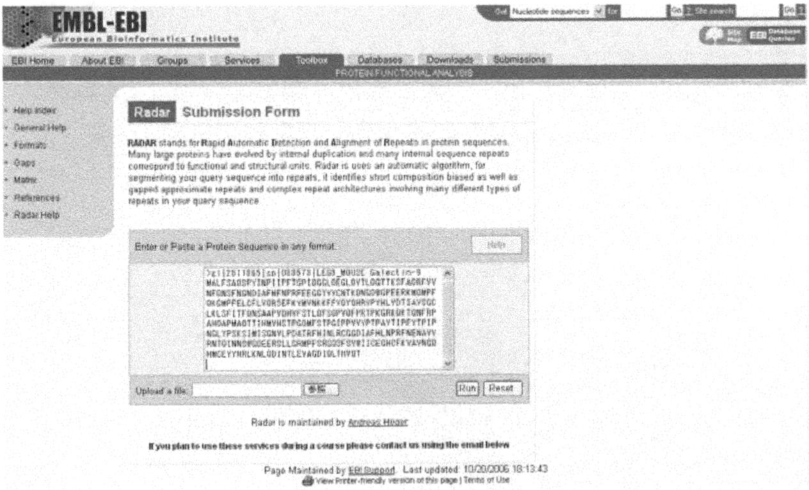

EMBL-EBI
European Bioinformatics Institute
EBI Home
About EBI
Groups
Services
Toolbox
Databases
Downloads
Submissions
PROTEIN FUNCTIONAL ANALYSIS
Help index
General Help
Formats
Gaps
Matrix
References
Radar Help
Radar Submission Form
RADAR stands for Rapid Automatic Detection and Alignment of Repeats in protein sequences. Many large proteins have evolved by internal duplication and many internal sequence repeats correspond to functional and structural units. Radar is uses an automatic algorithm, for segmenting your query sequence into repeats, it identifies short composition biased as well as gapped approximate repeats and complex repeat architectures involving many different types of repeats in your query sequence.
Enter or Paste a Protein Sequence in any format
Help
Upload a file:
Run
Reset

FIGURE 11.6 Submit mouse galectin-9 sequence in RADAR server.

3. Click "Run," and you will get the results (Figure 11.7). Galectin-9 is a tandem-repeat type of galectin gene family and consists of N-terminal and C-terminal carbohydrate-binding domains. These domains share 38.5% homology, and RADAR recognizes the repeat of the domains (Figure 11.7).

4. *PlotScale*

   1. Type the following address (http://us.expasy.org/tools/protscale.html) in your browser (Internet Explorer or others).

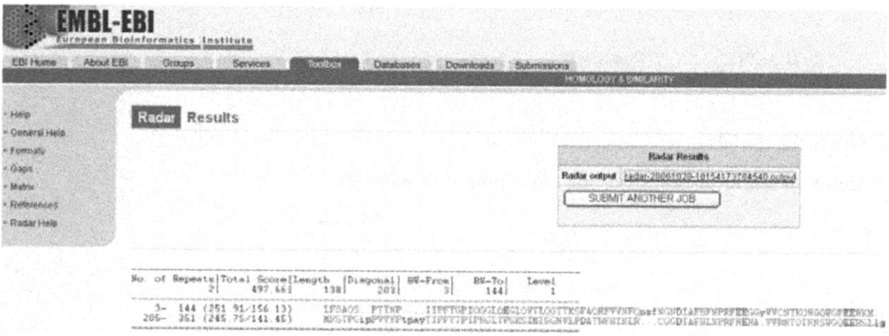
EMBL-EBI
European Bioinformatics Institute
Radar Results
Radar Results
SUBMIT ANOTHER JOB

FIGURE 11.7 Tandem repeat of carbohydrate domains in mouse galectin-9 sequence revealed by RADAR.

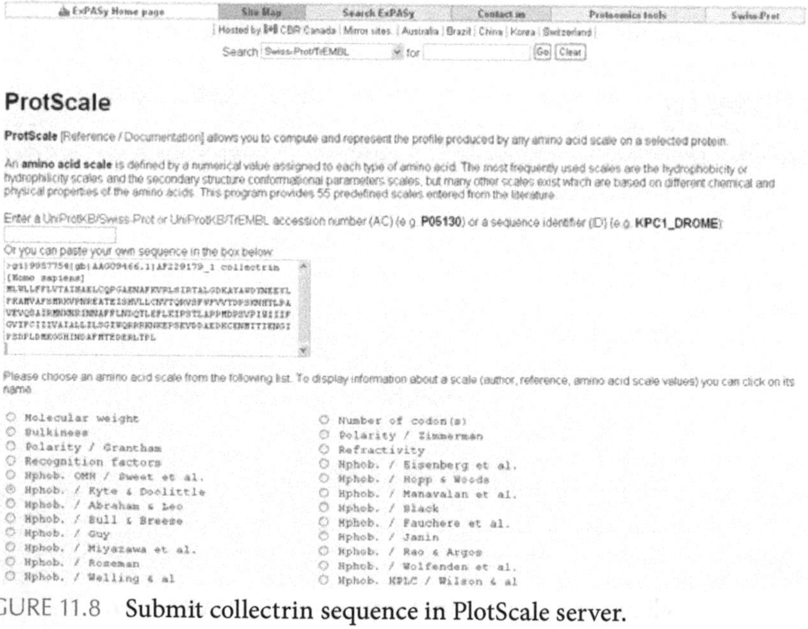
ProtScale
Molecular weight
Bulkiness
Polarity / Grantham
Recognition factors
Hphob. / Kyte & Doolittle
Number of codon(s)
Polarity / Zimmerman
Refractivity

FIGURE 11.8 Submit collectrin sequence in PlotScale server.

2. Enter UniProtKB/Swiss-Prot protein identifiers (ID) (e.g., *ALBU_HUMAN*) or UniProt Knowledgebase accession numbers. You can also cut and past the FASTA format amino acid sequences in the box (Figure 11.8).

3. Choose and click an amino acid scale from the list (Figure 11.8).

4. Click "Submit," and you will have the results (Figure 11.9). In the hydropathy plot, you can see the two hydrophobic domains of collectrin, signal peptide, and transmembrane domains (Figure 11.9).

Using the sale Hphob. / Kyte & Doolittle, the individual values for the 20 amino acids are:

Ala:	1.800	Arg:	−4.500	Asn:	−3.500	Asp:	−3.500	Cys:	2.500	Gln:	−3.500
Glu:	−3.500	Gly:	−0.400	His:	−3.200	Ile:	4.500	Leu:	3.800	Lys:	−3.900
Met:	1.900	Phe:	3.800	Pro:	−1.600	Ser:	−0.800	Thr:	−0.700	Trp:	−0.900
Tyr:	−1.300	Val:	4.200	Asx:	−3.500	Glx:	−3.500	Xaa:	−0.490		

Weights for window positions 1,..,9, using linear weight variation model:

1	2	3	4	5	6	7	8	9
1.00	1.00	1.00	1.00	1.00	1.00	1.00	1.00	1.00
edge				center				edge

MIN: −3.711
MIX: 3.756

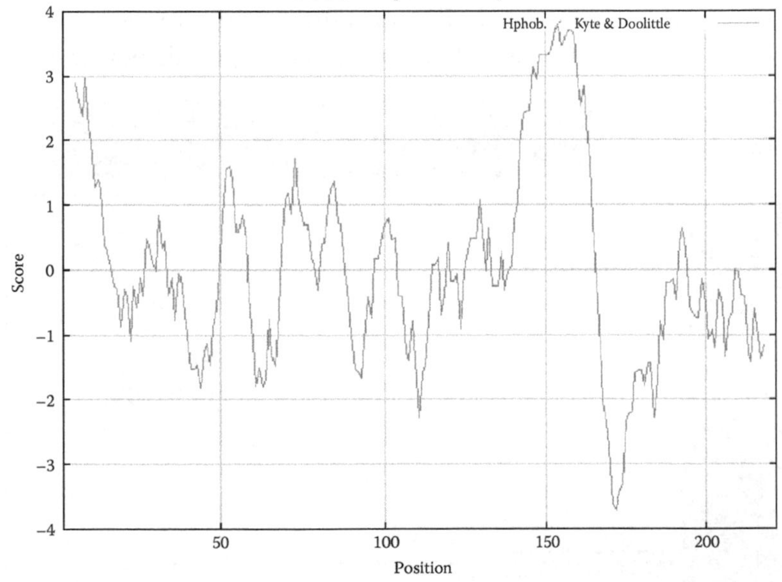
ProtScale output for user sequence
Hphob. / Kyte & Doolittle
Score
4
3
2
1
0
−1
−2
−3
−4
50
100
150
200
Position

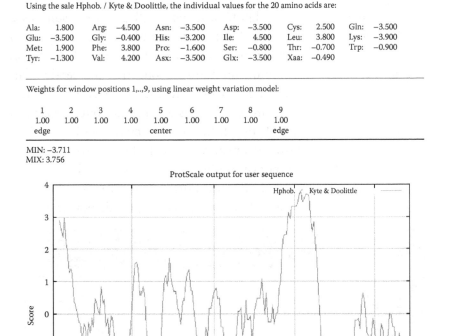

FIGURE 11.9 Hydropathy plot of collectrin sequence using PlotScale server.

5. Hydrophobic Cluster Analysis (HCA) of Proteins

1. Type the following address (http://bioserv.rpbs.jussieu.fr/RPBS/cgi-bin/Ressource.cgi?chzn_lg=an&chzn_rsrc=HCA) in your browser (Internet Explorer or others).
2. Enter a protein sequence in the box by cutting and pasting the single-letter code as plain text (Figure 11.10). This program does not accept FASTA format sequences.
3. Click "Submit" and wait until next page appears.
4. Click "Access to pdf file" and a new browser window will pop up.
5. Click "data_to_download.pdf," and you will get the results as a PDF file (Figure 11.11). HCA plots are useful for analyzing signal peptide and transmembrane segments. In Figure 11.11A, signal peptide and transmembrane domains of collectrin are clearly demonstrated. Globular domains, which are characterized by a typical thick distribution of hydrophobic clusters and are often separated from other domains by hydrophilic variable length of hinges. MCA is a useful tool in visual delineation of such domain structures. In Figure 11.11B, two carbohydrate-binding domains and link peptide are clearly demonstrated. The structural similarity of two globular domains is also seen. Finally, the HCA plot demonstrates the

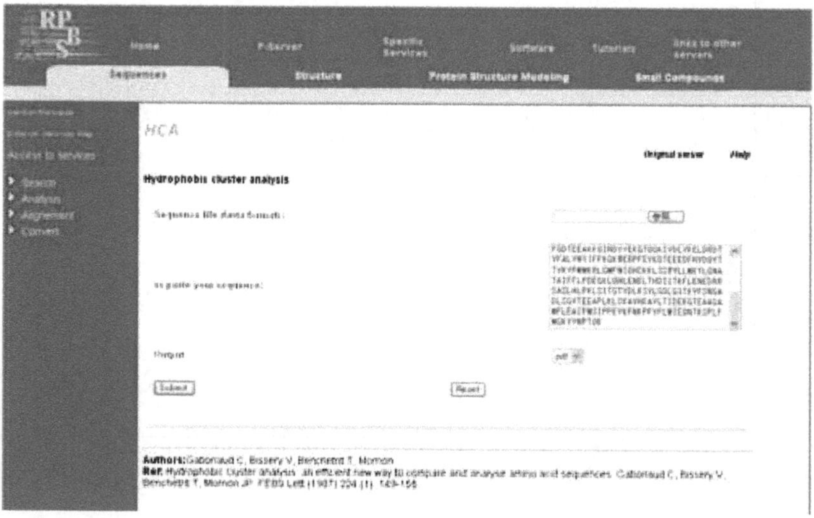
HCA
Hydrophobic cluster analysis

FIGURE 11.10 Submit human α1-antitrypsin sequence in Hydrophobic Cluster Analysis (HCA) server

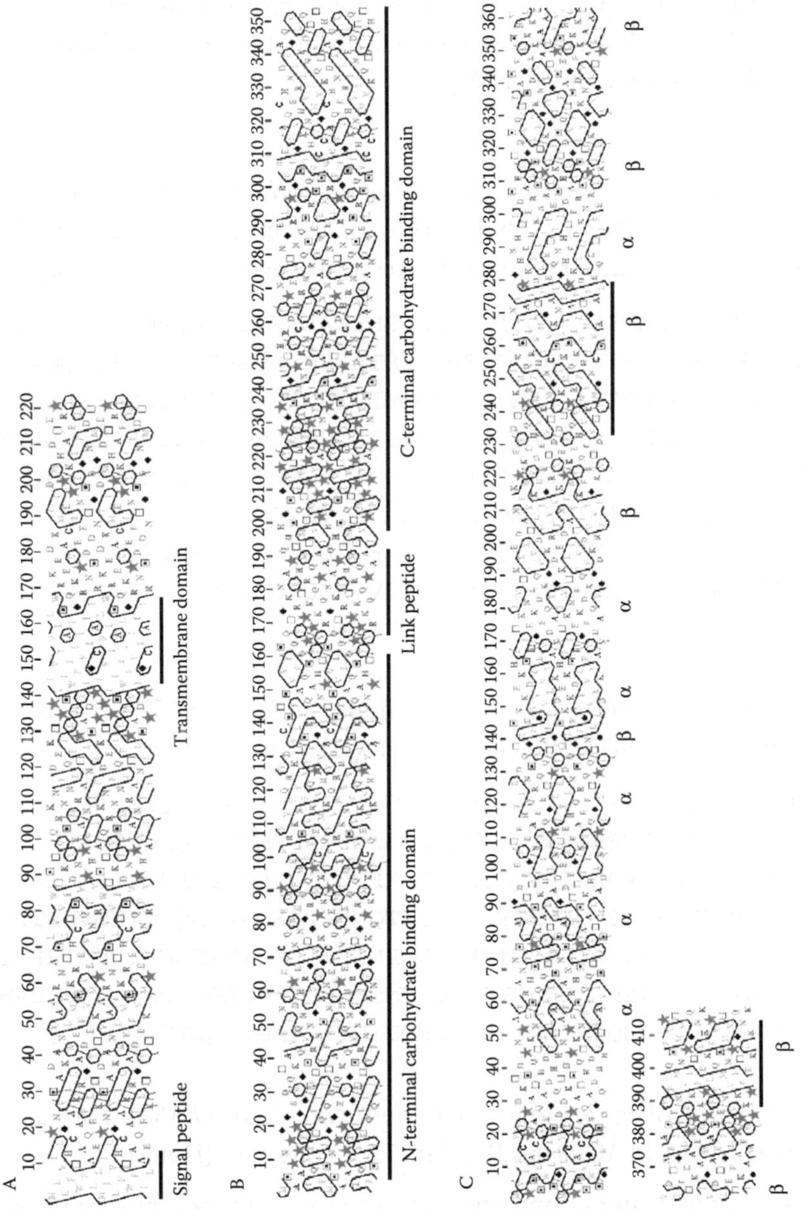
A
Signal peptide
Transmembrane domain
B
N-terminal carbohydrate binding domain
Link peptide
C-terminal carbohydrate binding domain
C
α
β

FIGURE 11.11 Hydrophobic cluster analysis (HCA) of human collectrin (A), mouse galectin-9 (B), and human α1 antitrypsin.

clusters of β-strand and α-helix secondary structures. In Figure 11.11C, an HCA plot of human α1-antitrypsin is indicated with α-helix and β-strand structures obtained from protein data bank (PDK; 1KCT). Typical shapes of hydrophobic amino acid clusters are demonstrated in an HCA plot.

Part III Sample Data

1. Human Vaspin Amino Acid Sequence in the FASTA Format

>gi|27777657|ref|NP_776249.1| serine (or cysteine) proteinase inhibitor, clade A (alpha-1 antiproteinase, antitrypsin), member 12 [*Homo sapiens*]

```
MNPTLGLAIFLAVLLTVKGLLKPSFSPRNYKALSEVQGWKQRMAAKELARQN-
MDLGFKLLKKLAFYNPGRNIFLSPLSISTAFSMLCLGAQDSTLDEIKQGFN-
FRKMPEKDLHEGFHYIIHELTQKTQDLKLSIGNTLFIDQRLQPQRKFLEDA-
KNFYSAETILTNFQNLEMAQKQINDFISQKTHGKINNLIENIDPGTVMLLANY-
IFFRARWKHEFDPNVTKEEDFFLEKNSSVKVPMMFRSGIYQVGYDDKLSC-
TILEIPYQKNITAIFILPDEGKLKHLEKGLQVDTFSRWKTLLSRRVVD-
VSVPRLHMTGTFDLKKTLSYIGVSKIFEEHGDLTKIAPHRSLKVGEAVH-
KAELKMDERGTEGAAGTGAQTLPMETPLVVKIDKPYLLLIYSEKIPSVLFLG-
KIVNPIGK
```

2. Mouse Galectin-9 Amino Acid Sequence in the FASTA Format

>gi|2811065|sp|O08573|LEG9_MOUSE Galectin-9

```
MALFSAQSPYINPIIPFTGPIQGGLQEGLQVTLQGTTKSFAQRFVVNFQNSF-
NGNDIAFHFNPRFEEGGYVVCNTKQNGQWGPEERKMQMPFQKGMPFEL-
CFLVQRSEFKVMVNKKFFVQYQHRVPYHLVDTIAVSGCLKLSFITFQNSAA-
PVQHVFSTLQFSQPVQFPRTPKGRKQKTQNFRPAHQAPMAQTTIHMVHSTP-
GQMFSTPGIPPVVYPTPAYTIPFYTPIPNGLYPSKSIMISGNVLPDATRFHIN-
LRCGGDIAFHLNPRFNENAVVRNTQINNSWGQEERSLLGRMPFSRGQSFSVWI-
ICEGHCFKVAVNGQHMCEYYHRLKNLQDINTLEVAGDIQLTHVQT
```

3. Human Collectrin Amino Acid Sequence in the FASTA Format

>gi|9957754|gb|AAG09466.1|AF229179_1 collectrin [Homo

```
MLWLLFFLVTAIHAELCQPGAENAFKVRLSIRTALGDKAYAWDTNEEYLFKAM-
VAFSMRKVPNREATEISHVLLCNVTQRVSFWFVVTDPSKNHTLPAVEVQ-
SAIRMNKNRINNAFFLNDQTLEFLKIPSTLAPPMDPSVPIWIIIFGVIF-
CIIIVAIALLILSGIWQRRRKNKEPSEVDDAEDKCENMITIENGIPSDPLD-
MKGGHINDAFMTEDERLTPL
```

4. Human α1-Antitrypsin Amino Acid Sequence in the FASTA Format

>gi|50363217|ref|NP_000286.3| serine (or cysteine) proteinase inhibitor, clade A (alpha-1 antiproteinase, antitrypsin), member 1 [Homo sapiens]

```
MPSSVSWGILLLAGLCCLVPVSLAEDPQGDAAQKTDTSHHDQDHPTF-
NKITPNLAEFAFSLYRQLAHQSNSTNIFFSPVSIATAFAMLSLGTKADTHDEI-
LEGLNFNLTEIPEAQIHEGFQELLRTLNQPDSQLQLTTGNGLFLSEGLKLVD-
KFLEDVKKLYHSEAFTVNFGDTEEAKKQINDYVEKGTQGKIVDLVKELDRDTV-
FALVNYIFFKGKWERPFEVKDTEEEDFHVDQVTTVKVPMMKRLGMFNIQHCK-
KLSSWVLLMKYLGNATAIFFLPDEGKLQHLENELTHDIITKFLENEDRRSASL-
HLPKLSITGTYDLKSVLGQLGITKVFSNGADLSGVTEEAPLKLSKAVHKAV-
LTIDEKGTEAAGAMFLEAIPMSIPPEVKFNKPFVFLMIEQNTKSPLFMGKVVNPTQK
```

SECTION 2 SECONDARY AND TERTIARY STRUCTURE ANALYSIS

Part I Introduction

1. What Is Secondary and Tertiary Structure Analysis?

A protein does not exhibit a full biological activity until it folds into a three-dimensional structure. Information on the secondary and three-dimensional (3D) structures of a protein is important for understanding its biological activity, because the shape and nature of the protein molecule surface account for the mechanisms of rational protein functions. It is also practically useful for rational molecular design of proteins with improved function as well as of drugs targeting the proteins. As of November 2006, about 40,000 protein structures have been deposited in the RCSB (Research Collaboratory for Structural Bioinformatics) Protein Data Bank (PDB) and the number is rapidly increasing, and more than 200,000 protein sequence entries are in the Swiss-Prot protein sequence database, a ratio of approximately one structure to five sequences. If a 3D structure of the queried sequence has already been solved and deposited in PDB, the structure of the protein itself as well as the proteins showing high sequence similarity with it can be obtained easily from PDB. Protein sequences whose structures have not yet been solved have to be analyzed with prediction tools for protein structure. A very large number of methods have been devised for secondary and tertiary protein structure prediction, and many servers provide them via Internet. The set of 3D coordinates of the protein thus obtained from PDB or from the prediction servers can be processed into intuitive graphics showing the 3D molecular models and the position of secondary structures of the protein by using molecular modeling and visualization tools. It should be noted that the results provided by the prediction servers are possible models: the results on the same sequence from one method (server) may be different from those from another method (server). The accuracy of the prediction results depends on the sequence submitted and the programs used. The prediction results will be relatively accurate for a sequence with high similarity to known protein structures, whereas it is still difficult to predict long sequences of unknown fold proteins (particularly, β-rich, multidomain, and complicated-topology proteins). Trends in the research on 3D structure prediction methods are published in reports of CASP (Critical Assessment of Techniques for Protein Structure Prediction), which provides valuation contests on protein structure prediction research and servers every two years.

2. What Is Involved in Secondary and Tertiary Structure Analysis?

A. Searching PDB for Protein Sequences Using a sequence search provided by the PDB servers is one of the fastest way to know if a protein sequence encodes a known protein structure or not.

RCSB PDB (http://www.rcsb.org/pdb/home/home.do). It has a menu that retrieves the PDB database for a sequence submitted via sequence search program using blastp from NCBI or the program of FASTA. If the sequence has significant homology to known proteins whose PDB data are available, the servers provide the list of PDB IDs of selected proteins with the results of sequence alignment. You can obtain the structure information on the selected PDB ID, such as its structural classification (according to SCOP, CATH, and PFAM), secondary structures, and geometry (bond length, bond angle, dihedral angle). The selected protein molecules will be easily displayed via viewers in the Web browser. The deposited structural data including the set of 3D coordinates of all the atoms can be downloaded as a PDB file for further analyses on your computer.

B. Secondary Structure Prediction Predicting secondary structure of a protein whose structure is unknown is an alternative or first step toward the much more difficult task of predicting the 3D structure. Several methods have been developed and improved for a secondary structure prediction of the protein structure from a sequence. In the 1970s, simple and basic methods using single sequences for prediction were developed initially.

Chou & Fasman (http://fasta.bioch.virginia.edu/fasta_www2/fasta_www.cgi?rm=miscl). It is one of the earliest methods that determined the frequency of occurrence of each amino acid in α-helices and β-sheets. This was calculated from a survey of fifteen proteins of known structure. Residues were assigned into strong formers, weak formers, formers, indifferent formers, strong breakers, and breakers, according to their ability to initiate or terminate these structures. In simple terms the method is executed as follows:

1. Assign residues as shown earlier for both helices and sheets.
2. For helices, locate a cluster of 4 formers or strong formers within 6 residues. Extend the helix in both directions until terminated by a tetrapeptide with an average alpha propensity equal to 1 i.e. indifferent. Disallow prolines in helices.
3. For sheets: locate a cluster with 3 out of 5 formers or strong formers. Extend in both directions. Terminate as for alpha helices.
4. Use similar method for turns. It gives 50% accuracy.

GOR (http://npsa-pbil.ibcp.fr/cgi-bin/npsa_automat.pl?page=npsa_gor4.html). The GOR method is based on information theory and was developed by J. Garnier, D. Osguthorpe, and B. Robson. The present version, GOR IV, uses all possible pair frequencies within a window of seventeen amino acid residues. After cross-validation on a data base of 267 proteins, version IV of GOR has a mean accuracy of 64.4%. The program gives two outputs, one giving the sequence and the predicted secondary structure in rows (H = helix, E = extended or beta strand, and C = coil) and the second gives the probability values for each secondary structure at each amino acid position. The predicted secondary structure is the one of highest probability compatible with a predicted helix segment of at least four residues and a predicted extended segment of at least two residues.

Recent programs for secondary structure prediction use multiple sequences and more information on protein nature combined with new computational algorithms (neural networks, nearest-neighbor methods, etc.) for better prediction. The following are some representative programs (servers) for secondary structure prediction:

PHD (http://www.predictprotein.org/). **PHDsec** in PHD (a suite of programs predicting from primary structure) predicts secondary structure from multiple sequence alignments. The secondary structure is predicted by a system of neural networks.

PROF (http://www.predictprotein.org/). **PROFsec** (in PROF) is an improved version of PHDsec and predicts secondary structure by using a profile-based neural network.

APSSP2 (http://www.imtech.res.in/raghava/apssp2/). This method uses the standard neural network and multiple sequence alignment generated by PSI-BLAST (Position Specific Iterated - BLAST) profiles, and a modified example-based learning (EBL) technique. The combination of the two is based on a reliability score.

PSIpred (http://bioinf.cs.ucl.ac.uk/psipred/psiform.html). PSIpred incorporates two feed-forward neural networks that perform an analysis on output obtained from PSI-BLAST. Version 2.0 of PSIpred includes a new algorithm that averages the output from up to four separate neural networks in the prediction process to further increase prediction accuracy.

SAM-T02 (http://www.soe.ucsc.edu/research/compbio/HMM-apps/T02-query.html). The Sequence Alignment and Modeling system (SAM) is a collection of flexible software tools for creating, refining, and using linear hidden Markov models (HMMs) for biological sequence analysis. SAM-T02 method for iterative SAM HMM construction and remote homology detection and protein structure prediction updates SAM-T99 by using predicted secondary structure information in its scoring functions.

SSpro (http://www.igb.uci.edu/tools/scratch/). It is a server for protein secondary structure prediction based on an ensemble of eleven BRNNs (bidirectional recurrent neural networks based on PSI-BLAST profiles). In SSpro version 2.0 a better algorithm to obtain multiple alignments of homologue sequences, based on PSI-BLAST, is exploited.

Jpred server (http://www.compbio.dundee.ac.uk/~www-jpred/): JNet predicts the secondary structure from multiple sequence alignments, PSI-BLAST profiles, and HMMER2 profiles. The JPred Server takes a single protein sequence or multiple sequence alignments and runs the JNet secondary structure prediction method as well as methods to predict coiled coil structures.

The accuracy of these secondary prediction methods is often evaluated with a value of a three-state prediction (Q3). The Q3 values of those recent methods are evaluated in two categories (sequence unique and similar 3D) to be about 70–80%. The evaluations of these programs can be seen in the Evaluation of automatic structure prediction servers for Critical Assessment of Fully Automated Structure Prediction (CAFASP/EVA) (http://cubic.bioc.columbia.edu/eva/cafasp/index.html) and Live Bench (http://bioinfo.pl/meta/livebench.pl).

C. Tertiary Structure Prediction A very large number of methods have been devised for tertiary protein structure prediction and are available via Internet freely for academic use. Tertiary structure prediction methods are categorized into mainly two types: modeling the 3D structure using a structure of a known protein as a template (fold-recognition methods) or modeling without templates (*Ab initio* modeling).

Fold-recognition (FR) programs first select appropriate protein structure suitable for a queried (target) sequence from PDB and then build models using the selected structure as a template by a restraint-based modeling program such as Modeller. Two types of methods are utilized to select template structures from PDB. Sequence-Only Methods utilize only sequence information available about the target and the templates to find similarities, based on the knowledge that proteins having similar sequences show similar 3D structures. Programs using this method, such as ESyPred3D, FFAS03, PDB-blast, and PSI-blast, employ so-called comparative modeling (CM) or homology modeling. On the other hand, it has long been recognized that proteins often adopt similar folds despite no significant sequence or functional similarity and that nature is apparently restricted to a limited number of protein folds. To find folds that are compatible with a target sequence from libraries of known protein folds, Threading Methods utilize sequence information and predicted structural features in their scoring functions. In many cases the predicted secondary structure of the target is compared with the observed secondary structure of the templates. Programs such as 3D-PSSM, FUGUE2, mGenThreader, and Sam-T02 use the threading method. The accuracy of these template-based methods is relatively high, only if the appropriate protein structure can be found and used as a template for modeling.

Ab initio programs use physicochemical simulations (e.g., energy minimization) combined with fragment assembly method to build models. They are used for a "New Fold (NR)" sequence whose fold cannot be assigned to any known folds. In the Rosetta algorithm, a representative program of this category, the distribution of conformations observed for each short sequence segment in known protein structures is taken as an approximation of the set of local conformations that sequence segment would sample during folding. The program then searches for the combination of these local conformations that has the lowest overall energy. Programs of this category are also referred as New Fold or *de novo* modeling. The accuracy of these programs is not so high, particularly for long sequences >100 aa.

There are several meta-servers available that integrate protein structure predictions performed by various methods described earlier. Some meta-servers, which submit a target sequence to several external prediction servers and evaluate the models provided by the servers, have higher accuracy than individual structure prediction servers (Figure 11.12). Some

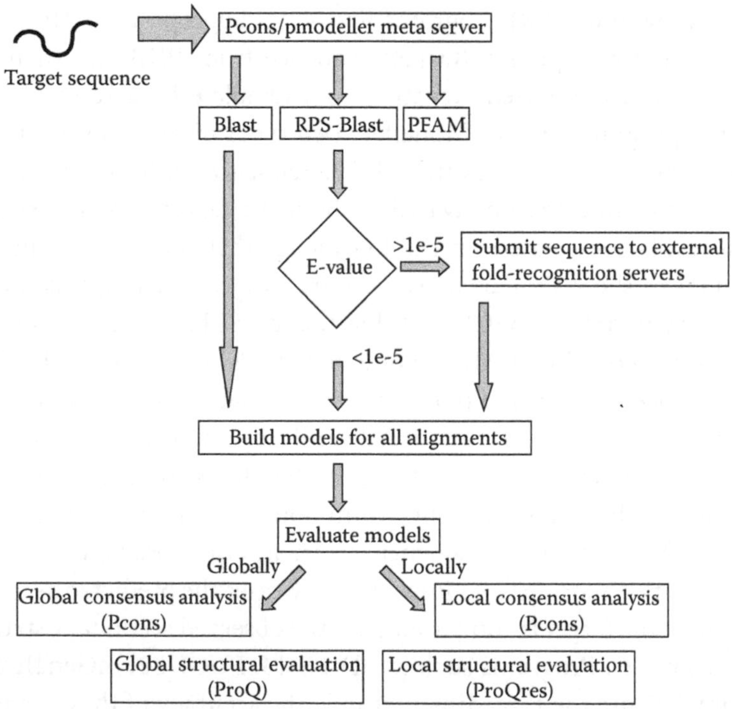
Target sequence
Pcons/pmodeller meta server
Blast
RPS-Blast
PFAM
E-value
>1e-5
Submit sequence to external fold-recognition servers
<1e-5
Build models for all alignments
Evaluate models
Globally
Locally
Global consensus analysis (Pcons)
Local consensus analysis (Pcons)
Global structural evaluation (ProQ)
Local structural evaluation (ProQres)

FIGURE 11.12 Three-dimensional structure prediction server (Pcons/Pmodeller Meta Server as an example).

representative programs and meta-servers for tertiary structure prediction are now described.

PredictProtein (http://www.predictprotein.org/). The meta-server in this site provides several tertiary structure prediction tools (four homology modeling and ten threading servers) as well as six interresidue contact predictions, fifteen secondary structure predictions, five membrane-helix predictions, and other methods predicting signal peptide, domains, o-glycosylation sites, and so on. This site provides the other services of basic protein structure analysis for a sequence such as Prosite motif search.

@Tome (@utomatic Threading Optimisation Modelling & Evaluation) (http://bioserv.cbs.cnrs.fr/HTML_BIO/frame_home.html). This meta-server allows one to submit an amino acid sequence to six remote servers (one similarity search, three threading, and two secondary prediction) dedicated to structural predictions and fold recognition.

GeneSilico Metaserver (http://genesilico.pl/meta). This server provides access to eleven FR servers. It aligns the target sequence to sequences of proteins with known structures, identified by protein FR methods. The Pcons consensus server will evaluate to what extent the FR alignments agree with each other and whether a particular fold can be singled out. Registration is required before use. From FR alignments you can automatically generate crude 3D models (without variable loops).

BioInfoBank (http://bioinfo.pl/). The Structure Prediction Meta Server in this site provides access to various fold recognition, function prediction, and secondary structure prediction methods. The 3D-Jury system generates meta-predictions from sets of models created using variable methods.

Pcons/Pmodeller Meta Server (http://www.cbr.su.se/pcons/index.php?about=pcons). Pcons5 integrates information from three different sources: consensus analysis, structural evaluation, and the score from the FR servers. It selects one of the models returned by these services. Pmodeller provides models optimized in the last step using the Modeller program.

Robetta (http://robetta.bakerlab.org/). Robetta produces full-chain models with the Rosetta *de novo* and comparative modeling methods. *De novo* models are built by fragment insertion simulated annealing. Comparative models are built by detecting a parent PDB with PSI-BLAST or Pcons2, aligning the query to the parent with the K*SYNC alignment method, and modeling variable regions with a modified version of the *de novo* protocol. Registration is required before use.

The servers CAFASP/ EVA and Live Bench also evaluate these 3D structure prediction programs and servers. The results of CASP can be seen in the Protein Structure Prediction Center (http://predictioncenter.org/).

Part II Step-By-Step Tutorial

The demo example is a sequence of rat betacellulin (BTC), whose tertiary structure itself has not been solved, whereas the structures of the EGF domains of its human homologue and other members of EGF family proteins have been available at PDB. Before submitting to prediction programs, it is recommended that the sequence be analyzed for protein motifs and topology using the tools described in the next section, in order to divide the sequence into short fragments corresponding to domains. Better and faster results will be obtained when domains are submitted separately. The time it takes before results are returned from each server depends on the length of the target sequence as well as the complexity of the algorithm. The sequence of rat BTC consists of five domains of signal sequence, N-terminal region whose structure is unknown, EGF-like domain, transmembrane helix, and cytoplasmic region whose structure is unknown, respectively.

1. Analyze the Sequence with the RCSB PDB

1. Submit a sequence to the prediction server.
    a. Type the following address (http://www.rcsb.org/pdb/) to get access to the RCSB PDB in your browser (Internet Explorer or others). Go to "Sequence Search" window by selecting "Search" tag and "Sequence" database in the menu at the left side of the window.
    b. Select "use Sequence" checkbox. Enter a protein sequence by cutting and pasting your sequence as a simple text (Figure 11.13). Click "Search" in your browser.

2. If there are structural data of a protein identical to or significantly similar to the queried sequence, their PDB IDs will be listed with the alignment statistics (Figure 11.14). In this sample sequence (rat BTC), there is no structural data of the queried sequence itself, but several PDB IDs showing significant similarity to a portion (K63–Y11) of the sequence, which corresponds to the EGF-like domain, are listed. Partial sequences of rat BTC containing the EGF domain will also give a similar list, but partial sequences without the EGF-like

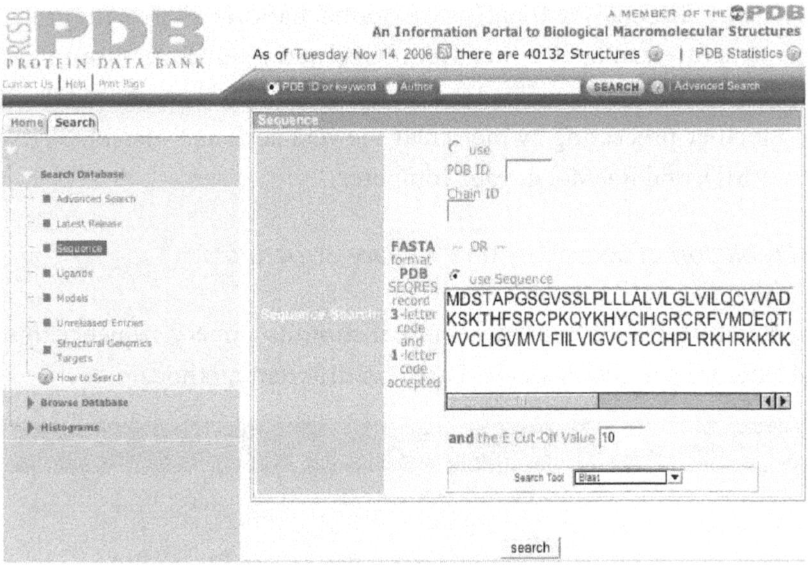

FIGURE 11.13 Search for a protein structure from the RCSB PDB database with a protein sequence.

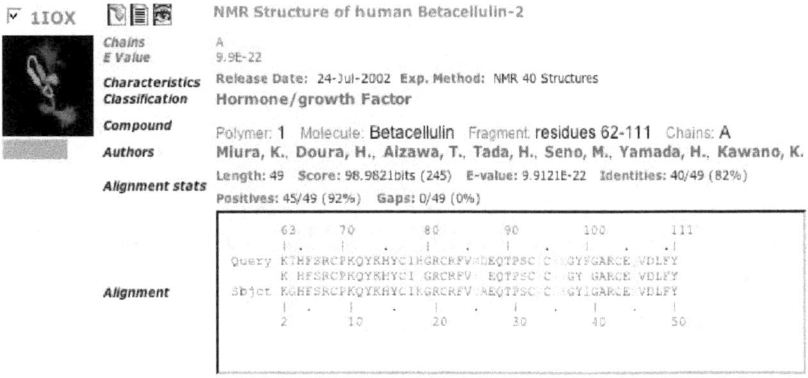

FIGURE 11.14 Display the PDB IDs of proteins with sequence homology.

domain will not, because only the structures of the EGF-like domain are available in PDB.

3. To get the structure information of a listed protein, click its PDB ID in the list. Structural summary of the selected PDB ID will be displayed. You can display and move the protein molecule by using several viewers listed in the "Display Molecule" column. The description of its secondary structure can be seen in the "Sequence Details"

menu. The analyzed data (such as Ramachandran Plot) of the protein's structure can be obtained from "Geometry" in "Structure analysis" menu. You can download the "PDB Files" from "Download Files" for further processing by molecular viewers such as Swiss-PDBViewer, VMD, and RasMol in your computer.

2. Prediction of Secondary and Tertiary Structure by Using PredictProtein Metaserver

The metaserver is convenient because it submits a query sequence to several external prediction servers based on different methods.

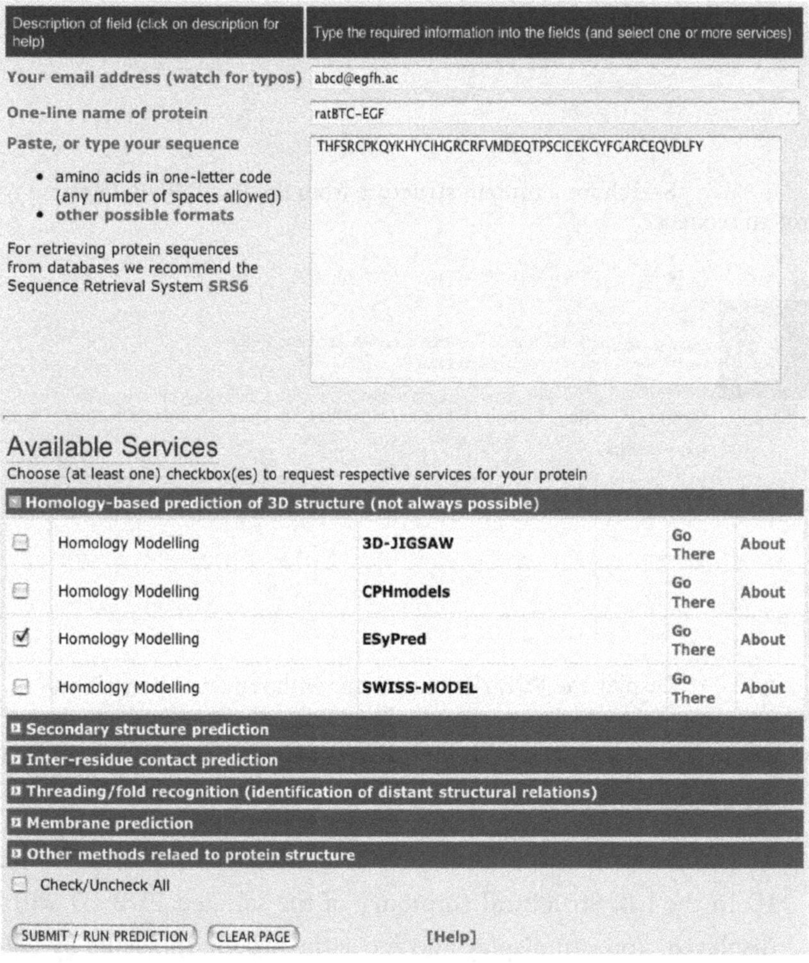

Description of field (click on description for help)	Type the required information into the fields (and select one or more services)
Your email address (watch for typos)	abcd@egfh.ac
One-line name of protein	ratBTC–EGF
Paste, or type your sequence • amino acids in one-letter code (any number of spaces allowed) • other possible formats For retrieving protein sequences from databases we recommend the Sequence Retrieval System SRS6	THFSRCPKQYKHYCIHGRCRFVMDEQTPSCICEKGYFGARCEQVDLFY

Available Services
Choose (at least one) checkbox(es) to request respective services for your protein

Homology-based prediction of 3D structure (not always possible)

☐	Homology Modelling	3D-JIGSAW	Go There	About
☐	Homology Modelling	CPHmodels	Go There	About
☑	Homology Modelling	ESyPred	Go There	About
☐	Homology Modelling	SWISS-MODEL	Go There	About

Secondary structure prediction
Inter-residue contact prediction
Threading/fold recognition (identification of distant structural relations)
Membrane prediction
Other methods relaed to protein structure
☐ Check/Uncheck All

[SUBMIT / RUN PREDICTION] [CLEAR PAGE]   [Help]

FIGURE 11.15 Submit a sequence for secondary and tertiary structure prediction programs via the metaserver in PredictProtein.

1. Submit a sequence to the prediction server.
   a. Type the following address (http://www.predictprotein.org/) to get access to the PredictProtein server in your browser. Click the "MetaPP" tag in the menu to access the metaserver.
   b. Type your e-mail address and name of your sequence in the corresponding boxes (Figure 11.15). Enter a protein sequence by cutting and pasting the single-letter code as plain text. Forty-nine amino acid sequences corresponding to the EGF-like domain of rat BTC is used as an example suitable for homology modeling. From the list of the "Available Services," choose checkboxes to request the programs for the sequence. Then click "SUBMIT/RUN PREDICTION."

2. Receive the results from the PredictProtein metaserver.
   a. If the request is submitted successfully, the "Submission Summary" will be displayed and the e-mail entitled "submission to MetaPP" will also come soon from the PredictProtein meta server.
   b. Prediction results will be sent by another e-mail from the PredictProtein metaserver, usually within a day. Access the address in the e-mail to see the prediction results. If there are some programs commented with "(No Response from Server)" at the time, access this site again another day, and you will find the updated results from the program servers. It may take several days until you receive all the results from all the program servers.

3. Obtain and process the prediction results.
   a. For example, responses received from secondary structure prediction servers of PSIpred and SAM-T99 are shown in Figure 11.16. PSIpred presents the prediction results in a simple style that gives the sequence, the predicted secondary structure, and the confidence value of the prediction at each amino acid position in rows (H = helix, E = extended or beta strand, and C = coil). The graphical output for the result can be obtained by accessing the address contained in the server response. Some program servers will only give the addresses to be accessed, as in the case of SAM-T99 (Figure 11.16). Access this SAM-T99 site, and the prediction results are provided in other styles, such as a list of the probability values for each secondary structure at each amino acid position (Figure 11.17) or a list of the most probable structure and its probability at each amino acid position (CASP format).

Results Recieved by PSIPRED

Server Response:

```
PSIPRED PREDICTION RESULTS

Key

Conf: Confidence (0=low, 9=high)
Pred: Predicted secondary structure (H=helix, E=strand, C=coil)
  AA: Target sequence


# PSIPRED HFORMAT (PSIPRED V2.5 by David Jones)

Conf: 976647724259006248984400270038727874234023032029
Pred: CCCCCCCHHHCCCCCCCEEEEEEECCEEEEEECCCCCCCCCCEEEECC
  AA: THFSRCPKQYKHYCIHGRCRFVMDEQTPSCICEKGYFGARCEQVDLFY
              10        20        30        40


Calculate PostScript, PDF and JPEG graphical output for this result using:
http://bioinf2.cs.ucl.ac.uk/cgi-bin/psipred/gra/nph-view2.cgi?id=7413f046fbbc0206.psi
```

Results Recieved by SAM-T99

Server Response:

```
Your query has completed processing. Please retrieve the results at:

    http://www.cse.ucsc.edu/~farmer/target99-query/query-1163567222-799/output/index.html

within 7 days.
```

FIGURE 11.16 The results of secondary structure prediction received from PSIpred and SAM-T99 servers via MetaPP server.

Pos	AA	E	H	L
10N	1S	5N	5N	5N
0	T	0.156	0.19	0.654
1	H	0.21	0.201	0.589
2	F	0.265	0.192	0.544
3	S	0.245	0.173	0.581
4	R	0.205	0.177	0.619
5	C	0.171	0.142	0.687
6	P	0.119	0.194	0.687
7	K	0.114	0.271	0.614
8	Q	0.141	0.281	0.578
9	Y	0.171	0.263	0.566
10	K	0.194	0.225	0.58

FIGURE 11.17 Secondary structure predicted by SAM-T99 (rdb format).

```
EXPDTA    MODEL, MODELLER Version  6v2 2006/11/15  04:01:26.921
REMARK MODELLER OBJECTIVE FUNCTION:          362.0733
ATOM      1  N   HIS     2      -1.146  15.971   0.164  1.00 52.87      1SG   2
ATOM      2  CA  HIS     2      -1.138  15.851   1.643  1.00 52.87      1SG   3
ATOM      3  ND1 HIS     2      -2.904  16.900   4.477  1.00 52.87      1SG   4
ATOM      4  CG  HIS     2      -1.719  17.063   3.796  1.00 52.87      1SG   5
ATOM      5  CB  HIS     2      -1.630  17.157   2.299  1.00 52.87      1SG   6
ATOM      6  NE2 HIS     2      -1.293  16.934   6.010  1.00 52.87      1SG   7
ATOM      7  CD2 HIS     2      -0.744  17.084   4.747  1.00 52.87      1SG   8
ATOM      8  CE1 HIS     2      -2.593  16.829   5.796  1.00 52.87      1SG   9
ATOM      9  C   HIS     2      -1.980  14.718   2.130  1.00 52.87      1SG  10
ATOM     10  O   HIS     2      -3.157  14.619   1.788  1.00 52.87      1SG  11
ATOM     11  N   PHE     3      -1.368  13.830   2.945  1.00116.68      1SG  12
ATOM     12  CA  PHE     3      -2.034  12.679   3.491  1.00116.68      1SG  13
                                          ⋮
ATOM    391  CE2 TYR    48       7.900 -18.438   3.578  1.00 52.09      1SG 392
ATOM    392  CZ  TYR    48       8.315 -19.727   3.337  1.00 52.09      1SG 393
ATOM    393  OH  TYR    48       7.410 -20.798   3.489  1.00 52.09      1SG 394
ATOM    394  C   TYR    48      13.262 -15.634   3.505  1.00 52.09      1SG 395
ATOM    395  O   TYR    48      13.657 -15.531   2.311  1.00 52.09      1SG 396
ATOM    396  OXT TYR    48      13.656 -14.883   4.440  1.00 52.09      1SG 397
END
```

FIGURE 11.18 Submission a sequence for 3D structure prediction programs via

b. Results from EsyPred are shown in Figure 11.18 as an example of 3D structure prediction. EsyPred will provide the predicted molecular structure as well as the template structure in PDB format. Copy the data corresponding to the target protein structure (from the row next to "EXPDTA" to that of "END," which contain the set of space coordinates of all atoms of the sequence) and past into an appropriate word file such as Microsoft Word to make a PDB file in a text format (with an extension of "pdb"). Open and analyze the file with a molecular viewer program such as RasMol or VMD.

3. *Prediction of 3D Structure by the Metaserver in BioInfoBank*
   1. Submit a sequence to the prediction server.
      a. Go to **BioInfoBank** Web site (http://bioinfo.pl/meta/livebench.pl), and access the "Meta server" at the site.
      b. Type your e-mail address and name of the sequence in the corresponding boxes (Figure 11.19). Enter a protein sequence by cutting and pasting the single-letter code as a simple text. Forty-two amino acid sequence corresponding to cytoplasmic domain of rat BTC is used as an example of sequences with little homology to known structures. Click checkboxes to skip the prediction methods if you like. Click "submit" in your browser, then "Request submitted job successfully" will be displayed.
   2. Obtain and process the prediction results.

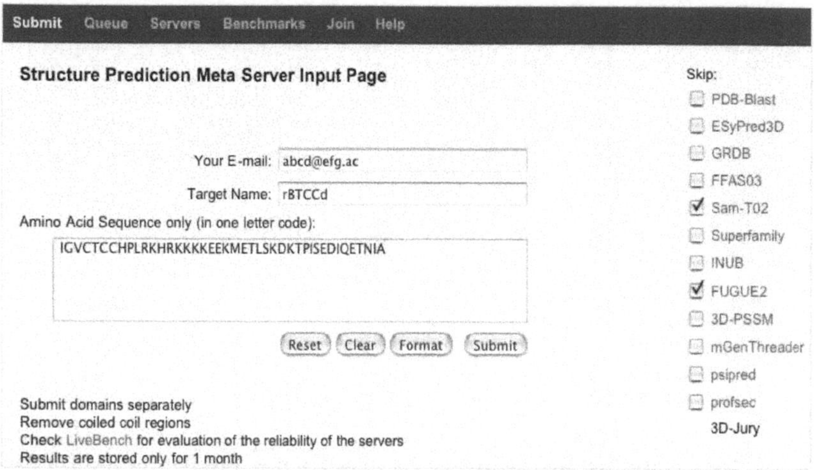
Submit Queue Servers Benchmarks Join Help
Structure Prediction Meta Server Input Page
Your E-mail: abcd@efg.ac
Target Name: rBTCCd
Amino Acid Sequence only (in one letter code):
IGVCTCCHPLRKHRKKKKEEKMETLSKDKTPISEDIQETNIA
Reset Clear Format Submit
Submit domains separately
Remove coiled coil regions
Check LiveBench for evaluation of the reliability of the servers
Results are stored only for 1 month
Skip:
PDB-Blast
ESyPred3D
GRDB
FFAS03
Sam-T02
Superfamily
INUB
FUGUE2
3D-PSSM
mGenThreader
psipred
profsec
3D-Jury

FIGURE 11.19 The results of 3D structure predictions provided by BioInfo-Bank metaserver.

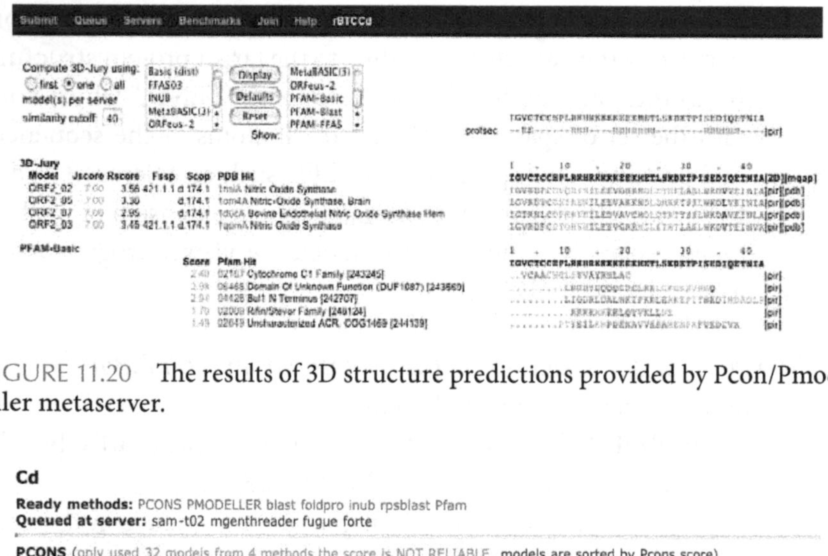
Submit Queue Servers Benchmarks Join Help rBTCCd
Compute 3D-Jury using:
first one all
model(s) per server
similarity cutoff 40
Display
Defaults
Reset
Show:
profsec
3D-Jury
Model Jscore Rscore Fssp Scop PDB Hit
PFAM-Basic
Score Pfam Hit

FIGURE 11.20 The results of 3D structure predictions provided by Pcon/Pmodeller metaserver.

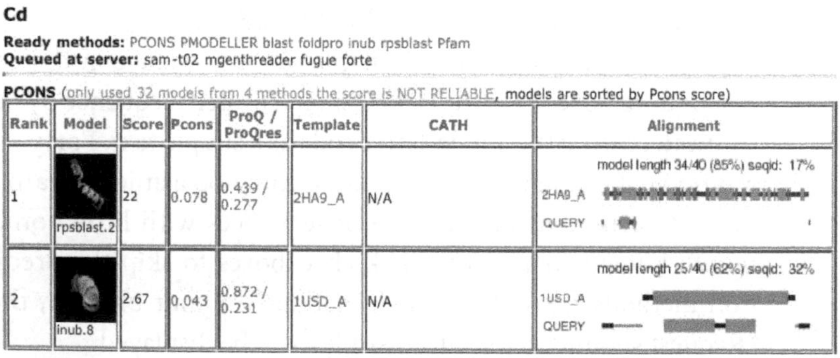
Cd
Ready methods: PCONS PMODELLER blast foldpro inub rpsblast Pfam
Queued at server: sam-t02 mgenthreader fugue forte
PCONS (only used 32 models from 4 methods the score is NOT RELIABLE, models are sorted by Pcons score)
Rank Model Score Pcons ProQ / ProQres Template CATH Alignment
1 rpsblast.2 22 0.078 0.439 / 0.277 2HA9_A N/A
model length 34/40 (85%) seqid: 17%
2HA9_A
QUERY
2 inub.8 2.67 0.043 0.872 / 0.231 1USD_A N/A
model length 25/40 (62%) seqid: 32%
1USD_A
QUERY

FIGURE 11.21 The InterProScan schema.

a. Click "Queue" in the menu, and you will find your sequence name in a list of submitted sequences. Click your sequence name or your submission Id to see the results. If calculation of the servers is not over, reaccess this page another day. (Bookmark the site for easy access.) It will take several days to receive all the results from all the servers selected. If the submitted sequence has a significant homology to a known PDB protein such as rat BTC-EGFd, the threading prediction programs may be stopped automatically and only the results based on sequence alignment will be displayed.
b. As in Figure 11.20, the results of prediction will be shown as lists of (1) alignment and threading programs used, (2) programs whose results are displayed in this window, (3) the secondary structure prediction results, (4) the 3D models processed by 3D-Jury, and (5) the results of fold recognition from selected programs. You will be able to see more results by selecting other programs followed by clicking the "Display" button. Using the structure data of "PDB hit" as a template, the 3D models are calculated and evaluated by 3D-July to give the 3D-Jury score (Jscore). The models that are most similar to others have a higher chance of being correct and obtaining higher J scores. When default settings are applied and results for most servers are available, the Jscore corresponds roughly to the number of C-alpha atoms of the model within 3.5Å from the native structure. Clicking "server" in the top menu can check the status of each program server. In the first row of 3D-Jury results in Figure 11.29 the model (ORF2_02) is built by using "PDB hit" (1nsiA) as a template. The alignment between the template and the target sequence will be shown at the right. Click "[pdb]" to obtain the set of 3D coordinates of the model, and the PDB file will be downloaded onto your computer. Open and analyze the file with a molecular viewer. The downloaded PDB file (ORF2_02) can be opened by RasMol but not by Swiss-PdbViewer 3.9 (for Macintosh), because the file contains only a Cα trace.
c. Figure 11.21 shows the results of the same sequence (rBTCCd) from the other metaserver, Pcon/Pmodeller. This server provides the models with their graphics in the summary list according to their evaluation score (Pcon score). You can download the PDB file by clicking the graphic of the model. The PDB file provided by

this server can be opened by Swiss-PdbViewer. Judging from their evaluation scores, reliability of the prediction models from both Pcon/Pmodeller and BioInfoBank metaservers is considered to be low. It is also suggested by the fact that the templates selected are quite different among the fold-recognition programs.

Part III Sample Data

1. *Amino Acid Sequence of Rat Betacellulin (BTC) in the FASTA Format Fetched from Genbank*

>AB028862 (genbank) 49.582/translation=

```
MDSTAPGSGVSSLPLLLALVLGLVILQCVVADGNTTRTPETNGSLCGAP-
GENCTGTTPRQKSKTHFSRCPKQYKHYCIHGRCRFVMDEQTPSCICEKGYF-
GARCEQVDLFYLQQDRGQILVVCLIGVMVLFIILVIGVCTCCHPLRKHRK-
KKKEEKMETLSKDKTPISEDIQETNIA
```

2. *A Partial Sequence Corresponding to EGF-Like Domain of Rat BTC*

>rat BTC-EGFd

```
THFSRCPKQYKHYCIHGRCRFVMDEQTPSCICEKGYFGARCEQVDLFY
```

3. *A Partial Sequence Corresponding to Cytoplasmic Domain of Rat BTC*

>rBTCCd

```
IGVCTCCHPLRKHRKKKKEEKMETLSKDKTPISEDIQETNIA
```

SECTION 3 PATTERN AND PROFILE SEARCH

Part I Introduction

1. *Why are Pattern and Profile Search Needed?*

The sequence alignments and blast search against the sequence database are the first steps in the characterization of the primary structure of your protein. Secondary and tertiary structure predictions are also available as described in Section 2, if you have your own sequence. However, further information would be important in order to know what your sequence looks like. It would be an easy way to see if we know just motifs that are commonly found in groups of proteins. Typically, many proteins are known to form many kinds of families with patterns of sequences localized in a restricted region of the primary structure even if they are composed of domains with high molecular weights. These domain structures are related to the higher order of structures and functional properties of the proteins. The comparison of the proteins with others, whose structures and functions are well known, is very useful to know the patterns and profiles of the protein of interest, to understand the function, and to design further experiments to test and modify the function of specific proteins. For the purpose of this pattern and profile search, many software packages coupled with unique databases have been developed recently. The protein signature databases are developing as vital tools to identify distant relationships in novel sequences and are used to classify protein sequences by inferring their functions.

2. What Is Pattern and Profile Search?

To know the pattern and profiles localized in your proteins, the easiest way would be if there are databases of protein domains, families, and functional sites. These kinds of databases are now being constructed at several sites as described in the following text:

PROSITE: PROSITE (http://www.expasy.ch/prosite/) is a database of protein families and domains. It consists of biologically significant sites, patterns, and profiles that help reliably identify to which known protein family (if any) a new sequence belongs. Profiles created by the Generalized Profile Syntax, which is a very sensitive method of finding motifs in a query sequence, from various protein domains are incorporated into the PROSITE library. Each signature is linked to a documentation that provides useful biological information on the protein family, domain, or functional site identified by the signature.

During the last two years, the documentation has been redesigned and the latest version of PROSITE (Release 20.0, of 15 November, 2006) contains 1449 documentation entries, 1331 patterns, 675 profiles, and 720 ProRule. Over the past two years more than 200 domains have been added, and now 52% of UniProtKB/Swiss-Prot entries (release 48.1 of September 27, 2005) have a cross-reference to a PROSITE entry. The ProRule section of PROSITE comprises manually created rules that increase the discriminatory power of PROSITE motifs (generally profiles) by providing additional information about functionally and structurally critical amino acids and can automatically generate annotation based on PROSITE motifs in the UniProtKB/Swiss-Prot format. Each ProRule is defined in the UniRule format.

Once a profile is found, wherever available, series of three dimensional structures are retrieved from the PDB database that share the motif, and you can examine them using such as RasMol program (http://www.umass.edu/microbio/rasmol/).

Pfam: Pfam (http://www.sanger.ac.uk/Software/Pfam/) is a large collection of protein families and domains. The database contains multiple sequence alignments and HMMs covering many common protein domains of these families. Pfam can be used to view the domain organization of proteins. A single protein can belong to several Pfam families. Seventy-four percent of protein sequences have at least one match to Pfam. The data in the Pfam database contains two parts: Pfam-A and Pfam-B. Pfam-A is the curated part of Pfam, containing over 8296 protein families (May 2006). Pfam-B is an automatically generated supplement to Pfam-A. It contains a large number of small families taken from the ProDom database that do not overlap with Pfam-A. In spite of lower quality, Pfam-B families is useful when no Pfam-A families are found. To search against Pfam database the program HMMER (http://hmmer.janelia.org/) is used.

ProDom: The ProDom (http://prodom.prabi.fr/prodom/current/html/home.php) database contains protein domain families generated from the SWISS-PROT database by automated sequence comparisons. The current version was built with a new improved procedure based on recursive PSI-BLAST homology searches. The BLAST2 program, the gapped version of the BLAST, is available to search a protein sequence against the ProDom database.

BLOCKS: BLOCKS (http://blocks.fhcrc.org/) are multiply aligned ungapped segments corresponding to the most highly conserved regions of proteins. The blocks for the BLOCKS Database are made automatically by looking for the most highly conserved regions in groups of proteins documented in the Prosite database. The Prosite pattern for a protein group is not used in any way to make the BLOCKS database, and the pattern may or may not be contained in one of the blocks representing a group. These blocks are then calibrated against the SWISS-PROT database to obtain a measure of the chance distribution of matches. It is these calibrated blocks that make up the BLOCKS database.

PRINTS: The PRINTS (Protein Fingerprint Database) (http://umber.sbs.man.ac.uk/dbbrowser/PRINTS/PRINTS.html) is a compendium of protein motif fingerprints. A fingerprint is a group of conserved motifs used to characterize a protein family; its diagnostic power is refined by iterative scanning of a SWISS-PROT/TrEMBL composite. Usually, the motifs do not overlap but are separated along a sequence, though they may be contiguous in 3D-space. Fingerprints can encode protein folds and functionalities more flexibly and powerfully than can single motifs, full diagnostic potency deriving from the mutual context provided by motif neighbors. It is derived by the excision of conserved motifs from sequence alignments and refined by iterative dredging of the OWL (http://umber.sbs.man.ac.uk/dbbrowser/OWL/), a nonredundant composite sequence database. Two types of fingerprint are represented in the database, either simple or composite, depending on their complexity: simple fingerprints are essentially single-motifs, whereas composite fingerprints encode multiple motifs.

HAMAP: The HAMAP profile (http://www.expasy.org/sprot/hamap/families.html) is a collection of orthologous microbial protein families, generated manually by expert curators. They are used for the high-quality automatic annotation of microbial proteomes in the framework of the Swiss-Prot protein knowledgebase. This database is only available on the Motif scan server.

TIGRFAMs: TIGRFAMs (http://www.tigr.org/TIGRFAMs/index.shtml) is a collection of protein families, featuring curated multiple sequence alignments, HMMs, and annotation, which provides a tool for identifying functionally related proteins based on sequence homology. Those entries that are "equivalogs" group homologous proteins which are conserved with respect to function.

Phospho.ELM: The Phospho.ELM database (http://phospho.elm.eu.org/) contains a collection of experimentally verified serine, threonine, and tyrosine sites in eukaryotic proteins. The entries, manually annotated and based on scientific literature, provide information about phosphorylated proteins and the exact position of known phosphorylated instances. Phospho.ELM version 5.0 (May

2006) contains 2540 substrate proteins from different species covering 1434 tyrosine, 4798 serine, and 974 threonine instances.

PIRSF: The PIRSF protein classification system (http://pir.georgetown.edu/iproclass/) is a network with multiple levels of sequence diversity from superfamilies to subfamilies that reflects the evolutionary relationship of full-length proteins and domains. The primary PIRSF classification unit is the homeomorphic family, whose members are both homologous (evolved from a common ancestor) and homeomorphic (sharing full-length sequence similarity and a common domain architecture).

SUPERFAMILY: SUPERFAMILY (http://supfam.mrc-lmb.cam.ac.uk/SUPERFAMILY/) is a library of profile HMMs that represent all proteins of known structure. The library is based on the SCOP classification of proteins: each model corresponds to a SCOP domain and aims to represent the entire SCOP superfamily that the domain belongs to. SUPERFAMILY has been used to carry out structural assignments to all completely sequenced genomes. The results and analysis are available from the SUPERFAMILY Web site.

InterPro: InterPro (http://www.ebi.ac.uk/interpro/index.html) is a database of protein families, domains, and functional sites in which identifiable features found in known proteins can be applied to unknown protein sequences. InterPro, an integrated documentation resource of protein families, domains, and functional sites, was created to integrate the major protein signature databases. Currently, it includes PROSITE, Pfam, PRINTS, ProDom, SMART, TIGRFAMs, PIRSF, and SUPERFAMILY. Signatures are manually integrated into InterPro entries that are curated to provide biological and functional information (8166 families, 201 repeats, 26 active sites, 21 binding sites, and 20 posttranslational modification sites). InterPro covers over 78% of all proteins in the Swiss-Prot and TrEMBL components of UniProt.

CATH: CATH (http://www.cathdb.info/latest/index.html) is a hierarchical classification of protein domain structures that clusters proteins at four major levels: Class, Architecture, Topology, and Homologous superfamily. Class, derived from secondary structure content, is assigned for more than 90% of protein structures automatically. Architecture, which describes the gross orientation of secondary structures, independent of connectivities, is currently assigned manually. The topology level clusters structures into fold groups according to their topological connections and numbers of secondary structures. The homologous superfamilies cluster proteins with highly similar structures and functions. The assignments of structures to fold groups and homologous superfamilies are made by sequence and structure comparisons.

3. Integrated Analyzing System Developed for the Databases

In recent years, scanning tools for the various databases have been developed and are being improved rapidly. Table 11.1 lists useful uniform resource locators (URLS). Some features of these programs are briefly described in the following text.

InterProScan: InterProScan is an integrated search in PROSITE, Pfam, PRINTS, and other family and domain databases. This is a tool that combines different protein signature recognition methods into one resource. The number of signature databases and their associated scanning tools, as well as the further refinement procedures, increases the complexity of the problem. InterProScan is more than just a simple wrapping of sequence analysis applications as it also performs a considerable amount of data lookup from various databases and program outputs. Results are shown in various formats such as "html," "xml," and so on.

FingerPRINTScan: FingerPRINTScan scans a protein sequence against the PRINTS Protein Fingerprint Database. The BLIMPS program, which scores a sequence against blocks or a block against sequences, is also available to search a protein sequence against the PRINTS database. The original PRINTS database will be converted to BLOCKS format for each release of PRINTS. "Composite" fingerprints were decomposed into motif components under the unique PRINTS code (such as "GLABLOOD" held in "gc" line, which is equivalent to "PR00001" in "gx" line in the PRINTS database.)

3of5: The 3of5 Web application enables complex pattern matching in protein sequences. 3of5 is named after a special use of its main feature, the novel n-of-m pattern type. This feature allows for an extensive specification of variable patterns in which the individual elements may vary in their position, order, and content within a defined stretch of sequence. The number of distinct elements can

TABLE 11.1 Several Useful URLS of Scanning Programs for Protein Profile Databases

Application	**URL**	**Databases**
InterProScan	http://www.ebi.ac.uk/InterProScan/	InterPro PROSITE Pfam PRINTS ProDom SMART TIGRFAMs PIRSF SUPERFAMILY
ScanProsite	http://kr.expasy.org/tools/scanprosite/	PROSITE Swiss-Prot/TrEMBL.
MotifScan	http://myhits.isb-sib.ch/cgi-bin/motif_scan	PROSITE Pfam HAMAP TIGRFAMs
Pfam HMM search	http://pfam.janelia.org/http://www.sanger.ac.uk/Software/Pfam/search.shtml	Pfam
BLIMPS	ftp://ftp.ncbi.nih.gov/repository/blocks/unix/blimps/	BLOCKS
FingerPRINTScan	http://umber.sbs.man.ac.uk/fingerPRINTScan/	PRINTS
3 of 5	http://www.dkfz.de/mga2/3of5/3of5.html	Not specified
ELM	http://elm.eu.org/	SMART Pfam
PRATT	http://www.ebi.ac.uk/pratt/http://kr.expasy.org/tools/pratt/	N/A
PPSEARCH	http://www.ebi.ac.uk/ppsearch/	PROSITE
PROSITE scan	http://npsa-pbil.ibcp.fr/cgi-bin/npsa_automat.pl?page=npsa_prosite.html	PROSITE
PATTINPROT	http://npsa-pbil.ibcp.fr/cgi-bin/npsa_automat.pl?page=npsa_pattinprot.html	A protein databases for one or several patterns at PBIL.
SMART	http://smart.embl-heidelberg.de/	Swiss-Prot SP-TrEMBL stable Ensembl proteomes
TEIRESIAS:	http://cbcsrv.watson.ibm.com/Tspd.html	N/A
Hits	http://myhits.isb-sib.ch/cgi-bin/index	MySQL
PANTHER	http://www.pantherdb.org/	UniProt
Gene3D	http://cathwww.biochem.ucl.ac.uk:8080/Gene3D/	BioMap UniProt CATH Pfam COG/KOG
SignalP	http://www.cbs.dtu.dk/services/SignalP/	N/A
TMHMM	http://www.cbs.dtu.dk/services/TMHMM/	N/A

be constrained by operators, and individual characters may be excluded. The n-of-m pattern type can be combined with common regular expression terms and thus also allows for a comprehensive description of complex patterns. 3of5 increases the fidelity of pattern matching and finds all possible solutions in protein sequences in cases of length-ambiguous patterns instead of simply reporting the longest or shortest hits. Grouping and combined search for patterns provides a hierarchical arrangement of larger patterns sets. This application offers an extended vocabulary for the definition of search patterns and thus allows the user to comprehensively specify and identify peptide patterns with variable elements. The n-of-m pattern type offers improved accuracy for pattern matching in combination with the ability to find all solutions, without compromising the user-friendliness of regular expression terms.

SMART: SMART is a Simple Modular Architecture Research Tool at EMBL. In Normal SMART, the database contains Swiss-Prot, SP-TrEMBL, and stable Ensembl proteomes. The protein database in Normal SMART has significant redundancy, even though identical proteins are removed. Genomic mode will enhance SMART to explore domain architectures, or to find exact domain counts in various genomes. The numbers in the domain annotation pages will be more accurate, and there will not be many protein fragments corresponding to the same gene in the architecture query results. A user must keep in mind exploring a limited set of genomes.

Hits: Hits is a database containing information about a protein, i.e., a sequence and annotations,and it is also a collection of tools for the investigation of the relationships between protein sequences and motifs. These motifs are defined by a heterogeneous collection of predictors, which currently includes regular expressions, generalized profiles, and HMMs. MyHits is an extension of Hits. It allows any registered user to manage its own private collections of protein sequences and motifs. The system relies on a MySQL database updated daily. Registration is free for academic use.

PANTHER: The PANTHER (Protein ANalysis THrough Evolutionary Relationships) Classification System is a unique resource that classifies genes by their functions, using published scientific experimental evidence and evolutionary relationships to predict function even in the absence of direct experimental evidence. Proteins are classified by expert biologists into families and subfamilies of shared function, which are then categorized by molecular function and biological process ontology terms. For an increasing number of proteins, detailed biochemical interactions in canonical pathways are captured and can be viewed interactively.

Gene3D: Gene3D is built upon the BioMap sequence database, which consists of UniProt (including the genome sets obtained from Integr8) and extra sequences from various functional resources (including KEGG and GO). These sequences are annotated using functional data from GO, COGS, and KEGG. This program scans the CATH domain database against the whole sequence database and adds Pfam domain family data for UniProt sequences. The protein–protein interaction data from BIND and MINT, where available, are added. The sequences are clustered into whole-chain protein families. These families should show good conservation of function and structural features. The complete genomes are obtained from Integr8 at the EBI. They are mostly correct, but for some genomes (i.e., rat) you may want to check on whether they are actually complete or not.

Part II Step-By-Step Tutorial

InterProScan is a metaserver integrated with different protein signature recognition methods into one resource with look up of corresponding InterPro and GO annotation and is best suitable for the demonstration purpose (Figure 11.22). This site is enhanced with a number of programs mentioned earlier. It scans PROSITE (patterns and profiles), PRINTS, PFAM, PRODOM, SMART, TIGRFAMs, PIR SuperFamily, and SUPERFAMILY and runs GENE3D, PANTHER, SignalP v3, and TMHMM v2 at the same time if you check boxes as many as you want.

The demo example shows how to run this InterProScan:

1. Obtain the protein sequences you want to analyze from the database or your own experiments. The sample data used in this demonstration are listed below in Part III titled "Sample Data."
2. Go to the InterProScan Web site at the address http://www.ebi.ac.uk/InterProScan/. Then you will reach the image shown in Figure 11.23.

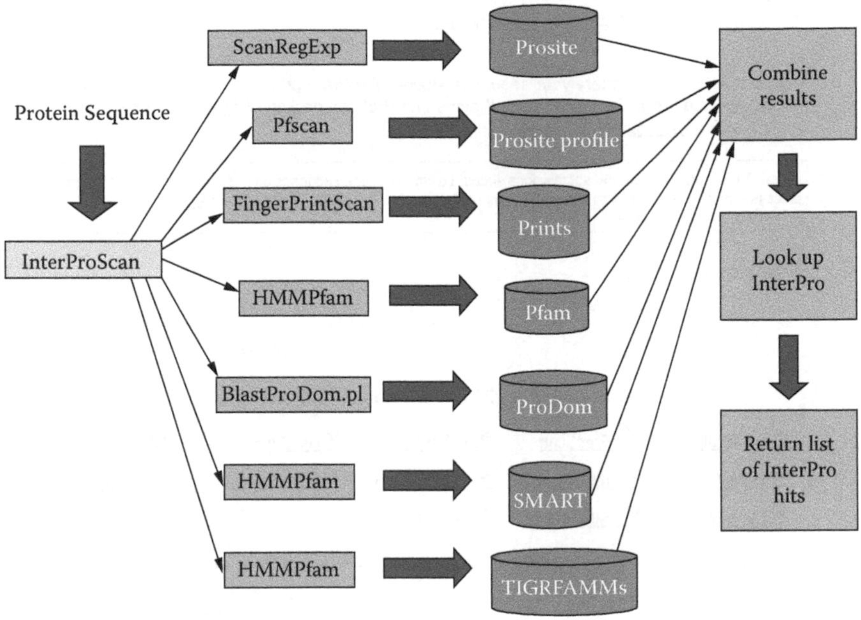
Protein Sequence
InterProScan
ScanRegExp
Pfscan
FingerPrintScan
HMMPfam
BlastProDom.pl
HMMPfam
HMMPfam
Prosite
Prosite profile
Prints
Pfam
ProDom
SMART
TIGRFAMMs
Combine results
Look up InterPro
Return list of InterPro hits

FIGURE 11.22 Search for protein profiles and patterns with a protein sequence through InterProScan.

3. Enter your e-mail address if you want to receive the results via e-mail.
4. Choose either an interactive run (when the results will be displayed online) or an e-mail run (when you will receive your results by e-mail).
5. Select the type of sequence you are going to input, i.e., protein or nucleotide.

You can copy and paste or type a nucleotide or protein sequence into the large text window. A free text (raw) sequence is simply a block of characters representing a protein or DNA/RNA sequence. You may also paste a sequence in Fasta, EMBL, Swiss-Prot, and GenBank format.

- To input a protein sequence, use free text/Raw, FASTA, or Swiss-Prot.
- To input a nucleotide sequence use free text/Raw, FASTA, EMBL, or GenBank.

Instead of putting your sequence into the window, you may want to attach a file containing your sequence.

InterProScan **Sequence Search**

This form allows you to query your sequence against InterPro. For more detailed information see the documentation for the perl stand-alone InterProScan package (Readme file or FAQ's), or the InterPro user manual or help pages.

Please Note: InterProScan job submissions should be limited to one sequence only. The system will no longer process 6 protein sequences simultaneously as of Monday Feb 13, 2006. Please contact support for help in submitting multiple sequences.

Download Software

YOUR EMAIL	RESULTS
	interactive

APPLICATIONS TO RUN ⌣ Clear all ⌣ Check all

☑ BlastProDom	☑ FPrintScan	☑ HMMPIR	☑ HMMPfam	☑ HMMSmart
☑ HMMTigr	☑ ProfileScan	☑ ScanRegExp	☑ SuperFamily	☑ SignalPHMM
☑ TMHMM	☑ HMMPanther	☑ Gene3D		

TRANSLATION TABLE (DNA/RNA only)	MIN. OPEN READING FRAME SIZE
None	100

Enter or Paste a PROTEIN Sequence in any format: Help

```
MDRAARCSGASSLPLLLALALGLVILHCVVADGNSTRSP
ETNGLLCGDPEENCAATTTQSKRKGHFSRCPKQYKHYC
IKGRCRFVVAEQTPSCVCDEGYIGARCERVDLFYLRGDR
GQILVICLIAVMVVFIILVIGVCTCCHPLRKRRKRKKKEEE
METLGKDITPINEDIEETNIA
```

Upload a file: ファイルを選択 ファイルが選...れていません   Submit Job   Reset

FIGURE 11.23 The results of InterProScan (Picture view).

6. When you are submitting a nucleotide sequence, you have to choose the minimum length of open reading frame (ORF) in your sequence. After submission, the sequence will be automatically translated into an amino acid sequence only when an ORF is present in the sequence. Copy and paste your sequence in the window appearing below the sentence. Figure 11.13 shows the image that has the amino acid sequence of human betacellulin precursor (No. 1 sequence in Part III titled "Sample Data") is input.

7. Check boxes appropriately depending on the assessment required. All boxes are checked as default. The checked protein sequence applications are launched. These applications search against specific databases and have preconfigured cutoff thresholds:

    - **BlastProDom** scans the families in the ProDom database.
    - **FprintScan** scans against the fingerprints in the PRINTS database.
    - **HMMPIR** scans the HMMs that are present in the PIR Protein Sequence Database (PSD) of functionally annotated protein sequences, PIR-PSD.
    - **HMMPfam** scans the HMMs that are present in the PFAM Protein families database.
    - **HMMSmart** scans the HMMs that are present in the SMART domain/domain families database.
    - **HMMTigr** scans the HMMs that are present in the TIGRFAMs protein families database.
    - **ProfileScan** scans against PROSITE profiles.
    - **ScanRegExp** scans against the regular expressions in the PROSITE protein families and domains database.
    - **SuperFamily** is a library of profile HMMs that represents all proteins of known structure.

    Refer **SignalPHMM** and **HMMPANTHER** to SignalP and PANTHER above, respectively.

8. Now you can perform an InterProScan query by clicking "Submit job." In this demonstration the result will appear in the computer screen but not via e-mail.

9. After a while, the result will appear as shown in Figure 11.24. This is a typical summary pattern as a picture view. Alternatively, this summary could be taken as a table by clicking the "Table View" button (Figure 11.25).

10. In either of these summaries, the left column shows the entry code of InterPro, i.e., IPR000742 (EGF-like, type3), IPR 001336 (EGF-like, type1), IPR 006209 (EGF-like), IPR 006210 (EGF), and IPR 013032 (EGF-like region).

11. The right column shows the entry code related with another database. For example, the line of IPR000742 (EGF-like, type3) shows the

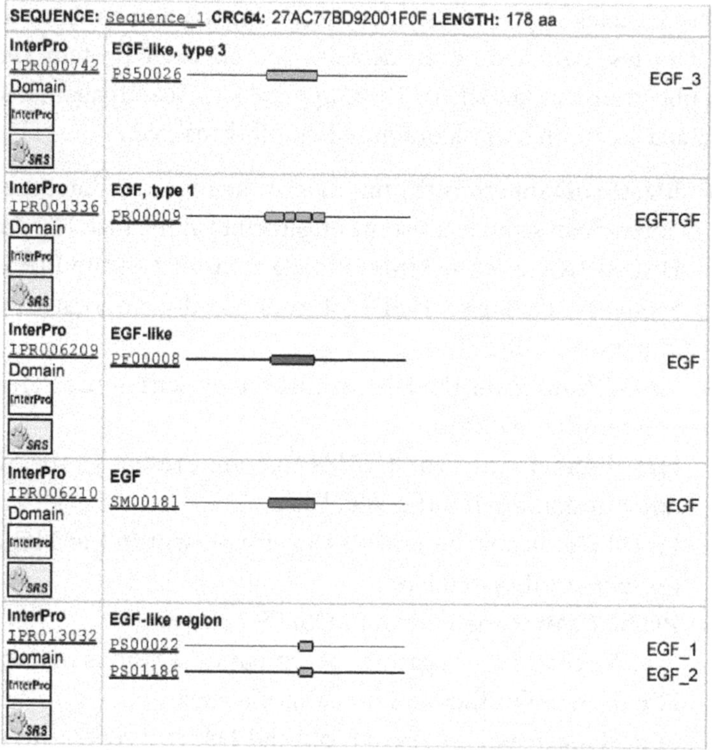
SEQUENCE: Sequence_1 CRC64: 27AC77BD92001F0F LENGTH: 178 aa
InterPro
IPR000742
Domain
EGF-like, type 3
PS50026
EGF_3
InterPro
IPR001336
Domain
EGF, type 1
PR00009
EGFTGF
InterPro
IPR006209
Domain
EGF-like
PF00008
EGF
InterPro
IPR006210
Domain
EGF
SM00181
EGF
InterPro
IPR013032
Domain
EGF-like region
PS00022
PS01186
EGF_1
EGF_2

FIGURE 11.24 The results of InterProScan (Table view).

category is related with PS50026 in PROSITE as for the amino acid residues from 65 to 105.

12. The entry code starting with IPR (e.g., IPR000742 assigned to the signature of "EGF-like, type 3") belongs to InterPro, and the contents of the entry can be seen by clicking the code.

13. Further information related to this entry code can be obtained by clicking the underlined character codes, which belong to the databases collaborating with InterProScan. The detailed tutorial for InterPro is found at (http://www.ebi.ac.uk/interpro/tutorial.html).

14. The scan should be carefully done to obtain more information that could not be shown with the first submission. This appears to occur owing to the priority of information automatically ordered by the program. In this case there is so much information on the EGF-motif that other information is removed from the final results. To obtain

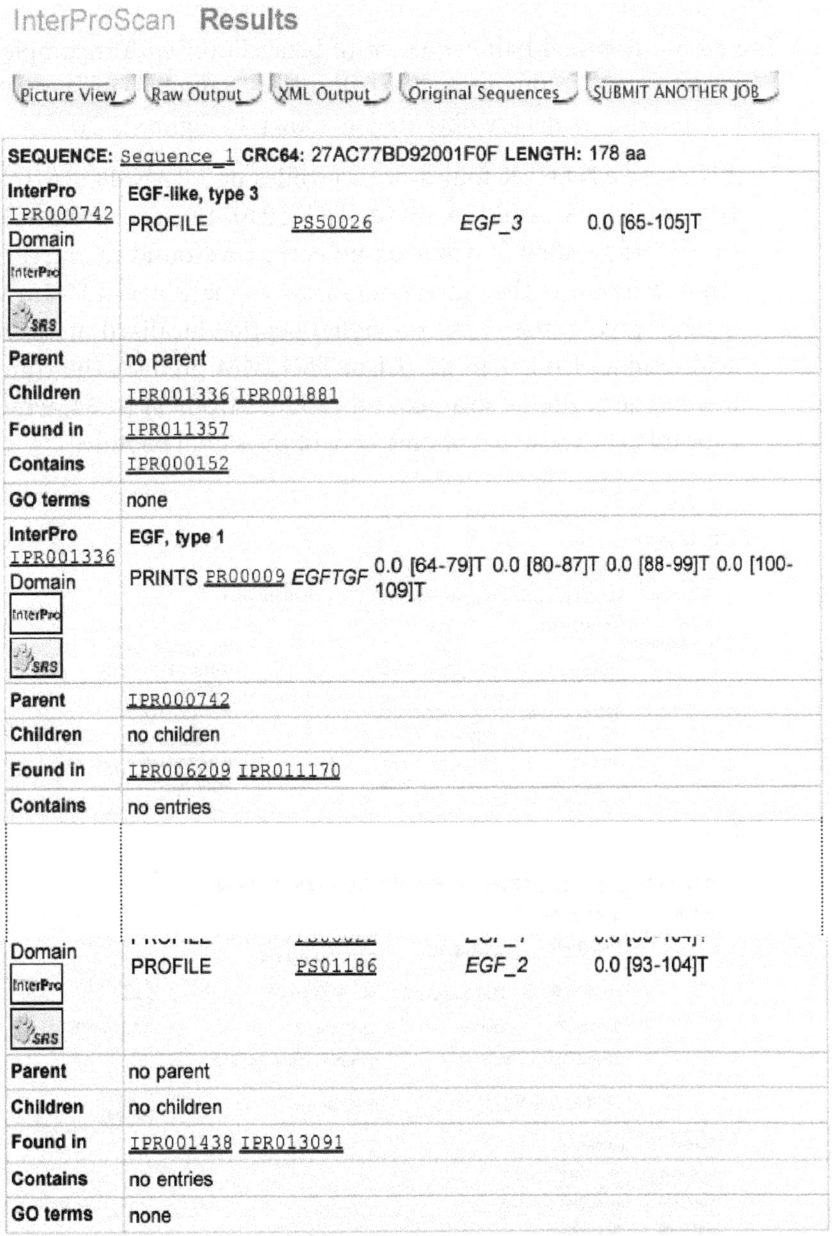

InterProScan **Results**

Picture View | Raw Output | XML Output | Original Sequences | SUBMIT ANOTHER JOB

SEQUENCE: Sequence_1 **CRC64:** 27AC77BD92001F0F **LENGTH:** 178 aa	
InterPro IPR000742 Domain	**EGF-like, type 3** PROFILE PS50026 *EGF_3* 0.0 [65-105]T
Parent	no parent
Children	IPR001336 IPR001881
Found in	IPR011357
Contains	IPR000152
GO terms	none
InterPro IPR001336 Domain	**EGF, type 1** PRINTS PR00009 *EGFTGF* 0.0 [64-79]T 0.0 [80-87]T 0.0 [88-99]T 0.0 [100-109]T
Parent	IPR000742
Children	no children
Found in	IPR006209 IPR011170
Contains	no entries
Domain	PROFILE PS01186 *EGF_2* 0.0 [93-104]T
Parent	no parent
Children	no children
Found in	IPR001438 IPR013091
Contains	no entries
GO terms	none

FIGURE 11.25 Results by the submission of amino-terminal sequence of betacellulin.

the missing information, it is strongly recommended to submit the sequence fragmented into short pieces.

15. The amino-terminal half sequence of betacellulin with incomplete EGF-motif (No. 2 sequence in Part III, titled "Sample Data") is submitted and the results are obtained as shown in Figure 11.26.

    In this case PANTHER found both profiles of TRANSFORMING GROWTH FACTOR ALPHA (PTHR10740) and Betacellulin (PTHR10740:SF3), whereas no entry was found in InterPro. Another point is that the results from SignalP and TMHMM. SignalP predict the secretion signal peptide localized at amino acid residues from 1 to 32. Also, TMHMM predicts the transmembrane region at amino acid residues from 14 to 34, which appears to be the hydrophobic core of the signal peptide.

A. Picture view

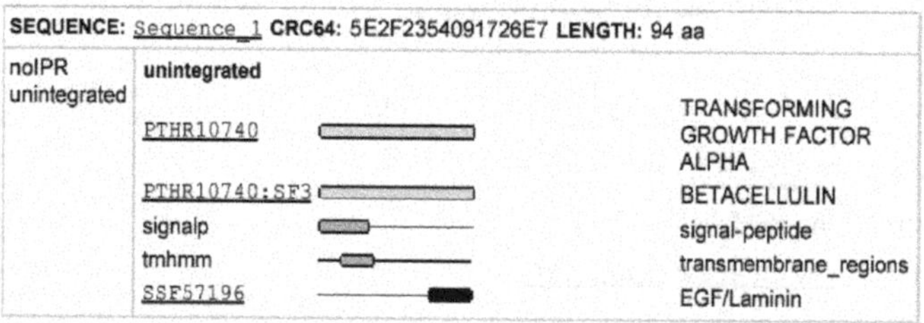
SEQUENCE: Sequence_1 CRC64: 5E2F2354091726E7 LENGTH: 94 aa
noIPR unintegrated
unintegrated
PTHR10740
TRANSFORMING GROWTH FACTOR ALPHA
PTHR10740:SF3
BETACELLULIN
signalp
signal-peptide
tmhmm
transmembrane_regions
SSF57196
EGF/Laminin

B. Table view

SEQUENCE: Sequence_1 CRC64: 5E2F2354091726E7 LENGTH: 94 aa				
noIPR unintegrated	**unintegrated**			
	PANTHER	PTHR10740	*TRANSFORMING GROWTH FACTOR ALPHA*	4.2e-72 [2-94]T
	PANTHER	PTHR10740:SF3	*BETACELLULIN*	4.2e-72 [2-94]T
	SIGNALP	signalp	*signal-peptide*	NA [1-31]?
	TMHMM	tmhmm	*transmembrane_regions*	NA [15-35]?
	SUPERFAMILY	SSF57196	*EGF/Laminin*	0.00028 [68-94]T
Parent	no parent			
Children	no children			
Found In	no entries			
Contains	no entries			
GO terms	none			

FIGURE 11.26 Results by the submission of carboxyl-terminal sequence of betacellulin.

16. Figure 11.27 shows the results when the carboxyl-terminal half sequence of betacellulin precursor without EGF motif (No. 3 sequence in Part III, titled "Sample Data") is submitted.

Again, PANTHER found both profiles of TRANSFORMING GROWTH FACTOR ALPHA (PTHR10740) and Betacellulin (PTHR10740:SF3). SignalP predicts the secretion signal peptide again; however, the submitted part of the sequence is not localized at the amino terminal of betacellulin, so that this is not a secretion signal but the transmembrane region as TMHMM predicts at amino acid residues from 14 to 34.

A detailed tutorial on InterProScan is available at "2Can Support Portal" (http://www.ebi.ac.uk/2can/tutorials/function/InterProScan.html).

A. Picture view

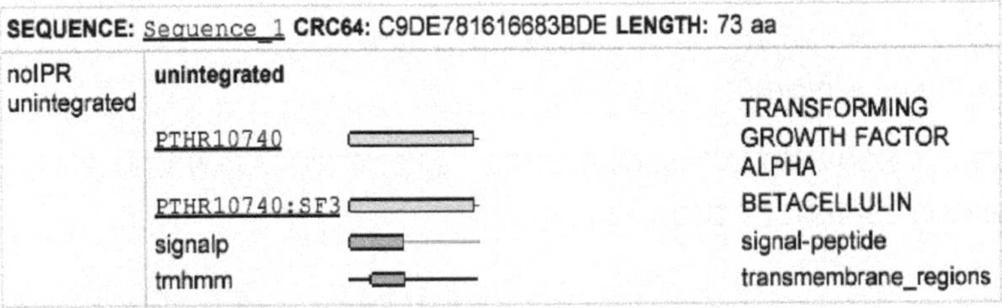
SEQUENCE: Sequence_1 CRC64: C9DE781616683BDE LENGTH: 73 aa
noIPR unintegrated
unintegrated
PTHR10740
TRANSFORMING GROWTH FACTOR ALPHA
PTHR10740:SF3
BETACELLULIN
signalp
signal-peptide
tmhmm
transmembrane_regions

B. Table view

SEQUENCE: Sequence_1 CRC64: C9DE781616683BDE LENGTH: 73 aa				
noIPR unintegrated	**unintegrated**			
	PANTHER	PTHR10740	*TRANSFORMING GROWTH FACTOR ALPHA*	1.2e-51 [1-70]T
	PANTHER	PTHR10740:SF3	*BETACELLULIN*	1.2e-51 [1-70]T
	SIGNALP	signalp	*signal-peptide*	NA [1-32]?
	TMHMM	tmhmm	*transmembrane_regions*	NA [14-34]?
Parent	no parent			
Children	no children			
Found in	no entries			
Contains	no entries			
GO terms	none			

FIGURE 11.27 Tertiary structure predicted by EsyPred.

Part III Sample Data

1. Human Betacellulin Precursor Amino Acid Sequence in FASTA Format

>gi|4502461|ref|NP_001720.1| betacellulin [Homo sapiens]

```
MDRAARCSGASSLPLLLALALGLVILHCVVADGNSTRSPETNGLLCGDPEEN-
CAATTTQSKRKGHFSRCPKQYKHYCIKGRCRFVVAEQTPSCVCDEGYIGAR-
CERVDLFYLRGDRGQILVICLIAVMVVFIILVIGVCTCCHPLRKRRKRK-
KKEEEMETLGKDITPINEDIEETNIA
```

2. The Amino-Terminal Half Amino Acid Sequence of Human Betacellulin

```
MDRAARCSGASSLPLLLALALGLVILHCVVADGNSTRSPETNGLLCGDPEEN-
CAATTTQSKRKGHFSRCPKQYKHYCIKGRCRFVVAEQTPSCV
```

3. The Carboxyl-Terminal Half Amino Acid Sequence of Human Betacellulin

```
RVDLFYLRGDRGQILVICLIAVMVVFIILVIGVCTCCHPLRKRRKRK-
KKEEEMETLGKDITPINEDIEETNIA
```

REFERENCES

1. Bjellqvist, B., Hughes, G.J., Pasquali, Ch., Paquet, N., Ravier, F., Sanchez, J.-Ch., Frutiger, S. & Hochstrasser, D.F. The focusing positions of polypeptides in immobilized pH gradients can be predicted from their amino acid sequences. *Electrophoresis* 1993; 14:1023–1031
2. Protein Identification and Analysis Tools on the ExPASy Server: Gasteiger, E., Hoogland, C., Gattiker, A., Duvaud, S., Wilkins, M.R., Appel, R.D., Bairoch, A.; In John M. Walker (ed): The Proteomics Protocols Handbook, Humana Press (2005). pp. 571–607
3. Heger, A., Holm, L. Rapid automatic detection and alignment of repeats in protein sequences. *Proteins*. 2000; 41(2):224–237.
4. Callebaut, I., Labesse, G., Durand, P., Poupon, A., Canard, L., Chomilier, J., Henrissat, B., Mornon, J.P. Deciphering protein sequence information through hydrophobic cluster analysis (HCA): current status and perspectives. *Cell Mol Life Sci*. 1997; 53(8):621–645.
5. Hida, K., Wada, J., Eguchi, J., et al.: Visceral adipose tissue-derived serine protease inhibitor: A unique insulin-sensitizing adipocytokine in obesity. *Proc Natl Acad Sci U S A* 2005; 102(30):10610–10615.

6. Wada, J., Ota, K., Kumar, A., Wallner, E.I., and Kanwar, Y.S.: Developmental regulation, expression, and apoptotic potential of galectin-9, a β-galactoside binding lectin. *J Clin Invest* 1997; (10):2452–2461.
7. Zhang, H., Wada, J., Hida, K., et al.: *Collectrin*, a collecting duct-specific transmembrane glycoprotein, is a novel homologue of ACE2 and is developmentally regulated in embryonic kidneys. *J Biol Chem* 2001; 276(20): 17132-17139.
8. Berman, H.M., Westbrook, J., Feng, Z., et al. The Protein Data Bank. *Nucleic Acids Research*, 2000; 28:235–242.
9. Chou, P.Y., Fasman, G.D. Prediction of the secondary structure of proteins from their amino acid sequence. Adv Enzymol Relat Areas Nol Biol, 1978;47:45-147.
10. Ouali, M., King, R.D. Cascaded multiple classifiers for secondary structure prediction. *Prot. Sci*, 2000; 9:1162–1176.
11. McGuffin, L.J., Bryson, K., Jones, D.T. The PSIPRED protein structure prediction server. *Bioinformatics*, 2000; 16:404–405.
12. Karplus, K., Karchin, R., Draper, J., et al. Combining local-structure, fold-recognition, and new fold methods for protein structure prediction. *Proteins*, 2003; 53 Suppl 6:491–496.
13. Rost, B., Yachdav, G., Liu, J. The PredictProtein Server. *Nucleic Acids Research*. 2004; 32(Web Server issue):W321–W326.
14. Ginalski, K., Elofsson, A., Fischer, D., et al. 3D-Jury: a simple approach to improve protein structure predictions. *Bioinformatics*, 2003; 19(8):1015–1018.
15. Wallner, B., Arne Elofsson, A. Identification of correct regions in protein models using structural, alignment and consensus information. *Protein Sci*, 2006; 15(4):900–913.
16. Chivian, D., Kim, D.E., Malmstrom, L., et al. Prediction of CASP6 structures using automated Robetta protocols. *Proteins*, 2005; 61 Suppl 7:157–66 (Supplementary Material)
17. Tada, H., Seno, M., Yamada, H., et al. Molecular cloning and expression of rat betacellulin cDNA. *Biochim Biophys Acta*, 2000; 1492(1):285–288.
18. Bucher, P. & Bairoch, A. (1994). A generalized profile syntax for biomolecular sequence motifs and its function in automatic sequence interpretation. In: Proceedings Second International Conference on Intelligent Systems for Molecular Biology. (Altman, R., Brutlag, D., Karp, P., Lathrop, R. & Searls, D., Eds.), pp 53–61, AAAI Press, Menlo Park, CA.
19. Bateman, A., Coin, L., Durbin, R., et al. The Pfam protein families database. *Nucleic Acids Res.* 2004 Database Issue 32:D138–D141.
20. Bru, C., Courcelle, E., Carrere, S., Beausse, Y., Dalmar, S., Kahn, D.. The ProDom database of protein domain families: more emphasis on 3D. Nucleic Acids Res. 33 (Database issue): D212-D215 (2005). *Bioinformatics,* 1999; 15:471–479.
21. Henikoff, J.G., Greene, E.A., Pietrokovski, S., Henikoff, S. Increased coverage of protein families with the blocks database servers. *Nucl. Acids Res.* 2000; 28(1):228–230.

22. Attwood, T.K., Blythe, M., Flower, D.R., Gaulton, A., Mabey, J.E., Maudling, N., McGregor, L., Mitchell, A., Moulton, G., Paine, K., and Scordis, P. PRINTS and PRINTS-S shed light on protein ancestry. *Nucleic Acids Res.* 2002; 30(1):239–241.
23. Wu, C.H., Huang, H., Nikolskaya, A., Hu, Z., Yeh, L.S., Barker, W.C. The iProClass Integrated database for protein functional analysis. *Computational Biology and Chemistry,* 2004; 28:87–96.
24. Gough, J., Karplus, K., Hughey, R., Chothia, C. Assignment of homology to genome sequences using a library of Hidden Markov Models that represent all proteins of known structure. *J. Mol. Biol.* 2001; 313:903–919.
25. Mulder, N.J., Apweiler, R., Attwood, T.K., et al. InterPro, progress and status in 2005. *Nucleic Acids Res.* 33 (Database Issue):D201–205 (2005).
26. Pearl, F., Todd, A., Sillitoe, I., et al. (2005) The CATH Domain Structure Database and related resources Gene3D and DHS provide comprehensive domain family information for genome analysis. *Nucleic Acids Res* Vol. 33 Database Issue D247-D251.

CHAPTER 12

Protein Function Analysis

Lydie Lane, Yum Lina Yip, and
Brigitte Boeckmann

CONTENTS

Proteins largely determine cellular events. Proteins may be classified into a variety of functional categories such as catalytic, transporting, signaling, regulatory, controlling, structural, or mechanical proteins.

One of the major challenges of the postgenomic era is to unravel the underlying molecular mechanisms of protein function and to elucidate how individual proteins interact in biological processes. As only a small fraction of known proteins have been experimentally characterized, the majority of uncharacterized proteins are annotated by the transfer of existing knowledge from homologous proteins.

In the first section of this chapter, the application of protein sequence analysis to protein function prediction for the purpose of database annotation is described. The second section is dedicated to the prediction of posttranslational protein modifications, as such modifications provide an important means of regulating protein function and increasing functional diversity. Although the *in silico* protein sequence analysis methods are valuable for the prediction of biochemical function, they may be unable to reliably predict the biological role of a protein. Such context-dependent functions can be analyzed by studying protein–protein interactions, a subject covered in the third section.

SECTION 1 PROTEIN FUNCTION ANNOTATION

Part I Introduction

1. Why Is Protein Function Annotation Needed?

An amino acid sequence without any function annotation is meaningless. What is more, the vast majority of proteins cannot perform their proper cellular function as such. As deduced from characterized proteins, it has been shown that most are chemically altered: amino acid modifications can either stabilize the protein's structure or influence a protein's function. More than 350 different types of amino acid modifications have already been described. What is more, most proteins assemble with other proteins to achieve a function that none of the components can perform on their own. The knowledge of such modifications is indispensable for understanding the molecular function of proteins and their contribution to the biological processes of cells.

Hundreds of raw sequences enter protein databases every day. The characterization of an individual protein requires a multitude of experimental studies, that are performed mostly on model organisms such as mouse, fruit fly, yeast, or *Escherichia coli*. Most other proteins will never be experimentally characterized; in contrast, their characterization will be based on the assumption that corresponding proteins of related organisms are likely to perform the same function. The knowledge obtained on one protein will thus be used to infer the possible function of related proteins. Consequently, the combination of biological knowledge, experimental findings, and *in silico* protein sequence analysis makes protein function annotation possible also for uncharacterized proteins.

2. What Is Involved in Protein Function Annotation?

Three basic steps are involved in protein function prediction: the reading of relevant literature, *in silico* function analysis, and the interpretation of the obtained results.

Access to experimental records. The only way to acquire knowledge on the function of proteins is through experiments. The results of such studies are published in scientific journals and review articles written by experts, who give a summarized interpretation of the available knowledge. Tools for finding relevant information in the literature are helpful, but cannot replace the careful reading of the individual articles. Reading is actually one of the main tasks of an annotator. Only a small fraction of protein function annotation goes back to direct author submissions. Since the introduction of large-scale experiments, databases have been created for the storage of experimental findings from numerous studies, among them gene expression and protein–protein interactions. The raw data are generally not used for protein function prediction without further validation.

***In silico* protein sequence analysis.** Numerous tools can be used to gain insight into a protein's function. A nonexhaustive list is presented in Table 12.1. In the following text, the successive steps of a protein sequence analysis are described, using one possible combination of software tools.

The first step in sequence analysis is generally a database similarity search. These often detect closely related sequences from other species, and their annotation can give a first clue regarding a possible function — if at least one of the sequences has already been characterized and curated. The function of most proteins is actually inferred from the known function of related sequences. Similar sequences originating from different species — and in an order consistent with evolutionary distances — may be considered potentially orthologous. The similarity of the orthologs will generally cover the full length of the protein.

Further, a multiple sequence alignment of the query sequence and its potential orthologs might help refine the sequence's primary structure. For example, initiation codons can be inferred from orthologs for which they have been experimentally determined. Moreover, locally divergent regions in the alignment are often caused by splice variants or frameshift errors, which can be checked by analyzing the coding region of the gene. At this stage of the study, the orthologous relationship of the proteins can be clarified by performing a phylogenetic analysis. Based on the assumption that orthologs share the same function, a literature search for all the orthologs will detect more relevant experimental findings than would be obtained for an individual protein.

Going deeper into the protein's function will generally require scanning the query protein against a family and domain database. Predicted domains might also indicate a structural and functional similarity between more distantly related proteins. Protein-function-relevant information is documented in family and domain databases that also report on functional sites and regions within domains. Such information helps accurately predict biologically important sites, e.g., active sites, binding sites, or posttranslational modifications (PTMs) (see next section).

Another aspect critical for the deduction of protein function is the prediction of the subcellular location of the protein. A number of methods have been developed to predict features relevant to the final destination

TABLE 12.1 List of Selected Programs Relevant to Protein Function Prediction

Database Similarity Search Against a Protein Sequence Database		
Blast UniProtKB	http://www.uniprot.org/blast/	Find similar entries in UniProtKB
Blast RefSeq	http://www.ncbi.nlm.nih.gov/BLAST/	Find similar entries in RefSeq
Multiple Sequence Alignment (see Chapter 3)		
Phylogenetic Analysis (see Chapter 3)		
Family and Domain Prediction (see Chapter 10)		
InterProScan	http://www.ebi.ac.uk/InterProScan/	Families, domains, and sites from ten databases, including Pfam, PROSITE, SMART, and CATH/SCOP
Pfam	http://www.sanger.ac.uk/Software/Pfam/search.shtml	Families and domains
ScanProsite	http://www.expasy.org/tools/scanprosite/	Families, domains, sites, and annotation rules
SMART	http://smart.embl-heidelberg.de/	Domains
HAMAP	http://www.expasy.org/sprot/hamap/families.html	Family prediction and annotation of microbes
COILS	http://www.ch.embnet.org/software/COILS_form.html	Coiled coils
Deduction of the Subcellular Location		
PSORT	http://www.psort.org/psortb/	Subcellular location of bacterial proteins
PSORTII	http://wolfpsort.seq.cbrc.jp/	Subcellular location of eukaryotic proteins
ChloroP	http://www.cbs.dtu.dk/services/ChloroP/	Chloroplast transit peptides
PTS1	http://mendel.imp.ac.at/mendeljsp/sat/pts1/PTS1predictor.jsp	Peroxisomal targeting signal 1
SignalP	http://www.cbs.dtu.dk/services/SignalP/	Secretory pathway signal peptide cleavage site
TargetP	http://www.cbs.dtu.dk/services/TargetP/	Chloroplast transit peptide, mitochondrial targeting peptide and secretory pathway signal peptide
DAS	http://www.sbc.su.se/~miklos/DAS	Transmembrane regions in prokaryotes
TMHMM	http://www.cbs.dtu.dk/services/TMHMM-2.0/	Orientation and location of transmembrane helices
big-PI	http://mendel.imp.ac.at/gpi/gpi_server.html	GPI-anchor
PTM Prediction (see Chapter 11, Section 2)		
Prediction of Protein–Protein Interactions (see Chapter 11, Section 3)		
Deduction of the Biological Role of a Protein		
String	http:/string.embl.de/	Protein–protein interactions

of a protein in the cell, e.g., targeting signals for the secretory pathway, mitochondria, peroxisomes and chloroplasts, transmembrane domains, and glycosyl phosphatidylinositol (GPI) anchors (GPI-anchors).

Combined analysis methods. The biological role of a protein can also be inferred from characterized proteins that are — in one way or another — associated with the protein of interest. A simple case would be the deduction of a protein function from known proteins encoded by the same operon. Functional evidence can also be obtained from cluster analysis of expression data or the analysis of protein–protein interaction networks, but both approaches have not yet achieved the accuracy needed for function prediction. Results seem to be more reliable when combining information from different sources, such as from genomics, proteomics, and high-throughput experiments.

Interpretation of the analysis results. Annotation is more than an accumulation of experimental findings and analysis results. All outcomes need checking for their logical coherence, and this requires biological knowledge. Whereas some inconsistencies in the predicted results are blatant, others are much less obvious. In turn, though analysis results may appear to be paradoxical, they could actually be correct. However, complex explanations are possible only when based on experimental results.

3. New Developments in Protein Function Annotation

Automated annotation. Protein function annotation is time consuming, and fortunately some of the analysis and annotation steps can be automated. Various approaches to automatic annotation have been developed over the past few years: procedures differ extensively in their extent of integrated analysis steps and quality checks, dependent on the scope of the project. Depending on the scope of the project, procedures differ extensively in their extent of integrated analysis steps and quality checks. Large-scale analysis — as for the detection of trends in comparative genomics — tends to maximize the coverage of function predictions. In contrast, systems for the proper characterization of individual proteins perform sequence analysis as extensively as with manual annotation. Validation steps implemented in the analysis pipeline filter sequences with unexpected characteristics for sequence correction or special annotation. Quality-oriented annotation procedures are not meant to replace manual annotation, but rather increase efficiency: instead of annotating an individual sequence entry, family-specific annotation rules are developed and applied to sequences, with coherent analysis results. Annotation rules are constantly maintained and updated as new scientific findings emerge.

Controlled vocabularies. Annotation is regularly used to access and compare data. This is possible when the database format is well structured and the information content is standardized. To this end, controlled vocabularies have been created; a typical example is keywords. In contrast, ontologies are a combination of controlled vocabularies with defined relationships. Building controlled vocabularies with precise definitions is a huge task, especially when they are used to characterize related data from distinct disciplines. Multiple research groups are involved in the Gene Ontology (GO) project, which provides controlled vocabularies to describe gene product attributes in a species-independent manner. So far, the GO project has created three ontologies for molecular functions, biological processes, and cellular components. GO terms are data independent and can thus be assigned to data of distinct nature, e.g., proteins, protein families, protein domains, protein complexes, and pathways. Thanks to the GO annotation, data derived from different species can be compared more easily.

Part II Step-By-Step Tutorial

This step-by-step tutorial pinpoints distinct aspects of protein function annotation: *in silico* protein sequence analysis (demo 1), exploration of existing database annotation (demo 2), and automatic annotation (demo 3).

1. Prediction of the Subcellular Location for a Protein

This demo example shows how to apply various methods for the prediction of the subcellular location by (1) retrieving information from the annotation of related data, (2) applying several programs for the prediction of the subcellular location of a protein, and (3) using InterProScan to complement the information obtained. The protein analyzed is an automatically annotated protein of *Drosophila melanogaster*. Besides a strong similarity to other probable insect orthologs, the strongest similarity to reviewed data of the UniProtKB database is detected for ALK tyrosine kinase receptors and leukocyte tyrosine kinase receptors. The sequence similarity covers the full length of these proteins.

The steps of the demo example follow:

1. Access a protein entry from the UniProtKB Web site by typing the address http://www.uniprot.org/ in your browser. In the "Query" field, type "P97793." The UniProtKB database entry for the mouse ALK tyrosine kinase receptor is displayed (Figure 12.1).
   a. The annotation indicates that the protein is expected (see non-experimental qualifier "Potential") to be a type-1 membrane protein. This implies, that (1) the protein possesses a signal peptide for the secretory pathway, (2) that it has one transmembrane domain, and (3) that the topology of the N-terminal is "outside" (e.g., extracellular) and the C-terminal of the protein remains in the cytosol. Three distinct types of domains have been predicted

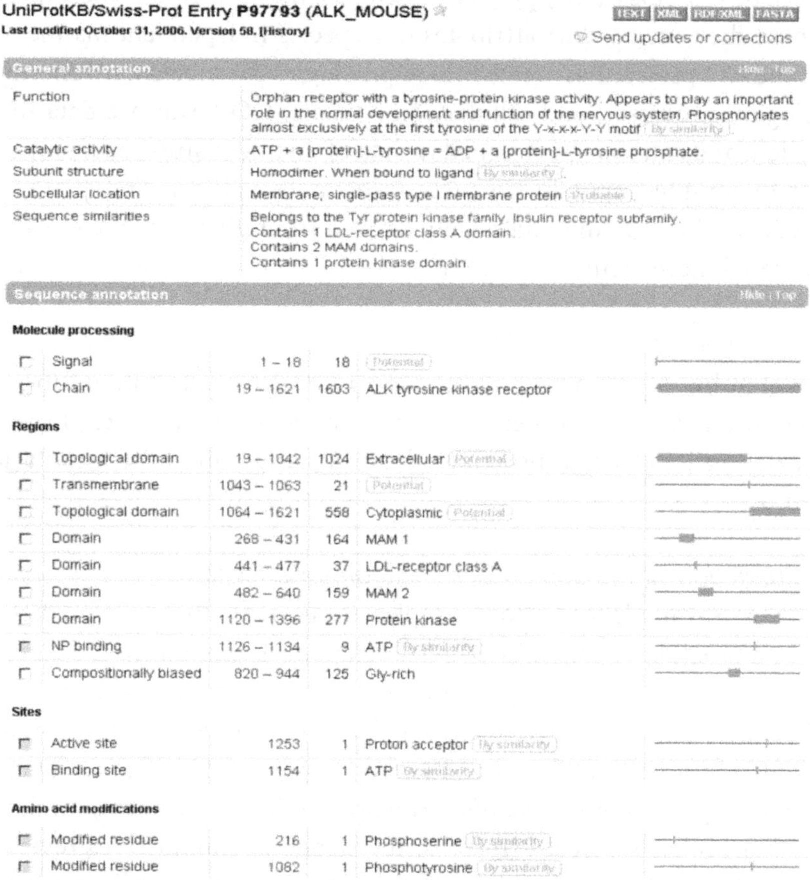

UniProtKB/Swiss-Prot Entry P97793 (ALK_MOUSE)

Last modified October 31, 2006. Version 58. [History]

TEXT XML RDF/XML FASTA

Send updates or corrections

General annotation Hide | Top

Function	Orphan receptor with a tyrosine-protein kinase activity. Appears to play an important role in the normal development and function of the nervous system. Phosphorylates almost exclusively at the first tyrosine of the Y-x-x-x-Y-Y motif (By similarity).
Catalytic activity	ATP + a [protein]-L-tyrosine = ADP + a [protein]-L-tyrosine phosphate.
Subunit structure	Homodimer. When bound to ligand (By similarity).
Subcellular location	Membrane; single-pass type I membrane protein (Probable).
Sequence similarities	Belongs to the Tyr protein kinase family. Insulin receptor subfamily. Contains 1 LDL-receptor class A domain. Contains 2 MAM domains. Contains 1 protein kinase domain.

Sequence annotation Hide | Top

Molecule processing

Signal	1 – 18	18	Potential
Chain	19 – 1621	1603	ALK tyrosine kinase receptor

Regions

Topological domain	19 – 1042	1024	Extracellular (Potential)
Transmembrane	1043 – 1063	21	Potential
Topological domain	1064 – 1621	558	Cytoplasmic (Potential)
Domain	268 – 431	164	MAM 1
Domain	441 – 477	37	LDL-receptor class A
Domain	482 – 640	159	MAM 2
Domain	1120 – 1396	277	Protein kinase
NP binding	1126 – 1134	9	ATP (By similarity)
Compositionally biased	820 – 944	125	Gly-rich

Sites

Active site	1253	1	Proton acceptor (By similarity)
Binding site	1154	1	ATP (By similarity)

Amino acid modifications

Modified residue	216	1	Phosphoserine (By similarity)
Modified residue	1082	1	Phosphotyrosine (By similarity)

FIGURE 12.1 Function-related annotation of mouse ALK tyrosine kinase receptor (UniProtKB/Swiss-Prot: P97793, version 58).

for this protein: two MAM domains and an LDL-receptor class A domain that are located within the extracellular terminal of the protein, whereas the protein kinase domain is located in the cytoplasmic terminal of the protein. The catalytic activity of the protein kinase domain is given, and the function annotation describes the possible biological role of the protein. Question: Does the topology of this protein also apply to the similar protein of *Drosophila*?

2. Access a protein entry from the UniProt Web site by typing the address http://www.uniprot.org/ in your browser. In the "Query" field, type "Q7KJ08." The UniProtKB database entry for an uncharacterized protein of *Drosophila melanogaster* is displayed. Click on the FASTA button to display the entry in FASTA format. Copy the entry.
   a. Type the address http://www.cbs.dtu.dk/services/SignalP/ in your browser. Paste the FASTA-formatted sequence data into the "Search" field and click "Submit." The result is presented in Figure 12.2a. A signal for the secretory pathway has been predicted with a high probability. The signal peptide is likely to be cleaved between sequence positions 23 and 24. The existence of a signal anchor is predicted to be improbable.
   b. Type the address http://www.cbs.dtu.dk/services/TMHMM/ in your browser. Paste the FASTA-formatted sequence data into the "Submission" field, select under "Organism group" "Eukaryotes" and click "Submit." The result is presented in Figure 12.2b. Two transmembrane domains are predicted: one is located in the N-terminal at positions 7 to 29, and the second one is located at sequence positions 1106 to 1128; the N-terminal region of the second predicted transmembrane domain is expected to be "outside," and the C-terminal "inside." The predicted N-terminal transmembrane domain clashes with the predicted signal peptide: both are hydrophobic, alpha-helical regions, and the difficulty of discrimination between the two regions is well known. In this case, we assume that the signal peptide is likely to be true, based on the outcome of SignalP and the global sequence similarity to the tyrosine kinase receptors. To verify if the *Drosophila* protein could be a possible protein kinase receptor, the *Drosophila* sequence is scanned against family and domain databases.

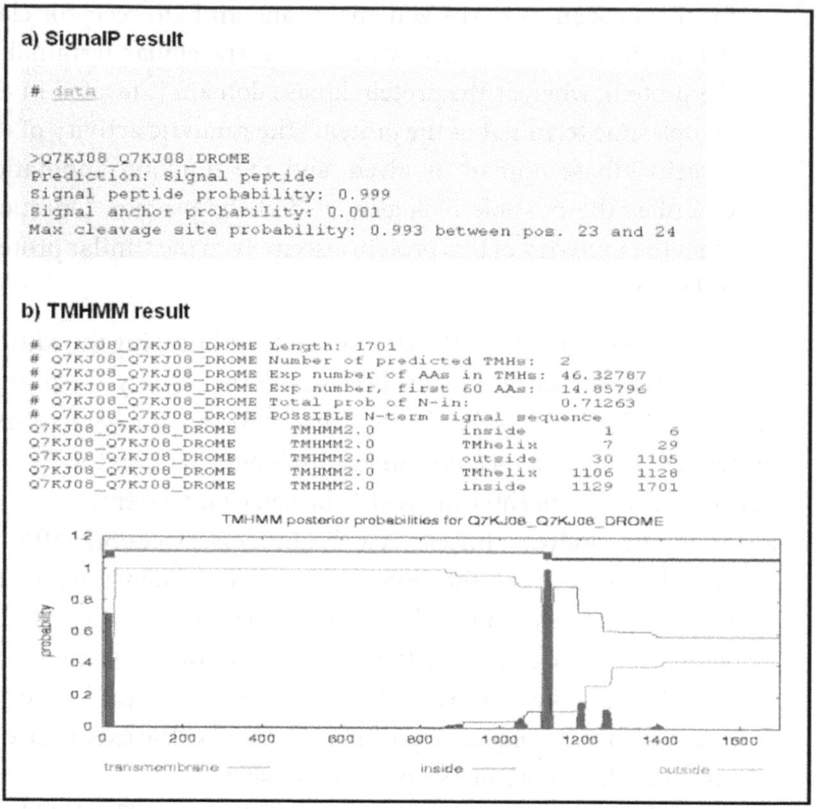
a) SignalP result
data
>Q7KJ08_Q7KJ08_DROME
Prediction: Signal peptide
Signal peptide probability: 0.999
Signal anchor probability: 0.001
Max cleavage site probability: 0.993 between pos. 23 and 24
b) TMHMM result
Q7KJ08_Q7KJ08_DROME Length: 1701
Q7KJ08_Q7KJ08_DROME Number of predicted TMHs: 2
Q7KJ08_Q7KJ08_DROME Exp number of AAs in TMHs: 46.32787
Q7KJ08_Q7KJ08_DROME Exp number, first 60 AAs: 14.85796
Q7KJ08_Q7KJ08_DROME Total prob of N-in: 0.71263
Q7KJ08_Q7KJ08_DROME POSSIBLE N-term signal sequence
Q7KJ08_Q7KJ08_DROME TMHMM2.0 inside 1 6
Q7KJ08_Q7KJ08_DROME TMHMM2.0 TMhelix 7 29
Q7KJ08_Q7KJ08_DROME TMHMM2.0 outside 30 1105
Q7KJ08_Q7KJ08_DROME TMHMM2.0 TMhelix 1106 1128
Q7KJ08_Q7KJ08_DROME TMHMM2.0 inside 1129 1701
TMHMM posterior probabilities for Q7KJ08_Q7KJ08_DROME
probability
1.2
1
0.8
0.6
0.4
0.2
0
0
200
400
600
800
1000
1200
1400
1600
transmembrane
inside
outside

FIGURE 12.2 Prediction of the subcellular location of an uncharacterized protein of *D. Melanogaster* (Q7AJ08). Results from the programs (a) SignalP, (b) TMHMM.

3. Type the address http://www.ebi.ac.uk/InterProScan/ in your browser, paste the FASTA-formatted sequence data into the "Enter or Paste..." field and click "Submit Job." The result is shown in Figure 12.3. Three distinct types of domains are predicted: two MAM domains, one low-density lipoprotein receptor class A domain, and a tyrosine protein kinase domain (more specific than the more general signature for the protein kinase domain that is also predicted). The four domains are predicted by distinct methods, which strengthens their plausibility. Both the MAM domain and the low-density lipoprotein receptor class A domain are typical extracellular domains: their structure is stringently stabilized by disulfide bonds and thus unstable in the oxidizing environment of the cytosol. Their N-terminal location in the probable transmembrane domain is consis-

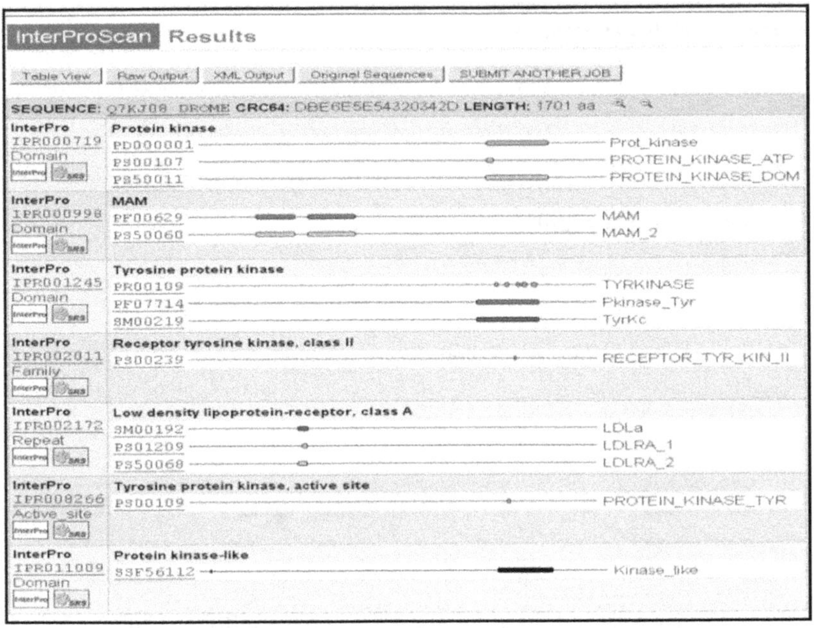
InterProScan Results
Table View
Raw Output
XML Output
Original Sequences
SUBMIT ANOTHER JOB
SEQUENCE: Q7KJ08_DROME CRC64: DBE6E5E543203420 LENGTH: 1701 aa
InterPro
IPR000719
Domain
Protein kinase
PD000001
PS00107
PS50011
Prot_kinase
PROTEIN_KINASE_ATP
PROTEIN_KINASE_DOM
InterPro
IPR000998
Domain
MAM
PF00629
PS50060
MAM
MAM_2
InterPro
IPR001245
Domain
Tyrosine protein kinase
PR00109
PF07714
SM00219
TYRKINASE
Pkinase_Tyr
TyrKc
InterPro
IPR002011
Family
Receptor tyrosine kinase, class II
PS00239
RECEPTOR_TYR_KIN_II
InterPro
IPR002172
Repeat
Low density lipoprotein-receptor, class A
SM00192
PS01209
PS50068
LDLa
LDLRA_1
LDLRA_2
InterPro
IPR008266
Active_site
Tyrosine protein kinase, active site
PS00109
PROTEIN_KINASE_TYR
InterPro
IPR011009
Domain
Protein kinase-like
SSF56112
Kinase_like

FIGURE 12.3 Result of the InterProScan for the *Drosophila* protein (Q7AJ08).

tent with the predicted topology of the protein. The protein kinase domain is located in the cytoplasmic region of the protein. Conclusion: The *Drosophila* protein is probably located in a membrane of the secretory pathway and/or in the cell membrane.

2. Exploring Disease-Related Annotation

Disease-related annotation can be descriptive (1a), include controlled vocabularies (1c), be position specific (1d), or give access to other resources (1b).

1. Search a protein entry from the UniProtKB Web site by typing the address http://www.uniprot.org/ in your browser. In the search field, type "Q15848." The database entry for human adiponectin is displayed (Figure 12.4).
   a. Scroll down to the section "General annotation" and find the topic "Involvement in disease." The enclosed annotation gives a brief description of the disease and includes a cross-reference to the corresponding MIM database entry (a disease database), which provides more details on the disease correlated to a deficiency in the protein.

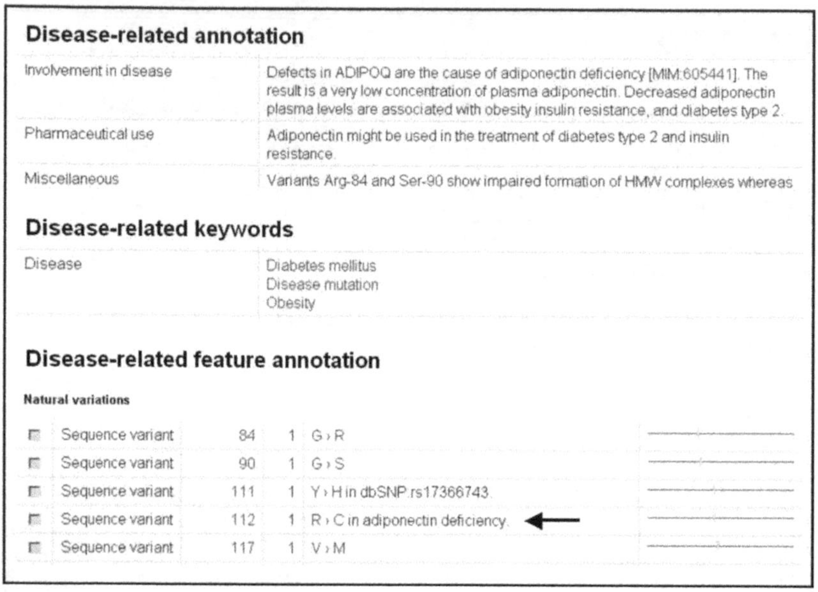

Disease-related annotation

Involvement in disease	Defects in ADIPOQ are the cause of adiponectin deficiency [MIM:605441]. The result is a very low concentration of plasma adiponectin. Decreased adiponectin plasma levels are associated with obesity insulin resistance, and diabetes type 2.
Pharmaceutical use	Adiponectin might be used in the treatment of diabetes type 2 and insulin resistance.
Miscellaneous	Variants Arg-84 and Ser-90 show impaired formation of HMW complexes whereas

Disease-related keywords

Disease	Diabetes mellitus
	Disease mutation
	Obesity

Disease-related feature annotation

Natural variations

□ Sequence variant	84	1	G › R	————————
□ Sequence variant	90	1	G › S	————————
□ Sequence variant	111	1	Y › H in dbSNP:rs17366743.	————————
□ Sequence variant	112	1	R › C in adiponectin deficiency. ←	————————
□ Sequence variant	117	1	V › M	————————

FIGURE 12.4 Disease-related annotation of human adiponectin (UniProtKB/Swiss-Prot: Q15848, version 74).

b. By clicking the MIM accession number, the MIM entry relevant to adiponectin is displayed. Click the "Back" button of your browser to retrieve the UniProtKB/Swiss-Prot entry Q15848. In the same section of the entry, there is the topic "Pharmaceutical use," which indicates that this protein is used or considered for disease treatment.

c. Scroll further down to the section "Ontologies," subsection "Keywords," topic "Disease." Several disease-related keywords are listed. Click the term "Diabetes mellitus." The uploaded Web site provides a definition for the usage of the keyword in UniProtKB. Click the term "Disease" in the "Category" section. All disease-related keywords used in UniProtKB are listed. Click the "Back" button of your browser to get back to the keyword entry. Click the term "UniProtKB" to get a list of all protein entries that are related to this disease. Click the "Back" button of your browser twice to get back to the UniProtKB/Swiss-Prot entry Q15848.

d. In the protein entry, scroll further down to the section "Sequence annotation." Under the subsection "Natural modifications" find the feature "Sequence variant" for amino acid position 112. It describes a disease mutation for "Adiponectin deficiency." In the description field of this feature, click "R->C." The Web site for the variant entry is uploaded and provides structural details for the specific protein variant.

3. Retrieve Function Annotation from the HAMAP Server

An uncharacterized prokaryotic protein is scanned against the HAMAP family database for microbial proteins. The resulting Web site provides functional annotation.

1. Fetch a protein entry in FASTA format.
   a. Search a protein entry from the UniProtKB Web site by typing the address http://www.uniprot.org/ in your browser. In the search field, type "Q0WAA3." The database entry for a currently uncharacterized protein from *Yersinia pestis* is displayed.
   b. Click on the box "FASTA" on the top right of the page to retrieve the entry in FASTA format. Copy the entry.
2. Run the program
   a. Access the HAMAP server by typing in your browser the address http://www.expasy.org/sprot/hamap/families.html. In the section "Scan your sequence against the HAMAP families," paste the entry in the sequence field and press "Run the scan."
   b. The resulting Web site lists a trusted hit for a protein family. Click on the family identifier "MF_01531." The resulting Web site contains automatically assigned annotation for this entry (Figure 12.5). In the section "Sets of member sequences," topic "All," click on the number at the end of the line, which indicates the number of family members in the UniProtKB/Swiss-Prot database. The resulting Web site gives access to all reviewed family members.

SECTION 2 PROTEIN POSTTRANSLATIONAL MODIFICATIONS

Part I Introduction

1. What Are PTMs?

During the late steps of protein synthesis, many proteins undergo modifications of their amino acids. The majority of these modifications occur posttranslationally, i.e., once the protein has undergone folding, and are typically catalyzed by specific enzymes found in the endoplasmic reticulum, the Golgi apparatus, the cytoplasm, or the nucleus. In the literature,

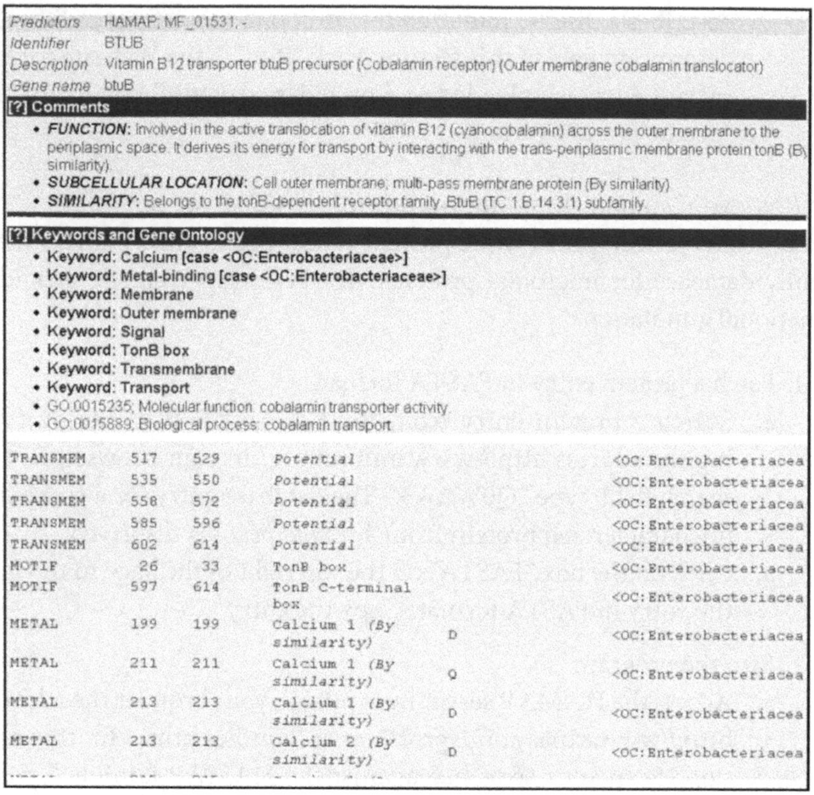

Predictors HAMAP; MF_01531; -
Identifier BTUB
Description Vitamin B12 transporter btuB precursor (Cobalamin receptor) (Outer membrane cobalamin translocator)
Gene name btuB

[?] Comments

- ***FUNCTION***: involved in the active translocation of vitamin B12 (cyanocobalamin) across the outer membrane to the periplasmic space. It derives its energy for transport by interacting with the trans-periplasmic membrane protein tonB (By similarity).
- ***SUBCELLULAR LOCATION***: Cell outer membrane; multi-pass membrane protein (By similarity).
- ***SIMILARITY***: Belongs to the tonB-dependent receptor family. BtuB (TC 1.B.14.3.1) subfamily.

[?] Keywords and Gene Ontology

- **Keyword: Calcium [case <OC:Enterobacteriaceae>]**
- **Keyword: Metal-binding [case <OC:Enterobacteriaceae>]**
- **Keyword: Membrane**
- **Keyword: Outer membrane**
- **Keyword: Signal**
- **Keyword: TonB box**
- **Keyword: Transmembrane**
- **Keyword: Transport**
- GO:0015235; Molecular function: cobalamin transporter activity
- GO:0015889; Biological process: cobalamin transport

TRANSMEM	517	529	*Potential*		<OC:Enterobacteriacea
TRANSMEM	535	550	*Potential*		<OC:Enterobacteriacea
TRANSMEM	558	572	*Potential*		<OC:Enterobacteriacea
TRANSMEM	585	596	*Potential*		<OC:Enterobacteriacea
TRANSMEM	602	614	*Potential*		<OC:Enterobacteriacea
MOTIF	26	33	TonB box		<OC:Enterobacteriacea
MOTIF	597	614	TonB C-terminal box		<OC:Enterobacteriacea
METAL	199	199	Calcium 1 *(By similarity)*	D	<OC:Enterobacteriacea
METAL	211	211	Calcium 1 *(By similarity)*	Q	<OC:Enterobacteriacea
METAL	213	213	Calcium 1 *(By similarity)*	D	<OC:Enterobacteriacea
METAL	213	213	Calcium 2 *(By similarity)*	D	<OC:Enterobacteriacea

FIGURE 12.5 Automatic HAMAP annotation. A protein of the bacterium *Yersinia pestis* (Q0WAA3) is scanned against the HAMAP family database and automatically annotated according the corresponding annotation rule.

the term PTM is often used in a rather general sense and includes both co- and posttranslational modifications. In general, PTMs are performed by enzymatic mechanisms that are present in a subset of organisms and/or subcellular compartments and are specific to certain amino acids.

There are three naturally occurring types of PTM. The first involves a change in the chemical nature of amino acids, for example, via deimination (arginine) or deamidation (glutamine or asparagine). The second type involves changes in the primary structure of the protein. These include proteolytic cleavages, or the formation of disulfide bridges by covalent linkage of two cysteines. The third type involves the addition of functional groups to amino acids. For example, phosphorylation adds a phosphate group to serine, threonine, or tyrosine; glycosylation adds a glycosyl group to either asparagine, serine, or threonine; acetylation adds an acetyl group, usually at the N-terminus of the protein; methylation adds a methyl group at lysine or arginine residues, and isoprenylation adds an isoprenoid group to a cysteine. PTMs can also consist of covalent linkages to another protein such as ubiquitin or an ubiquitin-like modifier.

In addition to specific proteolytic cleavages, more than 350 naturally occurring PTMs have been identified to date. The freely accessible RESID database (http://www.ebi.ac.uk/RESID) is a comprehensive collection of annotations and structures for these PTMs. For each PTM, it provides systematic and alternate names, the atomic formula and mass, enzymatic activities that generate the modification, keywords, literature citations, Gene Ontology (GO) cross-references, structure diagrams, and molecular models. In addition, it shows how PTMs are annotated in the UniProtKB/Swiss-Prot database. A complete list of PTMs currently annotated in the UniProtKB/Swiss-Prot is available on the ExPASy Web site (http://www.expasy.org/cgi-bin/ptmlist.pl). The Human Protein Reference Database (HPRD) (www.hprd.org) also contains high-quality PTM annotation for many human disease-related proteins. In addition, several manually curated PTM-specific databases, including O-GlycBase (http://www.cbs.dtu.dk/databases/OGLYCBASE/), Phospho.ELM (http://phospho.elm.eu.org/), and PhosphoSite (http://www.phosphosite.org) provide large amounts of experimental data.

2. Functional Impact of PTMs

One or more distinct PTMs can occur in a protein — and in various combinations — thus effectively extending the structural variety of a gene product. Hence, PTMs are a powerful mechanism to enhance the diversity of protein structures and to modify protein properties. Many sequence modifications such as proteolytic processing are irreversible, thus changing the property of the protein irreversibly too. Others are reversible and dynamically alter protein conformation, subcellular location, or interactions with other proteins; they are considered major regulatory mechanisms for metabolic enzymes and signaling pathways. In addition, competition for different PTMs at a single site in response to distinct upstream signals can provide a fine-tuning mechanism for signal integration. Therefore, it is not surprising that PTMs are now recognized as important targets in molecular medicine and pharmacology.

It is estimated that as many as a third of the eukaryotic proteins that enter the secretory pathway are glycosylated. The presence of carbohydrate chains has a profound influence on the physicochemical properties of glycoproteins. It can aid protein folding and quality control in the endoplasmic reticulum, and affect protein stability or aggregation properties. In addition, carbohydrate moieties are implicated as ligands in recognition phenomena, and can determine or affect the subcellular location, activity, or function of glycoproteins. Congenital disorders of glycosylation are inherited metabolic diseases caused by defects in the biosynthesis of glycoconjugates and hypoglycosylation of different glycoproteins. They cover a large variety of symptoms affecting multiple systems including, in most cases, statomotor and mental retardation.

In human cells, 518 protein kinases with different substrate specificities have been identified so far. They can phosphorylate cytosolic or nuclear proteins on serine, threonine, or tyrosine residues. As their activities are counterbalanced by a set of about 150 phosphatases, the resulting phosphoproteome can change in a highly dynamic way. Protein phosphorylation is considered to be the key event in many signal transduction pathways governing cell biology, including cell cycle control, membrane transport, cell adhesion, neurotransmission, and metabolism.

A good illustration of complex multisite PTMs is the "histone code," which regulates the structure and function of nucleosomes in chromatin. This code relies on a battery of enzymes that reversibly and dynamically methylate, acetylate, phosphorylate, and ubiquitinate distinct amino acids in the core histones, thereby modulating their interactions with DNA and with other factors involved in DNA replication, transcription, and repair. The ability to interfere with this code by using chemical compounds such as histone deacetylase inhibitors seems to be a promising strategy to block gene activation in cancer cells and check their proliferation.

3. How Are PTMs Experimentally Studied?

PTMs are usually present at substoichiometric levels, which means that a PTM at a given site is often present in only a small fraction of the protein molecules: for example, the occupancy of a phosphorylation site in 5% of a protein population may be sufficient to activate a signaling pathway. Thus, direct analysis of PTMs requires isolation of the correctly modified protein in a sufficiently large amount for biochemical study. For this reason, much work has been done on recombinant proteins expressed in systems thought to produce modification patterns similar to those of the organism of interest. For example, the baculovirus expression system is often used to mimic mammalian cell expression. Whatever the expression system used, there are still frequent significant differences between recombinant proteins and native proteins.

Two-dimensional gel electrophoresis separates protein populations on the basis of charge and molecular weight. For some PTMs such as phosphorylation, the resolution is sufficient to separate the different states of a protein directly. If sufficient information is available regarding the type and position of expected PTMs in the protein of interest, and if high-quality antibodies are available, Western blot on one- or two-dimensional gels is an easy and powerful method for PTM analysis. This technique is frequently coupled with site-directed mutagenesis.

Once a protein has been isolated, amino acid sequencing by the classical technique of Edman degradation is the method of choice to determine proteolytic processing. Edman degradation can also be performed on peptides obtained after enzymatic or chemical degradation of the protein. In this case, modified amino acids become apparent because of their absence or retention-time shift in the corresponding sequencing cycle, and can be mapped on the initial protein sequence.

In theory, any PTM can be detected by mass spectrometry (MS), provided that it leads to a difference in mass. The technique of peptide mass fingerprinting is widely used for high-throughput MS protein identification. This involves the digestion of the protein with an endoproteinase of known cleavage specificity, the measurement of the masses of resulting peptides by MS, and protein identification by matching the observed peptide masses against databases of theoretical masses of proteins and their derived peptides. Because unmatched peptides can be due to artifactual chemical modifications, to contamination with other proteins, but also to true PTMs, their inspection can sometimes directly provide useful PTM information.

However, in general, the mass shift of a modified peptide is not sufficient to determine confidently the nature of its modification. Peptides are usually refragmented at peptide bonds and reanalyzed by MS, which allows a precise determination of both their sequence and modifications. This type of approach is called "tandem mass spectrometry" or MS/MS. These MS techniques currently allow analysis of large sets of proteins at once. The development of affinity capture/enrichment techniques to examine subproteomes of proteins containing specific PTMs has allowed both increased sensitivity and simplified data analysis. For example, affinity enrichment of phosphopeptides by immobilized metal-affinity chromatography (IMAC) or antiphosphotyrosine antibodies and affinity enrichment of glycoproteins by lectins have been successfully performed to analyze protein phosphorylation and glycosylation in many model organisms (*Arabidopsis*, yeast, human, mouse, and *C. elegans*).

4. Which PTMs Can Be Predicted? How?

PTMs that are experimentally proved for a given protein are expected to occur in the same way in its orthologs from evolutionary close organisms. Therefore, PTMs are often inferred by similarity to a model organism. However, this relies on two main assumptions that always have to be checked first: the same PTM-performing enzymatic machinery is present in both organisms, and the sequence surrounding the PTM site is highly conserved in both organisms.

Even in the absence of PTM information available for ortholog proteins, it remains possible to predict several PTMs. As a matter of fact, PTMs are located at specific amino acid residues in proteins, usually in the context of particular sequence patterns. If they are sufficiently well defined and linear, these consensus motifs can be used to predict PTM occurrence. For example, N-glycosylation takes place at the Asn residue in the sequon Asn-X-Ser/Thr-X (where X is any amino acid but not Pro), which is easily revealed by computational sequence analysis. In the same vein, a wide range of computational approaches have been developed for prediction of PTMs, ranging from simple consensus motif searches to more complex methods such as artificial neural networks. Valuable tools are now available to predict proteolytic cleavages, tyrosine sulfation, N-terminal acetylation and myristoylation, O-glycosylation, or GPI-anchoring. Several algorithms have also been developed for prediction of kinase-specific phosphorylation sites. Although the rate of false-positive predictions is generally high for all these predictors, they may be useful to select potential substrates for further experimental analysis. In addition to PTM-related patterns from the PROSITE database (http://www.expasy.org/prosite/), a list of PTM prediction tools can be found at http://www.expasy.org/tools/#ptm.

The lack of highly curated PTM data sets makes it difficult to evaluate, compare, and improve PTM prediction tools in terms of sensitivity and specificity. Still, it is currently admitted that most PTM predictors tend to overpredict sites and therefore need more accurate filtering. For example, the taxonomic range of the organism studied should be considered, and protein subcellular location and topology should be taken into account in order not to predict N-glycosylation sites in cytosolic domains, or phosphorylation sites in transmembrane or extracellular domains. Finally, rapid developments in proteomics should increase the number of available PTM data and guide the improvement of PTM predictors.

Part II Step-By-Step Tutorial

1. Retrieve PTM Information on a Given Protein from General and More Specific Databases

The demo example is to illustrate what kind of PTM information is retrievable for a given protein from different databases. The chosen protein example is human crystallin alpha (A chain). The steps are as follows:

1. Go to UniProtKB by typing the address http://www.uniprot.org/ in your browser. In the dropdown list "Search in" select "Protein Knowledgebase (UniProtKB)" and type "human crystallin alpha" in the "Query" field. In the result list, click P02489 (CRYAA_HUMAN) to access the entry. Figure 12.6 presents part of the entry relevant to PTM. Modified amino acids are annotated in the section "Sequence annotation." The type of feature key used depends on the nature of the modification, and the exact name of the modified amino acid is indicated in the description field. Here, the sequence annotation shows one proteolytic cleavage, N-terminal acetylation, one glycosylation site, four phosphorylation sites, four internal acetylation sites, two methylation sites, one deamidation site, one disulfide bridge, and three sites that are susceptible to oxidation. Some details about deamidation, glycosylation, and phosphorylation are given in the section "General annotation" under the topic "Post-translational modification." In addition, five PTM-associated keywords are given under the topic "Ontologies." All the references from which the information was extracted are available under the topic "References," and for each of them a summary of the extracted information is provided.

2. Go to HPRD by typing the address http://www.hprd.org/ in your browser. Click on "Query" to obtain the form page. In the "Protein Name" field, type "crystallin alpha." Click on "Crystallin, alpha A" to access the entry. Click on "PTMs & SUBSTRATES" to access the PTM information (Figure 12.7). The disulfide bridge and the phosphorylation, methylation, glycosylation, and internal acetylation sites are annotated as in UniProtKB/Swiss-Prot. The HPRD entry also mentions two glycated sites that are not annotated in UniProtKB/Swiss-Prot, because they were obtained after chemical treatment of the protein and may be artifactual. The abstracts of the references

UniProtKB/Swiss-Prot Entry **P02489** (CRYAA_HUMAN) TEXT XML RDF/XML FASTA

Last modified November 28, 2006. Version 79. [History]

Send updates or corrections

General annotation

Function	May contribute to the transparency and refractive index of the lens.
Posttranslational modification	O-glycosylated; contains N-acetylglucosamine side chains. Deamidation of Asn-101 in lens occurs mostly during the first 30 years of age, followed by a small additional amount of deamidation (approximately 5%) during the next approximately 38 years, resulting in a maximum of approximately 50% deamidation during the lifetime of the individual. Phosphorylation on Ser-122 seems to be developmentally regulated. Absent in the first months of life, it appears during the first 12 years of human lifetime. The relative amount of phosphorylated form versus unphosphorylated form does not change over the lifetime of the individual.

Ontologies

Keywords

PTM	Acetylation Glycoprotein Methylation Oxidation Phosphorylation

Sequence annotation

Molecule processing

Chain	1 – 173	173	Alpha crystallin A chain
Chain	1 – 172	172	Alpha crystallin A chain, short form
Propeptide	173	1	Removed in short form

Sites

Site	18	1	Susceptible to oxidation
Site	34	1	Susceptible to oxidation
Site	138	1	Susceptible to oxidation

Amino acid modifications

Modified residue	1	1	N-acetylmethionine
Modified residue	13	1	Phosphothreonine
Modified residue	21	1	Omega-N-methylated arginine
Modified residue	45	1	Phosphoserine
Modified residue	70	1	N6-acetyllysine
Modified residue	78	1	N6-acetyllysine
Modified residue	88	1	N6-acetyllysine (alternate)
Modified residue	88	1	N6-methylated lysine (alternate)
Modified residue	101	1	Deamidated asparagine (partial)
Modified residue	122	1	Phosphoserine
Modified residue	140	1	Phosphothreonine
Modified residue	145	1	N6-acetyllysine
Glycosylation	162	1	O-linked (GlcNAc) By similarity
Disulfide	131, 142		

References

[9] **"Vertebrate lens alpha-crystallins are modified by O-linked N-acetylglucosamine."**
Roquemore E.P., Dell A., Morris H.R., Panico M., Reason A.J., Savoy L.-A., Wistow G.J., Zigler J.S. Jr., Earles B.J., Hart G.W.
J. Biol. Chem. 267:555-563(1992) [PubMed: 1730617] [Abstract]
Cited for: STRUCTURE OF CARBOHYDRATE.

[10] **"Post-translational modifications of water-soluble human lens crystallins from young adults."**
Miesbauer L.R., Zhou X., Yang Z., Sun Y., Smith D.L., Smith J.B.
J. Biol. Chem. 269:12494-12502(1994) [PubMed: 8175657] [Abstract]
Cited for: PHOSPHORYLATION AT SER-122, DISULFIDE BOND, C-TERMINAL PROCESSING, DEAMIDATION AT ASN-101, MASS SPECTROMETRY

[11] **"Differential phosphorylation of alpha-A crystallin in human lens of different age."**
Takemoto L.J.
Exp. Eye Res. 62:499-504(1996) [PubMed: 8759518] [Abstract] [Article from publisher]
Cited for: PHOSPHORYLATION AT SER-122.

[12] **"Quantitation of asparagine-101 deamidation from alpha-A crystallin during aging of the human lens."**
Takemoto L.J.
Curr. Eye Res. 17:247-250(1998) [PubMed: 9543632] [Abstract] [Article from publisher]
Cited for: DEAMIDATION AT ASN-101.

[13] **"In vivo acetylation identified at lysine 70 of human lens alphaA-crystallin."**
Lin P.P., Barry R.C., Smith D.L., Smith J.B.
Protein Sci. 7:1451-1457(1998) [PubMed: 9655350] [Abstract]
Cited for: C-TERMINAL PROCESSING, ACETYLATION AT LYS-70, PHOSPHORYLATION, DEAMIDATION AT ASN-101, MASS SPECTROMETRY

[14] **"The major in vivo modifications of the human water-insoluble lens crystallins are disulfide bonds, deamidation, methionine oxidation and backbone cleavage."**
Hanson S.R.A., Hasan A., Smith D.L., Smith J.B.
Exp. Eye Res. 71:195-207(2000) [PubMed: 10930324] [Abstract] [Article from publisher]
Cited for: PHOSPHORYLATION, SUSCEPTIBILITY TO OXIDATION, C-TERMINAL PROCESSING, MASS SPECTROMETRY

[15] **"Shotgun identification of protein modifications from protein complexes and lens tissue."**
MacCoss M.J., McDonald W.H., Saraf A., Sadygov R., Clark J.M., Tasto J.J., Gould K.L., Wolters D., Washburn M., Weiss A., Clark J.I., Yates J.R. III
Proc. Natl. Acad. Sci. U.S.A. 99:7900-7905(2002) [PubMed: 12060738] [Abstract] [Article from publisher]
Cited for: PHOSPHORYLATION [LARGE SCALE ANALYSIS] AT THR-13, SER-45, SER-122 AND THR-140, ACETYLATION AT LYS-70, LYS-78, LYS-88 AND LYS-145, METHYLATION AT ARG-21 AND LYS-88, SUSCEPTIBILITY TO OXIDATION, MASS SPECTROMETRY.

FIGURE 12.6 UniProtKB/Swiss-Prot PTM annotation on human crystallin alpha, A chain (Release 50.8 of 03 October, 2006).

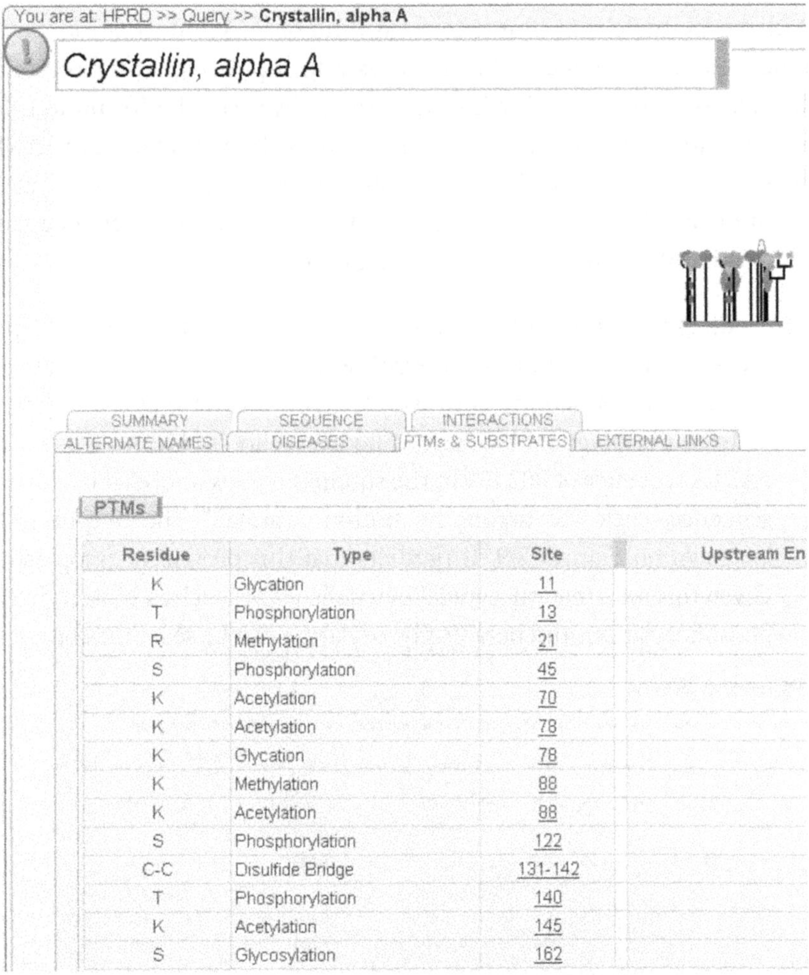
You are at: HPRD >> Query >> Crystallin, alpha A
Crystallin, alpha A
SUMMARY
SEQUENCE
INTERACTIONS
ALTERNATE NAMES
DISEASES
PTMs & SUBSTRATES
EXTERNAL LINKS
PTMs

Residue	Type	Site	Upstream En
K	Glycation	11	
T	Phosphorylation	13	
R	Methylation	21	
S	Phosphorylation	45	
K	Acetylation	70	
K	Acetylation	78	
K	Glycation	78	
K	Methylation	88	
K	Acetylation	88	
S	Phosphorylation	122	
C-C	Disulfide Bridge	131-142	
T	Phosphorylation	140	
K	Acetylation	145	
S	Glycosylation	162	

FIGURE 12.7 HPRD PTM information on crystallin alpha, A chain (03 October, 2006).

used for HPRD annotation are easily retrieved by clicking on the modified site positions.

3. Go to Phospho.ELM by typing the address http://phospho.elm.eu.org/ in your browser. Click on "Query" to obtain the form page. In the "Protein Name" field, type "crystallin alpha." Click on "Crystallin alpha A" to access the entry. For the human protein, the four phosphorylation sites are provided, with links on the corresponding references and on the UniProtKB/Swiss-Prot entry (Figure 12.8).

2. Prediction of PTMs from Sequence

The demo example is to illustrate what kind of PTM information can be predicted for the sequence of human crystallin alpha (A chain) from available tools, and to compare it with experimentally verified information. The first step is to get the human crystallin alpha (A chain) sequence in FASTA format. One way to proceed is to click on "P02489 in FASTA format" at the bottom of the UniProtKB/Swiss-Prot entry and copy the sequence from the uploaded Web site. The steps are as follows:

1. Prediction of subcellular location: Before performing any PTM prediction, it is important to determine where the protein is located in the cell. One possibility is to use BaCelLo predictor at the following address: http://gpcr.biocomp.unibo.it/bacello/pred.htm. Paste the FASTA sequence of P02489 in the submission box and, as it is a human sequence, click the taxonomy button "animal." The output result is shown on Figure 12.9. It predicts that this protein is cytoplasmic. Given this information, typical cytosolic PTMs such as N-acetylation, phosphorylation, and beta O-glycosylation can be searched for.

Phospho.ELM Currently 9090 instances in Phospho.ELM database New Search....

Summary of the phosphorylation sites (instances). Click on substrate name for additional information.

Substrate	Short Description	SWALL Id/Acc	Position	Sequence	Kinase	PubMed	Source	Binding Motif	Smart/Pfam	ELM	PDB/MSD
Crystallin alpha A	Heat shock protein	CRYAA_BOVIN	S122	REFHRRYRLPSNVDQSALSCS	-	92065884	LTP	-			-
Crystallin alpha A	Heat shock protein	CRYAA_HUMAN	T13	VTIQHPWFKRTLGPFYPSRLF	-	12060738	LTP	-	Crystallin		-
			S45	DLLPFLSSTISPYYRQSLFRT	-	12060738	LTP	-	Crystallin		-
			S122	REFHRRYRLPSNVDQSALSCS	-	12060738	LTP	-	HSP20		-
			T140	SCSLSADGMLTFCGPKIQTGL	-	12060738	LTP	-	HSP20		-

FIGURE 12.8 Phospho.ELM PTM information on crystallin alpha, A chain (Version 5.0 May 2006)

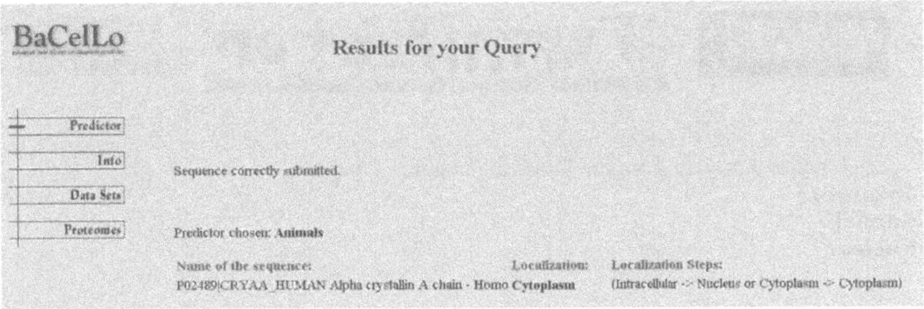
BaCelLo
Results for your Query
Predictor
Info
Data Sets
Proteomes
Sequence correctly submitted.
Predictor chosen: Animals
Name of the sequence:
P02489|CRYAA_HUMAN Alpha crystallin A chain - Homo
Localization:
Cytoplasm
Localization Steps:
(Intracellular -> Nucleus or Cytoplasm -> Cytoplasm)

FIGURE 12.9 Subcellular location prediction of human crystallin alpha chain A by BaCelLo.

2. Prediction of N-acetylation: Go to Terminator2 server by typing the following address: http://www.isv.cnrs-gif.fr/terminator2/. Select "Eukaryote" and then "Animal," "Nuclear" genome, and "No" predicted LPR cleavage in the selector fields. Paste the FASTA sequence of P02489 in the submission box and delete all characters before the first methionine of the sequence. Click on "Run prediction." The output result is shown on Figure 12.10. It predicts that this protein has a *N*-acetylmethionine, with a likelihood of 67%. This prediction is accurate, as N-acetylation of the methionine has been experimentally proven.

3. Prediction of phosphorylation and O-glycosylation sites: Because interplay of phosphorylation and O-glycosylation at the same site may result in the protein undergoing functional switches, it can be important to predict both O-glycosylation and phosphorylation on Ser/Thr residues of the protein. Go to YinOYang server by typing the address http://www.cbs.dtu.dk/services/YinOYang/ in your browser. Paste the FASTA sequence of P02489 in the "Sequence" field, select the "Output for all S/T residues" and "Yin-yang site predictions" buttons and click on the "Submit sequence" button. The resulting output is shown in Figure 12.11. NetPhos predicts eleven serine phosphorylation sites and three threonine phosphorylation sites. Among these sites, Ser-122 has one of the highest scores and is indeed shown to be phosphorylated. In contrast, Thr-13 and Ser-45, which are experimentally shown to be phosphorylated, had a very low probability of being phosphorylated according to the predictor. YinOYang predicts five sites with a high probability to be O-glycosylated. Among them, three also have a high probability of being phosphorylated. Among

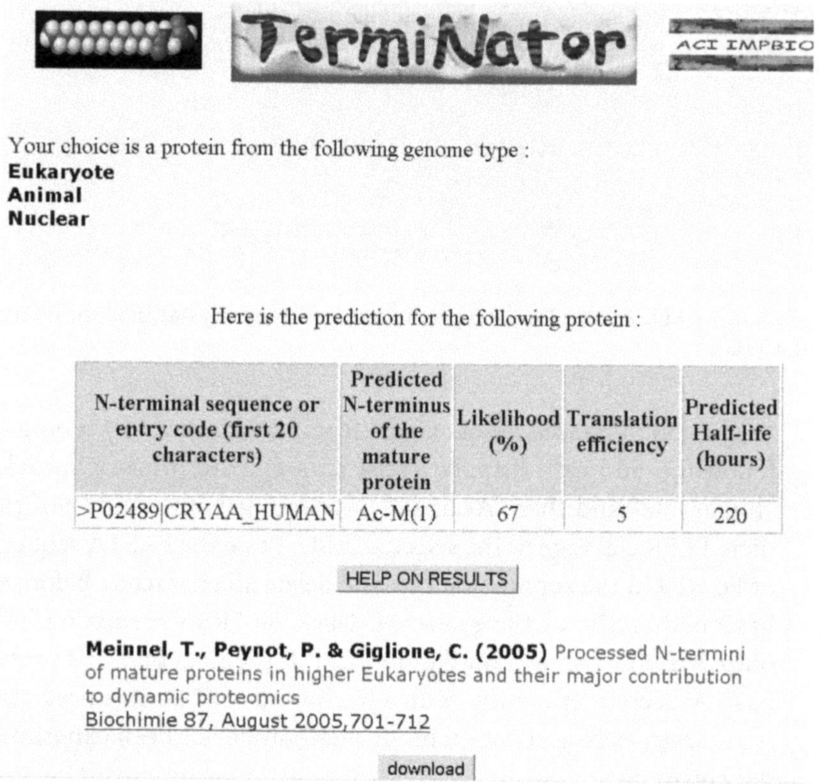
TermiNator
ACI IMPBIO
Your choice is a protein from the following genome type :
Eukaryote
Animal
Nuclear
Here is the prediction for the following protein :
N-terminal sequence or entry code (first 20 characters)
Predicted N-terminus of the mature protein
Likelihood (%)
Translation efficiency
Predicted Half-life (hours)
>P02489|CRYAA_HUMAN
Ac-M(1)
67
5
220
HELP ON RESULTS
Meinnel, T., Peynot, P. & Giglione, C. (2005) Processed N-termini of mature proteins in higher Eukaryotes and their major contribution to dynamic proteomics
Biochimie 87, August 2005,701-712
download

FIGURE 12.10 N-terminal acetylation prediction in human crystallin alpha chain A by TermiNator.

the predicted sites, Ser-162 was shown to be O-glycosylated in the sequence from the Rhesus macaque ortholog.

This example shows that PTM predictors can be efficiently used to predict some potential PTM sites. Depending on the considered protein, its taxonomy range, and subcellular location, many other prediction tools may be used. A (nonexhaustive) list can be found on http://www.expasy.org/tools/#ptm. However, these predictors may fail to predict real PTM sites (e.g., the two false-negative phosphorylation sites in the preceding example), and often predict many PTM sites that are probably not modified *in vivo* (e.g., the false-positive phosphorylation and glycosylation sites in the preceding example). Thus, results from PTM predictors should always be interpreted carefully, and each PTM site must be experimentally validated.

YinOYang 1.2 Prediction Results

(Click here for an explanation of the output)

The predictions for Yin-Yang sites in 1 sequence
(NetPhos threshold used: 0.5)

```
Name:  Sequence           Length:  173
MDVTIQHPWFKRTLGPFYPSRLFDQFFGEGLFEYDLLPFLSSTISPYYRQSLFRTVLDSGISEVRSDRDKFVIFLDVKHF      80
SPEDLTVKVQDDFVEIHGKHNERQDDHGYISREFHRRYRLPSNVDQSALSCSLSADGMLTFCGPKIQTGLDATHAERAIP     160
VSREEKPTSAPSS
...G..................................G....................Y..................      80
..........................................Y..........................................     160
.Y.....GY..GY
```

SeqName	Residue	O-GlcNAc result	Potential (o-glcnac)	Thresh. (1)	Thresh. (2)	NetPhos potential (Thresh=0.5)	YinOYang?
Sequence	4 T	++	0.5257	0.4070	0.4990		
Sequence	13 T	-	0.3263	0.3954	0.4834		
Sequence	20 S	-	0.2738	0.4440	0.5489		
Sequence	41 S	++	0.5842	0.4343	0.5358		
Sequence	42 S	-	0.4440	0.4440	0.5489		
Sequence	43 T	-	0.3071	0.4311	0.5314		
Sequence	45 S	-	0.2852	0.4315	0.5320		
Sequence	51 S	-	0.2799	0.4352	0.5370	0.817	
Sequence	55 T	-	0.4012	0.4567	0.5660		
Sequence	59 S	-	0.3472	0.4309	0.5312	0.950	
Sequence	62 S*	+	0.4403	0.4060	0.4977	0.931	*
Sequence	66 S	-	0.2363	0.3607	0.4366	0.997	
Sequence	81 S	-	0.1623	0.3684	0.4469	0.998	
Sequence	86 T	-	0.2529	0.3889	0.4746	0.813	
Sequence	111 S	-	0.2003	0.4401	0.5436	0.981	
Sequence	122 S	-	0.2598	0.3786	0.4607	0.997	
Sequence	127 S	-	0.3091	0.4050	0.4963		
Sequence	130 S*	++	0.5570	0.4328	0.5338	0.797	*
Sequence	132 S	-	0.3218	0.4316	0.5321		
Sequence	134 S	-	0.1841	0.4183	0.5142		
Sequence	140 T	-	0.2424	0.4719	0.5865	0.673	
Sequence	148 T	-	0.2468	0.3862	0.4709	0.740	
Sequence	153 T	-	0.3157	0.3969	0.4854		
Sequence	162 S*	+++	0.6382	0.3734	0.4537	0.997	*
Sequence	168 T	+++	0.6962	0.3040	0.3601		
Sequence	169 S*	+++	0.6494	0.3005	0.3555	0.857	*
Sequence	172 S	+++	0.5056	0.2794	0.3270		
Sequence	173 S*	+++	0.5651	0.2817	0.3301	0.755	*

Done

FIGURE 12.11 Prediction of O-glycosylation sites and phosphorylation sites in human crystallin alpha chain A by YinOYang.

Part III Sample Data

Human crystallin alpha (A chain) sequence in the FASTA format

```
>P02489|CRYAA_HUMAN Alpha crystallin A chain - Homo sapiens (Human).
MDVTIQHPWFKRTLGPFYPSRLFDQFFGEGLFEYDLLPFLSSTISPYYRQSLFRTVLDSG
ISEVRSDRDKFVIFLDVKHFSPEDLTVKVQDDFVEIHGKHNERQDDHGYISREFHRRYRL
PSNVDQSALSCSLSADGMLTFCGPKIQTGLDATHAERAIPVSREEKPTSAPSS
```

SECTION 3 PROTEIN–PROTEIN INTERACTIONS

Part I Introduction

1. Why Is Protein–Protein Interaction Important?

Regardless of function, proteins seldom act alone and are usually assembled into complexes and dynamic macromolecular structures to perform their task in the cell. Protein interactions are important for the smooth functioning of regulatory pathways that are crucial for cell survival. For example, the response to external stimuli through signaling cascades, the control of DNA replication and transcription, and the progression of a cell's cycle all require a dynamic interaction between proteins. Protein interactions are also essential for certain proteins to exert their particular cellular function. For example, human hemoglobin comprises a complex of four stable subunits to fix oxygen.

For a long time, it had only been possible to study proteins as isolated entities. More recently, genome-scale high-throughput techniques combined with novel computational tools have made it possible to study protein interactions on a network level.

2. How Do Proteins Interact?

Protein–protein interaction is characterized by a number of parameters. Apart from wondering which proteins interact together, protein–protein interaction itself is defined by its stoichiometry, its affinity, the kinetics of its formation and dissociation, and its particular structural features.

The stoichiometry of the interaction defines the number of each protein involved (e.g., monomer vs. dimer or multimer). Proteins of identical or nonidentical chains are usually named as homo- or heterooligomeric complexes, respectively. The affinity of the interaction defines the strength of binding. Quantitatively, the interaction strength between two proteins is characterized by the equilibrium constant Kd, with weak interactions in the mM range and strong interactions in the nM range or below. The kinetics of protein–protein interaction are defined by both the rate of association (k_{on}) and dissociation (k_{off}). Whereas the typical association rates are in the order of 10^5–10^6 M^{-1}sec^{-1} and do not vary much, the dissociation rate constant can vary considerably.

Structure-wise, protein–protein interactions are mediated mostly by physical contact made between protein domains. In the simplest scenario, two protein domains that are complementary in shape and charge may interact with no change in their conformation, as in the key-and-lock model. In other scenarios, interactions only occur following conformational changes (induced-fit model). Protein–protein interfaces appear to have a minimal contact surface of around 800 Å and an average surface of 1600 ± 400 Å. The interfaces have detectable properties such as biases in residue and motif composition. Several major forces are involved in the interaction. These include electrostatic forces, hydrogen bonds, van der Waals attractions, and hydrophobic effects. The energy of interaction is not evenly distributed between the two proteins at the interface, but is mostly contributed by hot spot residues.

Crudely, protein–protein interactions can be described either as stable or transient. Stable interactions are those associated with proteins that can be purified and eventually structurally analyzed as multisubunit complexes, e.g., ribosome, RNA polymerase II, and proteasome. Transient interactions, on the contrary, are temporary in nature and typically require a set of conditions that promote the interaction. They are responsible for the control of the majority of cellular processes, such as signal cascades, enzyme–substrate reactions, and transport. Both stable and transient interactions can be either strong or weak.

Inside the cell, a protein usually resides in a crowded environment with many potential binding partners with different surface properties. Apart from structural properties and energetic factors that determine which interactions may occur, other "environment"-dependent types of control are also present. These include (1) the colocalization of binding partners, in time and space; (2) local concentration of the proteins; and (3) the local physicochemical environment such as the presence of an effector molecule (e.g., ATP, Ca^{2+}) or a change in physiological conditions (e.g., changes in pH and temperature).

3. How Are the Interactions Determined Experimentally?

Traditionally, protein interactions have been studied in isolates using biochemistry-based methods such as protein–protein affinity chromatography, coimmunoprecipitation, and gel filtration. More recently, high-throughput techniques such as yeast two-hybrid and mass-spectrometry-based methods started to produce genome-size datasets. Protein interaction networks have already been proposed for yeast, *C. elegans*, fruit fly, and humans, using these techniques.

It is essential to note that each of these techniques (traditional or high-throughput) produces a distinct interaction dataset with respect to the nature and functional categories of interactions. Transient and direct binary interactions are best determined by yeast-two hybrid or fluorescence resonance energy transfer (FRET), which have the additional advantage of detecting actual *in vivo* interactions. Stable interactions, on the other hand, are best studied by coimmunoprecipitation and large-scale mass-spectrometry-based methods. These methods rely on the purification of complexes and cannot pinpoint direct physical interactions between the components constituting the complexes. The different experimental techniques also have their own limitations, and the interpretation of data is subject to caution. It is well known that yeast two-hybrid produces a high number of false-positives as well as false-negatives. An alternative method is usually required to confirm true interactions. Mass-spectrometry-based methods also have high false-positive rates, and are biased toward highly abundant stable complexes. They might also miss some complexes that are not formed under the experimental conditions.

Given this situation, it is hardly surprising that there are few overlaps between the datasets, and that it was estimated that genomewide screens might overlook between 20 and 80% of the interactions, depending on the species and the experiments. In addition to this incompleteness, it should also be noted that these experiments only report one aspect of protein–protein interaction, i.e., who interacts with who. What is more, it is a static and not a dynamic picture of the interactome that is given. Molecular details are also missing, and information such as the atomic description of the protein–protein interface can only be gathered by high-resolution 3D structures of interacting proteins and complexes.

4. In Silico Study of Protein Interaction: From Interaction Network to Structural Details

A protein–protein interaction network (or interactome) can be considered as the complete collection of all physical protein–protein interactions that can take place within a cell. Mathematically or computationally, this network is represented as a graph, where proteins are nodes and interactions between proteins are edges. Although such a representation does not reflect the true dynamic and complex nature of protein interactions, it serves to answer some fundamental questions, such as: "What is the global architecture of the network?"

Indeed, the major topological properties of protein interaction networks have been studied since the first large-scale experiments were published. It has been suggested by several groups that interaction networks are "scale-free" and present a "small-world" property. Scale-free networks are characterized by a few highly connected nodes (hubs) and many nodes with few connections. The distribution of the node degree k follows a power law. Although the exact biological consequences of these topological properties are not clear and there are still some controversies on this subject, the scale-free nature does explain why highly connected proteins (hubs) play an essential role in the cell and are vulnerable to attack by mutations. Network topology and connectivity have also been used in conjunction with other information (e.g., expression profiles) to derive methods to evaluate the accuracy of the experimentally determined interactions.

Apart from studying global network properties, an interaction network is useful to answer biological questions that were not addressable before. For example, how conserved is the protein–protein interaction network among the species? An answer to this could help identify evolutionarily conserved pathways, or even reconstruct networks of some less well-studied organisms by transferring experimentally verified interactions from one organism to another. Another interesting question to address is whether one can identify important subcomponents or functional modules within a network. Clearly, a protein interaction network as a whole can be too complex to study. Highly interconnected (or clustered) regions in networks have been suggested to represent groups of proteins with similar cellular functions. Most often they are stable protein complexes, but transient protein interactions involved in pathways can be present as well. Identification of these functional modules can prove useful for annotating uncharacterized proteins, studying evolution, and uncovering new pathways. A variety of supervised or unsupervised clustering methods have been employed to identify these modules. It remains to be seen if the different methods provide consistent results. Alternatively, it is possible to use current biological knowledge to map well-documented signaling, or metabolic pathways to the whole network. This approach can be useful for finding relationships between known pathways, although — currently — an exact match is not always found.

The global analysis of network features, although useful, is not sufficient to provide molecular details of individual constituents and the mechanistic aspects of their interactions. For this, knowledge of 3D structures is essential. Many physical interactions between domains are conserved through evolution, regardless of the proteins harboring the domains. Many studies have thus focused on identifying domain–domain interactions. Domain–domain interaction networks have been constructed, and this offers an alternative view on network organization. At an even more atomic level, several groups have recently exploited the increase in the number of high-resolution 3D structures in the Protein Data Bank (PDB) to systematically investigate the features of protein–protein interfaces. In general, binding sites share common properties that distinguish them from the rest of the proteins. Interfaces tend to be planar, and residue composition can be different between transient and obligate complexes. Numerous other geometrical, chemical, and energetic features such as shape complementarities and crystal packing have also been used to distinguish different types of interfaces. From an evolutionary point of view, residues in protein–protein interfaces appear to be more conserved, although this observation is not universally agreed upon. Apart from general interface analysis, various investigators have identified and studied energetic hot spots in protein interfaces using different approaches.

5. *How About Prediction?*

There are many prediction challenges in the field of protein–protein interactions. The most obvious one is to predict interacting partners. A number of different strategies have been proposed. They use information either from the genetic environment, phylogenetics, or protein structures. Genome-context-based methods propose interactions between proteins for which there is evidence of an association. This association can be either based on the relative position of one protein to the other in known genome sequences, or on similar expression profiles. For example, the gene fusion of "Rosetta stone" method relies on the observation that some pairs of interacting proteins have homologs fused to a single protein chain in another organism. For phylogenetic-based methods, such as phylogenetic profiling, tree similarity and clusters of orthologous proteins, the basic assumption is that proteins that interact are likely to coevolve. Structure-based methods — either domain-based or sequence/structural features-based — rely on the existent knowledge of protein domain interactions or specific characteristics of protein interfaces to perform predictions.

It has been shown that the accuracy of protein–protein interaction prediction is comparable to those of large-scale experiments. Prediction data can therefore complement high-throughput experimental data that are still far from complete.

Apart from predicting interacting partners, progress has been made in the prediction of actual binding sites involved in protein interactions. A communitywide evaluation of different methods used in the prediction of protein–protein interactions and protein docking (Critical Assessment of Prediction Interaction, or CAPRI) has been carried out annually since 2001. However, although significant advances have been observed in protein docking, the prediction of protein–protein interaction sites is still far from satisfactory and has been the subject of intense studies in recent years. In fact, no single property absolutely differentiates protein–protein interfaces from other surface patches. Therefore, most binding site prediction methods combine more than one physiochemical property. The most effective of these methods make extensive use of structural information, and evolutionary information is also sometimes included.

6. Can Protein–Protein Interaction Be Used to Predict Protein Function?

It has been observed that the majority of the interacting proteins (up to 70–80%) share at least one function. This property has been exploited to assign a function to an uncharacterized protein based on the functions of its characterized binding partners. The major disadvantage of this simple method is that predictions are limited to proteins that have at least one interaction partner with a known function. Recently, several groups have further explored the use of functional modules in protein networks and indirect interacting partners for function prediction.

Clearly, annotating protein function using information from protein–protein interaction is a promising technique that may become more useful when the available datasets are more complete and accurate. This approach can be integrated with other methods, such as those focusing on mRNA expression profiles or evolutionary data.

7. Resources on Protein–Protein Interaction

Resources on protein–protein interaction can be classified into three categories: databases, analysis, and visualization tools. It is more and more common, however, to find databases that also offer prediction, analysis, or visualization tools.

There is a plethora of databases that contain information on protein–protein interactions (Table 12.2). Most of the databases have user-friendly interfaces and provide good search engines. Interactions can usually be searched by gene or protein name. A list of interactors is usually given, together with their functional annotation or experimental details when available.

The databases are different in a number of ways. Databases can store either experimental (e.g., DIP, BIND, MINT, IntAct), predicted (e.g., InterDom, OPHID), or both types of interaction data (e.g., STRING). They

TABLE 12.2 Resources on Protein–Protein Interactions

Resource Name	URL	Remarks
	PPI Databases	
IntAct	www.ebi.ac.uk/intact/	Experimental data, curated
DIP	dip.doe-mbi.ucla.edu/	Experimental data
MINT	mint.bio.uniroma2.it/mint/Welcome.do	Experimental data, curated
BOND	bond.unleashedinformatics.com/Action?	This new database includes BIND
MIPS mammalian PPI db	mips.gsf.de/proj/ppi/	Curated, mammals
BioGRID	www.thebiogrid.org/	Experimental data, curated
OPHID	ophid.utoronto.ca/ophid/	Includes predicted data
HPRD	www.hprd.org/	Experimental data, curated, human specific
STRING	String.embl.de	Experimental data, prediction, functions
	Domain–Domain Interaction Database	
PIBASE	alto.compbio.ucsf.edu/pibase/	Information on interface
3did	gatealoy.pcb.ub.es/3did/	
PSIBase	psibase.kobic.re.kr/	
Ipfam	www.sanger.ac.uk/Software/Pfam/iPfam/	
InterDom	interdom.lit.org.sg/	
InterPare	interpare.net/	Information on interface
SCOWLP	www.scowlp.org/	Information on interface
PRISM	Gordon.hpc.eng.ku.edu.tr/prism/H	Information on interface
	Network Analysis	
NetAlign	www1.ustc.edu.cn/lab/pcrystal/NetAlign/	Pathways comparison
MAVisto	mavisto.ipk-gatersleben.de/	Exploration of network motifs
ToPNET	networks.gersteinlab.org/genome/interactions/networks/core.html	Topology analysis
FANMOD	www.minet.uni-jena.de/~wernicke/motifs/	Fast network motif detection
	Interface Analysis	
ProFace	www.boseinst.ernet.in/resources/bioinfo/stag.html	
NoxClass	Noxclass.bioinf.mpi-inf.mpg.de	Prediction of protein–protein interaction type
MolSurfer	Projects.villa-bosch.de/dbase/molsurfer	Compute parameters such as the distribution of electrostatic potential

TABLE 12.2 Resources on Protein–Protein Interactions

Resource Name	URL	Remarks
Computational alanine scanning	http://robetta.bakerlab.org/alascansubmit.jsp	Identify hot spot residues
	Visualization Programs	
Cytoscape	www.cytoscape.org	Open source, with lots of plug-ins offered
VisANT	visant.bu.edu	
Osprey	biodata.mshri.on.ca/osprey/servlet/Index	
WebInterViewer	interviewer.inha.ac.kr/	

can also record protein–protein interactions (e.g., DIP, BIND), or structural domain–domain interactions (e.g., 3did, PiBase, iPfam, PSIbase, InterPare, PRISM). Some databases contain manually curated data (e.g., IntAct), whereas others are species specific (e.g., HPRD). There are also databases that incorporate genomic data to provide novel insights into protein–protein interaction networks (e.g., STRING). In order to facilitate data exchange or comparison between this diverse set of databases, the HUPO Protein Standard Initiative (PSI) is developing a standard data format for the representation of protein interaction data. The standard has already been adopted by numerous databases belonging to the International Molecular Exchange Consortium (IMEX).

Apart from databases, protein–protein interaction data can also be found in the literature. PreBIND (http://prebind.bind.ca/), a data-mining tool, locates information on interactions in scientific articles.

In terms of analysis tools, a lot of novel methodologies have been published in recent years, covering areas from global network analysis to the detection of hot spots in protein interfaces. Not all these tools, however, are online and made available to the public. Table 12.2 lists some of the resources known to the authors. The list is not exhaustive. Users should also bear in mind that this field is still in active development. Therefore, tools may not all have been properly benchmarked or compared to one another.

For network visualization, most databases now provide their own tool to visualize protein interactions as a graph (e.g., IntAct, MINT). Several specialized visualization tools are also available and may offer additional features (Table 12.2). Among these tools, Cytoscape is an open-source software project for visualizing and analyzing biomolecular interaction network data. Besides its software core that provides basic functionalities to display and query the network, the tool is extensible through a plug-in architecture that allows contribution from the research community to add additional computational features. For example, it is possible to integrate gene expression profiles for interaction data analysis (Dynamic Expression Plugin), or to determine clusters present in the network (MCODE).

Part II Step-By-Step Tutorial

In this tutorial, we will retrieve information on protein–protein interaction from the open-source database IntAct. The data will then be visualized and analyzed using Cytoscape. It should be reminded that (1) the protein–protein interaction networks determined for different organisms are still incomplete, and (2) there are few overlaps between the datasets, depending on the experimental method used (e.g., yeast two-hybrid vs. mass-spectrometry-based methods). Moreover, many analysis tools are still currently under development, and their performance will certainly improve as novel algorithms emerge and the available datasets become more complete. For these reasons, it is not the intention of the authors to demonstrate a standard or best way to analyze protein–protein interactions, but to expose the users to the different possibilities while focusing on function annotation. Users, by knowing the resources available and their pros and cons, should not hesitate to explore the different tools and identify the most suitable resources for their specific needs.

1. How to Find out if the Protein of Interest Has PPIs.

In this part, we will try to gather information for the Swiss-Prot protein Q96CG3_HUMAN (UniProtKB AC: Q96CG3). At the time of the preparation of this chapter (Oct 2006), this protein was still largely uncharacterized, with no function and GO annotation.

A. Search the IntAct Database First, go to the IntAct Web site at http://www.ebi.ac.uk/intact/site/index.jsp. Perform the search using UniProtKB accession number Q96CG3.

A result table is shown (Figure 12.12a) with the query protein highlighted in red and its interacting proteins listed in the following text. You will find that protein Q96CG3 interacts with seven different proteins. The "Number of interactions" column indicates the number of reported interactions that involve the protein of interest. Seven interactions are stated here. Click on the hyperlink to obtain the list of the interactions. One can see that each interaction is characterized by a name, an accession number, a description, and an experiment. In this case, all seven interactions are

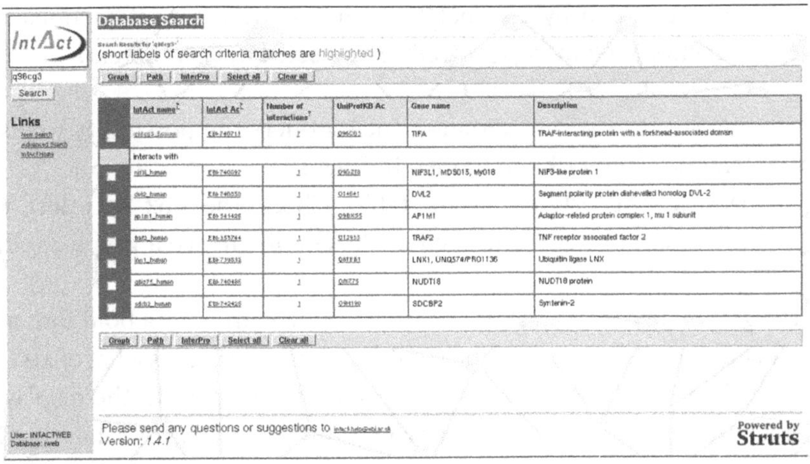
IntAct
Database Search
(short labels of search criteria matches are highlighted)
q96cg3
Search
Links
Please send any questions or suggestions to
Version: 1.4.1
Powered by Struts

FIGURE 12.12A IntAct search result for protein Q96CG3.

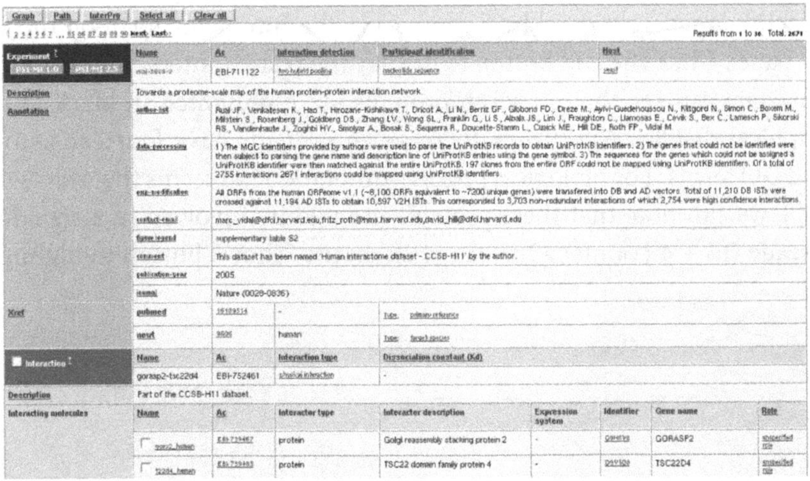

FIGURE 12.12B The experiment summary page for the experiment set rual-2005-2.

obtained from the experiment rual-2005-2. Click on one of the interaction names (e.g., ap1m1-tifa), and this will lead you to the experimental summary page (Figure 12.12b).

The experimental summary page provides all the annotated experimental data extracted from the source literature or submission. It can be clearly seen that IntAct stores an interaction in the context of the experiment that originally describes the interaction. For the experiment rual-2005-2, it records a large-scale yeast two-hybrid analysis with 2671 different interactions. All seven interactions involving Q96CG3 are detected in this context.

B. HierachView of the Interactions IntAct offers the possibility to visualize interaction networks contained in IntAct.

Go back to the original result table. Check the box to select the Q96CG3_human protein, and then click "Graph" to see a graphical display (Figure 12.13).

The graphical display shows the specified protein with a bold font and its immediate interactors. On the right-hand panel, there is a list of all GO and InterPro terms associated with the specified protein in the displayed interaction network. In this case, protein Q96CG3 does not have a GO annotation but contains an FHA domain.

To get an idea about the properties of all the seven interacting proteins of Q96CG3, we can expand the network by typing the list of seven protein names (separated by a comma) in the Interactor form on the top left panel. Click the add button to display the related interaction network. Now, one can see that all seven proteins are highlighted in black in the graphical display. Altogether, these seven proteins have fourteen associated GO terms. For biological process annotation (all terms that start with "P:"), we can note that these proteins are mainly involved in a signaling cascade (16 + 4 counts). One can click on the "show" button to highlight

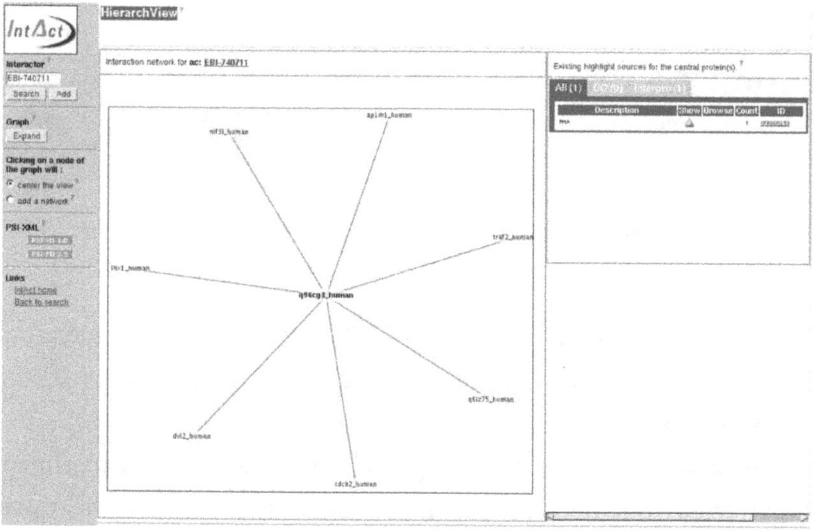
HierarchView
IntAct
Interactor
EBI-740711
Search
Add
Graph
Expand
Clicking on a node of the graph will :
center the view
add a network
PSI-XML
Links
IntAct home
Back to search
Interaction network for ac: EBI-740711
Existing highlight sources for the central protein(s)
All (1)
Description
Show
Browse
Count
ID
q96cg3_human
traf2_human
cdcb2_human
dvl2_human

FIGURE 12.13 Graphical display of proteins interacting with Q96CG3.

the proteins associated with one particular term in red. By clicking on the hierarchy icon, the related GO hierarchy will appear in the right lower panel. Try to browse and see which proteins share common GO or InterPro annotations.

C. Others IntAct offers a large range of other options to facilitate the analysis of protein interactions. Owing to space limitations, it is impossible to cover all these options here. Users are required to refer to their documentation for further details.

To download interaction data, users can either go to the download site (ftp://ftp.ebi.ac.uk/pub/databases/intact/current) or obtain data directly from the experiment summary page (Figure 12.12b). The latter option is especially useful for retrieving data from individual large-scale experiments.

2. Analyze a Protein–Protein Interaction Network

In this part, we will use Cytoscape version 2.3.2 to analyze global network architecture and detect functional modules in a protein–protein interaction network. This part is slightly more difficult, and more time will be required. For users interested in the structural aspect of the interaction, they can go directly to the section titled "Structural Analysis of Protein–Protein Interface."

A. Dataset A human protein–protein interaction network (Rual et al., 2005) will be used for demonstration. This dataset includes a set of interactions determined by a high-throughput yeast two-hybrid experiment, data collected from literature, and a number of coimmunoprecipitation results aiming to verify the yeast two-hybrid data. The dataset is available at http://www.cytoscape.org/cgi-bin/moin.cgi/Data_Sets/.

(Note: This combined dataset is different from the two datasets recorded in IntAct, rual-2005-1 (EBI-710837) and rual-2005-2 (EBI-711122), which contain the coimmunoprecipitation and the yeast two-hybrid data, respectively.)

To start, please install the latest version of Cytoscape onto your computer by going to their Web page (http://www.cytoscape.org/download_list.php) and following the installation instructions. Once you have verified that it works, proceed to obtain the three plug-ins at http://www.cytoscape.org/plugins2.php:

- NetworkAnalyzer plugin (Note: To use this plug-in in Cytoscape 2.3, you will need to download two additional files jfreechart-0.9.20.

jar and jfreechart-common-0.9.5.jar from http://med.bioinf.mpi-inf.mpg.de/netanalyzer/faq.php.)

- MCODE plugin
- BINGO plugin

All the plug-ins should be saved in the [Cytoscape_Home]/plugins directory. Cytoscape should now be closed and restarted. The installed plug-ins should be visible from the Plugins menu on top (Figure 12.14).

Cytoscape has a comprehensive online tutorial (http://www.cytoscape.org/tut/tutorial.php). Readers are encouraged to go through the "Getting started" section first. In the following, only major steps are described, and readers are referred to the relevant tutorial sessions provided by Cytoscape for more detailed step-by-step instruction.

B. Global Network Architecture In this part, we will analyze the global network architecture:

1. Load the RUAL.sif network and the node attribute file RUAL.na into Cytoscape.

   Hint: Please refer to the online "Basic tutorial."

2. Filter the dataset so as to exclude the interactions determined by coimmunoprecipitation (coAP), and keep only those determined by the yeast two-hybrid experiment (Y2H) and from the literature (core, noncore, and hypercore).

   Hint: Please refer to the online tutorial "Filters and Editor."

3. Save the filtered network as a new network.

   Hint: Select edges with the filter "Edge:interact ~*" and save the new network by using the File menu bar (File → New → Network → From selected nodes, selected edges).

4. Obtain information about the resultant network by using the plugin NetworkAnalyzer, from the Plugins menu bar (Plugins → Network Analyzer → Analyze Network).

You will obtain the result icon as depicted in Figure 12.14. Note that the human protein–protein interaction network displays a typical scale-free topology, i.e., a few highly connected nodes (nodes with a large degree)

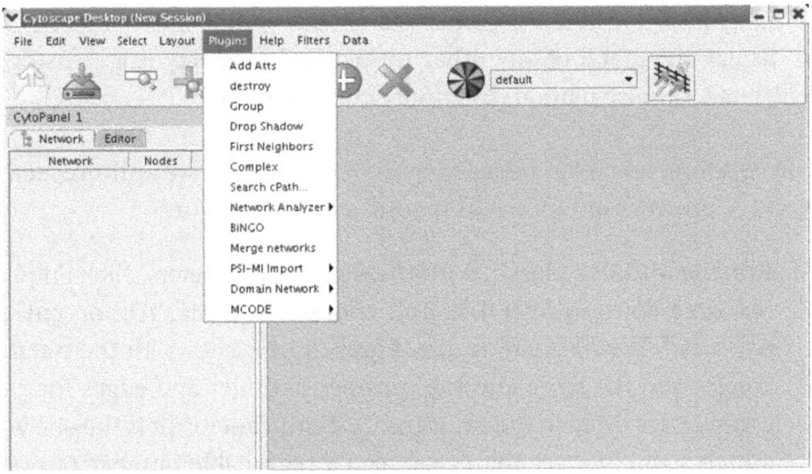
Cytoscape Desktop (New Session)
File Edit View Select Layout Plugins Help Filters Data
Add Atts
destroy
Group
Drop Shadow
First Neighbors
Complex
Search cPath...
Network Analyzer
BiNGO
Merge networks
PSI-MI Import
Domain Network
MCODE
default
CytoPanel 1
Network Editor
Network Nodes

FIGURE 12.14A Visualizing network properties using Cytoscape. The main Cytoscape frame.

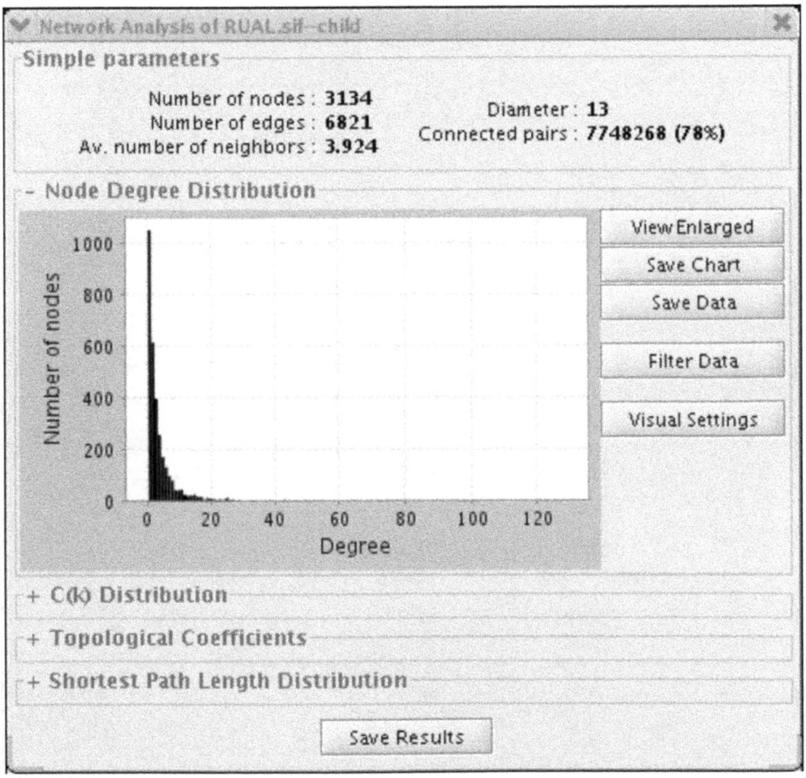
Network Analysis of RUAL.sif--child
Simple parameters
Number of nodes : 3134
Number of edges : 6821
Av. number of neighbors : 3.924
Diameter : 13
Connected pairs : 7748268 (78%)
- Node Degree Distribution
Number of nodes
1000
800
600
400
200
0
0 20 40 60 80 100 120
Degree
View Enlarged
Save Chart
Save Data
Filter Data
Visual Settings
+ C(k) Distribution
+ Topological Coefficients
+ Shortest Path Length Distribution
Save Results

FIGURE 12.14B Visualizing network properties using Cytoscape. Network statistics provided by the NetworkAnalyzer plugin.

and many nodes with few connections (nodes with a small degree). One can display the graph of any other topological property (e.g., clustering coefficient C(k) distribution) by simply clicking on it.

C. Detect Functional Modules Using MCODE Now, we will proceed to detect the presence of functional modules in this network:

1. Run the MCODE plug-in with the default parameters: click on plug-ins, drag down to MCODE, and choose "Run MCODE on current network." The MCODE result (Figure 12.15) shows all the putative complexes. The score and the number of nodes and edges for each complex are listed in the columns. A significant result is the one with a high score (greater than one) and a reasonable number of nodes

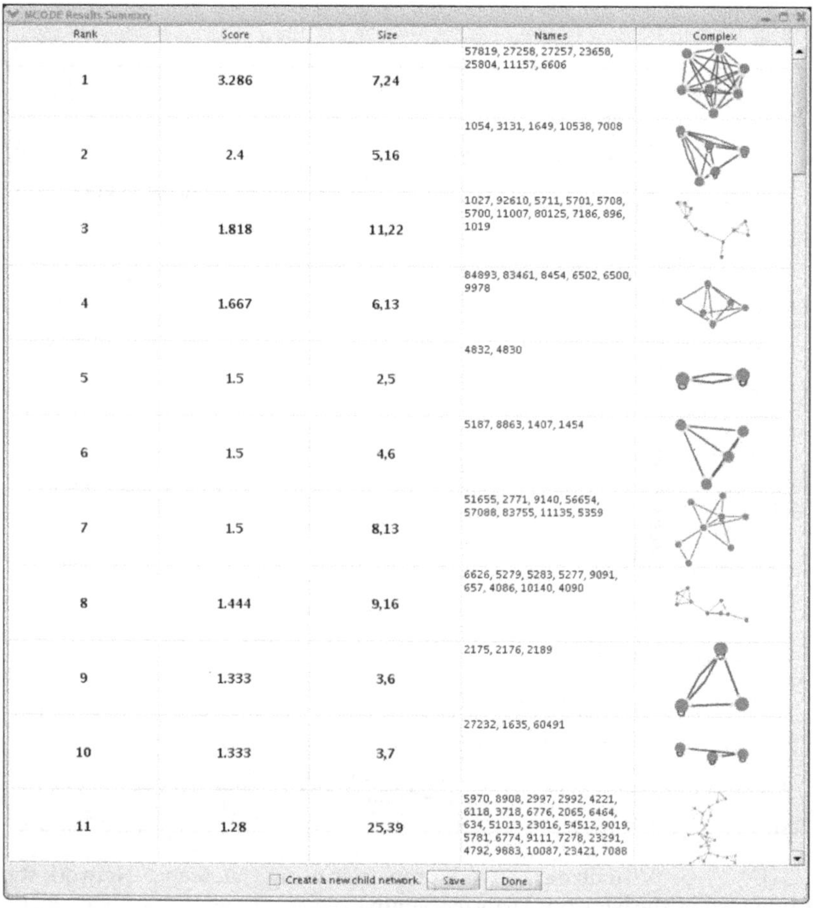
MCODE Results Summary
Rank
Score
Size
Names
Complex
1
3.286
7,24
57819, 27258, 27257, 23658, 25804, 11157, 6606
2
2.4
5,16
1054, 3131, 1649, 10538, 7008
3
1.818
11,22
1027, 92610, 5711, 5701, 5708, 5700, 11007, 80125, 7186, 896, 1019
4
1.667
6,13
84893, 83461, 8454, 6502, 6500, 9978
5
1.5
2,5
4832, 4830
6
1.5
4,6
5187, 8863, 1407, 1454
7
1.5
8,13
51655, 2771, 9140, 56654, 57088, 83755, 11135, 5359
8
1.444
9,16
6626, 5279, 5283, 5277, 9091, 657, 4086, 10140, 4090
9
1.333
3,6
2175, 2176, 2189
10
1.333
3,7
27232, 1635, 60491
11
1.28
25,39
5970, 8908, 2997, 2992, 4221, 6118, 3718, 6776, 2065, 6464, 634, 51013, 23016, 54512, 9019, 5781, 6774, 9111, 7278, 23291, 4792, 9883, 10087, 23421, 7088
Create a new child network.
Save
Done

FIGURE 12.15 Functional modules identified by the MCODE plugin.

and edges. One of the properties of a functional module is that most of the time, all members of the complex share a similar function. To analyze the members of a complex and to see if they share common GO annotations, we will use the plug-in BINGO.

2. In the MCODE results summary frame, check the box "Create a new child network" and click on the first cluster. A new frame will appear that shows only proteins belonging to the first clusters.
3. Select all the nodes in this frame, and select BINGO from the Plugins menu. A dialog box appears. Fill in a cluster name (e.g., cluster 1), check "Get Cluster from Network." Leave the parameters at their default value. Select organism "*Homo sapiens*" and check "Entrez Gene ID." This is because we are analyzing a human network and the ID provided for the proteins is EntrezGeneID. Start BINGO. A graph (Figure 12.16) appears. Note that GO terms related to RNA splicing are significantly enriched (red color) in this cluster. Repeat the operation for clusters 2 to 4. You will find that cluster 2 has no

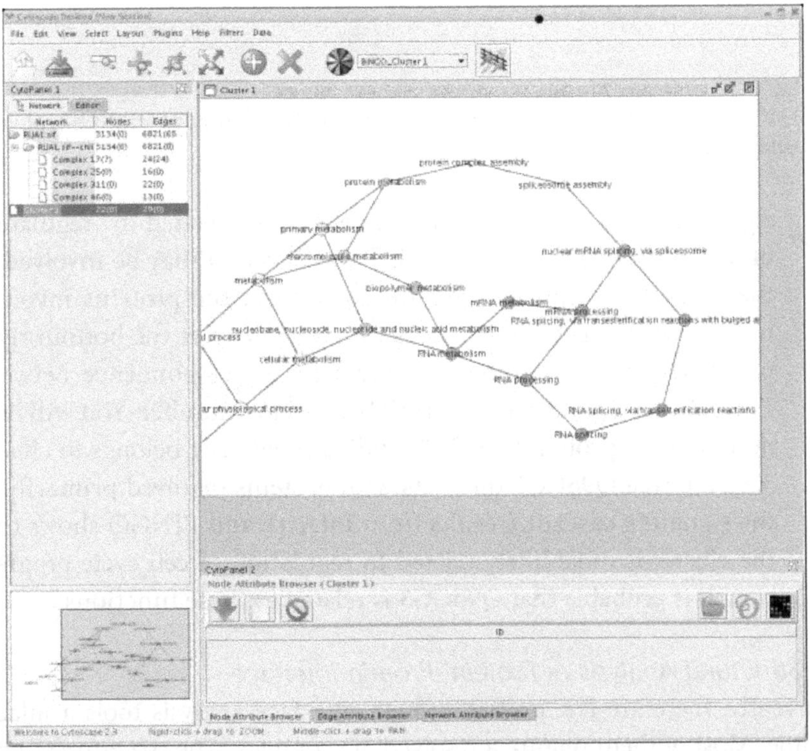
Cluster 1
protein complex assembly
protein metabolism
spliceosome assembly
primary metabolism
macromolecule metabolism
nuclear mRNA splicing, via spliceosome
metabolism
biopolymer metabolism
mRNA metabolism
mRNA processing
RNA splicing, via transesterification reactions with bulged a
nucleobase, nucleoside, nucleotide and nucleic acid metabolism
cellular metabolism
RNA metabolism
RNA processing
RNA splicing, via transesterification reactions
RNA splicing

FIGURE 12.16A Enriched GO annotations identified by the BINGO plugin Results for cluster 1.

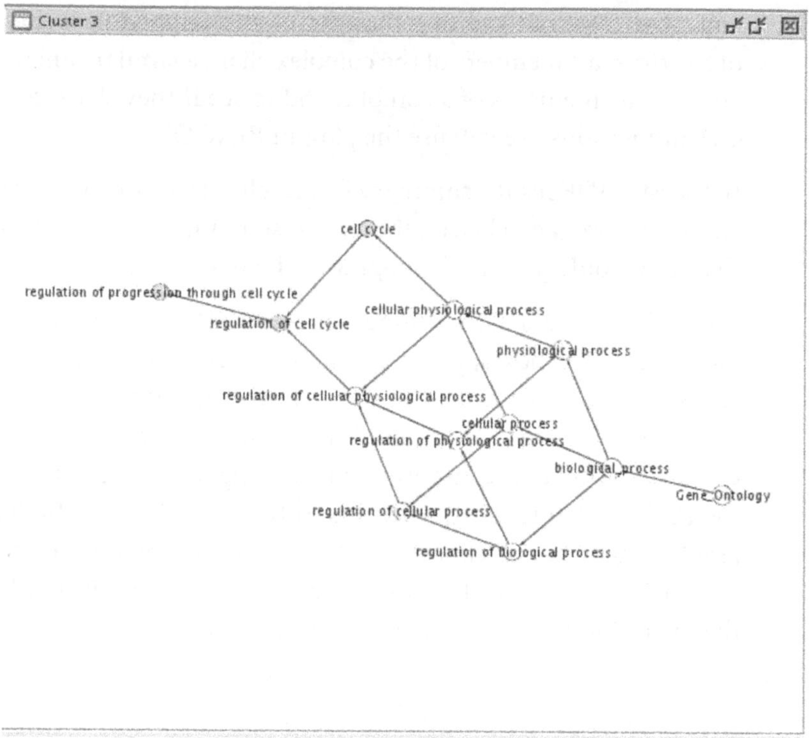
Cluster 3
cell cycle
regulation of progression through cell cycle
regulation of cell cycle
cellular physiological process
physiological process
regulation of cellular physiological process
cellular process
regulation of physiological process
biological process
Gene Ontology
regulation of cellular process
regulation of biological process

FIGURE 12.16B Enriched GO annotations identified by the BINGO plugin. Results for cluster 3.

significant GO terms, cluster 3 appears to be related to "regulation of progression through cell cycle," and cluster 4 may be involved in the ubiquitin cycle. To learn the identity of the proteins involved in each cluster, use the Node Attribute Browser at the bottom right panel (Cytopanel2). Table 12.3 gives the correspondence between EntrezGene ID and the UniProtKB accession number. You will now note that the protein Q96CG3 studied previously belongs to cluster 3. Given that Q96CG3 interacts with proteins involved primarily in the signaling cascade (results from IntAct), and BINGO shows that the cluster is most likely related to regulation of cell cycle progression, it is probable that Q96CG3 is related to these functions.

3. Structural Analysis of Protein–Protein Interface

Currently, there are few protein-protein interface analysis tools available online. Most of them require a resolved 3D complex. The structure of protein Q96CG3 is not solved. Although a homology model is available in Mod-

TABLE 12.3 Mapping of EntrezGene ID to UniProtKB ID and AC for the Four Best-Scoring Clusters Identified by MCODE

EntrezGene ID	UniProtKB ID	UniProtKB AC
	Cluster 1	
6606	SMN_HUMAN	Q16637
25804	LSM4_HUMAN	Q9Y4Z0
27258	LSM3_HUMAN	P62310
11157	LSM6_HUMAN	P62312
27257	LSM1_HUMAN	O15116
57819	LSM2_HUMAN	Q9Y333
23658	LSM5_HUMAN	Q9Y4Y9
	Cluster 2	
3131	HLF_HUMAN	Q16534
10538	BATF_HUMAN	Q16520
1054	CEBPG_HUMAN	P53567
1649	DDIT3_HUMAN	P35638
7008	TEF_HUMAN	Q10587
	Cluster 3	
7186	TRAF2_HUMAN	Q12933
11007	DIPA_HUMAN	Q15834
5711	PSMD5_HUMAN	Q16401
5701	PRS7_HUMAN	P35998
1019	CDK4_HUMAN	P11802
1027	CDN1B_HUMAN	P46527
5708	PSMD2_HUMAN	Q13200
896	CCND3_HUMAN	P30281
80125	Q8TAX6_HUMAN	Q8TAX6
92610	Q96CG3_HUMAN	Q96CG3
5700	PRS4_HUMAN	P62191
	Cluster 4	
6500	SKP1_HUMAN	P63208
83461	Q99618_HUMAN	Q99618
84893	FBX18_HUMAN	Q8NFZ0
9978	RBX1_HUMAN	P62877
8454	CUL1_HUMAN	Q13616
6502	SKP2_HUMAN	Q13309

Base (http://modbase.compbio.ucsf.edu/), the sequence homology between Q96CG3 and the template used (PDB code: 1UHT) is obviously too low (13%) for the model to be trusted. Therefore, in this section, we will use the complex between TRAF2 and TRADD (PDB code: 1F3V) as an example:

1. Go to Protein Data Bank (PDB) (www.rcsb.org/pdb) to download the coordinate file for the structure 1F3V.

This structure is composed of two chains. Chain A represents the TRAF2 binding domain of TRADD (residues 7–163), and chain B represents the TRAF domain of human TRAF2 (residues 331–501).

2. Go to ProFace (www.boseinst.ernet.in/resources/bioinfo/stag.html), and upload the PDB file 1F3V.

   A result page will appear that contains four sections: (1) graphical plots of interface residues; (2) statistics of interface parameters, including interface area, residues and atom composition; (3) downloadable files, including the coordinates of interface atoms as well as the PDB files in which the interface residues are tagged, a list of neighboring residues (within 4.5 Å); and (4) a structure viewer, either Rasmol or Chime.

3. Click on "To simply identify interface residues." Choose to display interface information for Chain A by clicking "Chain A." A new window will open, showing a graphical plot of interface residues and secondary structure (Figure 12.17a).

4. Display the information for Chain B. Choose To identify interface residues, dissected into patches and core/rim" this time (Figure 12.17b). If we compare the result for Chain A (TRAF2 binding domain of TRADD) to the published data (Park et al., 2000), we will find that the program successfully identifies all the residues involved

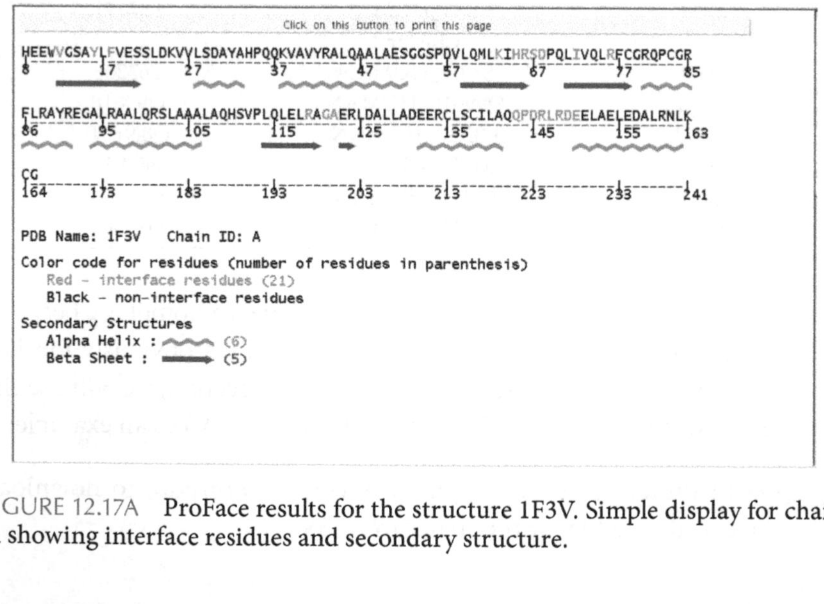
Click on this button to print this page
HEEWVGSAYLFVESSLDKVVLSDAYAHPQQKVAVYRALQAALAESGGSPDVLQMLKIHRSDPQLIVQLRFCGRQPCGR
FLRAYREGALRAALQRSLAAALAQHSVPLQLELRAGAERLDALLADEERCLSCILAQQPDRLRDEELAELEDALRNLK
CG
PDB Name: 1F3V Chain ID: A
Color code for residues (number of residues in parenthesis)
Red - interface residues (21)
Black - non-interface residues
Secondary Structures
Alpha Helix : (6)
Beta Sheet : (5)

FIGURE 12.17A ProFace results for the structure 1F3V. Simple display for chain A showing interface residues and secondary structure.

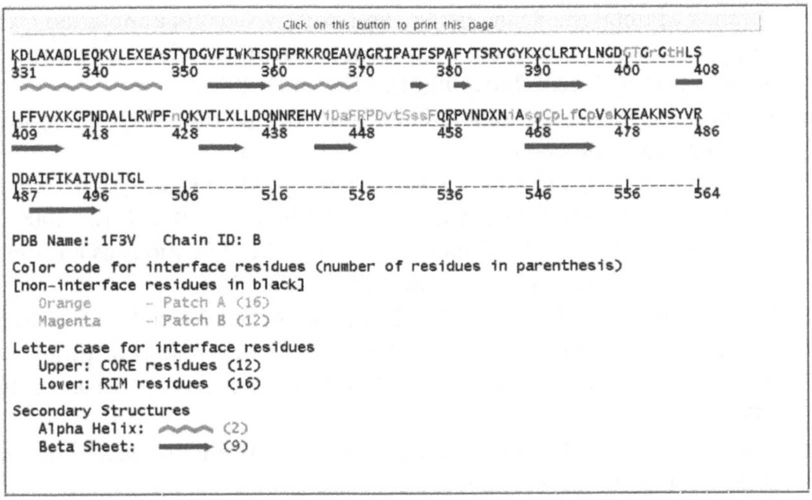
Click on this button to print this page
PDB Name: 1F3V Chain ID: B
Color code for interface residues (number of residues in parenthesis)
[non-interface residues in black]
Orange - Patch A (16)
Magenta - Patch B (12)
Letter case for interface residues
Upper: CORE residues (12)
Lower: RIM residues (16)
Secondary Structures
Alpha Helix: (2)
Beta Sheet: (9)

FIGURE 12.17B ProFace results for the structure 1F3V. Alternative display for chain B where interface residues were further divided into rim and core residues. Distinct patches were shown as well.

in interactions. Three residues, namely, V12, R119, and E150, were not mentioned in the paper as belonging to the interface. They are, however, in close proximity to the interface.

Part III Sample Data

All the sample data are available online.

The human protein–protein interaction network, RUAL.sif and RUAL.na: http://www.cytoscape.org/cgi-bin/moin.cgi/Data_Sets.

The PDB file 1F3V: www.rcsb.org/pdb.

REFERENCES

1. Boeckmann, B., Blatter, M.C., Famiglietti, L., Hinz, U., Lane, L., Roechert, B., and Bairoch, A. Protein variety and functional diversity: Swiss-Prot annotation in its biological context. *C R Biol* 328(10–11): 882–899, 2005.
2. Gardy, J.L. and Brinkman, F.S. Methods for predicting bacterial protein subcellular localization. *Nat Rev Microbiol* 4(10): 741–751, 2006.
3. Watson, J.D., Laskowski, R.A., and Thornton, J.M. Predicting protein function from sequence and structural data. *Curr Opin Struct Biol* 15(3): 275–284, 2005.
4. Suresh, S., Sujatha Mohan, S., Mishra, G., Hanumanthu, G.R., Suresh, M., Reddy, R., and Pandey, A. Proteomic resources: integrating biomedical information in humans. *Gene* 364: 13–18, 2005.

5. Ofran, Y., Punta, M., Schneider, R., and Rost, B. Beyond annotation transfer by homology: novel protein-function prediction methods to assist drug discovery. *Drug Discov Today* 10(21): 1475–1482, 2005.
6. Valencia, A. Automatic annotation of protein function. *Curr Opin Struct Biol* 15(3): 267–274, 2005.
7. Jensen, L.J., Saric, J., and Bork, P. Literature mining for the biologist: from information retrieval to biological discovery. *Nat Rev Genet* 7(2): 119–129, 2006.
8. Garavelli, J.S. The RESID Database of Protein Modifications as a resource and annotation tool. *Proteomics* 4(6): 1527–1533, 2004.
9. Farriol-Mathis, N., Garavelli, J.S., Boeckmann, B., Duvaud, S., Gasteiger, E., Gateau, A., Veuthey, A.L., and Bairoch, A. Annotation of post-translational modifications in the Swiss-Prot knowledge base. *Proteomics* 4(6): 1537–1550, 2004.
10. Peterson, C.L. and Laniel, M.A. Histones and histone modifications. *Curr Biol* 14(14): R546–R551, 2004.
11. Santos-Rosa, H. and Caldas, C. Chromatin modifier enzymes, the histone code and cancer. *Eur J Cancer* 41(16): 2381–2402, 2005.
12. Hjerrild, M. and Gammeltoft, S. Phosphoproteomics toolbox: computational biology, protein chemistry and mass spectrometry. *FEBS Lett* 580(20): 4764–4770, 2006.
13. Blom, N., Sicheritz-Ponten, T., Gupta, R., Gammeltoft, S., and Brunak, S. Prediction of post-translational glycosylation and phosphorylation of proteins from the amino acid sequence. *Proteomics* 4(6): 1633–1649, 2004.
14. Wodak, S.J. and Janin, J. Structural basis of macromolecular recognition. *Adv Protein Chem* 61: 9–73, 2002.
15. Nooren, I.M.A. and Thornton, J.M. Diversity of protein-protein interactions. *EMBO J* 22(14): 3486–3492, 2003.
16. Causier, B. Studying the interactome with the yeast two-hybrid system and mass spectrometry. *Mass Spectrom Rev* 23(5): 350–367, 2004.
17. Chen, Y. and Xu, D. Computational analyses of high-throughput protein-protein interaction data. *Curr Protein Peptide Sci* 4: 159–181, 2003.
18. Titz, B., Schlesner, M., and Uetz, P. What do we learn from high-throughput protein interaction data? *Expert Rev Proteomics* 1(1): 111–121, 2004.
19. Russell, R.B., Alber, F., Aloy, P., Davis, F.P., Korkin, D., Pichaud, M., Topf, M., and Sali, A. A structural perspective on protein-protein interactions. *Curr Opin Struct Biol* 14: 313–324, 2004.
20. Hishigaki, H., Nakai, K., Ono, T., Tanigami, A., and Takagi, T. Assessment of prediction accuracy of protein function from protein–protein interaction data. *Yeast* 18(6): 523–531, 2001.
21. Bader, G.D. and Hogue, C.W.V. An automated method for finding molecular complexes in large protein interaction networks. *BMC Bioinf* 4(2), 2003.
22. Hermjakob, H., Montecchi-Palazzi, L., Bader, G. et al. The HUPO PSI's molecular interaction format — a community standard for the representation of protein interaction data. *Nat Biotechnol* 22(2): 177–183, 2004.

CHAPTER 13

Functional Annotation of Proteins in Murine Models

Shui Qing Ye

CONTENTS

The lowly mouse is highly regarded as a useful model for human diseases. As sequences of the human and mouse genomes indicate that the two genomes are 85% identical, a multidisciplinary collaboration of basic and clinical scientists worldwide has undertaken genomewide mutagenesis to functionally annotate the mouse genome and develop new mouse models relevant to human disease. In this chapter, Section 1 will present various murine models of human diseases and relevant online resources, and Section 2 will demonstrate how to fetch some online murine model information from selected databases.

SECTION 1 VARIOUS MURINE MODELS OF HUMAN DISEASES

1. Why Are Mouse Models of Human Diseases Valuable?

The house mouse, *Mus musculus*, has been linked with humans since the beginning of civilization — wherever farmed food was stored, mice would be found. Many of the advances in 20th century biology featured significant contributions from the mouse, which has become the favored model animal in most spheres of research. *Mus musculus* has played a prominent role in the study of human disease mechanisms throughout the rich, 100-year history of classical mouse genetics, evidenced by the insight gained from naturally occurring mutants such as agouti, reeler, and obese. Now, with the completion of the human and mouse genome sequences, increasing attention has been focused on elucidating human gene functions and their pathophysiological roles in mouse models. The large-scale production and analysis of induced genetic mutations in mice have greatly accelerated the understanding of human gene functions. Among the animal models of human diseases, the mouse offers particular advantages for the study of human biology and disease: (1) the development, body plan, physiology, behavior, and diseases of the mouse have much in common with those of humans; (2) most or nearly all (99%) mouse genes have homologues in humans; and (3) the mouse genome is relatively easy to be manipulated for the mutagenesis of its genes by homologous recombination in embryonic stem (ES) cells. A number of databases on gene knockout or overexpressing mouse models are now available on the Internet. Efficiently leveraging this knowledge and other resources in mouse models of human diseases will be crucial in harnessing the power of the genome to drive biomedical discovery.

2. Transgenic Mouse

A transgenic mouse is simply a mouse that has had foreign DNA introduced into one or more of its cells artificially. Transgenic mice are powerful tools for studying gene functions and testing drugs. Many human genetic diseases can be modeled by introducing the same mutation into a mouse. Although similar genetic manipulations can be performed in tissue culture, the interaction of transgenes with proteins, hormones, neurotransmitters, and other components of an intact organism provides a much more complete and physiologically relevant picture of the transgene's function than could be achieved in any other way. In 1982, a team led by Richard Palmiter and Ralph Brinster prepared the first transgenic mouse. They fused elements of a gene that can be regulated by dietary zinc to a rat growth-hormone gene, and injected it into fertilized mouse embryos. The resulting mice, when fed with extra zinc, grew to be huge, and the technique paved the way for a wave of genetic analysis using transgenic mice.

There are two common strategies to introduce foreign DNA into mouse — (1) microinjection and (2) targeted insertion:

1. Microinjection: DNA can be microinjected into the pronucleus of a fertilized ovum. Following injection, the DNA is incorporated into the genome of the cell. The transformed fertilized eggs are then injected back into pregnant females and brought to term. The injected DNA can integrate anywhere in the genome, and multiple copies often integrate in a head-to-tail fashion. There is no need for homology between the injected DNA and the host genome. Pronuclear injection of DNA is often used to characterize the ability of a promoter to direct tissue-specific gene expression. Another usage is to examine the effects of overexpressing and misexpressing endogenous or foreign genes at specific times and locations in the animal. A major drawback of this technique is that researchers could not predict where in the genome the foreign genetic material would be inserted. Because a gene's location in the genome is important for its expression pattern, mouse lines carrying the same transgene could display wildly varying phenotypes.

2. Targeted insertion: Targeted insertion is to insert the DNA into embryonic stem (ES) cells and select for cells with homologous recombinants. In 1987–1989, teams led by Martin Evans, Oliver Smithies, and Mario Capecchi created the first *knockout* mice, by selectively disabling a specific target gene in embryonic stem cells. The three received the Lasker Award in 2001 for this achievement. Knockout mice have become one of the most widely used tools in helping understand the human gene functions and their roles in disease. Embryonic stem cells are used as the target cells for the insertion of gene-targeting vectors because they are pluripotent, and thus able to generate all the different types of cells in the adult body. Scientists interested in examining a specific gene will remove or "knock out" the gene in an embryonic stem cell as described earlier, then put the cell into a recently fertilized embryo. Typically, homologous recombination is used to insert a selectable gene (e.g., *neo*mycin) driven by a constitutive promoter (e.g., PGK) into an essential exon of the gene one wishes to disrupt. Typically, the *neo* gene (or other gene) is flanked by large stretches of DNA (on the order of 2–7 kb) that exactly match the genomic sequences surrounding the desired insertion point. Once this construct is transfected into ES cells, the cells' own machinery performs the homologous recombination. To make it possible to select against ES cells with nonhomologous recombinants, it is common to include a negatively selectable gene *outside* the region intended to undergo homologous recombination in the targeting constructs. A commonly used gene for negative selection is the herpes virus thymidine kinase gene, which confers sensitivity to the drug gancyclovir, when a targeting vector is randomly incorporated into the targeting genome. Following positive selection (e.g., G418, to select for *neo*) and negative selection if desired, ES cell clones need to be screened for the right homologous recombinants. Because ES cells are diploid, only one allele is usually altered by the recombination event. When appropriate targeting has occurred, one usually sees bands representing both wild-type and targeted alleles. The ES cells are derived from the inner cell masses of blastocysts (early mouse embryos). These cells are pluripotent. They must be maintained on a layer of feeder cells, typically mouse embryo fibroblasts that have been irradiated to prevent them from dividing. ES cells must be passaged every 2 to 3 d to keep them from differentiating and losing pluripotency. It is important that the genomic DNA used

in making a targeting construct be derived from the ES cells of the same strain of mouse that one intends to target. Even small gaps in homology due to sequence polymorphisms between mouse strains can dramatically reduce the efficiency of homologous recombination. Once positive ES clones have been grown and frozen, the production of transgenic animals can begin. Donor females are mated, blastocysts are harvested, and 10–15 ES cells are injected into each blastocyst. Eight to ten blastocysts are then implanted into a uterine horn of each pseudopregnant recipient. By choosing the appropriate donor strain, the identification of chimeric offspring (i.e., those in which some fraction of tissue is derived from the transgenic ES cells) can be as simple as observing the change in hair and/or eye color. If the transgenic ES cells do not contribute to the germ line (sperm or eggs), the transgene cannot be passed on to offspring. Generally, those commonly used ES lines, if maintained properly, often yield the germ line.

Gene trapping is a high-throughput method of creating mutagenized ES cells for use in generating knockout and other mutant mouse strains for research in functional genomics. Major scientific initiatives are currently under way to knock out every mouse gene in ES cells in order to characterize gene function and provide insight into molecular mechanisms of human diseases. Gene trapping is an attractive strategy to generate gene mutations in mice on a large scale. Gene trapping is based on random integration of a gene-trap vector into the mouse genome. A promoterless reporter gene following a splice acceptor will produce a fusion transcript between the trapped gene and the reporter gene when the vector inserts into an intron. This allows the identification of the trapped genes easily by 5' rapid amplification of cDNA ends (RACE) and also to investigate both the *in vitro* and *in vivo* expression patterns of trapped genes. Because the reporter recapitulates the expression pattern of the trapped gene, *in vitro* expression screening in differentiating ES cells before *in vivo* expression analysis is useful to identify specific reporter expression in restricted cell types such as neuronal or hematopoietic lineage cells. If the insertion of a gene trap vector disrupts the endogenous gene structure, phenotypic analysis can be carried out in mice generated from gene-trapped ES cell lines. A number of international groups have used this approach to create sizeable public cell line repositories available to the scientific community for the generation of mutant mouse strains. The major gene trapping groups worldwide have recently joined together to centralize access to all publicly available gene trap lines by developing a user-oriented Web site for the International Gene Trap Consortium (IGTC). This collaboration provides an valuable public informatics resource comprising approximately 45,000 well-characterized ES cell lines, which currently represent approximately 40% of known mouse genes, all freely available for the creation of knockout mice on a noncollaborative basis. To standardize annotation and provide high-confidence data for gene trap lines, a rigorous identification and annotation pipeline has been developed combining genomic localization and transcript alignment of gene trap sequence tags to identify trapped loci. This information is stored in a new bioinformatics database accessible through the IGTC Web site interface. The IGTC Web site (www.genetrap.org) allows users to browse and search the database for trapped genes, BLAST sequences against gene trap sequence tags, and view trapped genes within biological pathways. In addition, IGTC data have been integrated into major genome browsers and bioinformatics sites to provide users with outside portals for viewing this data. The development of the IGTC Web site marks a major advance by providing the research community with the data and tools necessary to effectively use public gene trap resources to facilitate the characterization of mammalian gene function.

3. Usage of Knockout Mice

Knockout mice are used in a variety of ways:

1. They allow testing of the specific functions of any gene and observation of the processes that this gene could regulate. What actions does this gene turn off and on? By examining what is happening in an *in vivo* model, investigators are able to determine the effects a particular gene may have. These effects would be impossible to observe in a culture dish. However, to completely establish and assign an action to a particular gene is very challenging and requires lots of work.

2. They are models to study human diseases at the molecular level. The objective is that by examining what role a gene may play in the development of a particular disease in gene knockout mouse models, investigators can better understand molecular mechanisms on how that gene contributes to the pathogenesis of a counterpart human disease. Researchers can also take the knowledge a step further and search for drugs that act on that gene.

Although knockout technology is highly advantageous for both biomedical research and drug development, it also suffers from a number of limitations. For example, because of developmental defects, many knockout mice are embryonically lethal before the researcher has a chance to use the model for experimentation. Even if a mouse survives, several mouse models have somewhat different physical or other phenotypic traits than their human counterparts. An example of this phenomenon is that the p53 knockout mice develop a completely different range of tumors than do humans. In particular, mice develop lymphomas and sarcomas, whereas humans tend to develop epithelial-cell-derived cancers. Because such differences exist, it cannot be assumed that a particular gene will exhibit identical function in both mouse and human, and thus limits the utility of knockout mice as models of human disease.

4. Conditional Knockout Mouse

Although conventional gene targeting technology has been a powerful tool for studying gene function *in vivo* and has shed light on many developmental biological questions, it has not achieved the same success in explaining physiological and pathophysiological processes in mature animals. The reasons for such relatively poor results are largely twofold. First, it was initially not possible to control the timing of gene disruption. Originally, gene targeting typically involved insertion, using homologous recombination in mouse embryonic stem (ES) cells, of an exogenous DNA fragment into an exon critical for target gene function, resulting in gene knockout. Animals derived from these stem cells are affected by mutant gene dysfunction throughout ontogenesis, often yielding undesired effects. For example, endothelins-1 and -3 (ET-1 and ET-3) were initially implicated in blood pressure regulation; however, homozygous ET-1 knockout mice die at birth from first pharyngeal arch malformation, and homozygous ET-3 knockout mice die shortly after birth owing to failure to develop a myenteric plexus. In these cases, the biological roles of ET-1 and ET-3 could not be studied in mature mice. The second major reason why traditional gene targeting has had limited success is that the targeted gene is affected in all cell types. Thus, if one wanted to examine the biological significance of a targeted gene in a particular cell type, then this would be precluded by the confounding and potentially injurious effects of gene dysfunction throughout the body. Hence, it becomes clear that conditionally regulating the gene targeting is needed. Unlike the goal of conventional knockout technology, which is to knock out both alleles so that the gene is entirely absent from all cells, the purpose of conditional knockouts, in contrast, is to delete a gene in a particular organ, cell type, or stage of development. Researchers can use the technique to knock out certain portions of specific genes at particular times when those genes are important. Conditional knockout mice have several benefits over the conventional type. Not only do they typically survive longer than traditional knockout mice, but conditional knockout methods are more precise as well.

There are several different ways to make conditional knockout models; however, the most widely used method is the site-specific recombinase (SSR) system (Cre-loxP, Flp/frt and φC31/att). SSR systems are transforming both forward and reverse genetics in mice. By enabling high-fidelity DNA modifications to be induced *in vitro* or *in vivo*, these systems have initiated a wave of new biology, advancing our understanding of gene function, development, and disease.

Cre–loxP system. The Cre–loxP system mediates site-specific DNA recombination and was originally described in bacteriophage P1. Two components are involved: (1) a 34-bp DNA sequence containing two 13-bp inverted repeats and an asymmetric 8-bp spacer region termed loxP ("locus of X-over in P1") that targets recombination, and (2) a 343 amino acid monomeric protein termed Cre (Cause recombination of the bacteriophage P1 genome) recombinase that mediates the recombination event. Any DNA sequence will be excised by Cre if it is flanked by two loxP sites in the same orientation. On the other hand, any DNA sequence will be inverted by Cre if it is flanked by two loxP sites in opposite orientation.

A major advantage of the Cre–loxP system lies in its relative simplicity. First, no cofactors are required for Cre activity. Second, loxP target sites are short and easily synthesized. Third, there are no apparent external energy requirements. This is because the Cre–loxP complex provides the necessary energy through formation of phosphotyrosine intermediates at the point of strand exchange. Fourth, Cre is a very stable protein. Finally, and most important, it is easy to generate DNA constructs with any promoter of interest driving Cre expression. This permits controlling the tissue site, and possibly the timing, of Cre expression and resultant gene disruption. It is not surprising, therefore, that this system has been increasingly employed in manipulating eukaryotic genes *in vivo*.

Utilization of the Cre–loxP system for gene targeting *in vivo* involves two lines of mice. The first mouse line is generated using ES cell technology. Typically, the target gene is altered, by homologous recombination in

ES cells, such that genomic regions critical for protein activity are flanked by loxP sites ("floxed" gene). Mice derived from these ES cells should ultimately contain floxed alleles in all cells. These mice should be phenotypically normal because the loxP sites were inserted into introns, where they theoretically do not affect gene function. The second line of transgenic mice is generated by standard oocyte injection techniques. These mice express Cre under the control of a "particular" (time- or tissue-specific) promoter. Mating of the two mouse lines should result in Cre-mediated gene disruption only in those cells in which the promoter is active.

The power and versatility of the Cre–loxP system is largely a function of promoter activity. Such activity can be regulated to achieve tissue-specific Cre expression in mice by either (1) endogenous cell-specific elements or (2) exogenously administered regulatory (inducing) factors. Cre expression may also be temporally regulated using inducible promoters. This has the theoretical advantage of restricting gene targeting events to a particular time in the animal's life, thereby avoiding potentially adverse consequences of defective gene function during earlier developmental stages. A few inducible systems have been coupled to Cre expression, such as the interferon-responsive Mx1 promoter, a tamoxifen-dependent mutated estrogen receptor promoter, and the tetracycline-regulated transactivator/tet operator system.

Flp–FRT system. The Flp–FRT recombination system is essentially the eukaryotic homologue of the Cre–loxP system. Flp was named for its ability to invert, or flip, a DNA fragment in *S. cerevisiae.* It is a 423 amino acid monomeric peptide encoded within the 2-μm plasmid of *Saccharomyces cerevisiae.* Flp is similar to Cre in that it requires no cofactors, uses a phosphotyrosine intermediate for energy, and is relatively stable. FRT (Flp recombinase recognition target) is also very similar to loxP in that it is composed of three 13-bp repeats surrounding an 8-bp asymmetric spacer region. The asymmetric region dictates whether excision (FRT sites in same orientation) or inversion (FRT sites in inverted orientation) of an intervening DNA sequence occurs after recombination.

Although not as widely used as Cre–loxP, the Flp–FRT system has been shown to cause site-specific DNA recombination in ES cells and transgenic mice. Interestingly, despite similar mechanisms of action and DNA recognition sites, the Cre–loxP and Flp–FRT systems do not exhibit significant cross-reactivity. The uniqueness of these two recombination systems may allow them to be used in concert to simplify the gene targeting process.

φC31–att system. The *Streptomyces*-phage-derived φC31 site-specific recombinase (SSR) has been established for use in ES cells. The φC31 SSR mediates recombination only between the heterotypic sites *attB* (34 bp in length) and *attP* (39 bp in length). *attB* and *attP*, named for the attachment sites for the phage integrase on the bacterial and phage genomes, respectively, both contain imperfect inverted repeats that are likely bound by φC31 homodimers. The product sites, *attL* and *attR*, are effectively inert to further φC31-mediated recombination, making the reaction irreversible. For catalyzing insertions, it has been found that *attB*-bearing DNA inserts into a genomic *attP* site more readily than an *attP* site into a genomic *attB* site. Thus, typical strategies position by homologous recombination an *attP*-bearing "docking site" into a defined locus, which is then partnered with an *attB*-bearing incoming sequence for insertion. Importantly, expression of φC31 in ES cells (like Cre and Flp) is compatible with germ line competence.

5. Chemical Mutagenesis in Mouse

The alkylating agent *N*-ethyl *N*-nitrosourea (ENU), considered one of the most potent mutagens in mice, is estimated to induce a point mutation (most often AT-to-TA transversions or AT-to-GC transitions) at a given locus every 700 gametes. In contrast to null mutations generated by gene targeting, ENU mutagenesis may result in hypomorphic or hypermorphic alleles. Repetitive intraperitoneal administration of ENU to male mice (G0, generation 0) efficiently mutagenizes spermatogonial stem cells; G1 mice are subsequently generated through breeding of treated males with wild-type females. Dominant or recessive mutations are identified through phenotypic screening for variation in the traits under study, using different breeding strategies. Dominant screens of G1 mice are logistically simpler and more rapid then recessive screens that require two- or three-generation breeding schemes. In two large-scale screens carried out to date, 2% of all G1 mice displayed a heritable phenotypic abnormality. Mutations of many disease phenotypes are identifiable with robust screening methods and, to date, dysmorphological mutants, circadian rhythm abnormalities, clinical chemical abnormalities, immunological phenotypes, and neurobehavioral phenotypes have all been successfully identified. Confirmation of the heritable nature of these mutations is obtained through repetitive screening of subsequent generations of mice derived from affected animals.

Chromosomal localization of stable, heritable mutations is achieved by linkage analysis of backcross or intercross populations derived from breeding of mutagenized mice with another strain that is phenotypically divergent for the trait of interest, using a panel of informative markers spanning the genome.

Chemical mutagens have contributed to large-scale generation of mouse mutants. Conventional germ-cell mutagenesis with *N*-ethyl-*N*-nitrosourea (ENU) is compromised by an inability to monitor mutation efficiency, strain, and interlocus variation in mutation induction, and extensive husbandry requirements. New methods for generating mouse mutants were devised to generate germ line chimaeric mice from ES cells heavily mutagenized with ethylmethanesulphonate (EMS). Germ line chimeras were derived from cultures that underwent a mutation rate of up to 1 in 1200 at the Hprt locus (encoding hypoxanthine guanine phosphoribosyl transferase). The spectrum of mutations induced by EMS and the frameshift mutagen ICR191 was consistent with that observed in other mammalian cells. Chimeras derived from ES cells treated with EMS transmitted mutations affecting several processes, including limb development, hair growth, hearing, and gametogenesis. This technology affords several advantages over traditional mutagenesis, including the ability to conduct shortened breeding schemes and to screen for mutant phenotypes directly in ES cells or their differentiated derivatives.

6. Using Mouse Models to Identify Quantitative Trait Loci (QTL)

Definition of QTL. A QTL is a region of DNA that is associated with a particular phenotypic trait. A quantitative trait is one that has measurable phenotypic variation. Generally, quantitative traits are multifactorial and are influenced by several polymorphic genes and environmental conditions. One or many QTLs can influence a trait or a phenotype. Two classic examples of quantitative traits are height and weight. Loci that modulate these traits are therefore called *QTLs*. These traits can also be influenced by loci that have large discrete effects (often called *Mendelian loci*). The distinction between Mendelian loci and QTLs is artificial, as the same mapping techniques can be applied to both. In fact, the classification of genetic (and allelic) effects should be considered as a continuum. At one end of the spectrum is the dichotomous Mendelian trait with only two detectable and distinct phenotypes, which are governed by a single gene. At the other end are traits, such as growth, which are likely to be affected by many genes that each contribute a small portion to the overall phenotype. Although they present challenges of discovery and analysis, QTL is important for us to understand disease processes because they are responsible for most of the genetic diversity in human disease susceptibility and severity. With the development of new genetic techniques and with the completion of human, mouse and other mammalian genomes, a new wave of optimism is permeating scientific circles that QTLs will become easier to identify and will provide valuable information about normal development and disease processes.

Mouse-to-human strategy. Finding QTL genes in mice is much more cost-effective, less time consuming, and less fraught with ethical issues than finding them first in humans, especially now that many new genetic, genomic, and bioinformatics tools for the mouse are available. A QTL gene can be first identified in the mouse, and then tested for its validity in human association studies in human diseases. The completion of the human HapMap project has greatly simplified such studies by facilitating the selection of representative SNPs of a haplotype block. In addition, dense SNPs throughout the human genome have facilitated whole-genome association studies in which the candidacy of all the genes in human diseases are systematically studied. The success of this mouse-to-human strategy has been documented in the identification of the proopiomelanocortin as a gene influencing obesity, cytotoxic T-lymphocyte-associated protein 4 as a gene contributing to autoimmune disorders, and engrailed 2 as a gene involved in autism-spectrum disorder. This strategy has also been used to the discovery of an atherosclerosis-susceptibility gene, a tumor necrosis factor superfamily member 4, a polymorphism of *which* was significantly associated with the risk of myocardial infarction and coronary artery disease in humans. Among nearly 30 human atherosclerosis QTLs reported, more than 60% are concordant to mouse QTLs, suggesting that these mouse and human atherosclerosis QTLs have the same underlying genes. Therefore, genes regulating human atherosclerosis will be most efficiently discovered by first identifying their orthologues in concordant mouse QTLs. Once QTL genes are identified in mice, they can be tested in human association studies for their relevance in human atherosclerotic disease. This paradigm can be applied to identify genes regulating QTLs in any other complex diseases.

Bioinformatics tools for dissecting rodent QTLs. A number of bioinformatics tools for dissecting rodent QTLs are available, and they are continuously updated and improved. A nutshell of those tools are presented in the following text. Comparative genomics can identify regions of chromosomal synteny in QTLs that are concordant across species. Experimental evidence suggests that causal genes underlying rodent QTLs are often conserved as disease genes in humans. Therefore, aligned concordant QTLs by comparative genomics analysis are likely to contain the causal gene. Generally, comparative genomic analysis only modestly narrows a QTL interval in most cases. Combined cross-analysis combines multiple crosses, detecting a shared QTL into one susceptibility and one resistance genotype into a single QTL analysis. Combining multiple crosses increases the number of recombination events, leading to better resolution of the QTL interval, based on the assumption that the same causal gene underlies the QTL in each cross. Haplotype analysis is useful to refine further a QTL interval to several Mb. There are interval-specific and genomewide haplotype analyses. Interval-specific haplotype analysis is often used to identify small regions (<5 Mb) within a QTL that are likely to contain the causal gene. Genomewide haplotype association analyzes haplotype patterns across the whole genomes of inbred strains surveyed for a phenotype to associate that phenotype with conserved haplotype patterns. Strain-specific sequence and gene expression comparisons are effective for focusing on a few strong candidate genes that may underlie relevant QTLs. Each strategy has its advantages, limitations and pitfalls. Interested readers may refer to an excellent review on the subject by DiPetrillo et al. (2005) for the details.

7. The Mouse Genome Informatics (MGI)

MGI provides integrated access to data on the genetics, genomics, and biology of the laboratory mouse in order to facilitate the use of the mouse as a model system for understanding human biology and disease processes. A core component of the MGI effort is the acquisition and integration of genomic, genetic, functional, and phenotypic information about mouse genes and gene products. MGI works within the broader bioinformatics community to define referential and semantic standards to facilitate data exchange between resources, including the incorporation of information from the biomedical literature. MGI is also a platform for computational assessment of integrated biological data with the goal of identifying candidate genes associated with complex phenotypes. MGI is web accessible at http://www.informatics.jax.org, hosted at The Jackson Laboratory (http://www.jax.org). Its core component is the Mouse Genome Database

(MGD). Other projects and resources that are part of the MGI system include the Gene Expression Database (GXD) (http://www.informatics.jax.org/mgihome/GXD/aboutGXD.shtml) and the Mouse Tumor Biology Database (MTB) (http://tumor.informatics.jax.org). All MGI component groups participate actively in the development and application of the Gene Ontology (GO) (http://www.geneontology.org). MGI continues to evolve, expanding its data coverage, improving data access, and providing new data query, analysis, and display tools.

SECTION 2 SEARCH FOR MURINE DATABASES

In this part, step-by-step tutorials are presented on the search for the following:

1. Mouse strains and stocks available worldwide, including inbred, mutant, and genetically engineered mice from International Mouse Strain Resources (IMSR, http://www.informatics.jax.org/imsr/IMSRSearchForm.jsp)

2. Mutant mouse embryonic stem (ES) cell lines from The International Gene Trap Consortium (IGTC)

3. The candidate gene in a QTL region.

1. Search for International Mouse Strain Resources

International Mouse Strain Resources include sixteen repositories worldwide. All regions and repositories are selected by default (ANY). One can limit the search to a specific region or more specific repositories from the selection list. Here, searching for Apolipoprotein E (Apoe) gene mutant mouse models is demonstrated:

1. In the Internet browser, type the following address: http://www.informatics.jax.org/imsr/IMSRSearchForm.jsp; then, under Gene or Allele Symbol/Name in IMSR Search Form, type gene symbol Apoe as displayed in Figure 13.1.

2. Click "Search," and query summary of "Apoe" appears as displayed in Figure 13.2. There are nineteen matching items found. In each item listed in the IMSR, users can obtain information about: Where a strain is available (Holder Site); in what states a strain is available (e.g. live, frozen, germplasm, ES cells); links to further information

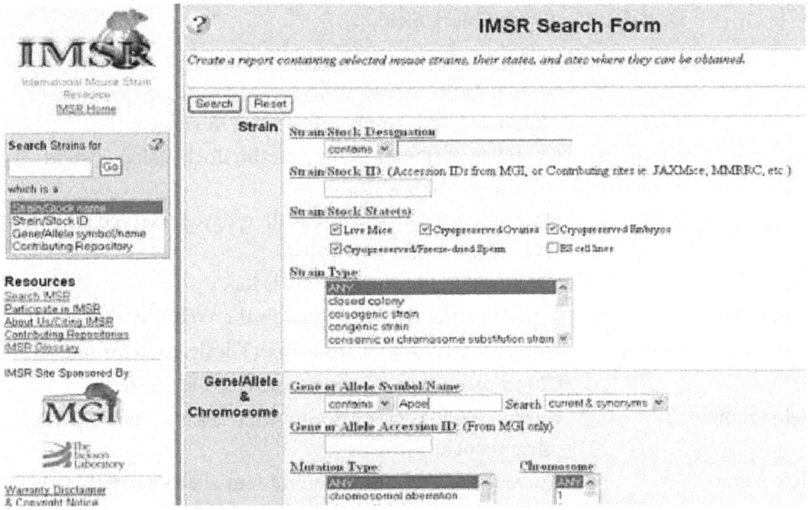
IMSR
IMSR Search Form
Search
Reset
Strain
Search Strains for
Go
which is a
Strain/Stock ID
Gene/Allele symbol/name
Contributing Repository
Resources
IMSR Site Sponsored By
MGI
closed colony
coisogenic strain
congenic strain
Gene/Allele & Chromosome
contains
Apoe
chromosomal aberration

FIGURE 13.1 Search for Apoe mutant mouse from the International Mouse Strain Resources.

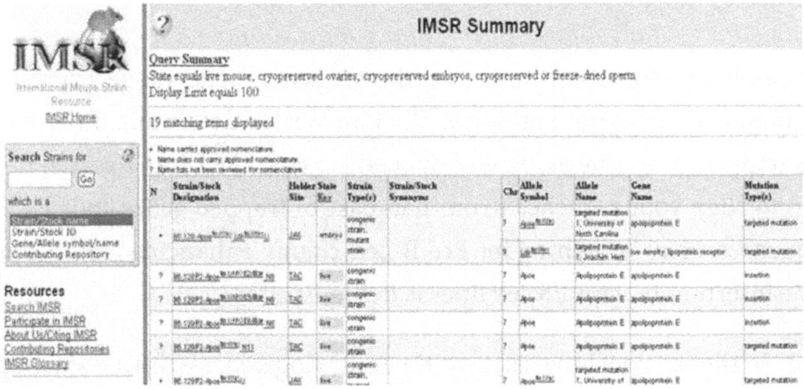
IMSR
IMSR Summary
Query Summary
Display Limit equals 100
19 matching items displayed
Search Strains for
which is a
Resources

FIGURE 13.2 Query summary of the "Apoe" search. Nineteen matching items are displayed. Only the partial list is shown.

about a strain; and links for contacting the holder to order a strain or send a query. Descriptions of each field are presented in Table 13.1.

2. Search for Mutant Mouse Embryonic Stem (ES) Cell Lines from The International Gene Trap Consortium (IGTC)

IGTC represents all publicly available gene trap cell lines, which are available on a noncollaborative basis for nominal handling fees. Researchers can search and browse the IGTC database for cell lines of interest using

TABLE 13.1 Description of Each Query Field

Field	Description
N	Nomenclature status of the Strain/Stock Designation: Approved, Not Approved, or Not Reviewed.
Strain/stock designation	Complete designation for a strain/stock, using current official nomenclature.
State	State of the strain/stock (e.g., live, cryopreserved embryos, etc.) at the holder site listed.
Holder site	Site at which the strain/stock is held.
Strain types	Type of strain as enumerated in the Web site.
Strain/stock synonyms	Alternative (unofficial or former) designated for that strain.
Chr	Chromosomes containing the designated alleles.
Allele symbol	Current allele symbols for mutant alleles carried by the strain/stock.
Allele name	Current allele names for all mutant alleles carried by the strain/stock.
Gene name	Current gene names for all genes with mutant alleles carried by the strain/stock.
Mutation types	Mutant type indicates how the allele was created.

accession numbers or IDs, keywords, sequence data, tissue expression profiles, and biological pathways. The best way to determine if the IGTC has trapped your gene or locus of interest is to use the BLAST search function to align your sequence to our database of trapped genes and cell line sequences. Here, searching for Pre-B-cell colony enhancing factor (Pbef) gene insertion mutant ES cell lines is demonstrated:

1. In the Internet browser, type the following address: http://www.genetrap.org/dataaccess/blast.html; then paste the query Pbef cDNA sequence (Sample data B) in the FASTA format into search window as displayed in Figure 13.3. Click on the "Quick Search" button; the default BLAST search for the NM_021524 sequence (NM_021524) yields the result displayed in Figure 13.4. There are twenty-one blast hits on the query sequences. Five of significant hits is presented. The user can record cell line numbers and further search the IGTC by clicking hyperlinked cell lines to get additional information on the supplier site, and availability status from the cell line annotation page.

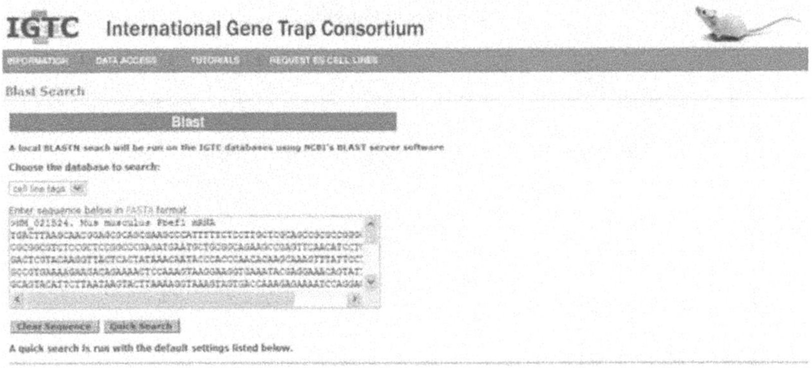
IGTC International Gene Trap Consortium
Blast Search
Blast
Choose the database to search:
A quick search is run with the default settings listed below.

FIGURE 13.3 Search for mutant mouse ES cell lines from IGTC. The screenshot shows pasting the query Pbef cDNA sequence in the FASTA format into the search window.

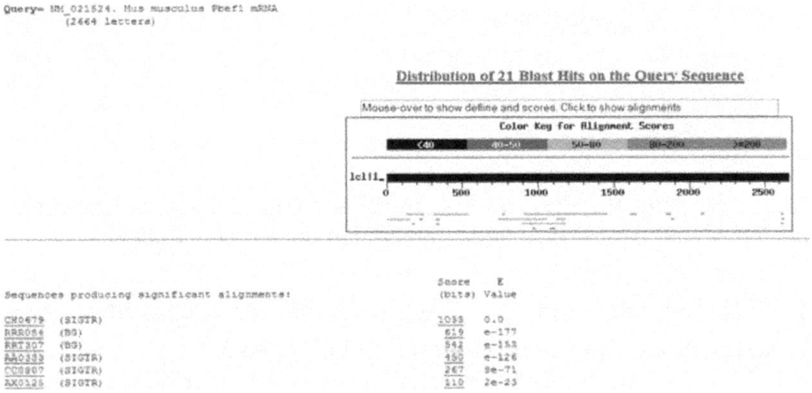
Distribution of 21 Blast Hits on the Query Sequence
Color Key for Alignment Scores
Sequences producing significant alignments:

FIGURE 13.4 Output of the query PBEF1. Top panel shows the distribution of 21 blast hits on the query sequence. Lower panel shows 6 out of 21 hits producing the significant alignments.

3. Search for the Candidate Gene in a QTL Region

A. *Find a QTL*

1. Go to the MGI homepage (http:www.informatics.jax.org), click “Genes and Markers” under “Search Menus,” click again “Genes and Markers” under “Searches,” select QTL under “Type” in “Gene/marker,” type “HDL cholesterol” under “Phenotype/Human Disease” in “Mouse phenotypes & mouse models of human disease,” and sort by Nomenclature under “Sorting and output format.” The end screen shot of all these steps is presented in Figure 13.5.

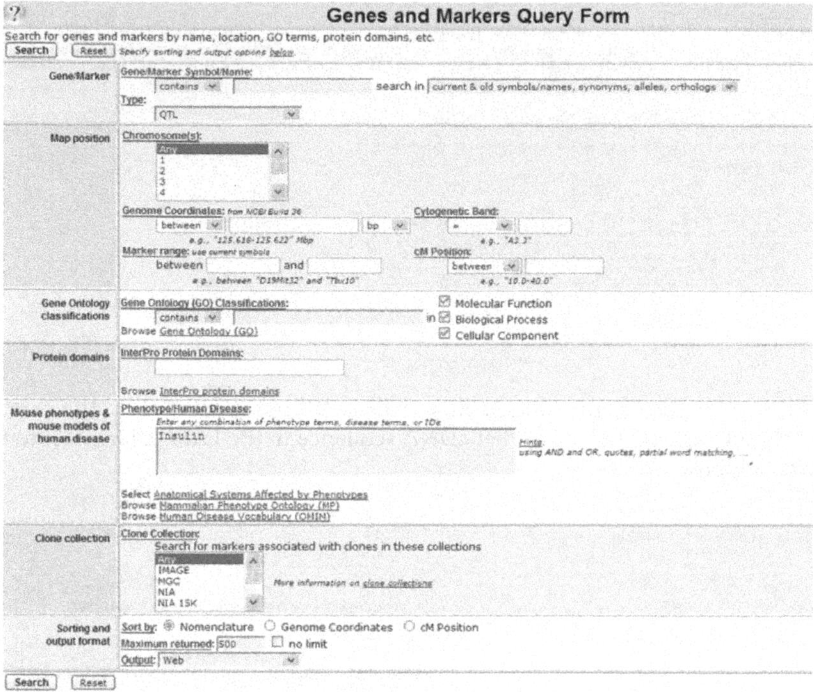
Genes and Markers Query Form
Search for genes and markers by name, location, GO terms, protein domains, etc.
Search
Reset
Gene/Marker
Gene/Marker Symbol/Name:
search in current & old symbols/names, synonyms, alleles, orthologs
Type:
QTL
Map position
Chromosome(s):
Genome Coordinates:
Cytogenetic Band:
Marker range:
cM Position:
Gene Ontology classifications
Gene Ontology (GO) Classifications:
Molecular Function
Biological Process
Cellular Component
Protein domains
InterPro Protein Domains:
Mouse phenotypes & mouse models of human disease
Phenotype/Human Disease:
Insulin
Clone collection
Clone Collection:
Search for markers associated with clones in these collections
Sorting and output format
Sort by: Nomenclature Genome Coordinates cM Position
Maximum returned: 500 no limit
Output: Web

FIGURE 13.5 Find the QTL on insulin. This screenshot shows where search terms "QTL" and "Insulin" should be entered in the search window.

2. Click "Search," and 217 matching items are displayed. A partial screenshot of the list is presented in Figure 13.6.

B. Find the Genes in the QTL Region

Here, we use an example of the QTL region: Chromosome 2: 117803890-174126009bp, the last list of Figure 13.6 to search for the genes in this QTL region.

1. Go to the MGI homepage (http:www.informatics.jax.org), click "Genes and Markers" under "Search Menus," click again "Genes and Markers" under "Searches," select Gene under "Type" in "Gene/marker," select Chromosome 2 and Genome Coordinates: 117803890-174126009bp under "Map position," type "Insulin" under "Phenotype/Human Disease" in "Mouse phenotypes & mouse models of human disease," and sort by Genome Coordinates under "Sorting and output format." The end screenshot of all these steps is presented in Figure 13.7.

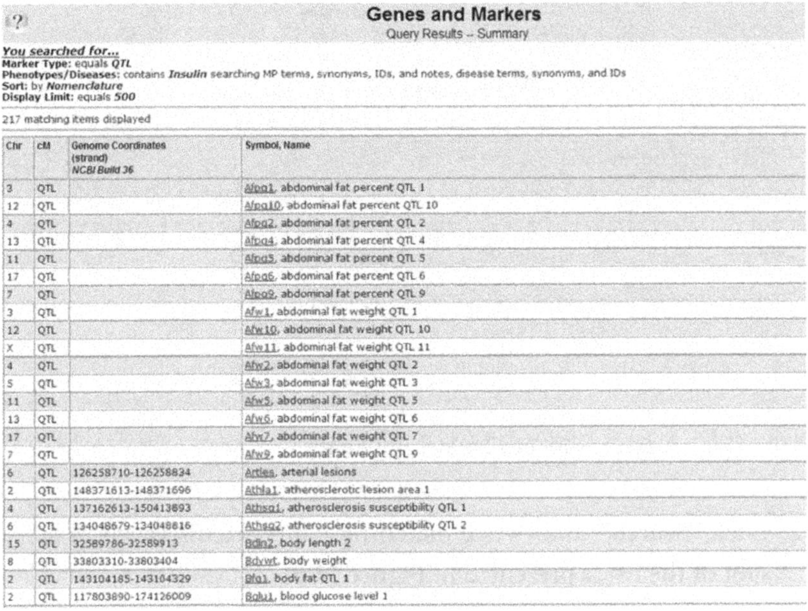

Genes and Markers

Query Results -- Summary

You searched for...
Marker Type: equals *QTL*
Phenotypes/Diseases: contains *Insulin* searching MP terms, synonyms, IDs, and notes, disease terms, synonyms, and IDs
Sort: by *Nomenclature*
Display Limit: equals *500*

217 matching items displayed

Chr	cM	Genome Coordinates (strand) NCBI Build 36	Symbol, Name
3	QTL		Afpq1, abdominal fat percent QTL 1
12	QTL		Afpq10, abdominal fat percent QTL 10
4	QTL		Afpq2, abdominal fat percent QTL 2
13	QTL		Afpq4, abdominal fat percent QTL 4
11	QTL		Afpq5, abdominal fat percent QTL 5
17	QTL		Afpq6, abdominal fat percent QTL 6
7	QTL		Afpq9, abdominal fat percent QTL 9
3	QTL		Afw1, abdominal fat weight QTL 1
12	QTL		Afw10, abdominal fat weight QTL 10
X	QTL		Afw11, abdominal fat weight QTL 11
4	QTL		Afw2, abdominal fat weight QTL 2
5	QTL		Afw3, abdominal fat weight QTL 3
11	QTL		Afw5, abdominal fat weight QTL 5
13	QTL		Afw6, abdominal fat weight QTL 6
17	QTL		Afw7, abdominal fat weight QTL 7
7	QTL		Afw9, abdominal fat weight QTL 9
6	QTL	126258710-126258834	Artles, arterial lesions
2	QTL	148371613-148371696	Athla1, atherosclerotic lesion area 1
4	QTL	137162613-150413893	Athsq1, atherosclerosis susceptibility QTL 1
6	QTL	134048679-134048816	Athsq2, atherosclerosis susceptibility QTL 2
15	QTL	32589786-32589913	Bdln2, body length 2
8	QTL	33803310-33803404	Bdywt, body weight
2	QTL	143104185-143104329	Bfq1, body fat QTL 1
2	QTL	117803890-174126009	Bglu1, blood glucose level 1

FIGURE 13.6 A partial screenshot of 217 matching items.

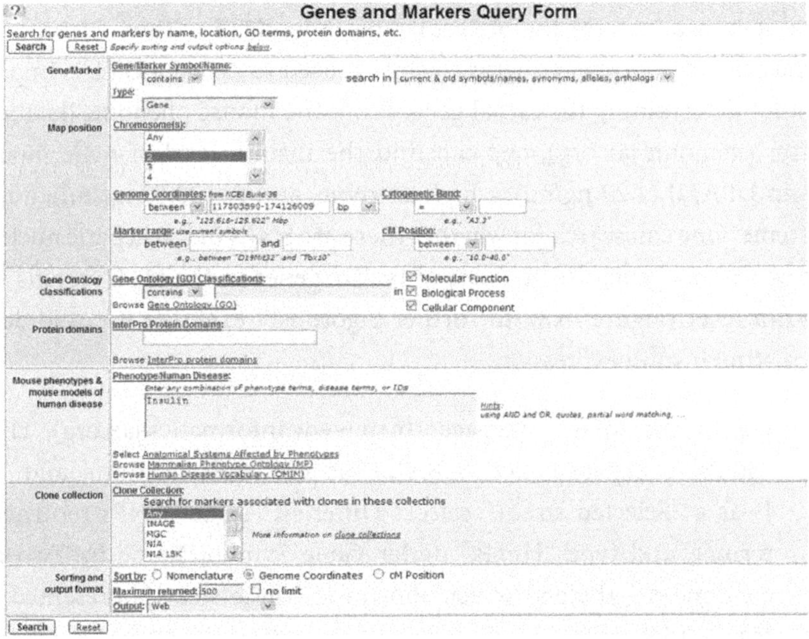

Genes and Markers Query Form

Search for genes and markers by name, location, GO terms, protein domains, etc.

[Search] [Reset] *Specify sorting and output options below.*

Gene/Marker — Gene/Marker Symbol/Name: [contains ▼] [____] search in [current & old symbols/names, synonyms, alleles, orthologs ▼]
Type: [Gene ▼]

Map position — Chromosome(s): [Any / 1 / 2 / 3 / 4 ▼]
Genome Coordinates: from NCBI Build 36 [between ▼] [117803890-174126009] [bp ▼]
e.g. "125.618-128.602" Mbp
Cytogenetic Band: [= ▼] e.g. "A3.3"
Marker range: use current symbols [between ▼] [and ____] e.g. between "D1Mit32" and "Tbx10"
cM Position: [between ▼] e.g. "10.0-40.0"

Gene Ontology classifications — Gene Ontology (GO) Classifications: [contains ▼] [____]
☑ Molecular Function
in ☑ Biological Process
☑ Cellular Component
Browse Gene Ontology (GO)

Protein domains — InterPro Protein Domains: [____]
Browse InterPro protein domains

Mouse phenotypes & mouse models of human disease — Phenotype/Human Disease:
Enter any combination of phenotype terms, disease terms, or IDs
[Insulin]
Hints:
using AND and OR, quotes, partial word matching, ...
Select Anatomical Systems Affected by Phenotypes
Browse Mammalian Phenotype Ontology (MP)
Browse Human Disease Vocabulary (OMIM)

Clone collection — Clone Collection:
Search for markers associated with clones in these collections
[Any / IMAGE / MGC / NIA / NIA 15K ▼]
More information on clone collections

Sorting and output format — Sort by: ○ Nomenclature ⦿ Genome Coordinates ○ cM Position
Maximum returned: [500] ☐ no limit
Output: [Web ▼]

[Search] [Reset]

FIGURE 13.7 Find the genes in the QTL region. Fill in genome coordinates and find how many genes are in the region.

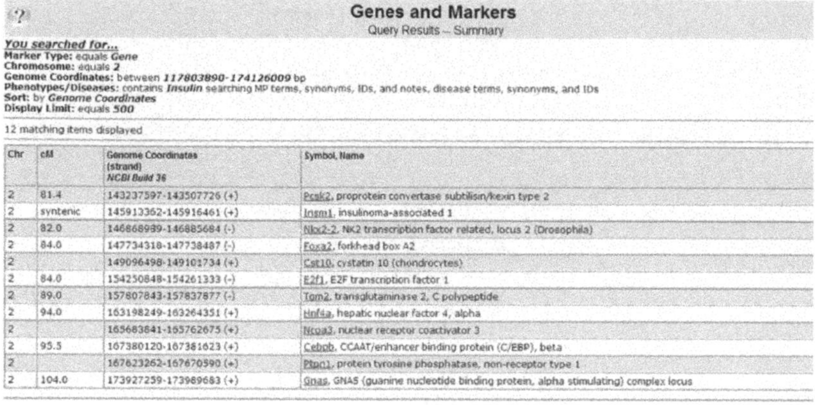

Genes and Markers
Query Results – Summary

You searched for...
Marker Type: equals Gene
Chromosome: equals 2
Genome Coordinates: between 117803890-174126009 bp
Phenotypes/Diseases: contains Insulin searching MP terms, synonyms, IDs, and notes, disease terms, synonyms, and IDs
Sort: by Genome Coordinates
Display Limit: equals 500

12 matching items displayed

Chr	cM	Genome Coordinates (strand) NCBI Build 36	Symbol, Name
2	81.4	143237597-143507726 (+)	Pcsk2, proprotein convertase subtilisin/kexin type 2
2	syntenic	145913362-145916461 (+)	Insm1, insulinoma-associated 1
2	82.0	146868989-146885684 (-)	Nkx2-2, NK2 transcription factor related, locus 2 (Drosophila)
2	84.0	147734318-147738487 (-)	Foxa2, forkhead box A2
2		149096498-149101734 (+)	Cst10, cystatin 10 (chondrocytes)
2	84.0	154250848-154261333 (-)	E2f1, E2F transcription factor 1
2	89.0	157807843-157837877 (-)	Tgm2, transglutaminase 2, C polypeptide
2	94.0	163198249-163264351 (+)	Hnf4a, hepatic nuclear factor 4, alpha
2		165683841-165762675 (+)	Ncoa3, nuclear receptor coactivator 3
2	95.5	167380120-167381623 (+)	Cebpb, CCAAT/enhancer binding protein (C/EBP), beta
2		167623262-167670590 (+)	Ptpn1, protein tyrosine phosphatase, non-receptor type 1
2	104.0	173927259-173989683 (+)	Gnas, GNAS (guanine nucleotide binding protein, alpha stimulating) complex locus

FIGURE 13.8 A screenshot of 12 matching items.

2. Click "Search," and twelve matching items are displayed. A screenshot of the list is presented in Figure 13.8.

C. Search for SNPs in Candidate Gene

DNA sequences polymorphisms affecting either expression or function of a gene product are the molecular basis for QTLs. Thus, identifying sequence polymorphisms between strains used to detect a QTL is important for determining the causal gene. From the mouse phenome database (http://phenome.jax.org), one can find the insulin level of male mouse strain DBA/2J (4.29 ng/ml), a high extreme, and A/J (0.53 ng/ml), a low extreme. One can search for whether there are SNPs in the hepatic nuclear factor 4 alpha gene, which is one of twelve genes in the QTL regions for insulin level (Figure 13.8), to further rigorously examine its candidacy regulating insulin expression.

1. Go to the MGI homepage (http:www.informatics.jax.org), click "SNPs" query form, add "DBA/2J" as a "Reference strain" and "A/J" as a "Selected strain," select "Different" as an "SNPs returned" format, and type "Hnf4a" under Gene Symbol/Name in "Associated genes." The end screen shot of all these steps is presented in Figure 13.9.

2. Click "Search," and forty matching items are displayed. A screenshot of the partial list is presented in Figure 13.10.

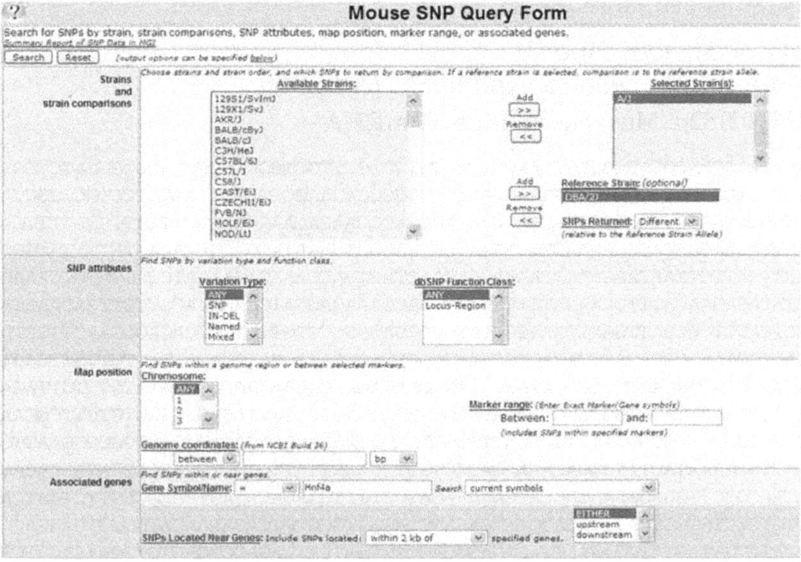
Mouse SNP Query Form
Search for SNPs by strain, strain comparisons, SNP attributes, map position, marker range, or associated genes.
Strains and strain comparisons
SNP attributes
Map position
Associated genes

FIGURE 13.9 Search for SNPs in the Hnf4a gene from the MGI Web site.

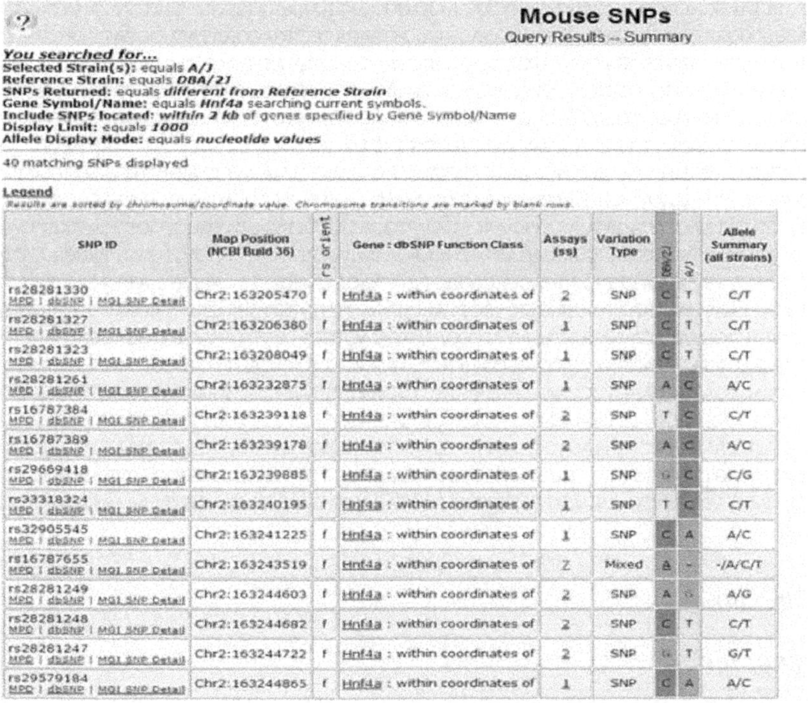
Mouse SNPs
Query Results – Summary
You searched for...
Selected Strain(s): equals A/J
Reference Strain: equals DBA/2J
SNPs Returned: equals different from Reference Strain
Gene Symbol/Name: equals Hnf4a searching current symbols.
Include SNPs located: within 2 kb of genes specified by Gene Symbol/Name
Display Limit: equals 1000
Allele Display Mode: equals nucleotide values
40 matching SNPs displayed
Legend

FIGURE 13.10 A screenshot of 40 matching items.

4. Sample Data for Section 2

A. Gene Symbol "Apoe"

B. Pbef mRNA Sequence in the FASTA Format

>NM_021524. Mus musculus Pbef1 mRNA

```
TGACTTAAGCAACGGAGCGCAGCGAAGCCCATTTTTCTCCTTGCTCGCAGCCGCGCCGGGCAGCTCGTGG
CGCGGCGTCTCCGCTCCGGCCCGAGATGAATGCTGCGGCAGAAGCCGAGTTCAACATCCTGCTGGCCACC
GACTCGTACAAGGTTACTCACTATAAACAATACCCACCCAACACAAGCAAAGTTTATTCCTACTTTGAAT
GCCGTGAAAAGAAGACAGAAAACTCCAAAGTAAGGAAGGTGAAATACGAGGAAACAGTATTTTATGGGTT
GCAGTACATTCTTAATAAGTACTTAAAAGGTAAAGTAGTGACCAAAGAGAAAATCCAGGAGGCCAAAGAA
GTGTACAGAGAACATTTCCAAGATGATGTCTTTAACGAAAGAGGATGGAACTACATCCTTGAGAAATACG
ATGGTCATCTCCCGATTGAAGTAAAGGCTGTTCCCGAGGGCTCTGTCATCCCCAGAGGGAACGTGCTGTT
CACAGTGGAAAACACAGACCCAGAGTGCTACTGGCTTACCAATTGGATTGAGACTATTCTTGTTCAGTCC
TGGTATCCAATTACAGTGGCCACAAATTCCAGAGAACAGAAGAGAATACTGGCCAAATATTTGTTAGAAA
CCTCTGGTAACTTAGATGGTCTGGAATACAAGTTACATGACTCTGGTTACAGAGGAGTCTCTTCGCAAGA
GACTGCTGGCATAGGGGCATCTGCTCATTTGGTTAACTTAAAAGGAACAGATACTGTGGCGGGAATTGCT
CTAATTAAAAAATACTATGGGACAAAAGATCCTGTTCCAGGCTATTCTGTTCCAGCAGCAGAGCACAGTA
CCATAACGGCTTGGGGGAAAGACCATGAGAAAGATGCTTTTGAACACATAGTAACACAGTTCTCATCAGT
GCCTGTGTCTGTGGTCAGCGATAGCTATGACATTTATAATGCGTGTGAGAAAATATGGGGTGAAGACCTG
AGACATCTGATAGTATCGAGAAGTACAGAGGCACCACTAATCATCAGACCTGACTCTGGAAATCCTCTTG
ACACTGTATTGAAGGTCTTAGATATTTTAGGCAAGAAGTTTCCTGTTACTGAGAACTCAAAAGGCTACAA
GTTGCTGCCACCTTATCTTAGAGTCATTCAAGGAGATGGCGTGGATATCAATACTTTACAAGAGATTGTA
GAGGGAATGAAACAAAAGAAGTGGAGTATCGAGAATGTCTCCTTCGGTTCTGGTGGCGCTTTGCTACAGA
AGTTAACCCGAGACCTCTTGAATTGCTCCTTCAAGTGCAGCTATGTTGTAACCAATGGCCTTGGGGTTAA
TGTGTTTAAGGACCCAGTTGCTGATCCCAACAAAAGGTCAAAAAAGGGCCGGTTATCTTTACATAGGACA
CCAGCGGGGAACTTTGTTACACTTGAAGAAGGAAAAGGAGACCTTGAGGAATATGGCCATGATCTTCTCC
ATACGGTTTTCAAGAATGGGAAGGTGACAAAAAGCTACTCATTTGATGAAGTCAGAAAAAATGCACAGCT
GAACATCGAGCAGGACGTGGCACCTCATTAGGCTTCATGTGGCCGGGTTGTTATGTGTGCAGTGTGTGTA
TACATACATGCACGTATGTGTGCGCCTGTGCGTATGTACTAACATGTTCATTGTACAGATGTGTGGGTTC
GTGTTTATGATACACTGCAGCCAGATTATTTGTTGGTTTATGGACATACTGCCCTTTTTATTTTTCTCCC
AGTGTTTAGATGATCTCAGATTAGAAAACACTTACAACCATGTACAAGATTAATGCTGAAGCAAGCTTTT
CAGGGTCCTTTGCTAATAGATAGTAATCCAATCTGGTGTTGATCTTTTCACAAATAACAAACCAAGAAAC
TTTTATATATAACTACAGATCACATAAAACAGATTTGCATAAAATTACCATGCCTGCTTTATGTTTATAT
TTAACTTGTATTTTTGTACAAACGAGATTGTGTAAGATATATTTAAAGTTTCAGTGATTTACCAGTCTGT
TTCCAACTTTTCATGATTTTTATGAGCACAGACTTTCAAGAAAATACTTGAAATAAATTACATTGCCTTT
TGTCCATTAATCAGCAAATAAAACATGGCCTTAACCAAGTTGTTTGTGGTGTTGTACATTTGCAAATTAT
GTCAGGACAGACAGACACCCAACAGAGTTCCGAACATCACTGCCCTTGTAGAGTATGCAGCAGTCATTCT
CCGTGGAAGAGAAGAATGGTTCTTACGCGAATGTTTAGGCATTGTACAGTTCTGTGCCCTGGTCAGTGTA
TGTACCAGTGATGCCAAATCCCAAAGGCCTGTTCTGCAATTTTATATGTTGGATATTGCCTGTGGCTCTA
ATATGCACCTCAAGATTTTAAGAAGATAATGTTTTTAGAGAGAATTTCTGCTTCCACTATAGAATATATA
CATAAATGTAAAATATTGAAAGTGGAAGTAGTGTATTTTAAAGTAATTACACTTCTGAATTTATTTTTCA
TATTCTATAGTTGGTATGTCTTAAATGAATTGCTGGAGTGGGTAGTGAGTGTACTTATTTTAAATGTTTT
GATTCTGTTATATTTTCATTAAGTTTTTTAAAAATTAAATTGGATATTAAACTGTAAAAAAAAAAAAAAA
AAAA
```

C. Insulin

REFERENCES

1. The life history of the mouse in genetics. *Nature* 420: 510, 2002.

2. Walinski, H. Studying gene function: create knockout mice. *Sci Creat Q* September–November 2006 (http://www.scq.ubc.ca/?p=264).
3. Austin, C.P., Battey, J.F., Bradley, A. et al. The knockout mouse project. *Nat Genet* 36(9): 921–924, September 2004.
4. Stricklett, P.K., Nelson, R.D., and Kohan, D.E. The Cre/loxP system and gene targeting in the kidney. *Am J Physiol* 276(5 Pt. 2): F651–F657, 1999.
5. Branda, C.S. and Dymecki, S.M. Talking about a revolution: the impact of site-specific recombinases on genetic analyses in mice. *Dev Cell* 6(1): 7–28, 2004.
6. Blake, J.A., Eppig, J.T., Bult, C.J., Kadin, J.A., and Richardson, J.E. Mouse genome database group. The mouse genome database (MGD): updates and enhancements. *Nucl Acids Res* 34(Database issue): D562–D567, 2006.
7. Nord, A.S., Chang, P.J., Conklin, B.R. et al. The international gene trap consortium Website: a portal to all publicly available gene trap cell lines in mouse. *Nucl Acids Res* 34(Database issue): D642–D648, 2006.
8. Stanford, W.L., Cohn, J.B., and Cordes, S.P. Gene-trap mutagenesis: past, present and beyond. *Nat Rev Genet* 2, 756–768, 2001.
9. Chen, Y., Yee, D., Dains, K., Chatterjee, A., Cavalcoli, J., Schneider, E., Om, J., Woychik, R.P., and Magnuson, T. Genotype-based screen for ENU-induced mutations in mouse embryonic stem cells. *Nat Genet* 24(3): 314–317, March 2000.
10. Abiola, O., Angel, J.M., Avner, P. et al. The nature and identification of quantitative trait loci: a community's view. *Nat Rev Genet* 4(11): 911–916, 2003.
11. Wang, X., Ishimori, N., Korstanje, R., Rollins, J., and Paigen, B. Identifying novel genes for atherosclerosis through mouse-human comparative genetics. *Am J Hum Genet* 77(1):1–15, 2005.
12. DiPetrillo, K., Wang, X., Stylianou, I.M., and Paigen, B. Bioinformatics toolbox for narrowing rodent quantitative trait loci. *Trends Genet* 21(12): 683–692, 2005.
13. Rual, J.F., Venkatesan, K., Hao, T., Hirozane-Kishikawa, T., Dricot, A., Li, N., Berriz, G.F., Gibbons, F.D., Dreze, M., Ayivi-Guedehoussou, N., Klitgord, N., Simon, C., Boxem, M., Milstein, S., Rosenberg, J., Goldberg, D.S., Zhang, L.V., Wong, S.L., Franklin, G., Li, S., Albala, J.S., Lim, J., Fraughton, C., Llamosas, E., Cevik, S., Bex, C., Lamesch, P., Sikorski, R.S., Vandenhaute, J., Zoghbi, H.Y., Smolyar, A., Bosak, S., Sequerra, R., Doucette-Stamm, L., Cusick, M.E., Hill, D.E., Roth, F.P., Vidal, M. Towards a proteome-scale map of the human protein-protein interaction network. *Nature* 2005; 437(7062):1173-8.

CHAPTER 14

Application of Programming Languages in Biology

Hongfang Liu

CONTENTS

With a large number of prokaryotic and eukaryotic genomes completely sequenced and more forthcoming, the use of genomic information and accumulated biomedical knowledge for the discovery of new knowledge has become the central theme of modern biomedical research. The exploration of the Internet and the use of the World Wide Web (WWW) have enabled various data sources like GenBank and a variety of bioinformatics tools like BLAST accessible to biomedical researchers. However, to truly exploit various sources and bioinformatics tools, biologists need to understand and be able to use programming languages for managing, presenting, and analyzing large amounts of data or results obtained from many different sources or tools.

Similar to human languages, programming languages need to follow syntax rules that specify the basic vocabulary of the language and how programs can be constructed using control structures such as loops, branches, and subroutines. A syntactically correct program is one that can be successfully compiled or interpreted; programs that have syntax errors will be rejected. The program also needs to be semantically right. This chapter introduces you to three different open-source programming languages with some basic information about their syntax and semantics. Step-by-step tutorials are provided for the installation and the use of existing bioinformatics code for one operation system (OS), i.e., Window XP. We also provide sample applications. In the first section, we introduce Perl, which is the most commonly used language in bioinformatics for data manipulation and analysis. Next, we discuss JAVA, which has strengths in graphical user interface (GUI) development. R, a programming language and software environment for statistical computing and graphics, is described in the last section.

SECTION 1 APPLICATION OF PERL IN BIOLOGY

Part I Introduction

1. What Is Perl Programming?

Perl is a programming language that was created in 1986 as the result of one man's frustration with a task involved generating reports from a lot of text files with cross-references. Being tired of writing specific utilities to manage similar tasks dealing with extracting information from text and generating reports, Larry Wall invented a new language and wrote an interpreter in C for it that later became known as Perl (stands for Practical Extraction Report Language). The initial emphasis of Perl was on system management and text handling. More features were added during later revisions, including regular expressions, signals, and network sockets.

Programs written in Perl are usually called Perl scripts. Perl can be used for a large variety of tasks. A typical simple use of Perl would be extracting information from a text file and printing out a report, or converting a text file into another form. But Perl has evolved to provide a large number of tools for quite complicated systems.

Perl is implemented as an interpreted language (i.e., turned into binary instruction on fly) rather than a compiled language (i.e., turned into binary instructions at once). The execution of a Perl script may take more time than a program written in traditional programming languages such as C or JAVA. On the other hand, computers tend to get faster and faster, and writing something in Perl instead of languages such as C tends to save you time.

2. Why Is Perl Popular in Biology?

The Perl programming language is probably the most commonly used language in bioinformatics, partly for the following reasons. First, Perl is less structured than traditional programming languages, and it is easy to learn for biologists. Perl takes care of many of the low-level tasks in traditional programming, such as memory allocation, and developers can concentrate on solving the problem at hand. Perl is interpreted, which enables the quick and dirty creation of portable (i.e., platform-independent) analysis programs. Perl is also free software and available for a variety of operating systems, including Windows and Mac, as well as Linux and Unix platforms. It is often the case that a problem can be solved using Perl programming with a few lines of code while requiring several times more lines of code in C or Java. For example, the following several lines of code will enable an application to read a file "example.txt" and perform some task line by line.

```
open(FILE,"example.txt");
while(<FILE>){
 some task
}
```

Additionally, the Comprehensive Perl Archive Network (CPAN, http://www.cpan.org) provides an impressive collection of Perl code (mostly Perl modules), which makes the development of complex systems possible. For example, the ability to manipulate data stored in commercial or free database management systems (DBMSs) and the text manipulation capability have made Perl the language of choice to create Common Gateway Interface (CGI) scripts for handling data submission and information retrieval tasks (shown in Figure 14.1).

The approach used by Perl to access DBMSs is a two-tier approach:

1. The top level (DBI module) is visible to application scripts and presents an abstract interface for connecting to and using database engines. The interface does not depend on details specific to particular engines and is designed to provide a simple interface to send database queries in Structured Query Language (SQL) and retrieve the results.

2. The lower level (DBDs) consists of drivers for individual DBMS engines. Each driver handles the details necessary to map the abstract interface onto operations that a specific engine will understand. The

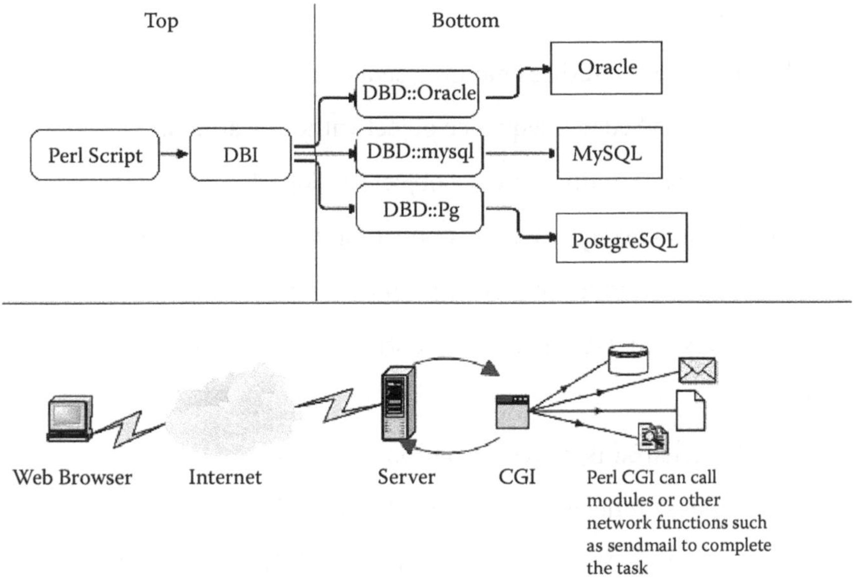
Top
Bottom
Perl Script
DBI
DBD::Oracle
DBD::mysql
DBD::Pg
Oracle
MySQL
PostgreSQL
Web Browser
Internet
Server
CGI
Perl CGI can call modules or other network functions such as sendmail to complete the task

FIGURE 14.1 Overviews of the two-tier architecture of database applications using Perl and the server-client architecture of Web applications using Perl Common Gateway Interface (CGI) scripts.

> DBD modules usually have the libraries for specific DBMSs in them and know how to communicate with the real databases; there is one DBD module for every database.

The Perl module CGI can be used to create CGI scripts such as querying databases using HTML Web forms (Figure 14.1).

Furthermore, the existence of a collection of Perl modules specific to bioinformatics applications, Bioperl, enables the development of complex bioinformatics applications in Perl. There are many reusable Perl modules in Bioperl that facilitate writing Perl scripts for sequence manipulation, accessing of databases using a range of data formats, and parsing of the results of various molecular biology programs including Blast, ClustalW, TCoffee, genscan, ESTscan, and HMMER. Bioperl is freely (under a very unrestrictive copyright) available at http://bioperl.org. Also, tutorials and documents on how to use them are also accessible at http://bioperl.org.

The following is a brief overview of the main types of modules in Bioperl, collected in a few broadly defined groups:

Sequences

Bio::LiveSeq::* handles changing sequences.

Bio::PrimarySeq is a sequence object without features.

Bio::Seq is the main sequence object in Bioperl.

Bio::Seq::LargeSeq provides support for very large sequences.

Bio::SeqIO provides sequence file input and output.

Bio::Tools::SeqStats provides statistics on a sequence.

Databases

Bio::DB::GenBank provides GenBank access.

Bio::Index::* indexes and accesses local databases.

Bio::Tools::Run::StandAloneBlast runs BLAST on your local computer.

Bio::Tools::Run::RemoteBlast runs BLAST remotely.

Bio::Tools::BPlite parses BLAST reports.

Bio::Tools::BPpsilite parses psiblast reports.

Bio::Tools::HMMER::Results parses HMMER hidden Markov model results.

Alignments

Bio::SimpleAlign manipulates and displays simple multiple sequence alignments.

Bio::UnivAln manipulates and displays multiple sequence alignments.

Bio::LocatableSeq is for sequence objects with start and end points for locating relative to other sequences or alignments.

Bio::Tools::pSW aligns two sequences using the Smith-Waterman algorithm.

Bio::Tools::BPbl2seq is a lightweight BLAST parser for pairwise sequence alignment using the BLAST algorithm.

Bio::AlignIO aligns two sequences using BLAST.

Bio::Clustalw is an interface to the multiple sequence alignment package Clustalw.

Bio::TCoffee is an interface to the multiple sequence alignment package TCoffee.

Bio::Variation::Allele handles sets of alleles.

Bio::Variation::SeqDiff handles sets of mutations and variants.

Features and genes on sequences

Bio::LocationI provides an interface to location information for a sequence.

Bio::Location::Fuzzy provides location information that may be inexact.

Bio::Location::Simple handles simple location information for a sequence, both as a single location and as a range.

Bio::Location::(Tisdall) provides location information where the location may encompass multiple ranges, and even multiple sequences.

Bio::SeqFeature is the sequence feature object in Bioperl.

Bio::Tools::OddCodes rewrites amino acid sequences in abbreviated codes for some specific statistical analyses.

Bio::Tools::EPCR parses the output of ePCR program.

Bio::Tools::ESTScan is an interface to the gene-finding program ESTScan.

Bio::Tools::MZEF is an interface to the gene-finding program MZEF.

Bio::Tools::Grail is an interface to the gene-finding program Grail.

Bio::Tools::Genemark is an interface to the gene-finding program Genemark.

Bio::Tools::Genscan is an interface to the gene-finding program Genscan.

Bio::Tools::RestrictionEnzyme identifies restriction sites in sequence.

Bio::Tools::SeqPattern provides support for regular expression descriptions of sequence patterns.

Bio::Tools::Sim4::Results (and Exon) is an interface to the gene exon finding program.

Bio::Tools::Sigcleave finds amino acid cleavage sites.

3. Some Perl Basics

Most operating systems (OSs), such as Linux and Mac, include Perl in their releases. In case Perl is not installed or you need the latest Perl release, you can go to http://www.perl.com/download.csp or http://www.cpan.org/ports/index.html and follow the instructions for downloading and installing it on your system. Detailed steps for installation of Perl for Windows will be provided in the tutorial section.

Writing and executing a Perl script can be broken into several steps: creating a script, running and debugging the scripts, and reading the output. An integrated development environment (IDE), a type of computer software that assists programmers in developing software, can be used to simplify and accelerate the development process. Most IDEs consist of a source code editor, a compiler and/or interpreter, build-automation tools, and usually a debugger. IDEs for Perl can also be found at http://www.cpan.org/ports/index.html and will not be discussed in detail.

A. Creating the Script A Perl script is just a text file that contains instructions written in the Perl language. The scripts can be created using a text editor such as vi, emacs, pico, or TextPad. By convention, Perl script files end with the extension .pl. Some text editors (e.g., TextPad, Emacs) have a Perl mode that will autoformat your Perl scripts and highlight keywords. Perl mode will be activated automatically if the script has a name ending with .pl. A Perl script consists of a series of statements and comments. Each statement is a command that is recognized by the Perl interpreter and executed. Statements are terminated by the semicolon character (;). They are also usually separated by a newline character to enhance readability. A comment begins with the # sign and can appear anywhere. Everything from the # to the end of the line is ignored by the Perl interpreter. The Perl interpreter will start at the top of the script and execute all the statements, in order from top to bottom, until it reaches the end of the script. This order of execution can be modified by loops and control structures.

The following is a script consisting of two lines of comments that indicate what the script is and what it does:

```
# time.pl
# print out the local time
$time = localtime;
print "The current time is $time \n";
```

There are three building blocks for statements: variables, operators, and functions. A variable is a named reference to a memory location. Variables provide an easy handle for programmers to keep track of data stored in memory. There are two basic types of variables in Perl. Scalar variables hold the basic building blocks of data: numbers and characters. Array variables hold lists. Perl has a rich set of operators (over fifty of them) that perform operations on string and numeric values. Some operators will be familiar from algebra (like "+", to add two numbers together), whereas others are more esoteric (like the "." string concatenation operator). In addition to its operators, Perl has many built-in functions. Functions usually have a human-readable name, such as `print`, and take one or more arguments passed as a list. A function may return no value, a single value, or a list. The argument list can be optionally included in parentheses. In the previous example, a scalar variable `$time` was assigned with a result from a build-in function `localtime` and "." is an operator used by Perl to concatenate strings. Besides built-in functions, users can also define functions, which are usually termed *subroutines* to differentiate them from built-in functions.

B. Running the Script There are two options to execute the script. The first option is to run it from the command line, giving it the name of the script file to run:

```
$ perl time.pl
The current time is Mon Nov 27 11:40:49 2006.
```

The second option is to run the script as a command by putting the comment `#!/usr/bin/perl` at the top of the script and make it executable by typing `chmod +x time.pl`. The comment directs the system to locate the Perl interpreter; you need to substitute `/usr/bin/perl` with the full path to your Perl installation, which can be found by typing `which perl` at the command line in Unix or Linux platforms.

Every script usually goes through a few iterations before the programmer gets it right. The common errors include syntax errors such as forgetting the "$" sign in the name of a scalar variable, or runtime errors such as dividing a number by zero. Other common errors include forgetting to make the script executable or putting the path to Perl wrong on the "#!" line. You can call Perl with a few command-line options to help catch errors:

```
-c - Perform a syntax check, but don't run.
-w - Turn on verbose warnings.
-d - Turn on the Perl debugger.
```

Usually, you will invoke these from the command line, as in `perl -cw time.pl` (syntax-check time.pl with verbose warnings). You can also put them in the top line: `#!/usr/bin/perl -w`.

Part II Step-By-Step Tutorial

The step-by-step tutorial provides an example of how to set up a computer for testing Bioperl scripts available at http://bioperl.open-bio.org/. The tutorial assumes that you have a personnel computer (PC) with Window XP as the operating system and that Perl is not installed. Also, we assume that Firefox is your Internet browser. Note that Firefox is available at http://www.mozilla.org/download.html) for download and installation.

1. Download and Install Perl

For Windows, we can download the latest Perl distribution from the ActiveState Web site, which is the leading provider of tools and services for languages such as Perl, PHP, Python, Ruby, and Tcl. Figure 14.2 shows a series of screenshots guiding the download and installation of ActivePerl:

1. Open a Firefox window and type in the address http://www.activestate.com/Products/ActivePerl.
2. Click on the link "Get ActivePerl."
3. Click "Free Download."
4. Select appropriate operating systems and download the MSI package.
5. Open the download file in the Firefox download window, and follow the installation directions to install the program.

2. Download and Install TextPad

Perl scripts can be created using text editors. Here, we use the trial version of TextPad. It can be downloaded from http://www.textpad.com/download/index.html. The software can be installed by following the instruction provided in the Web page. Figure 14.3 shows a series of screenshots guiding the download and installation of TextPad:

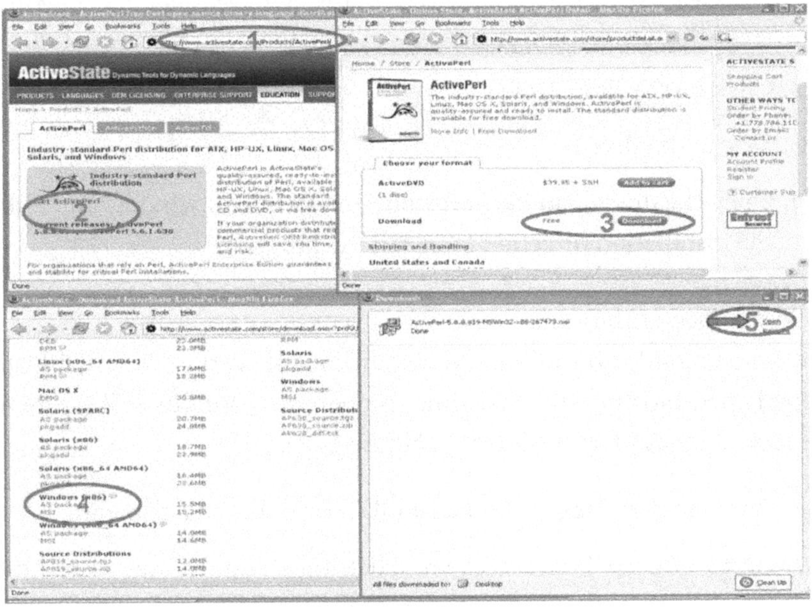
ActiveState
ActivePerl
1
2
3
4
5

FIGURE 14.2 Overview of the installation of Perl in Window XP.

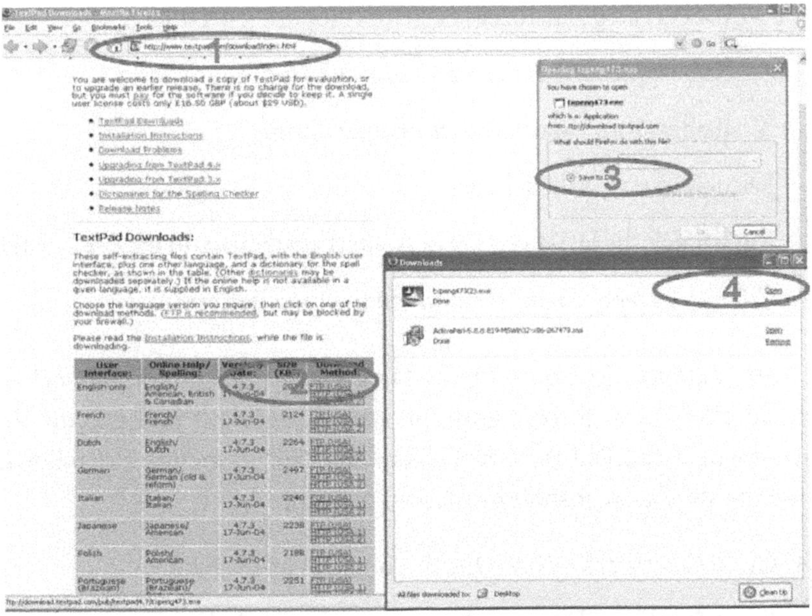
TextPad Downloads:
1
2
3
4

FIGURE 14.3 Overview of the installation of TextPad in Window XP.

1. Open a Firefox window, and type in the address http://www.textpad.com/download/index.html.
2. Click on the link "FTP."
3. Save to the disk.
4. Open the download file in the Firefox download window and follow the installation direction to install the program.

3. Install Bioperl Module

The instruction for installing the Bioperl module can be found at http://bioperl.open-bio.org/wiki/Installing_Bioperl_on_Windows. We simplify the procedure as follows (shown in Figure 14.4):

1. Start the Perl Package Manager GUI from the "Start" menu.
2. Go to Edit >> Preferences and click the "Repositories" tab. Add a new repository for each of the entries listed in the following with the format "Name:Location":
   - BioPerl-Release Candidates: http://bioperl.org/DIST/RC
   - BioPerl-Regular Releases: http://bioperl.org/DIST
   - Kobes: http://theoryx5.uwinnipeg.ca/ppms
   - Bribes: http://www.Bribes.org/perl/ppm
3. Select View >> All Packages.
4. Right-click the latest version of Bioperl available and choose "install."
5. Click the green arrow to complete the installation.

4. Start TextPad, Configure for Syntax Highlight and Running Perl

TextPad allows you to configure for syntax highlight and execute Perl. After starting TextPad (accessible from the "Start" menu), the following shows the steps (screenshots are shown in Figure 14.5):

1. Go to Configure>>New Document Class and click.
2. Enter "perl" as "Document class name" and click next.
3. Enter "*.pl" as "Class members" and click next.

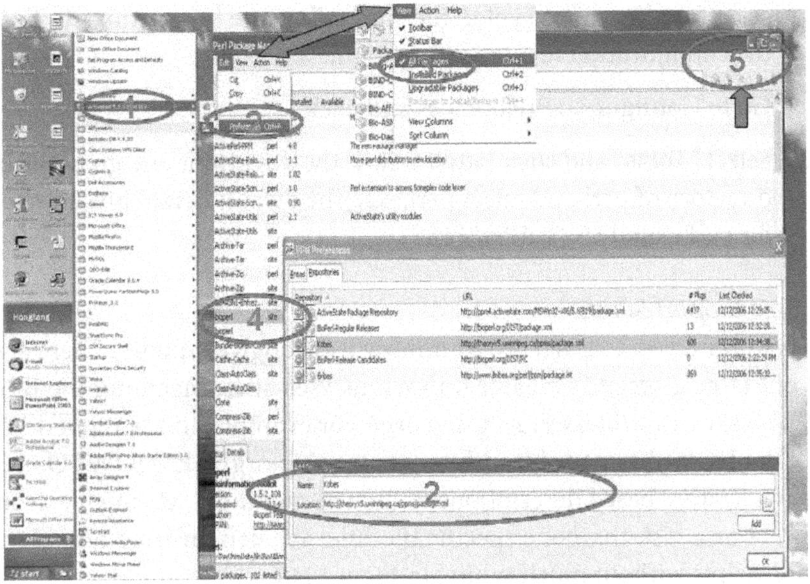
View Action Help
Toolbar
Status Bar
Upgradable Packages
View Columns
Sort Column
5
4
2

FIGURE 14.4 Overview of the installation of BioPerl through Perl Package Manager (PPM).

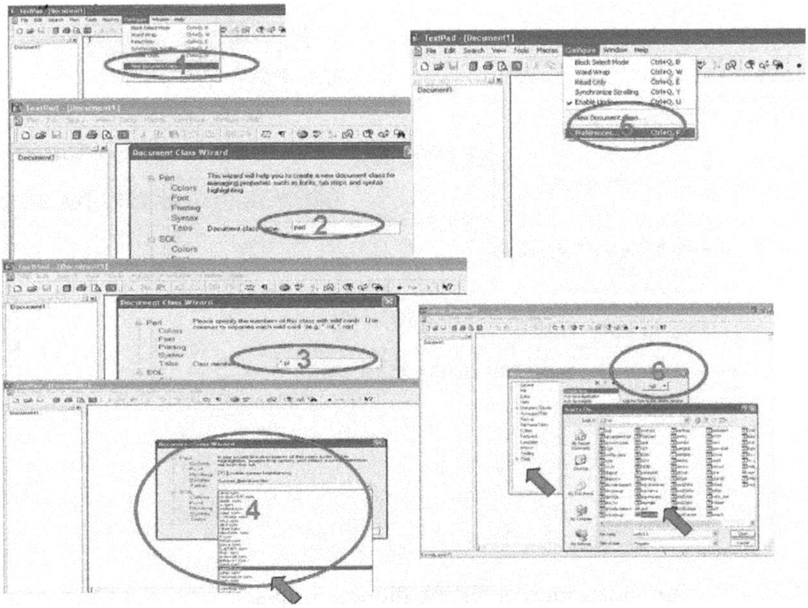
TextPad - [Document1]
Document Class Wizard
1
2
3
4
5
6

FIGURE 14.5 Set up TextPad for Perl Syntax Highlight.

4. Select "Enable syntax highlighting" and from the dropdown menu, select "perl5.syn" as syntax file, and follow the instructions to finish the configuration for syntax highlight.

5. Go to Configure>>Preferences and click.

6. Select "Tools" and click "add," follow the directions and go to the bin directory of perl and select perl.exe. The default install bin directory for perl usually is "C:\Perl\bin."

5. Test Bioperl Scripts

A very useful way to learn programming is through examples. There are a lot of example scripts available on various Web sites, including the Bioperl Web site http://bioperl.open-bio.org/. For example, in the Bioperl Web site, the HOWTOs page http://bioperl.open-bio.org/wiki/HOWTOs provides narrative-based descriptions of Bioperl modules focusing more on a concept or a task than one specific module. We demonstrate how to test those scripts without much typing in Figure 14.6:

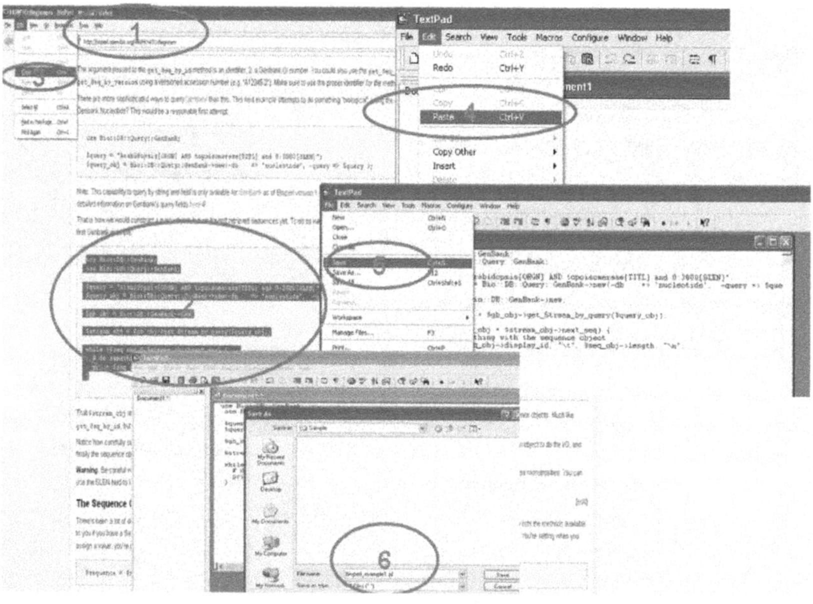
1
2
3
4
5
6
TextPad
File Edit Search View Tools Macros Configure Window Help
Redo Ctrl+Y
Copy Other
Insert

FIGURE 14.6 An illustration of testing Bioperl scripts.

1. Open Firefox and go to the Web site http://bioperl.open-bio.org/wiki/HOWTO:Beginners.
2. Browse the window until you identify the script that retrieves GenBank records, press the mouse, and select the script.
3. Go to Edit>>Copy in Firefox to copy the script.
4. Start TextPad and go to Edit>>Paste in Textpad to paste the example.
5. Go to File>>Save and click "Save."
6. Save the file as a local file ending with .pl (e.g., "bioperl_example1.pl").
7. Go to the left window of Textpad, right-click the mouse, and select "Tile horizontally."
8. Go to Tools>>Perl in TextPad and click.
9. The results will be shown in the "Command Results" area.

You can test any of the example scripts available at the Bioperl Web site, by using the previous nine steps.

Part III An Application Example

We introduce here a stand-alone Perl application that (1) processes a file containing official human gene names and symbols downloaded from HUGO Gene Nomenclature Committee (HGNC) in text format, (2) stores the information in a local MySQL database, and (3) retrieves the official human gene name for a given official symbol. We also introduce a Web application for querying the underlying MySQL database. The example is to demonstrate the usefulness of Perl for applications using DBMSs and the Web.

1. Background

We first provide a short introduction to HGNC, MySQL, and Web applications.

HGNC is an organization that approves one gene name and one gene symbol for each known human gene. Providing a unique symbol for each gene resolves ambiguity and facilitates electronic data retrieval from publications. All approved gene names and symbols are stored in the HGNC database, with a total of 24,044 approved entries. The HGNC database is downloadable from the Web site (http://www.gene.ucl.ac.uk/nomenclature/data/gdlw_index.html). Let us download the Core Database in text format and name the file hgnc.txt (see Figure 14.7).

MySQL is a popular database server that is used in various bioinformatics applications. SQL stands for Structured Query Language, which is what MySQL uses to communicate with other programs. On top of that, MySQL has its own expanded SQL functions to provide additional functionality to users. Please refer to http://www.mysql.com on how to do the initial MySQL installation, set up databases and tables, and create new users. Assume a user named `perlbio`, a password `perlbio`, and a database named `perlbio` have been created. In order to use Perl to communicate with MySQL, we need to install the Perl DBI and DBD::mysql modules (refer to the step-by-step tutorial and http://www.cpan.org about the installation of modules).

For the development of Web applications, a HTTP Web server, such as Apache Web Server (available at http://www.apache.com), needs to be installed. If you can design, develop, and post your own Web pages in a computer, then a Web server has already been installed in that computer. The details on how to set up a Web server are beyond the scope of this chapter.

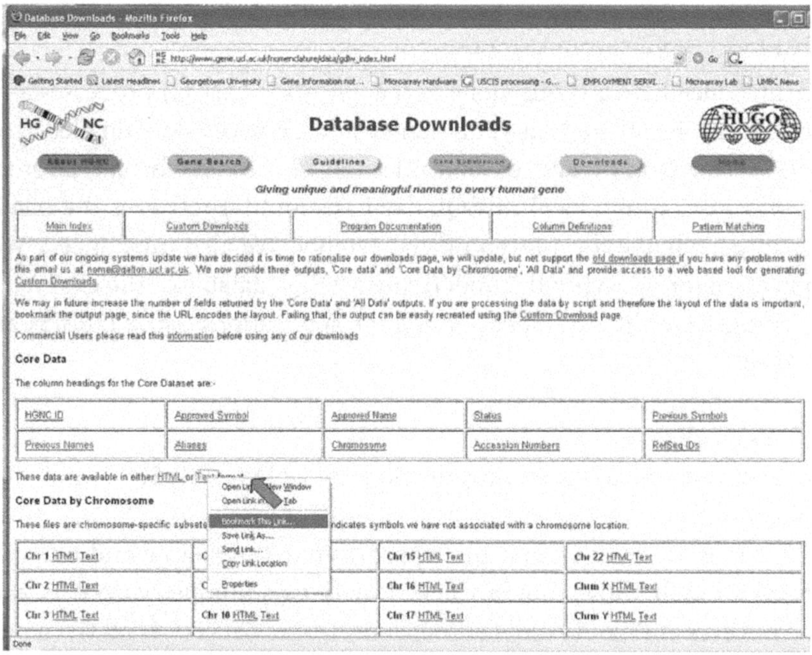
Database Downloads
Giving unique and meaningful names to every human gene
Core Data
Core Data by Chromosome

FIGURE 14.7 Screenshot on obtaining the HGNC core table.

2. Read the File

There are nine fields in the HGNC core table (see Figure 14.8 for a screenshot). We first start with a script that reads the file, and provides statistics of the number of records with nonempty values for each field:

```
#! /usr/bin/perl
#count.pl
#provide statistics for the HGNC core table
use strict; # a module that forces every variable
#needs to be declared as local or global variables
my @fieldnames; # store field names
my @fields; # store field values
my %count; # the counter hash,
 # keys will be field names, and
 # values will be the number of rows with non-empty
#values for the corresponding field
open(HGNC, "hgnc.txt"); # open the file to read
my @rows=<HGNC>; # read in lines into an array @rows
chomp(@rows); # remove the end-of-line characters from
#each row
@fieldnames=split(/\t/,$rows[0]);
# the first row contains field names, separate them
#into fields according to "\t".
for(my $i=1; $i<=$#rows; $i++){
# iterate through the number of rows, starting from
#the second row
@fields= split(/\t/,$rows[$i]); # separate the row
#into fields
for(my $j=0; $j<=$#fieldnames; $j++){ # iterate
#through the number of fields
      if($fields[$j] ne ""){ $count{$fieldnames[$j]}++
  ;} #increment the counter by 1
  }
}
foreach my $field (keys %count){
print $count{$field}." out of ".($#rows)." are non-
empty values at field $field! \n";
}
```

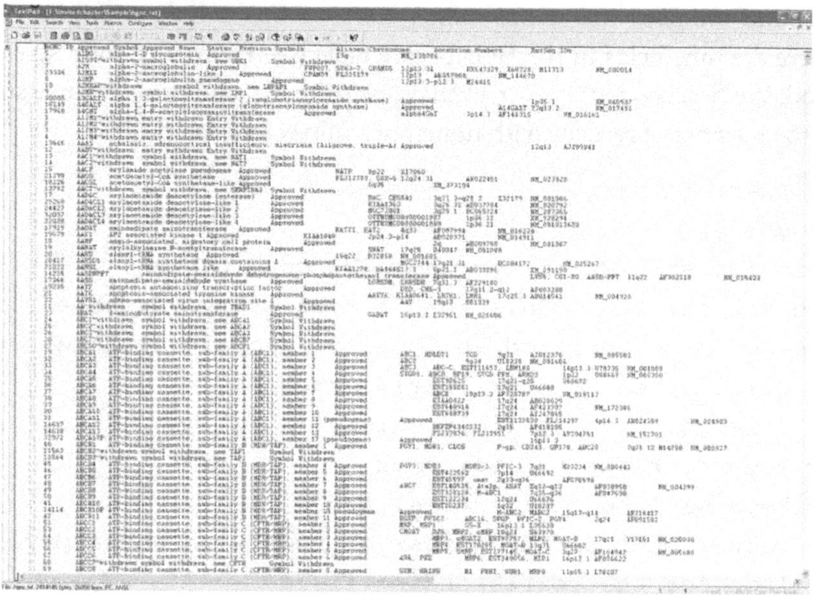

FIGURE 14.8 RECORDS IN THE HGNC CORE TABLE.

In the `count.pl` script, the code `use strict;` is a statement to tell the script to use the "strict" module. A module can be called by:

```
use modulename;
```

There are two types of variables in Perl: one is local (defined using my) and the other is global (by default). The use of the module "strict" will force every local variable in the script to be declared. Variables `@fieldnames` (to store field names) and `@fields` (to store field values) are arrays, and the variable `%count` is a hash that is a special type of array consisting of (key, value) pairs. The keys for `%count` will be field names, and the value will be the number of rows with nonempty values for the corresponding field.

The function open is used to open a file to read or write. There are generally two parameters: a file handler (e.g., HGNC) and a filename (e.g., `hgnc.txt`). All lines in the file can be read to an array. (e.g., `@rows=<HGNC>;`). They can also be read line by line, using the while loop as follows:

```
 while(<HGNC>){ chomp;
# The parameter to chomp is the default scalar $_.
......
}
```

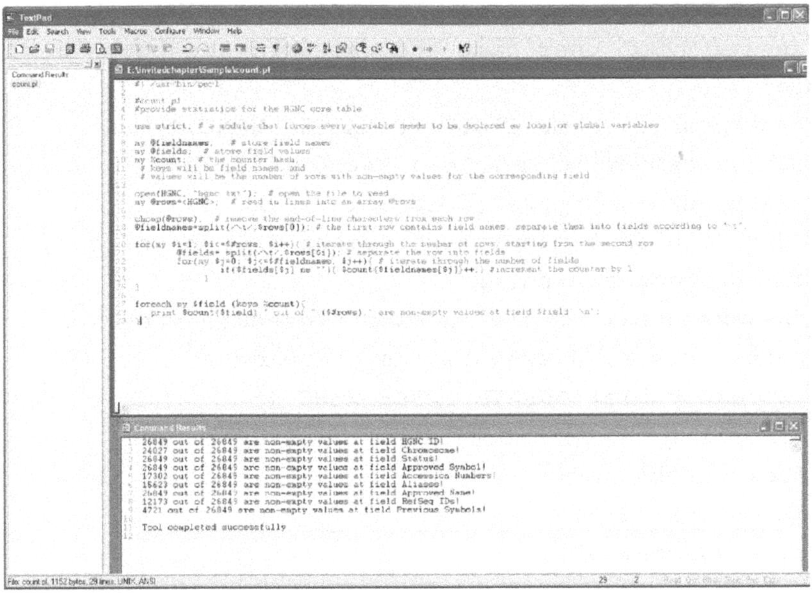

FIGURE 14.9 Screenshot for using TextPad to edit and run count.pl.

The function `chomp` is a built-in function that removes the end-of-line character from each element in the array `@rows`. Another function, `split`, is used to split a string to an array according to a delimiter. A nested `for` loop is used to iterate through each row and each field. A `if` condition block is to check if the value of a field is empty or not; if it is not empty, the counter is incremented by one for the corresponding field. Besides `for`, another way to iterate an array or an hash is `foreach`.

The result of running the script together with the script (i.e., the statistics of the HGNC core table (downloaded Nov 27, 2006)) is shown in Figure 14.9.

3. Store the Data into MySQL

The script `store.pl` stores the HGNC core table into the MySQL database `perlbio`:

```
# store.pl
# A script to process the HGNC core table and store in
#MySQl database perlbio
use DBI; # require DBI module
my $dsn="DBI:mysql:database=perlbio";
#assume MySQL database as "perlbio"
```

```
my $dbh=DBI->connect($dsn,"perlbio","perlbio");
# assume MySQL user name and password are both "perl
#bio"
my $droptable='DROP TABLE IF EXISTS HGNC;';
# SQL statement to delete a table if exists
my $sth=$dbh->prepare($droptable)
or die "Couldn't prepare statement ".$dbh->errstr;
# the statement handler
$sth->execute() or die "Couldn't execute statement
".$sth->errstr; # execute the statement
my $createtable='CREATE TABLE HGNC (
id int(11),
symbol varchar(50),
name varchar(255),
status varchar(50),
psymbols varchar(255),
aliases varchar(255),
chromosome varchar(50),
accession varchar(100),
refseq varchar(100),
PRIMARY KEY (id) );';
# SQL statements for creating a table
# The number inside parentheses is the maximum number
# of characters allowed
$sth=$dbh->prepare($createtable) or die "Couldn't
prepare statement ".$dbh->errstr;
$sth->execute() or die "Couldn't execute statement
".$sth->errstr;
my @fieldnames; # store field names
my @fields; # store field values
open(HGNC, "hgnc.txt"); # open the file to read
my @rows=<HGNC>; # read in lines into an array @rows
chomp(@rows);
# remove the end-of-line characters from each row
@fieldnames=split(/\t/,$rows[0]);
# the first row contains field names, separate them
into fields
# according to "\t".
```

```
$insertrecord='insert into HGNC (id, symbol, name,
status, psymbols, aliases, chromosome, accession,
refseq) values ('."?, " x (@fieldnames-1).'?)';
 # use of string operator "x", will repeat the string
#"?, " one less than the number of fields
$sth=$dbh->prepare($insertrecord) or die "Couldn't
prepare statement ".$dbh->errstr;
for(my $i=1; $i<=$#rows; $i++){
# iterate through the number of rows, starting from
#the second row
@fields= split(/\t/,$rows[$i]);
# separate the row into fields
$sth->execute(@fields[0..8]);
# ".." here defines an integer array from 0 to 8.
}
```

The line `use DBI;` indicates that the script will utilize the DBI module to interact with DBMSs. The actual connection is achieved by the following two lines:

```
my $dsn="DBI:mysql:database=perlbio";
#assume MySQL database as "perlbio"
my $dbh=DBI->connect($dsn,"perlbio","perlbio");
```

After making the connection, Perl sends SQL statements to MySQL to perform SQL operations such as deletion of a table, creation of a table, and insertion.

Execution of the script will generate a table called HGNC in MySQL database `perlbio`.

4. Retrieve the Data from MySQL

The script `retrieve.pl` provides a console interface to the HGNC core table. The complete code is shown in Figure 14.10. The execution of the program generates a console window and allows the users to query the MySQL database (see Figure 14.11). Note that here we can execute the script only from the console window. If executing from Textpad, a console window will not be brought up.

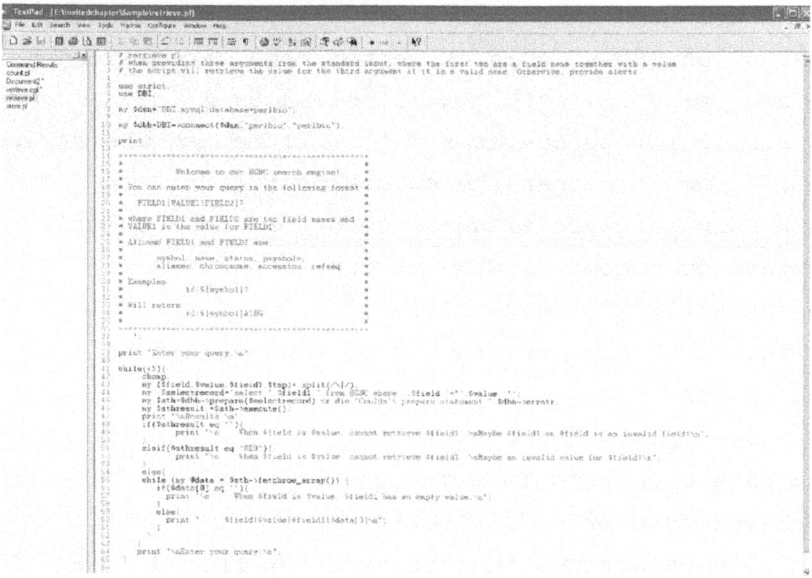

FIGURE 14.10 Screenshot displaying code in the script retrieve.pl that provides a command-line interface for the HGNC core table.

5. Web Application

Assume MySQL and the Web server are in the same computer and the Web server can be accessed as http://www.perlbio.org (replace the link with the address of your Web server). Under the Web server root Web directory, we create a test directory as `perlbio` and under the root cgi-bin directory, we create a test cgi-bin directory as `perlbio-cgi`. Under `perlbio` directory, we create a HTML file `hgnc _ query.htm` (see Figure 14.12 for htm text and screenshot). After making the directory and file readable by anyone (i.e., with `chmod -R 744 perlbio` in Linux/Unix or right-click to change the property of the directory and file to "readable by anyone"). Under `perlbio-cgi` directory, we create a file `retrieve.cgi`, which uses the CGI module to retrieve query parameters, submits the query to MySQL database using DBI, and generates a Web page to display the query result:

```
#!/usr/bin/perl
# retrieve.cgi
use strict;
use CGI;
use DBI;
```

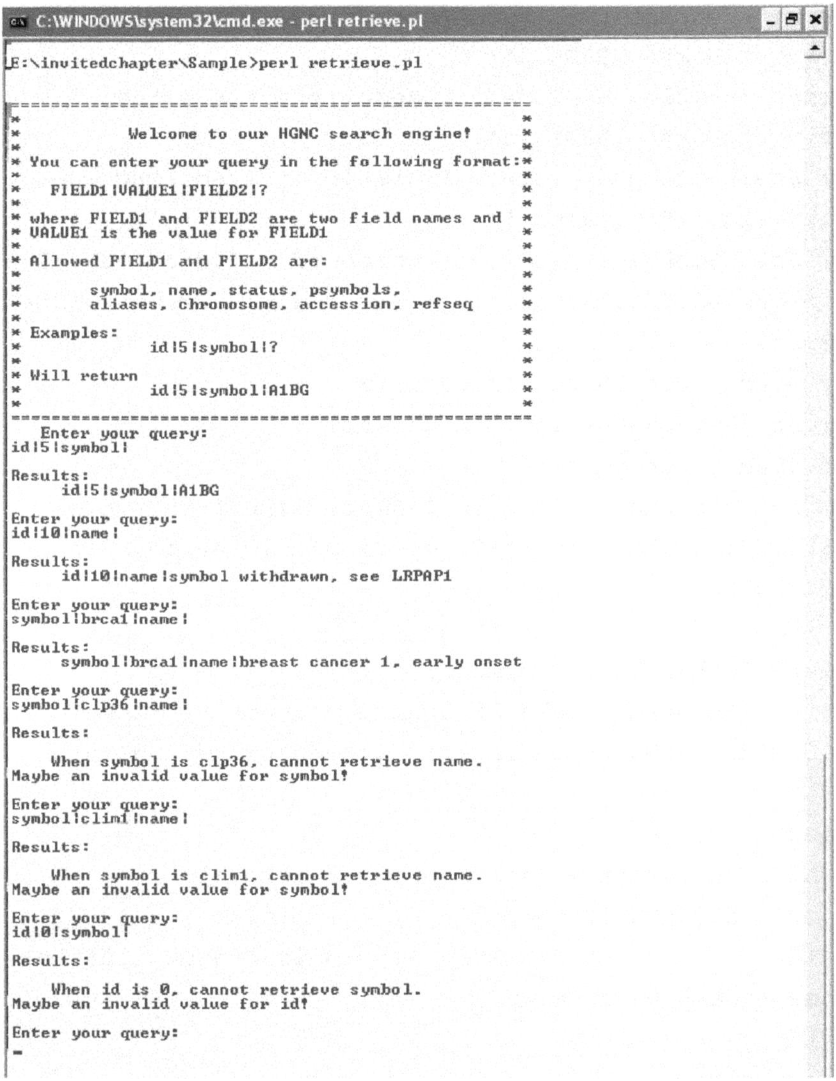
C:\WINDOWS\system32\cmd.exe - perl retrieve.pl
E:\invitedchapter\Sample>perl retrieve.pl
Welcome to our HGNC search engine!
You can enter your query in the following format:
FIELD1|VALUE1|FIELD2|?
where FIELD1 and FIELD2 are two field names and
VALUE1 is the value for FIELD1
Allowed FIELD1 and FIELD2 are:
symbol, name, status, psymbols,
aliases, chromosome, accession, refseq
Examples:
id|5|symbol|?
Will return
id|5|symbol|A1BG
Enter your query:
id|5|symbol|
Results:
id|5|symbol|A1BG
Enter your query:
id|10|name|
Results:
id|10|name|symbol withdrawn, see LRPAP1
Enter your query:
symbol|brca1|name|
Results:
symbol|brca1|name|breast cancer 1, early onset
Enter your query:
symbol|clp36|name|
Results:
When symbol is clp36, cannot retrieve name.
Maybe an invalid value for symbol!
Enter your query:
symbol|clim1|name|
Results:
When symbol is clim1, cannot retrieve name.
Maybe an invalid value for symbol!
Enter your query:
id|0|symbol|
Results:
When id is 0, cannot retrieve symbol.
Maybe an invalid value for id!
Enter your query:

FIGURE 14.11 SCREENSHOT OF THE COMMAND-LINE INTERFACE FOR THE HGNC CORE TABLE.

```
my $dsn="DBI:mysql:database=perlbio";
my $dbh=DBI->connect($dsn,"perlbio","perlbio");
my $query = new CGI;
my $field = $query->param('field');
my $value = $query->param('value');
my $field1 = $query->param('field1');
```

```
print "Content-type: text/html\n\n";
print "<HTML><HEAD>\n";
print "<title> HGNC Search Results </title>\n";
print "</HEAD><BODY>\n";
my $selectrecord='select '.$field1.' from HGNC where '
 .$field.'="'.$value.'"';
my $sth=$dbh->prepare($selectrecord) or die '<H3><font
 color="#CC0000"> Server is not ready! </font></H3></
 BODY>';
my $sthresult =$sth->execute();
print "\n<H3> Results:\n</H3><hr>";
if($sthresult eq ""){
 print "\n When $field is $value, cannot retrieve
 $field1. \nMaybe $field1 or $field is an invalid
 field!\n";
}
elsif($sthresult eq "0E0"){
print "\n When $field is $value, cannot retrieve
 $field1. \nMaybe an invalid value for $field!\n";
}
else{
 while (my @data = $sth->fetchrow_array()) {
      if($data[0] eq ""){
            print "\n When $field is $value, $field1
 has an empty value.\n";
      }
      else{
            print " When $field is $value, $field1 is
 $data[0]\n";
      }
 }
}
print '<hr>';
print '<a href="/test/hgnc_query.htm"> Try Again </
 a>';
print "</BODY>\n";
```

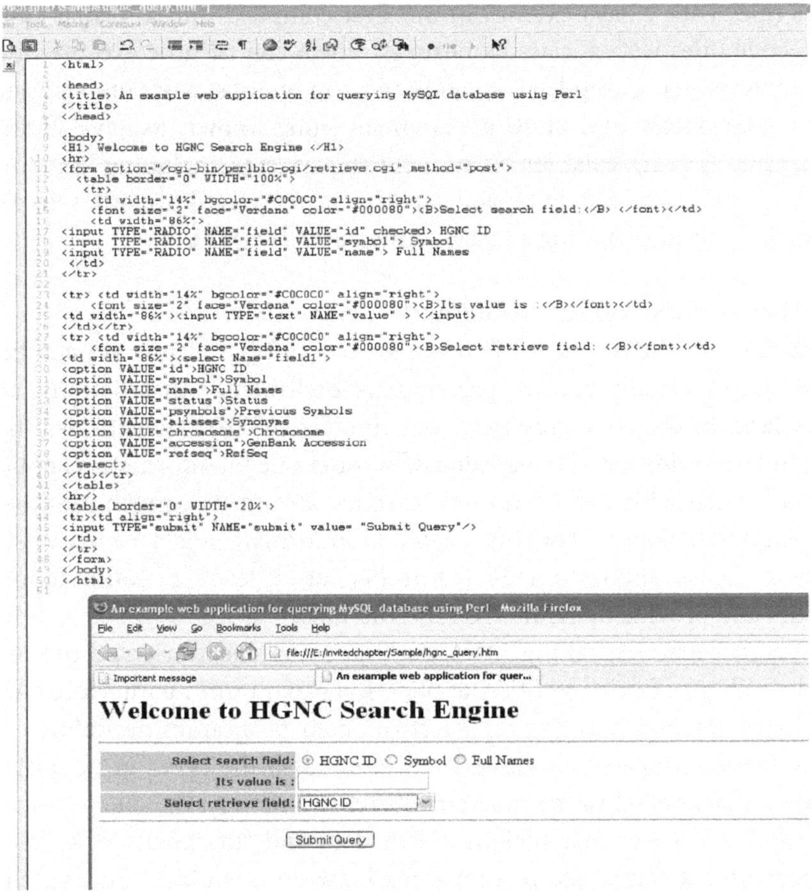
<html>
<head>
<title> An example web application for querying MySQL database using Perl
</title>
</head>
<body>
<H1> Welcome to HGNC Search Engine </H1>
<hr>
<form action="/cgi-bin/perlbio-cgi/retrieve.cgi" method="post">
<table border="0" WIDTH="100%">
<tr>
<td width="14%" bgcolor="#C0C0C0" align="right">
<font size="2" face="Verdana" color="#000080"><B>Select search field:</B> </font></td>
<td width="86%">
<input TYPE="RADIO" NAME="field" VALUE="id" checked> HGNC ID
<input TYPE="RADIO" NAME="field" VALUE="symbol"> Symbol
<input TYPE="RADIO" NAME="field" VALUE="name"> Full Names
</td>
</tr>
<tr> <td width="14%" bgcolor="#C0C0C0" align="right">
<font size="2" face="Verdana" color="#000080"><B>Its value is :</B></font></td>
<td width="86%"><input TYPE="text" NAME="value" > </input>
</td></tr>
<tr> <td width="14%" bgcolor="#C0C0C0" align="right">
<font size="2" face="Verdana" color="#000080"><B>Select retrieve field: </B></font></td>
<td width="86%"><select Name="field1">
<option VALUE="id">HGNC ID
<option VALUE="symbol">Symbol
<option VALUE="name">Full Names
<option VALUE="status">Status
<option VALUE="psymbols">Previous Symbols
<option VALUE="aliases">Synonyms
<option VALUE="chromosome">Chromosome
<option VALUE="accession">GenBank Accession
<option VALUE="refseq">RefSeq
</select>
</td></tr>
</table>
<hr/>
<table border="0" WIDTH="20%">
<tr><td align="right">
<input TYPE="submit" NAME="submit" value= "Submit Query"/>
</td>
</tr>
</form>
</body>
</html>
An example web application for querying MySQL database using Perl   Mozilla Firefox
File Edit View Go Bookmarks Tools Help
file:///E:/invitedchapter/Sample/hgnc_query.htm
Important message
An example web application for quer...
Welcome to HGNC Search Engine
Select search field: HGNC ID Symbol Full Names
Its value is :
Select retrieve field: HGNC ID
Submit Query

FIGURE 14.12 The HTML page that provides a Web form interface for the HGNC core table.

After making the directory executable such as using `chmod -R 755 perlbio-cgi` in Unix/Linux, the search engine can then be accessed at http://www.perlbio.org/perlbio/hgnc_query.htm.

SECTION 2 APPLICATION OF JAVA IN BIOLOGY

Part I Introduction

1. What Is JAVA?

Another popular language used in Biology is JAVA, which was developed by Sun Microsystems. Java is an object-oriented language similar to C++, but simplified to eliminate language features that cause common program errors. Similar to Perl, JAVA is an interpreted language. Java source code files (files with a .java extension) are first compiled into a format called bytecode (files with a .class extension), which can then be executed by a Java interpreter. Compiled Java code can run on most computers because Java interpreters and runtime environments, known as Java Virtual Machines (JVMs), exist for most operating systems, including Windows, Linux/Unix, and Mac. Bytecode can also be converted directly into executable applications using a just-in-time compiler (JIT).

2. Why Is JAVA Popular in Biology?

Traditionally, the language of choice for bioinformaticians has been Perl. Perl scripts usually require prerequisite dependency installations, and they lack the dynamic graphical user interface (GUI) interactions inherent in Java. Additionally, the value of stand-alone bioinformatics applications created with Perl is limited in scope and in its contribution back to research biologists. For this reason, bioinformaticians have been using Java to deliver applications to researchers at all levels of computational ability and provide dynamic GUI interactions.

Another feature making JAVA popular is its object-oriented property and a rich set of existing JAVA application programming interface (API) packages for bioinformatics. Object-oriented programming defines not only the data type of a data structure, but also the types of operations that can be applied to the data structure. In this way, the data structure becomes an object that includes both data and functions. In addition, programmers can create relationships between one object and another. For example, objects can inherit characteristics from other objects. One of the principal advantages of object-oriented programming techniques is that they enable programmers to create modules that do not need to be changed when a new type of object is added. A programmer can simply create a new object that inherits many of its features from existing objects. This makes object-oriented programs easier to modify.

Furthermore, the existence of many bioinformatics-based JAVA applications and APIs motivates bioinformaticians to use them as resources for acquiring or manipulating data. The development of applications and APIs that couple biological and computational knowledge to formally describe complex biological data types significantly reduces the number of conflicting formats, and the time required to access as well as analyze biological data. The following provides several bioinformatics-based applications or APIs that aid in acquiring, processing, and defining biological data, each in its own unique way:

- TIGR MultiExperiment Viewer (MEV) integrates large numbers of experiments to facilitate researchers in analyzing patterns of gene expression. MEV is an important example of a bioinformatics application that aims at integrating bioinformatics experimentation.
- WebMol analyzes molecular structure information. This program uses Java3D to allow researchers to visualize and manipulate complex protein structures.
- Apollo is a genome annotation tool designed to aid in the annotation of genes in various genomes, and has been used to annotate the fruit fly genome and parts of the human genome. Apollo is a good example of a Java bioinformatics tool that integrates biological data to aid in human-facilitated annotation activities.
- Sockeye facilitates comparative genome analysis (the science of comparing genomic-level similarities across species). It allows users to view and compare annotation and sequence from several genomes simultaneously. Sockeye uses Java3D to display genomes and their respective annotations.
- Spice is a browser for protein sequences, structures, and their annotations. It can display annotations for PDB, UniProt, and Ensembl Peptides. All data are retrieved from different sites on the Internet. Annotations are available using the DAS protocol. It is possible to add new annotations to Spice, and to compare them with information already available.
- BioJava is more of an API than an application. Many bioinformatics projects exist for different languages. These APIs try to organize the semantics of working with and manipulating biological data.
- caBIO (Cancer Bioinformatics Infrastructure Objects) is one component of the National Cancer Institute's Centre for Bioinformatics (NCICB) caCORE research management system. caBio is a great solution for bioinformatics developers who want the benefits of defined biomedical objects, coupled with search criteria objects capable of cross-platform data exchange and retrieval.
- PAL (Phylogenetic Analysis Library) is dedicated to the subset of bioinformatics analysis that pertains to the evolutionary development of genomes (DNA and protein sequence). Some of the more

advanced features include phylogenetic tree manipulation, amino acid substitution modeling, and several tree-construction methods.

- MAGE-stk facilitates the loading of MAGE-ML, an XML-based format based on the MIAME standards. Although that may not seem impressive, the Java component of MAGE-stk comprises more than 300 classes, making it one of the heavyweights when compared to the other APIs mentioned here.
- The Knowledge Discovery Object Model (KDOM) is a bioinformatics-based API designed to represent and manage biological knowledge during application development. The goal of KDOM is to create a framework for managing the acquisition and implementation of biological objects. Developers can further define relationships between biological objects to develop a knowledge ontology that is persistent through the system. KDOM also facilitates context-dependent display of biological objects. For instance, a gene can be displayed differently in the context of a chromosome or an exon.
- LSID (The Life Science Identifier) project is I3C URN specification that is being implemented by IBM. The concept boils down to creating a worldwide unique ID for life sciences data that includes the information required to resolve this ID. The LSID project is being developed for Java and Perl as freely downloaded open-source software. Basic LSID-handling capabilities are also appearing in BioJava under org.biojava.utils.lsid.

The effort to compartmentalize bioinformatics tasks into discrete APIs is a fundamental component of standardizing knowledge and creating reliable bioinformatics systems. User adoption and adherence to code-reuse strategies will significantly aid in the construction and delivery of complex analysis systems to researchers with varying computational proficiencies. Here, community efforts must be undertaken to increase the visibility and usability of JAVA APIs among developers. The majority of the existing Java APIs are the products of dedicated individuals or relatively small groups of developers. In many cases, there is little documentation or guarantee that the software will work out of the box. The measure of success has been related to the proportion of the bioinformatics community that utilizes the API. Unfortunately, most novice programmers are overwhelmed when they take their first steps into an object-oriented environment, and abandon or side-step the effort. Furthermore, senior programmers, who are more likely to use these APIs, frequently do not; they are either not aware of them, or the short-term tradeoff between time and the challenge of doing it themselves is not that appealing. Both reactions are understandable, as it is frequently unclear where to go to find these APIs or how to start using one once identified (even if it is bug free).

3. Some JAVA Basics

The latest JAVA API development kit (JDK) and JAVA Runtime Environment (JRE) can be downloaded from http://java.sun.com. The Java Tutorial series (available at http://java.sun.com/docs/books/tutorial/) provides practical guides for programmers who want to use the Java programming language to create applications, with hundreds of complete, working examples, and dozens of lessons.

Writing and executing a JAVA program can be broken into several steps: creating a program, compiling, and executing the program. Similar to Perl, IDEs can be used to simplify and accelerate the development process.

A. Creating the Program A JAVA source code file is just a text file that can be created using a text editor such as vi, emacs, pico, or TextPad. We will illustrate JAVA programming through the famous "Hello World" application.

```
/**
 * The HelloWorldApp class implements an application
 that *
simply prints "Hello World" or "Hello World from ?"
* to standard output.
 */
class HelloWorldApp {
 public static void main(String[] args) {
 if(length(args)>0){
  System.out.println("Hello World from " + args[0]+"\
 n"); // Display the string.
 }
 else{
 System.out.println("Hello World!\
n"); // Display the string.
 }
 }
}
```

An object in JAVA is usually a class that can have other classes as variables and have a collection of methods. It can be an extension of an existing class or can be implemented from several interfaces. The basic form of a class definition is:

```
class name {
  . . .
}
```

The keyword `class` begins the class definition for a class named `name`, and the code for each class appears between the opening and closing curly braces marked in bold above. The file needs to be named after its main class and ended with `.java`. For the above example, the main class is `HelloWorldApp`, so the program needs to be saved as `HelloWorldApp.java`.

Every JAVA program must have the main method, whose signature is:

```
public static void main(String[] args)
```

This array is the mechanism through which the runtime system passes information to the application. Each string in the array is called a *command-line argument*. Command-line arguments let users affect the operation of the application without recompiling it.

Java programming language supports three kinds of comments:

```
/* text */
```

The compiler ignores everything from /* to */.

```
/** documentation */
```

This indicates a documentation comment (`doc comment`, for short). The compiler ignores this kind of comment, just like it ignores comments that use /* and */. The `javadoc` tool uses doc comments when preparing automatically generated documentation.

```
// text
```

The compiler ignores everything from // to the end of the line.

Finally, the line:

```
System.out.println("Hello World!");
```

uses the `System` class from the core library to print the "Hello World!" message to standard output.

B. Compiling and Executing the Program The java program usually compiles using the command `javac` and runs using the command `java`. Usually, you may need to set up your computer environment so that the compilation and execution run smoothly. Let us assume a fresh installation of JDK. Figure 14.13 shows how to set up the computer for compiling and executing java programs:

1. From the start menu or Desktop, right-click "My Computer," and select "Properties."
2. Click on button "Advanced."
3. Click "Environment Variables."

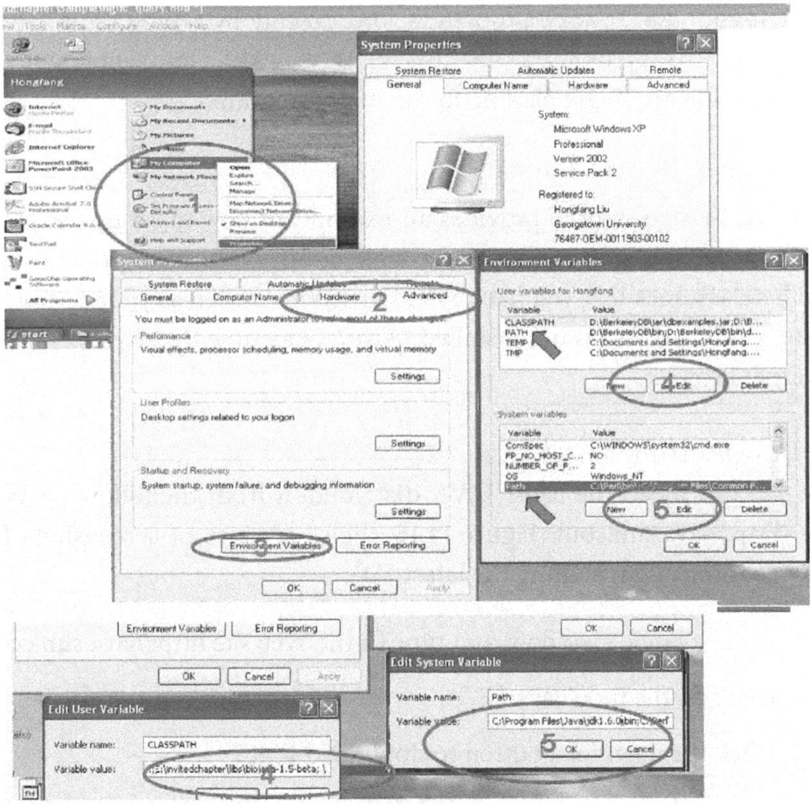
System Properties
System Restore
Automatic Updates
Remote
General
Computer Name
Hardware
Advanced
System:
Microsoft Windows XP
Professional
Version 2002
Service Pack 2
Registered to:
Hongfang Liu
Georgetown University
1
2
Performance
Settings
Environment Variables
Error Reporting
OK
Cancel
Apply
3
Environment Variables
Variable
Value
CLASSPATH
PATH
TEMP
TMP
New
Edit
Delete
4
System variables
ComSpec
OS
Windows_NT
Path
5
Edit System Variable
Variable name:
Path
Variable value:
Edit User Variable
Variable name:
CLASSPATH
Variable value:

FIGURE 14.13 STEPS FOR SETTING UP WINDOW XP'S ENVIRONMENT VARIABLES FOR COMPILING AND EXECUTING JAVA APPLICATIONS.

4. If variable `CLASSPATH` is shown as one of the User variables, select `CLASSPATH` and add ".;" at the beginning. If `CLASSPATH` is not shown, click "New" and create CLASSPATH.
5. Modify System variables `Path` and append the JDK bin directory (e.g., `C:\Program Files\Java\jdk1.6.0\bin`) at the beginning.

Now, we can compile and execute the program by restarting the computer and bringing up a command window (see Figure 14.14):

1. From the start menu select "run."
2. Type `cmd in` the popup run window.
3. Go to the directory where the java program saved, compile by typing `javac HelloWorldApp.java` and execute by typing `java HelloWorldApp` or `java HelloWorldAPP PERL`, etc., where `PERL` is the first argument passing to the main program.

Part II Step-By-Step Tutorial

The step-by-step tutorial provides an example of how to set up the computer for testing Biojava scripts available at http://biojava.org/. The tutorial here assumes that you have a PC with Window XP as your operation system and that java is not installed. Also, we assume that Firefox is your Internet browser.

1. Download and Install Java

We can download the latest JAVA distribution from the following Web site: http://java.sun.com. Figure 14.15 shows a series of screenshots for downloading and installing the latest JDK:

1. Open a Firefox window, and type in the Web site http://java.sun.com.
2. Click on the link "Java SE."
3. Click the download button to download JDK6.
4. Accept the agreement and click Windows online installation, Multi-language, and start the download.
5. Open the download file in the Firefox download window and follow the installation directions to install the program.

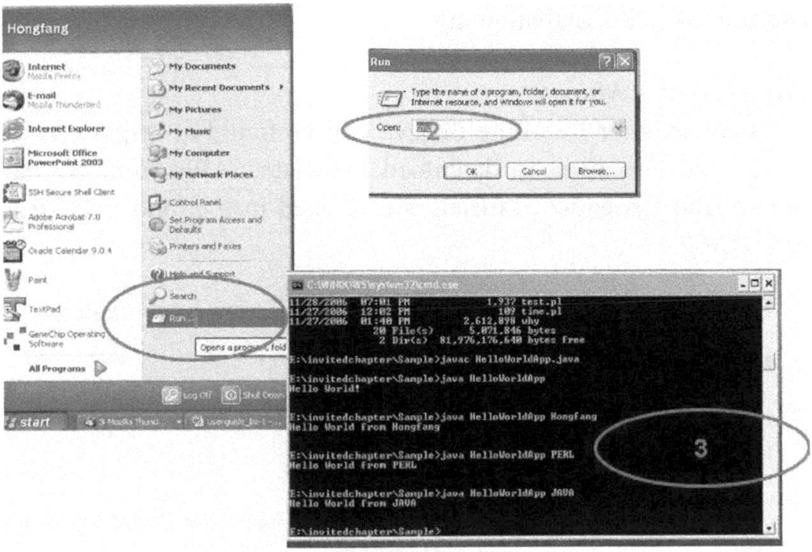
Hongfang
Internet
E-mail
Internet Explorer
Microsoft Office PowerPoint 2003
SSH Secure Shell Client
Adobe Acrobat 7.0 Professional
Oracle Calendar 9.0.4
Paint
TextPad
GeneChip Operating Software
All Programs
My Documents
My Recent Documents
My Pictures
My Music
My Computer
My Network Places
Control Panel
Set Program Access and Defaults
Printers and Faxes
Help and Support
Search
Run...
Log Off
Shut Down
start
Run
Type the name of a program, folder, document, or Internet resource, and Windows will open it for you.
Open:
OK
Cancel
Browse...
2
E:\invitedchapter\Sample>javac HelloWorldApp.java
E:\invitedchapter\Sample>java HelloWorldApp
Hello World!
E:\invitedchapter\Sample>java HelloWorldApp Hongfang
Hello World from Hongfang
E:\invitedchapter\Sample>java HelloWorldApp PERL
Hello World from PERL
E:\invitedchapter\Sample>java HelloWorldApp JAVA
Hello World from JAVA
E:\invitedchapter\Sample>
3

FIGURE 14.14 Run Java from the Command Window.

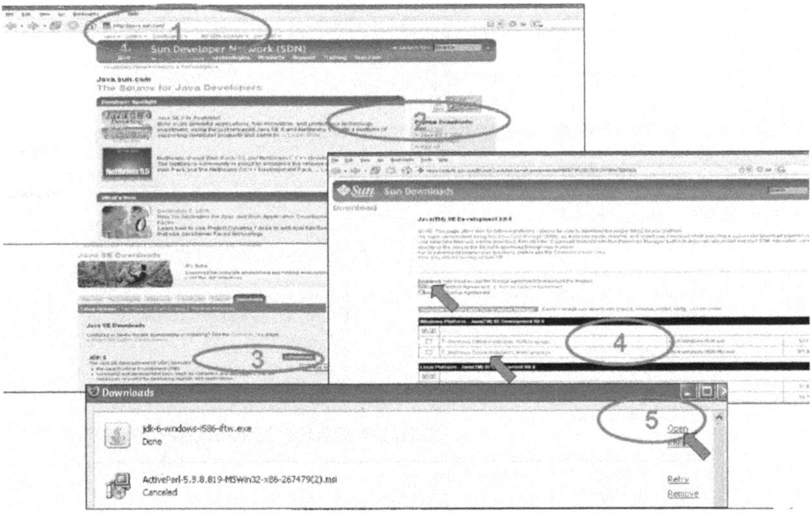
1
Sun Developer Network (SDN)
2
3
Sun Downloads
4
Downloads
5
jdk-6-windows-i586-iftw.exe
Done
Open
ActivePerl-5.8.8.819-MSWin32-x86-267479(2).msi
Canceled
Retry
Remove

FIGURE 14.15 Overview of the installation of JAVA in Window XP.

2. Download and Install TextPad

This step is the same as shown in Section 1. We need to reinstall TextPad to capture the JDK installation information.

3. Install Biojava Module

The instructions for installing Biojava can be found at http://www.biojava.org. You also need to download associated APIs from the same Web site. The procedure to install is described in the following list (see Figure 14.16):

1. Go to "C:" and right-click the mouse, and create a new folder.
2. Name the folder as `javalibs` under "C":.
3. Open a Firefox window and type http://www.biojava.org/wiki/BioJava:Download.

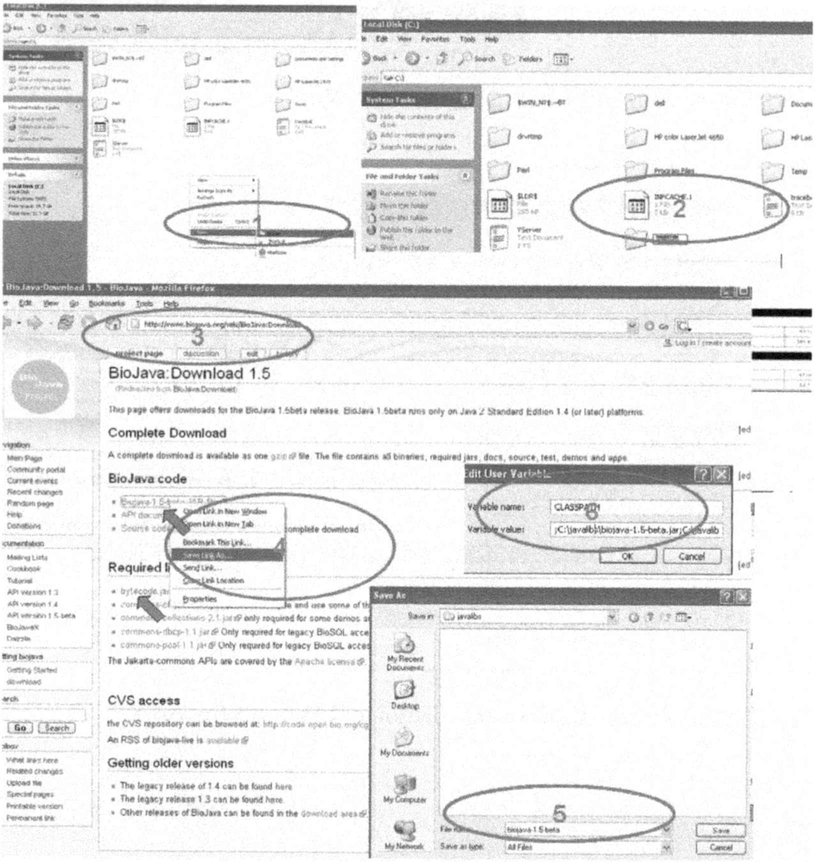

Local Disk (C:)
System Tasks
File and Folder Tasks
2
3
BioJava:Download 1.5
Complete Download
BioJava code
Required l
CVS access
Getting older versions
Edit User Variable
Variable name:
CLASSPATH
Variable value:
OK
Cancel
6
Save As
Save in:
javalibs
My Recent Documents
Desktop
My Documents
My Computer
My Network
5
Save as type:
All Files
Save
Cancel

FIGURE 14.16 Overview of the installation of Biojava in Window XP.

4. Point to `Biojava-1.5-beta.Jar file`, right-click, and select "Save the Link as."

5. Save the file to `C:\javalibs` (repeat 4 and 5 for bytecode.jar).

6. Configure `ClassPath` to include those two jar files.

4. Start TextPad, Configure for Syntax Highlight as well as Compiling and Running JAVA

By default, TextPad is configured for syntax highlight and java execute. Otherwise, follow the step shown in the tutorial of the previous section.

5. Program Using Biojava

Biojava contains 12 core biological packages (see Figure 14.17 for a screenshot of Biojava1.4 API). There are a lot of example programs available through the Biojava Web site. We will demonstrate one program using Biojava, which transfers GenBank-format DNA sequences to Fastaformat. Before introducing the code, we need to have a file that stores sequence records in GenBank format. One way to get such a file is to use the NCBI Entrez system (Figure 14.18).

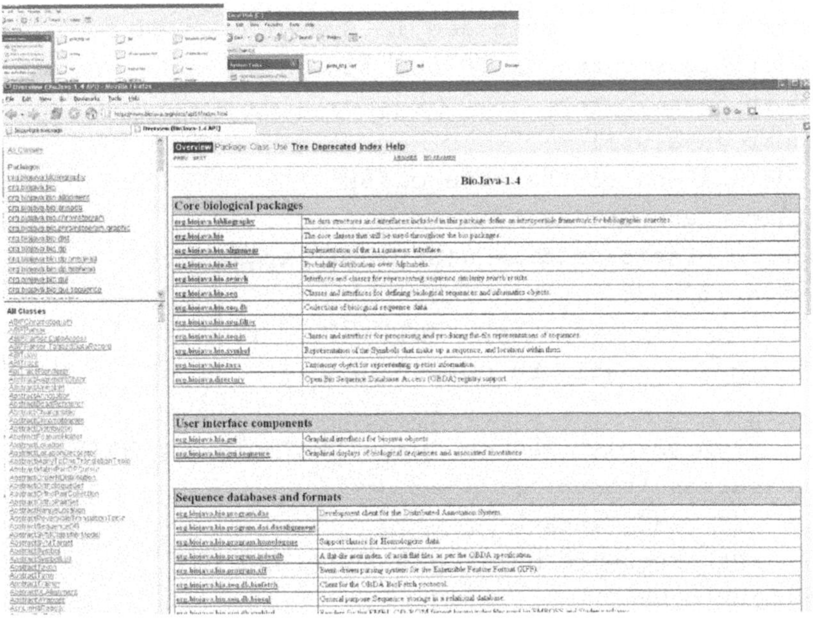
Overview
BioJava-1.4
Core biological packages
All Classes
User interface components
Sequence databases and formats

FIGURE 14.17 Core biological packages in Biojava.

1. Go to http://www.ncbi.nih.gov and search for “Nucleotide.”
2. Type CLIM1 (or any other gene symbol).
3. Select Display and GenBank.
4. Show to format as File.

A file named `sequences.gb` will be generated on the desktop. Let us place it in the same directory where you work on the code. Let us name the code file `GB2FASTA.java`.

```
//GB2FASTA.java
//An example program to transfer DNA sequences in
//GenBank format//to FASTA format
//import API packages
import org.biojavax.bio.seq.*;
import org.biojavax.bio.seq.io.*;
import org.biojavax.Namespace;
import org.biojavax.SimpleNamespace;
import org.biojavax.RichObjectFactory;
import org.biojava.bio.seq.DNATools;
import org.biojava.bio.seq.io.SymbolTokenization;
import org.biojava.bio.BioException;
import java.io.*;
import java.util.*;
// definition of the main class
public class GB2FASTA{
 public static void main(String[] args) {
 BufferedReader input=null;
 FileOutputStream output=null;
 // sequences will be DNA sequences
 try{
      SymbolTokenization dna = DNATools.getDNA().getTo
 kenization("token");
      // read Genbank
      RichSequenceFormat genbank = new
 GenbankFormat();
      // write FASTA
      RichSequenceFormat fasta = new FastaFormat();
      // compress only longer sequences
```

```
RichSequenceBuilderFactory factory =
RichSequenceBuilderFactory.THRESHOLD;
      // read/write everything using the 'perlbio'
//namespace
      Namespace perlbioNS = (Namespace) RichObjectFac-
 tory.getObject(
SimpleNamespace.class, new Object[]{"perlbio"});
// read seqs from file "sequences.gb"
// write seqs to file "sequences.fasta"
      try{
            input = new BufferedReader(new
 FileReader("sequences.gb"));
            output = new FileOutputStream("sequences.
 fasta");
}
catch(IOException e){
            e.printStackTrace();
}
RichStreamReader seqsIn = new
RichStreamReader(input,genbank,dna,factory,perlbioNS);
RichStreamWriter seqsOut = new RichStreamWriter(output
 ,fasta);
// one-step Genbank to Fasta conversion!
try{
            seqsOut.writeStream(seqsIn,perlbioNS);
      }
catch (IOException ex) {
            ex.printStackTrace();
}
}
catch (BioException ex) {
//not in GenBank format
ex.printStackTrace();
}
}
}
```

The program GB2FASTA.java can be compiled and executed by bringing up a Command window or in TextPad.

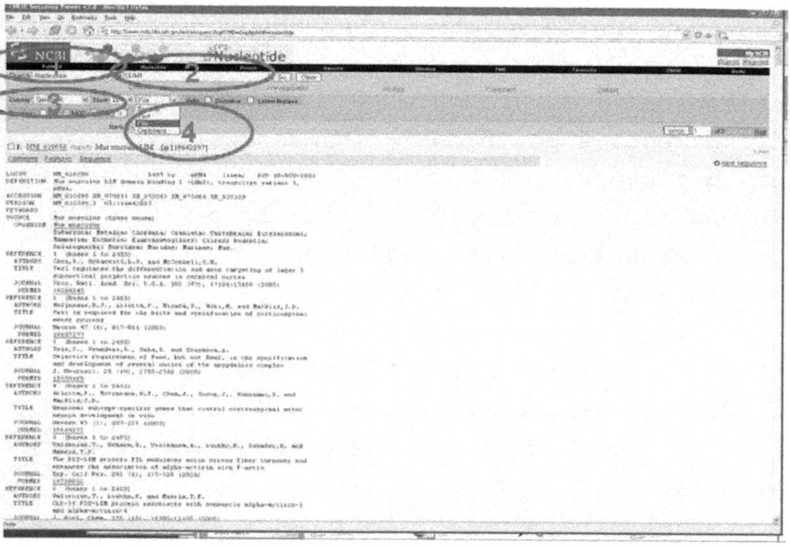
NCBI
Nucleotide
2
4

FIGURE 14.18 Process to obtain a Genbank format sequence file.

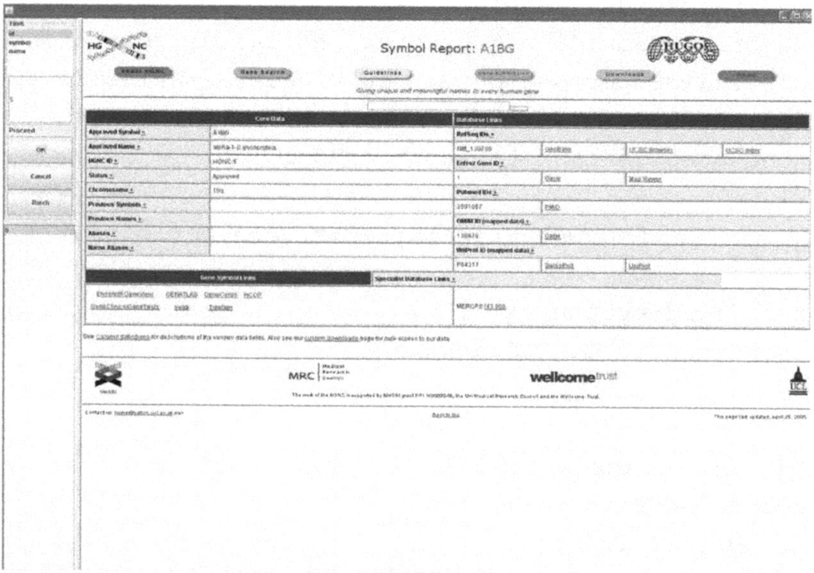
Symbol Report: A1BG
MRC
wellcome trust

FIGURE 14.19 The JAVA graphical user interface (GUI) for the HGNC core table.

Part III An Application Example

In the previous section, we have shown two different ways (i.e., console or Web) of querying the HGNC table. In the following text, we introduce another way to query the table, i.e., a stand-alone JAVA GUI application. The approach used by JAVA to interact with DBMSs is the same as Perl's (i.e., a two-tier approach), in which the top level is the JDBC interface, and the bottom level consists of various database driver packages. The JDBC interface allows developers to write applications that can be used with different databases with a minimum of porting effort. Once a driver for a given server engine is installed, JDBC applications can communicate with any server of that type.

In order to use JDBC to access MySQL, we need to download a database driver for MySQL. There are several drivers available. We recommend MySQL Connector/J, which can be downloaded at (available at http://www.mysql.com/products/connector/j/). Similar to the download and installation of Biojava, we copy the jar file to the directory `C:/javalibs/` and modify the `CLASSPATH` environment accordingly.

The graphical user interface (see Figure 14.19) consists of three panels: Search, Result, and Web site, where

1. Search panel includes three components: a list of fields that can be used to search, a textfield to enter query, and a group of buttons to execute the query.
2. Result panel is simply a list that displays a list if HGNC identifiers retrieved.
3. Web site panel is the same as the search result page of the HGNC Web site http://www.gene.ucl.ac.uk/.

We explain the code step by step.

The beginning of the code imports APIs used in the code.

```
import javax.swing.*;
import java.awt.event.*;
import javax.swing.event.*;
import javax.swing.border.*;
import java.awt.*;
import java.util.*;
import java.io.*;
```

```
import java.net.URL;
import java.sql.*;
```

JAVA controls the accessibility of a member, including methods, variables, and classes. There are three levels of accessibility: public, private, and protected. Public members allow other members to access them, private members only allow members in the same class to access them, and protected members can be accessed only within their own package. The main class `HGNCBrowser` here is a public class, and it extends an object called `JFrame` and implements two interfaces: `ListSelectionListener` and `HyperlinkListener`.

```
public class HGNCBrowser extends JFrame implements
 ListSelectionListener, HyperlinkListener{
```

The following are class-level variables, and all of them are private variables. They can only be accessed by members in the same class. `JEditorPane`, `JSplitPane`, `JScrollPane`, and `JList` are JAVA Swing classes, which provide different kinds of graphical user interface. `Connection` and `Statement` are JAVA JDBC classes that serve as top-level interfaces.

```
private JEditorPane definition=new JEditorPane();
//A panel to display webpage
private JSplitPane leftPane;
private JScrollPane rightPane;
private DefaultListModel listModel=new DefaultList-
 Model();
//store a list of HGNC identifiers
private JList list=null; //a swing object for list
private String field=""; //the search field
private Connection conn=null; //JDBC connection
private Statement s=null;
```

Every JAVA program must have the `main` method in order to be executable. The `main` program here just brings up the window and specifies the window's property including size and visibility.

```
public static void main(String[] args) {
 HGNCBrowser frame = new HGNCBrowser();
 Dimension screenSize = Toolkit.getDefaultToolkit().
 getScreenSize();
 frame.setBounds(5, 5,
```

```
 screenSize.width - 20,
 screenSize.height-20);
frame.addWindowListener(new WindowAdapter() {
      public void windowClosing(WindowEvent e) {
      System.exit(0);
      }
 });
 frame.setVisible(true);
}
```

The following defines a constructor that initializes class `HGNCBrowser` with no parameters.

```
public HGNCBrowser(){
```

The first task when initializing the constructor is to set up the connection to MySQL database and hook up the `Statement` with the `Connection`. Exceptions need to be caught here.

```
try{
 Class.forName("com.mysql.jdbc.Driver").newInstance();
 conn = DriverManager.getConnection("jdbc:mysql:///
 perlbio", "perlbio", "perlbio");
s=conn.createStatement();
}
catch(Exception e){
 e.printStackTrace();
}
```

The following code sets up the components in the `JFrame` and adds them to the `JFrame's` content panel. Note that we set `SelectionMode` and add `ListSelectionListener` to the `JList` list that displays a list of HGNC identifiers returned by the query. `mySearchLabel` is a class inside `HGNCBrowser` that sets up the Search panel.

```
list = new JList(listModel); //define JList to display
//the list of HGNC identifiers
list.setSelectionMode(ListSelectionModel.SINGLE_SELEC-
 TION); //only allow single selection
list.addListSelectionListener(this);
mySearchLabel msl=new mySearchLabel();
```

```
//combine the search panel annd right panel into the
//left pane
leftPane=new JSplitPane(JSplitPane.VERTICAL_SPLIT, new
 JScrollPane(msl), new JScrollPane(list));
leftPane.setDividerLocation(420);
//a scrollpane to display the Web site
rightPane=new JScrollPane(definition);
JSplitPane splitPane = new JSplitPane(JSplitPane.
 HORIZONTAL_SPLIT, leftPane,rightPane);
splitPane.setDividerLocation(140);
//add the splitpane into the Frame
getContentPane().add(splitPane);
pack();
}
```

The valueChanged method is to change the Web site display according to the selection of the list in the result panel.

```
public void valueChanged(ListSelectionEvent e) {
 if (e.getValueIsAdjusting()) return;
 JList theList = (JList)e.getSource();
 if (theList.isSelectionEmpty()) {
      theList.setSelectedIndex(0);
}
else {
 int index=theList.getSelectedIndex();
int intval=(Integer) listModel.elementAt(index);
//display the HGNC webpage
String def="http://www.gene.ucl.ac.uk/nomenclature/
 data/get_data.php?hgnc_id="+intval;
try {
      definition.setEditable(false);
      definition.addHyperlinkListener ( this ) ;
      definition.setPage(new URL(def));

 definition.setSize(definition.getPreferredSize());
 }
 catch(Exception ex){
      definition.setText ("Could not load
 page:"+def+"\n"+"Error:"+ex.getMessage());
```

```
 }
 }
}
```

The method hyperlinkUpdate allows users to activate links in the Web panel.

```
public void hyperlinkUpdate ( HyperlinkEvent e ) {
if (e.getEventType( ) == HyperlinkEvent.EventType.
 ACTIVATED ) {
try {
URL url = e.getURL ( ) ;
definition.setPage ( url ) ;
}
catch ( Exception ex ) {
definition.setText ("Could not load page!\
 n"+"Error:"+ex.getMessage());
 }
}
}
```

The following is the code for `mySearchLabel`. It sets up the search panel and associates actions with buttons and list selections.

```
public c lass mySearchLabel extends JPanel implements
 ActionListener, ListSelectionListener{
//varibles
JTextField tf = new JTextField(10);
// the textfield for entering query terms
DefaultListModel fieldList=new DefaultListModel();
// to hold the list of fields
mySearchLabel(){ //constructor
 //allow to search according to id, symbol, and name
 fieldList.addElement(new String("id"));
 fieldList.addElement(new String("symbol"));
 fieldList.addElement(new String("name"));
 JList fieldlist=new JList(fieldList);
fieldlist.setSelectionMode(ListSelectionModel.SINGLE_
 SELECTION);
fieldlist.addListSelectionListener(this);
fieldlist.setSelectedIndex(0);
```

```
//three buttons used to control the behavior of
//searching
JButton b1 = new JButton("OK");
 JButton b2 = new JButton("Cancel");
 JButton b3 = new JButton("Batch");
 b1.addActionListener(this);
 b2.addActionListener(this);
 b3.addActionListener(this);
 JPanel tfp=new JPanel();
 tfp.add(fieldlist);
 tfp.add(tf);
 tfp.setLayout(new GridLayout(2,1));
 tfp.setBorder(new TitledBorder(new EtchedBorder(),
 "Find: "));
tfp.setSize(tfp.getMinimumSize());
 add(tfp);
 JPanel p2=new JPanel(new GridLayout(3,1));
p2.add(b1);
 p2.add(b2);
 p2.add(b3);
 p2.setBorder(new TitledBorder("Proceed"));
add(p2);
setLayout(new BoxLayout(this,BoxLayout.Y_AXIS));
}
//the following modifies the field used for search
public void valueChanged(ListSelectionEvent e) {
if (e.getValueIsAdjusting()) return;
JList theList = (JList)e.getSource();
int index=theList.getSelectedIndex();
field=(String) fieldList.elementAt(index);
}
//the following lists actions when pressing buttons
public void actionPerformed(ActionEvent e){
if(e.getActionCommand().equals("OK")){
String str=tf.getText();
 if(str.equals("")){
 definition.setText("Please input text for search");
 //if it is empty, doing nothing
 }
```

```
 else{
      try{
            //remove previous results
listModel.removeAllElements();
s.executeQuery("SELECT id from HGNC where "+field+"=\
 ""+str+"\"");
      ResultSet rs=s.getResultSet();
      while(rs.next()){
            int idval = rs.getInt("id");
            listModel.addElement(idval);
            }
      //select the first result to display the //
corresponding webpage on the right panel
      list.setSelectedIndex(0);
      }
      catch (Exception ex){
            ex.printStackTrace();
            }
      }
}
else if (e.getActionCommand().equals("Batch")){
      //open a file chooser to select a file
//with query terms
      JFileChooser fc = new JFileChooser();
      fc.showOpenDialog((HGNCBrowser) SwingUtilities.
 getRoot(this));
      File selFile = fc.getSelectedFile();
try{
listModel.removeAllElements();
FileInputStream fstream = new FileInputStream(selFile)
 ;
      BufferedReader in = new BufferedReader(new Input
 StreamReader(fstream));
      String str=in.readLine();
      while (str!=null) {
s.executeQuery("SELECT id from HGNC where "+field+"
 = \""+str+"\"");
            ResultSet rs=s.getResultSet();
            while(rs.next()){
```

```
                    int idval = rs.getInt("id");
                    listModel.addElement(idval);
                    }
                    str=in.readLine();
                    }
        }
        //select the first result to display the
    //corresponding webpage on the right panel
        list.setSelectedIndex(0);
        }
        catch (Exception ex){
              ex.printStackTrace();
        }
    }
    else if(e.getActionCommand().equals("Cancel")){
        tf.setText("");
    }
}
} // the class end for mySearchLabel
} // the class end for HGNCBrowser
```

SECTION 3 APPLICATION OF R IN BIOLOGY

Part I Introduction

1. What Is R Programming?

R (www.r-project.org) is a widely used open-source language and environment for statistical computing and graphics — GNU's S-Plus. It provides a high-level programming environment together with a sophisticated packaging and testing paradigm for statistical analysis and presentation. It also has a number of mechanisms that allow it to interact directly with software that has been written in many different languages. The main object types in R include the following:

- Vectors: ordered collection of numeric, complex, logical and character values
- Factors: vector object used to specify grouping of its components
- Arrays and matrices: multidimensional arrays of vectors
- Lists: general form of vectors with different types of elements

- Data frames: matrixlike structures with different types of columns
- Functions: piece of code

R contains most arithmetic functions such as `mean`, `median`, `sum`, `prod`, `sqrt`, `length`, `log`, etc. An extensive list of R functions can be found on http://cran.r-project.org/manuals.html. Many R functions and datasets are stored in separate packages, which are only available after loading them into an R session.

2. Why Is R Popular in Biology?

Since its creation, R has revolutionized the statistical data analysis for most bioscience and chemistry disciplines. The R environment is completely free and runs on all common operating systems. The required time to learn the R software is well invested, as the R environment covers an unmatched spectrum of statistical tools, including an efficient programming language for automating time-consuming analysis routines. Specifically, the BioConductor project provides many additional R packages for statistical data analysis in biosciences, such as tools for the analysis of SNP and transcriptional profiling data derived from SAGE, cDNA microarrays, and Affymetrix chips. Because of their popularity, R and BioConductor are continuously updated and extended with the latest analysis tools that are available in the different research fields. BioConductor updates twice a year. Figure 14.20 shows the increasing trend of the number of packages in BioConductor. Packages in BioConductor can be browsed at the BioConductor Web site (http://www.bioconductor.org.) Figure 14.21 shows packages available in BioConductor for one-channel microarray data analysis.

3. Some R Basics

Precompiled binary distributions of the base system and contributed packages for various platforms can be downloaded at http://cran.r-project.org/. Online documentation for most of the functions and variables in R exists, and can be printed onscreen by typing help(*name*) (or ?*name*) at the R prompt, where *name* is the name of the topic help sought for. A very informative introduction to R can be found at http://cran.r-project.org/doc/manuals/R-intro.html. Different from Perl and JAVA, R is an expression language that interacts with users directly. When R is invoked, it issues a prompt and expects input commands. The default prompt is ">".

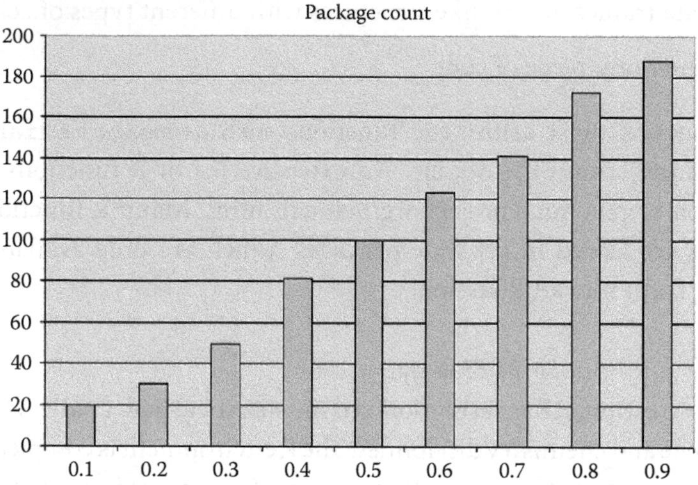
Package count
200
180
160
140
120
100
80
60
40
20
0
0.1
0.2
0.3
0.4
0.5
0.6
0.7
0.8
0.9

FIGURE 14.20 The number of packages for each Bioconductor release.

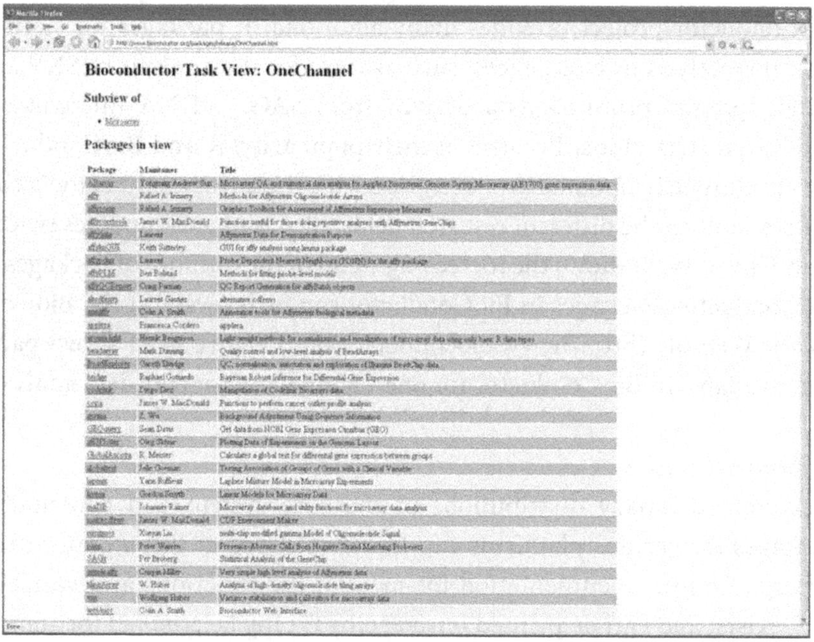
Bioconductor Task View: OneChannel
Subview of
Packages in view
Package
Maintainer
Title

FIGURE 14.21 An overview of the packages available in Bioconductor for one-channel microarray.

Input commands can be directly typed or read from an external file using the function "source." To quit the R program, the command is `q()`.

Part II Step-By-Step Tutorials

Bioconductor is an open-source and open development software project to provide tools for the analysis of SNP and transcriptional profiling data (SAGE, microarrays or Affymetrix chips) and the integration of genomic meta data. In the following, we demonstrate how to get R and Bioconductor and work on an example.

1. Download and Install R

We can download the latest R distribution from the following Web site: http://cran.r-project.org/. Figure 14.22 shows a series of screenshots:

1. Open a Firefox window and type in the Web site http://cran.r-project.org.
2. Click on the link "Windows (95 or later)."
3. Click the link "R-2.4.0-win32.exe" (the version number can change from time to time).

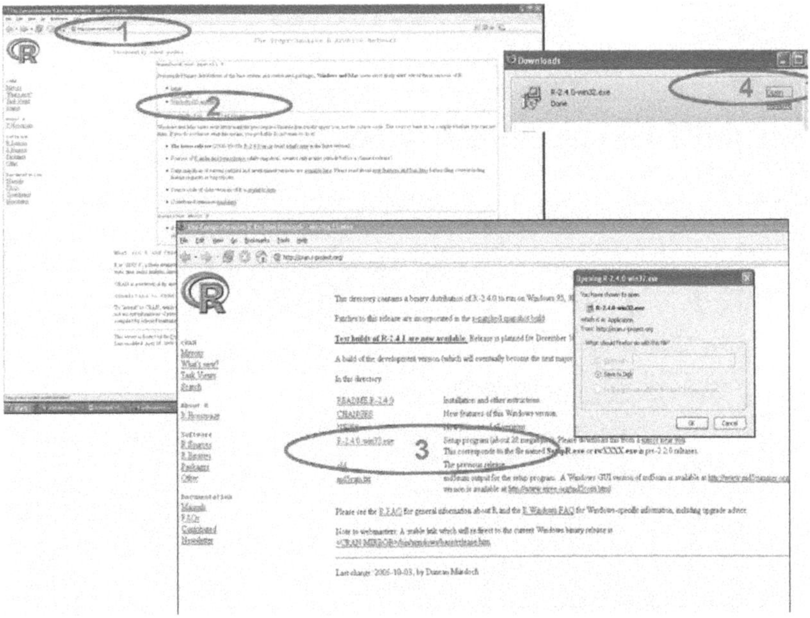
1
2
Downloads
4
3

FIGURE 14.22 Overview of the installation of R in Window XP.

4. Open the download file in the Firefox download window and follow the installation directions to install the program.

2. Install BioConductor

The instructions for installing Bioconductor can be found at http://www.bioconductor.org/. The procedure to install Bioconductor is shown in the following (see Figure 14.23):

1. Go to "Start" in Windows and hold the mouse "All Programs" and invoke R.
2. Type `source(http://www.bioconductor.org/getBioC.R)` and hit enter.
3. Type `getBioC()` and hit enter, and a list of default packages in BioConductor will be installed.
4. Use the library function to load packages needed for the data analysis. For example, `library(affy)` will load the `affy` package, a package for analyzing Affymetrix chips.

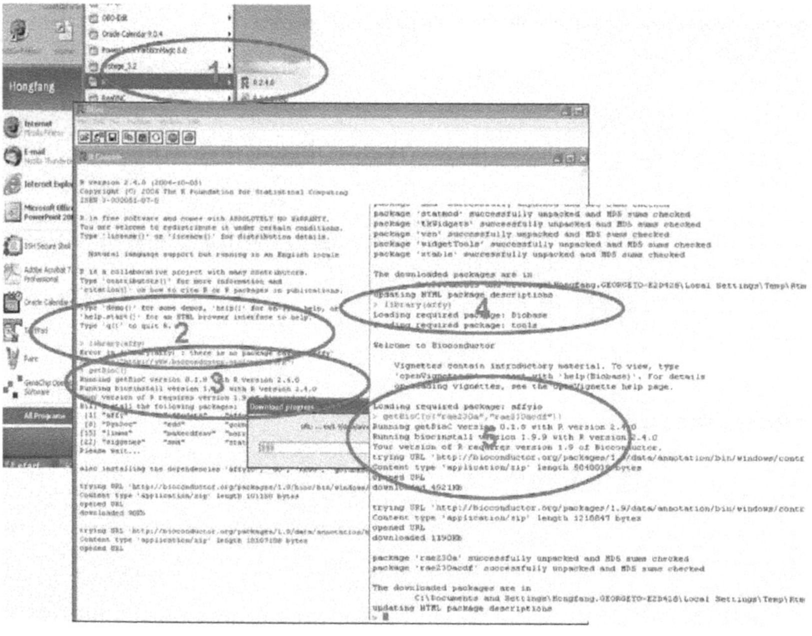
Hongfang
Download progress
1
2
3
4
5

FIGURE 14.23 Overview of the installation of Bioconductor.

5. Install packages that are not in the default list. Note that the command `getBioC()` only installs a list of default packages in BioConductor. In case a package is not included in the list, you can use `getBioC(PackageList)` to install other packages. For example, `getBioC("rae230a","rae230acdf")` will install two packages: the annotation package `rae230a` and the Chip Definition File (CDF) package `rae230acdf`, which will be used in Step 3.

3. Perform Microarray Data Analysis Using BioConductor

BioConductor provides extensive resources for the analysis of Affymetrix data. Here we introduce the `affy` package, which provides basic methods for analyzing Affymetrix chips. Let us first derive some CEL files from GEO from one of the GEO deposited data set ftp://ftp.ncbi.nih.gov/pub/geo/DATA/supplementary/series/GSE2275/. After unzipping and unpacking the data, let us move all the CEL files into a newly created directory under "C:" in the computer called `biochip` (shown in Figure 14.24). The following steps outline the usage of the `affy` package and associated packages.

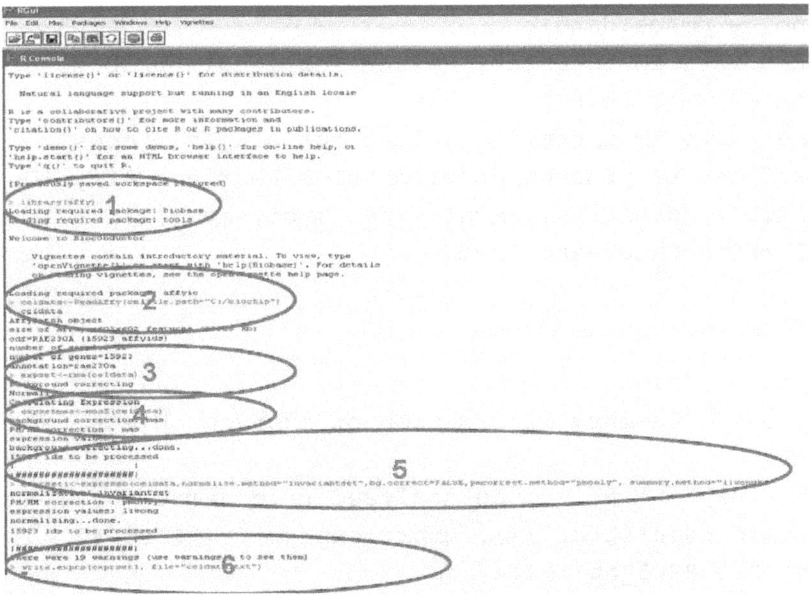
R Console
1
2
3
4
5
6

FIGURE 14.24 An illustration of using Bioconductor for microarray data analysis.

1. Use `library(affy)` to load the `affy` package.
2. Use `ReadAffy` to read `Cel` files and store them into affybatch object `celdata`.
3. Obtain expression values using `RMA` method and store in an object `expset` in standard `exprSet` format.
4. Obtain expression values using `MAS5` method and store in an object `expsetmas` in standard `exprSet` format.
5. Generate expression calls similar to `dChip (MBEI)` method from Li and Wong and store in an object `expset1`.
6. Export data from `expset1` object to text file `celdata.txt` in `C:\biochip` directory.

There are more functions in the `affy` package. The following provides an overview of them; each block can be copied and pasted to the R console for execution (after performing the previous six steps). Note that anything following the sign "#" is a comment.

A. Obtain Single Probe-Level Data From `affybatch` Objects

```
mypm <- pm(celdata)
# Retrieve PM intensity values for single probes
mymm <- mm(celdata)
# Retrieve MM intensity values for single probes
myaffyids <- probeNames(celdata) # Retrieve Affy IDs
result <- data.frame(myaffyids, mypm, mymm)
# Combine into data frame
```

B. Work with `exprSet` Objects

```
expset
# Provide summary information of exprSet object 'exp-
 set'
pData(expset) #List the analyzed file names
exprs(expset)[1:2,1:4]; exprs(expset)[c("1398982_
 at","1398983_at"),1:4]
# Retrieve specific rows and fields of exprSet object.
test <- as.data.frame(exprs(expset))
covN <- list(sample="arbitrary numbering")
```

```
geneCov <- data.frame(row.names=names(test), sample=1:
 length(names(test)))
phenoD <- new("phenoData", pData=geneCov,
 varLabels=covN);
eset2 <- new("exprSet", exprs=as.matrix(test),
 phenoData=phenoD)
# Example for creating an exprSet object from a data
#frame or a matrix.
data.frame(expset)
#Print content of 'expset' as data frame to STDOUT
```

C. Retrieve Annotation Data For `affy` IDs

```
library(rae230a) # Open library with annotation data.
library(help=rae230a)
# Show availability and syntax for annotation data.
library(rae230acdf)
ls(rae230acdf)
# Retrieve all Affy IDs for a chip in
# vector format.
x <- c("1398982_at", "1398983_at")
# Generate sample data set of Affy ID numbers.
mget(x, rae230aACCNUM, ifnotfound=NA)
# Retrieve locus ID numbers for Affy IDs.
mget(x, rae230aCHR)
#Retrieve chromosome numbers
mget(x, rae230aCHRLOC)
# Retrieve chromosome locations of Affy IDs.
mget(x, rae230aGO)
# Retrieve GO information for Affy IDs.
```

D. Access Probe Sequence Data

```
source("http://www.bioconductor.org/getBioC.R");getBio
 C(c("rae230aprobe")); library(rae230aprobe)
#Install and open library with probe sequence data.
print.data.frame(rae230aprobe[1:22,])
# Print probe sequences and their position for first
#twenty two Affy IDs.
pm <- rae230aprobe$sequence
```

```
# Assign sequence component of list object
#'rae230aprobe' to vector 'pm'.
mm <- complementSeq(pm, start = 13, stop = 13)
# Create mismatch sequences with complementSeq
# function by flipping the middle base at position 13.
cat(pm[1], mm[1], sep = "\n")
# The generic 'cat' function produces output in user
# defined format. Here: pm aligned above mm.
reverseSeq(complementSeq(pm[1])) # Command to generate
#reverse and complement of a sequence.
```

E. Visualization and Quality Controls

```
deg <- AffyRNAdeg(celdata)
# Performs RNA degradation analysis.
# It averages on each chip the probes relative to the
 5'/3' position on the target genes
summaryAffyRNAdeg(deg)
# A summary list and a plot are returned
plotAffyRNAdeg(deg)
hist(celdata[ ,1:2]) # Plots histogram of PM
# intensities for 1st and 2nd array
hist(log2(pm(celdata[,1])), breaks=100, col="blue")
# Plot bar histogram of the PM ('pm') or MM ('mm') log
# intensities of 1st array
boxplot(celdata,col="red")
# Generates boxplot of unnormalized log intensity
# values
boxplot(data.frame(exprs(expset)),col="blue",
 main="Normalized Data")
# Generates boxplot of normalized log intensity values
mva.pairs(pm(celdata)[,c(1,4)])
# Creates MA-plot for un-normalized data
# A MA-plot is a plot of log-intensity ratios
# (M-values) versus log-intensity averages
# (A-values) between selected chips (here '[1,4]')
>mva.pairs(exprs(expset)[,c(1,4)])
# Creates MA-plot for normalized data.
```

REFERENCES

1. Gautier, L., Cope, L., Bolstad, B.M. and Irizarry, R.A. affy—analysis of Affymetrix GeneChip data at the probe level, *Bioinformatics*, 20, 3.
2. Gentleman, R.C., Carey, V.J., Bates, D.M., Bolstad, B., Dettling, M., Dudoit, S., Ellis, B., Gautier, L., Ge, Y., and Gentry, J. (2004) Bioconductor: open software development for computational biology and bioinformatics, *Genome Biol*, 5, R80.
3. Jamison, D.C. *Perl Programming for Biologists*. Hoboken, NJ: John Wiley & Sons, 2003.
4. Loy, M. and Eckstein, R. *Java Swing*. O'Reilly, 2003.
5. Sun Microsystems. The Java Tutorial: A practical guide for programmers, 2002, Obtido de http://java.sun. com/docs/books/tutorial.
6. Permann, C. BioPerl and BioJava Tutorial.
7. Pocock, M., Down, T. and Hubbard, T. BioJava: open source components for bioinformatics, *ACM SIGBIO Newsletter*, 20, 10–12, 2000.
8. Stajich, J.E., Block, D., Boulez, K., Brenner, S.E., Chervitz, S.A., Dagdigian, C., Fuellen, G., Gilbert, J.G.R., Korf, I., and Lapp, H. *The Bioperl Toolkit: Perl Modules for the Life Sciences*. Cold Spring Harbor Lab, 2002.
9. Tisdall, J. *Beginning Perl for Bioinformatics*. O'Reilly, 2001.
10. Tisdall, J. *Mastering Perl for Bioinformatics*. O'Reilly, 2003.

CHAPTER 15

Web Site and Database Design

Jerry M. Wright

CONTENTS

This section covers two seemingly disparate topics. However, use of Web sites integrated with database functions is becoming increasingly common and is important in information presentation and sharing in the scientific world. This chapter provides a practical overview of concepts and technical limitations behind the creation of Web sites and reviews principles of how to organize data in a formal manner for efficient storage, retrieval, and analysis.

SECTION 1 PERSONAL WEB SITE DESIGN

Part I Introduction

This section describes what a Web site is, how it functions across a network, and how to create and publish your own site. Selected types of Web page editors are discussed, and some online resources are pointed out. The hands-on portion demonstrates how to create a basic page with graphics, tables, text, and sound, using readily available editors and templates.

1. Background

A personal Web site is generally regarded as a noncommercial posting of information of interest to the creator. A Web site is a collection of documents that are designed to be hosted on remote computers in a network and viewed in a Web browser. There are a number of interconnected computer standards and technologies behind the creation and presentation of Web sites. The network may be the Internet, a local network with no external connections, or a local network with limited access to the Internet. A Web browser is software designed to interact with and display a Web-based document. The term *Internet* refers to a publicly accessible network of connections to numerous local networks that can carry a wide variety of data and services.

The World Wide Web is a global information space for sharing documents that is accessible on the Internet; it is not the Internet. The standards used for document sharing will work on the Internet or on properly configured local networks because they address basic issues of how documents are formatted, how network resources are located, and how markup language controls the document display. This takes place in a client-server model in which a client computer with a browser application requests documents from a server.

Web-oriented documents functionally consist of two parts: the content to be displayed and the instructions on how it is to be displayed. The browser interprets the display commands and renders an image of the content on the client computer. The process starts with the browser program requesting a document. The document is located by entering a URL (uniform resource locator) such as http://www.jhu.edu, which is resolved to a physical address on the network where the documents are located. The documents are sent to the browser, which then renders them for display or action as dictated by the markup language included with the documents.

The first markup language for documents on the World Wide Web was HTML (hypertext markup language), and it is still the principal language. HTML specifies the layout of a page but leaves many of the details to the client machine. The result is that a page will display differently on different machines, depending on the local resources available. The HTML standard has been revised several times with extensions to address limitations, and supplementary standards have also been developed.

To address difficulties in sharing and displaying documents uniformly on a variety of client machines, XML (eXtensible Markup Language), XHTML (eXtensible HTML), and CSS (Cascading Style Sheets) standards were developed. CSS is designed to address the presentation of content specified by a markup language. CSS describes the style of elements in a document such as fonts, type size, and other layout details that cannot be specified in HTML. XML is a general markup language designed for sharing data across different computer systems and is capable of both describing and containing data. Currently, the field is moving toward extensive use of XML for data sharing. JavaScript is not a markup language but a programming scripting language developed to provide additional functional capability not possible in HTML alone. DHTML (dynamic HTML) is not a specific standard; rather, it is a combination of protocols such as HTML and JavaScript that together enable a browser to alter a Web page's look and style after the page is loaded.

2. *Creating a Web Page*
    1. **Editors:** There are three general classes of editing programs for creating a Web page: document editors, which can save pages in HTML format; full-featured WYSIWYG (what you see is what you get) graphic-oriented professional editors; and HTML text editors. Specialized editors are not addressed here.
        a. **Document editors:** Microsoft Word and Corel WordPerfect each can save a document in HTML format suitable for use as a Web page. The HTML export functionality is being adopted in

other programs as well. For instance, Microsoft Excel can save a spreadsheet in HTML format for Web posting. The exported HTML Web page may have accompanying directories with XML files, images, and ancillary information necessary for proper rendering of the page in a browser. They can be used in a Web site, but you may have problems matching style if are using a Web site template to standardize the appearance.

b. **Full-featured editors:** The two dominant full-featured editors for professional Web page design are Microsoft FrontPage and Macromedia Dreamweaver. Fortunately, novices can create good-looking Web sites by choosing a default approach, using recommended templates and allowing the program to generate the computer code. Although extensive knowledge of the various Web-specific standards is not necessary, the programs can present so many options that it becomes overwhelming to a novice. FrontPage is the easier of the two for a beginner to use in production of a personal Web site; Dreamweaver is decidedly geared for technically proficient professional developers. FrontPage will be fully supported until June 2008 but is currently in the process of being replaced by Microsoft Expression Web Designer.

c. **Text editors:** HTML can be edited in a pure text editor such as Notepad. Free HTML-specific editors are abundant on the Web; a recent search at download.com produced a list of 153 editors. A posted link to the developer's Web site may be a condition for noncommercial use. Many of these editors have both graphic and text editing modes. Capability to switch between graphic and HTML editing modes is an important consideration because it is tedious to edit in text mode then use a separate browser to display the result. The graphic editing mode considerably speeds up the process because you have immediate feedback on layout and appearance.

2. **Web page templates:** Templates are used to standardize Web site layout and simplify site creation. Basically, templates take care of HTML coding and graphics are needed for a polished presentation that allows the developer to focus on adding content. Dreamweaver and FrontPage both have numerous templates for single page layouts for commercial sites. FrontPage, however, has a specific template for a personal Web site that includes suggestions for and examples

of content. Also, there are numerous free templates available from many sources on the Web. Usually, they are created by a Web graphics company and are free for use as long as there is an easily used link to the company site.

Typically, a personal Web site will include a home page, a link page, a personal interest or hobby page, a photo page, and a means to provide feedback to the author via e-mail or postings. The home page presents basic information and brief summaries of what is on the other pages. The link page contains pointers to other Web sites the author finds interesting. Usually, the links are grouped by category with some commentary. Large photo pages may take considerable time to be loaded onto the client computer and are difficult to navigate. If the photo page becomes extensive, it is best to create additional photo pages by topic and provide links between the pages.

3. *Basic Steps to Setting up a Web Site*

1. **Hosting a Web site:** Numerous companies provide Web site hosting for a wide variety of prices, depending on disk space, connectivity, and user requirements. A quick online search will turn up hundreds. A simple solution for many users is to use the Web space available from their ISP (Internet Service Provider). Many providers, whether dial-up, DSL, or cable, will provide several megabytes of disk space for personal Web pages as part of the account. Also check with your employer; some companies and colleges will host a personal site if it meets their use standards. If you decide to register a domain name, many of the domain name registration companies also provide Web site hosting and management for additional fees.

2. **Domain registration:** Domain registration is not necessary for a Web site to be functional. However, name registration can make it easier for users to find a site later because the URL to a Web site can become rather long and difficult to remember.

   Domain name registration is a separate issue from Web site registration with search engines. Registering a domain name provides a direct link to your site through DNS (Domain Name System) resolution, in which the text, such as www.mysite.org, is translated to a physical address where the information resides. Without a domain name specific for your site, the pathway becomes indirect, usually going through the hosting company name space such as www.bighostcompany.com/personalsites/user12345/mywebsite.

Domain name registration is controlled by ICANN (Internet Corporation for Assigned Names and Numbers). The InterNIC Web site (http://www.internic.net/) established by ICANN has extensive listings of accredited registrar companies worldwide. Charges for registration vary among companies, and many offer additional services such as e-mail accounts and Web site hosting for additional fees. Registration of a domain name is for a limited period of time usually starting at two years. If you do not renew, the domain name becomes available for reuse by anyone paying the registration fee. Be cautious when paying renewal fees. Some companies may send a notice that your registration needs to be renewed, but it may not be from the company you originally registered with. This may result in paying fees twice or losing Web site hosting paid for with the first company.

With or without a site-specific domain name, most Web users will initially find the site through links from other sites or topic searches in one of the Web search engines such as Google. Search engine registration enters your Web site into the list of sites to be indexed. Search engine registration is not necessary, but it will speed up the process as the various companies have automated searching and indexing programs. Judicious use of keywords and document descriptors will help users find your Web site in a search.

3. **Loading Web pages on the host site:** FTP (file transfer protocol) is used to upload the Web site pages to the host. Check with host site technical support for details because the site index files must be correctly named and placed in specific directory locations. There are many FTP programs available at a wide variety of costs; many are free. CuteFTP is a low-cost, easily configured, general-purpose FTP program, and there are many others. Searching at www.download.com produces extensive listings. Usually, your hosting company will have detailed instructions on how to connect via FTP and may have specific instructions for a particular FTP program.

4. **Maintaining a site:** It is necessary to periodically review Web sites and update material. Even if the content you created has not changed, presentation styles become outdated, and you may have published

additional papers or changed hobbies. Links are subject to "link rot," where they are no longer functional because the target sites have changed. A page full of bad links usually indicates a site that is no longer maintained.

4. Basic Web Page Editing

An *element* is the basic unit of a Web document in HTML and has two basic properties: attributes and content. HTML uses attribute tags to indicate how content is to be displayed. Attributes are usually dictated by paired starting and ending tags, whereas content is between the tags; together they make a display element. Content may be text, images, links, or other files. Whereas HTML indicates *what* is to be displayed, the client system determines many details of *how* it is to be displayed. For example, a block of text may be tagged as a paragraph, but details such as font and font size are determined by what is available on the user's computer. The result may radically alter appearance of the page and not be what the Web site designer intended; this is frequently a problem with controlling spacing between images. The basic elements of Web page editing are described in the following text:

1. **How to hyperlink a text file:** HTML requires anchor tags to indicate a hyperlink in a block of text. The attribute portion of the tag "href=" is set to the URL, as an example <a href=http://www.jhu.edu> Johns Hopkins </a>. The first part within < and > is the opening tag, indicating that the following content is hyperlinked to this URL. The text to be linked is between the opening tag and the closing tag "</a>". Forgetting to properly place closing tags is a common source of problems for novice programmers.

   Links can be made to other places in the same document, other local files, files on other computers in a local network, or files on other machines accessible through the Internet.

   Link names are case sensitive under some circumstances. There are differences among operating systems as to how case differences are handled; some systems will accept either case for the same file. However, to many Web servers, "SomeFile" is not the same name as "Somefile" and a "file not found" error will be returned. Letter case errors can be very difficult to find.

2. **Adding a picture to your Web page:** There is a multitude of image file formats that can be displayed by various versions of browser but not all browsers can display all formats, and older browsers cannot display some of the newer formats. There are two that are common enough to be reliable: GIF and JPEG. Both use image compression to reduce the amount of data that needs to be transferred. JPEG loses some detail during compression, whereas GIF does not. However, GIF is limited to 256 colors, and JPEG is not. Try to avoid using uncompressed files, as they can add considerably to the time necessary for downloading and viewing. If it is a large image file, it is recommended to make a thumbnail image of it for display on the Web page, and provide a link to the full-size image.

   With WYSIWYG editors and Microsoft Word, just drag and drop the image onto the page. Many HTML editors do not have drag-and-drop capability, so you have to use the insert command from the menu edit bar. However, position control on the page is crude in HTML alone. In either case, the editor uses spaces, tabs, and paragraph positioning to move the image where you want it. The program then generates the underlying HTML code to position the image on the page. Exact positioning on a page and control of text flow around the image can be difficult in any editor. It is best to assume the image location will vary on different computers, browsers, and display hardware, so try not to make your page dependent on exact layout of images. Although tedious, it helps to view your Web page in different browsers with different screen resolutions.

   Steps to precisely control image placement and text flow on pages under all circumstances are beyond the scope of a novice tutorial. However, there is one crude solution that works. Position the images and text as you want it on a page in your computer and make an image of the display. You can use the print screen key to copy, and then paste the picture into a graphics editor. Alt–print screen copies the active page, and print screen copies the entire display. Use the image in the Web page instead of trying to control a variety of displays.

3. **Tables:** There are specific HTML tags for creation and arrangement of tables and lists. Although table tags are precise in meaning, the actual HTML code can become quite convoluted; the tutorial will

show an example from the template. Most HTML editing programs have table creation functions that automate the process and are a substantial asset if the page requires a table. The individual cells of a table can hold text or images; consequently, tables can be extremely useful in controlling placement of images.

4. **Background color and images:** Background colors and images are used to provide texture, contrast, enhance interest, or reinforce a theme. A color setting floods the display field with a single color over the entire page or just a section, depending on how it is placed in the document. A background image is repeated to fill the page both across and down. Any content text or image appears on top of the background. Portions of an image can be made to blend with the background by specifying a transparency mode in the image file; this feature is supported in GIF (Graphics Interchange Format) images. If the image is as wide as the page, the repetition can be used to produce interesting linear effects (Figure 15.1). (Note: When using either background images or background color, it is important to check the visibility of text. Some combinations are very difficult or impossible to read.)

5. **Adding animation and sound:** To add animation, use an animated image in GIF format. The image format is limited to 256 colors, so it is unsuitable for photos but does very well with drawn images. The single file is made up of a series of compressed images that are displayed sequentially in the same location to produce an animated effect. Numerous free animated images and background images are available at http://www.gifanimations.com.

   Original HTML did not natively support sound or animation, and there are limited means to add functional elements in a pure HTML Web page. HTML allows media files to be embedded in a document using the OBJECT command. However, it has numerous limitations in editing and display. Upon

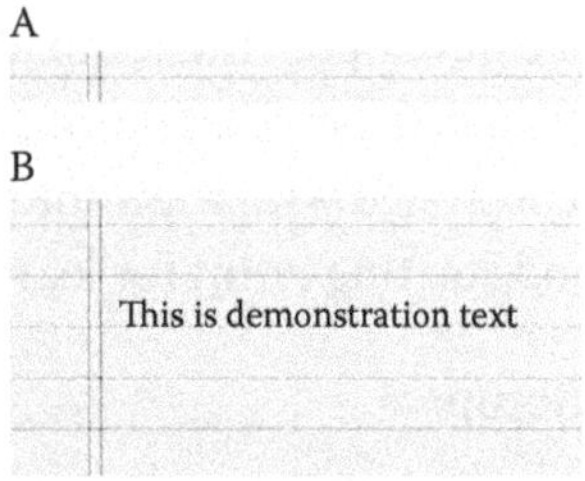
A
B
This is demonstration text

FIGURE 15.1 A background image as wide as the Web page (A) is repeated to produce the appearance of a page of legal paper (B). Graphics or text appears on top of the background image.

loading the Web page, a copy of the program that will play the file will open. Security on the Web browser may present a popup window asking if the user will allow the program to run or block it without notification.

A solution to adding audio is to use the JavaScript tag <BGSOUND>. JavaScript instructions are intended to be used in an HTML document and rendered by the browser. This command was originally introduced in Internet Explorer 3 and is commonly supported now. A client browser may be configured to prevent audio files from playing and ignore the command. Also, keep in mind that large audio or video files may take considerable time to load over a slow network connection, and frustrate the user. Overall, it is best to become proficient in JavaScript and other techniques to add anything more than simple media elements.

Part II Step-By-Step Tutorials

For the demonstration we will use the HTML editor PageBreeze (http://www.pagebreeze.com), which is free for personal or nonprofit use, and a template downloaded from the Internet. We will modify a template and save the changed version as a new template. From the new template we will create links to other pages and create a personal interest page with text, graphics, and a table.

The editor is limited to running in Windows and with Internet Explorer for previews. However, it has tabbed displays, which makes it very easy to set up a page, view and edit source code, set page properties, and view the result in a browser. These combined functions make the program very useful for a novice developer. The template selected is a single page HTML document with minimal use of advanced features. The graphics are simple, and placement controlled with a few interesting HTML tricks using tables; this makes it a good starting point for learning. The detailed steps are:

1. **Setup**:

Install and start PageBreeze. The program opens to an instructional page. First, examine the HTML code for a blank page. Create a new page using the blank template (File>New Page or press Ctrl-N), and name it demonstration. Switch to the HTML source tab (Figure 15.2). (Note: On first use, you may need to exit the program

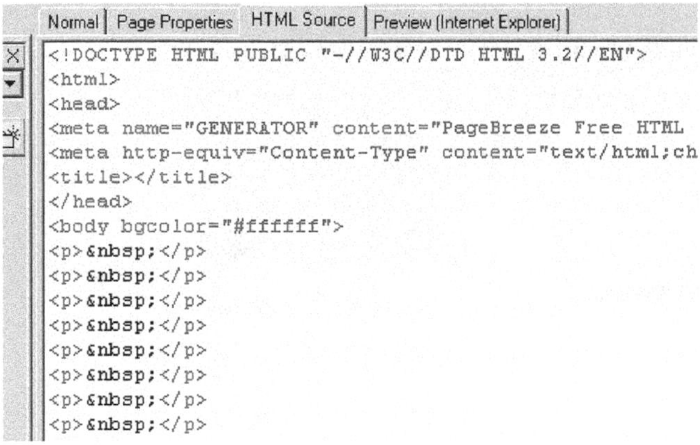
Normal | Page Properties | HTML Source | Preview (Internet Explorer)
<!DOCTYPE HTML PUBLIC "-//W3C//DTD HTML 3.2//EN">
<html>
<head>
<meta name="GENERATOR" content="PageBreeze Free HTML
<meta http-equiv="Content-Type" content="text/html;ch
<title></title>
</head>
<body bgcolor="#ffffff">
<p> </p>
<p> </p>
<p> </p>
<p> </p>
<p> </p>
<p> </p>
<p> </p>
<p> </p>

FIGURE 15.2 HTML coding of a blank page showing header and body regions.

then reload the blank page to see the code.) The document header between the <head> and <\head> tags is not displayed. The header section contains metadata, which is information about the document such as keywords, page titles for search engines, as well as refresh rates for the browser. This document is written compliant to HTML 3.2 standards. The start of the document to be displayed is indicated by the <body> tag. Background color is set to white (ffffff); nonbreaking spaces ( ) within paragraph tags <p> </p> are used to create a blank space to write on.

Download a template. From the Menu bar, select File > "Get website templates…" and navigate to Steve's Templates (http://www.steves-templates.com/templates.html). We will be working with template number 38; 39 and 40 are color variations of the same template and will work just as well. Download the file and save it to the My Webs directory in the My Documents folder. The file is zipped and can be easily extracted in Windows XP. Open the zipped file using Windows Explorer, and copy the contents to the My Webs directory. Keeping the files associated in a single root directory simplifies Web site creation and later transfer of files to a Web host.

2. **Overview of the template:** Open the template by selecting the index.html file. There are a number of generic features that can be easily modified by the user. All displayed graphics are in the folder img and easily modified in most graphics programs (Figure 15.3A). We will

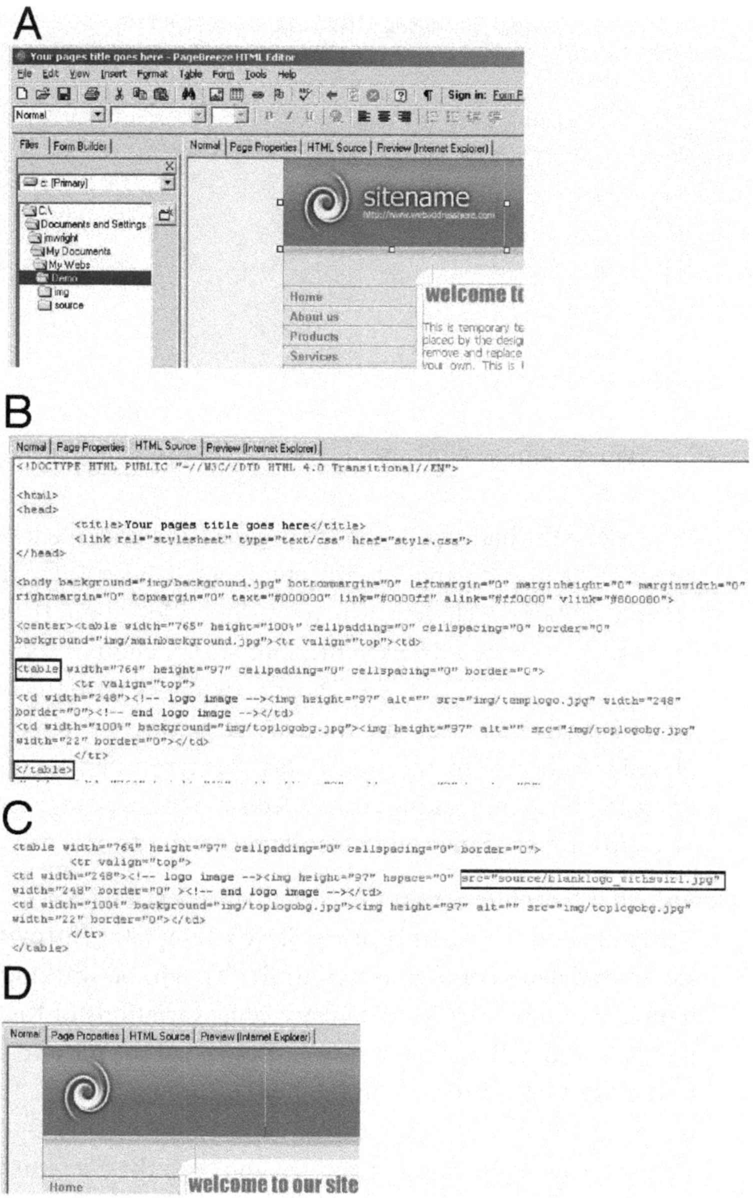
A
Your pages title goes here - PageBreeze HTML Editor
File Edit View Insert Format Table Form Tools Help
Sign in:
Normal
Files
Form Builder
Normal
Page Properties
HTML Source
Preview (Internet Explorer)
c: [Primary]
C:\
Documents and Settings
jmwright
My Documents
My Webs
Demo
img
source
sitename
Home
About us
Products
Services
welcome t
This is temporary te
placed by the desig
remove and replace
your own. This is
B
Normal
Page Properties
HTML Source
Preview (Internet Explorer)
<!DOCTYPE HTML PUBLIC "-//W3C//DTD HTML 4.0 Transitional//EN">
<html>
<head>
<title>Your pages title goes here</title>
<link rel="stylesheet" type="text/css" href="style.css">
</head>
<body background="img/background.jpg" bottommargin="0" leftmargin="0" marginheight="0" marginwidth="0" rightmargin="0" topmargin="0" text="#000000" link="#0000ff" alink="#ff0000" vlink="#800080">
<center><table width="765" height="100%" cellpadding="0" cellspacing="0" border="0" background="img/mainbackground.jpg"><tr valign="top"><td>
<table width="764" height="97" cellpadding="0" cellspacing="0" border="0">
<tr valign="top">
<td width="248"><!-- logo image --><img height="97" alt="" src="img/templogo.jpg" width="248" border="0"><!-- end logo image --></td>
<td width="100%" background="img/toplogobg.jpg"><img height="97" alt="" src="img/toplogobg.jpg" width="22" border="0"></td>
</tr>
</table>
C
<table width="764" height="97" cellpadding="0" cellspacing="0" border="0">
<tr valign="top">
<td width="248"><!-- logo image --><img height="97" hspace="0" src="source/blanklogo_withswirl.jpg" width="248" border="0" ><!-- end logo image --></td>
<td width="100%" background="img/toplogobg.jpg"><img height="97" alt="" src="img/toplogobg.jpg" width="22" border="0"></td>
</tr>
</table>
D
Normal
Page Properties
HTML Source
Preview (Internet Explorer)
Home
welcome to our site

FIGURE 15.3 Template customization: changing the default logo image. (A) Default page displayed in editing mode with the logo image selected. (B) HTML source code for logo image placement using a table to control position. (C) HTML source code for changing the image file while maintaining placement. (D) Page display with replaced image.

change the sitename image later. (Note: The name of the opening page on a Web site is usually `index.html`, but some Web hosts use the name `default.html`.)

The pink sidebar with white page space is created with a simple pink and white linear background image repeated multiple times in a designated space (Figure 15.3A; see also Figure 15.1). The page space, navigation bars on the left, and placement of images in the site name header at the top are controlled with tables. (Note: Gray layout lines appear in the Normal [editing] mode, but are not visible in the Preview [viewer] mode.)

On the HTML page, the source code between the highlighted tags shows the complexity of instructions necessary to create a defined space that holds the two slate gray images of the header bar (Figure 15.3B). Without use of tables, the images would shift in positioning, depending on the browser and display resolution of the client.

3. **Template customization**

   Change a template image: To customize the template, we need to change the default logo image in the upper left corner. Double-click on the sitename image indicated by the selection box in Figure 15.3A. An `Insert Graphic` screen appears and the selected image file `templogo.jpg` in the `img` folder is highlighted. Navigate to the `source` folder. In the source folder, select `blank-logo _ withswirl.jpg`.

   The HTML code shows the source of the image in the table cell has been changed (Figure 15.3C). Changing the image could have been done manually by typing in the selected text.

4. **Setting page properties:** Select the `Page` properties tab. The page title, keywords, and description are information that goes in the header section and is used by search engines to index the page for searches. The page title is the default label used when saving a link in the browser. Failure to change the title, or leaving it blank, results in Web pages being bookmarked as `index` or other strange names. In this case, the bookmark would read "Your pages' title goes here." Although PageBreeze is called an HTML editor, it can also edit CSS files.

5. **Setting the background and link color scheme:** While in the page properties tab, click on the "Set page Background …" to bring up the background properties window. In this case we want to use the default colors for the link color highlights, so check the appropriate boxes. Without specifying the link color, the links are not highlighted and not easily found. The page background image has already been selected by the template. These are document master settings used by default; areas within a page can have different settings.

   Exercise. Uncheck the image box and select a highly contrasting background color and see how that changes the page display. The margins in an expanded view will have a new color scheme; the colors inside the tables do not change. Restore the selections to continue.

6. **Creating a template and new page:** First, customize the navigation bar by changing the labels. Highlight the `Products` label and change it to `Interests`, change `Bookmark` to `Links`, and change `Services` to `Photos`.

   From the edit bar select File > "Save Copy of Page as Template …" and give it an easily identifiable label such as `My38template`. The customized page can now be used to generate new pages with the new logo image and navigation bar labels.

   Select "New page" from the file menu then select the template you just created. The new page will be loaded in the editor. Label the new page "Links" and save it as `Links.html`. Close the page. Create another page from the template and label it `Interests`.

7. **Adding text and images:**

   Text. Entering text is as simple as typing or using cut and paste from the clipboard. Positioning text can be a challenge. We will stick to simple text entry. Clear the place-holding text in the opening page. The two items "welcome to our site" and "temporary headline" are both graphic images; delete them as well.

   At the top of the page use the menu function insert>horizontal line. Type in introductory text from the Sample Data section, and then press `Enter` to end the paragraph. End with another horizontal line.

   Images. Download an image of the ship *Endurance* trapped in ice from the Wikipedia entry on the Ernest Shackleton 1914 expedition by right-clicking on the image and selecting save to file (http://en.wikipedia.org/wiki/Imperial_Trans-Antarctic_Expedition). Use the menu item insert>graphic or press F5 to bring up the menu. Select the image file to be used and type in an appropriate alternate text label (Figure 15.4). The alternative text will be used in browsers designed for visually impaired users, by search engines to index a site and is displayed when the browser is set to not download graphics. Click "OK."

Use the paragraph centering layout to position the image in the center of the page. Press return several times, and enter any additional text you want to fill out the page.

8. **Linking:** Highlight the `Links` item on the navigation bar, and press the create hyperlink icon (Figure 15.5A) or double-click on the word "Links" in the navigation bar; because it already has a link, an editing window pops up. The links associated with the words in the navigation bar words contain an "http://" prefix, which indicates this is a hypertext transfer protocol and the browser should request the page from the network destination, not the local disk (Figure 15.5B). In the template these links are nonfunctional place holders and need to be modified or removed in the final page. Editing can be manually

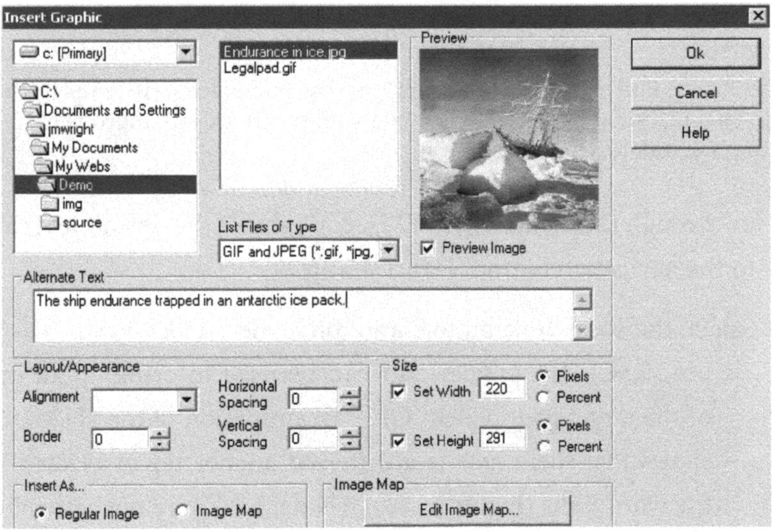
Insert Graphic
c: [Primary]
C:\
Documents and Settings
jmwright
My Documents
My Webs
Demo
img
source
Endurance in ice.jpg
Legalpad.gif
List Files of Type
GIF and JPEG (*.gif, *jpg,
Preview
Preview Image
Ok
Cancel
Help
Alternate Text
The ship endurance trapped in an antarctic ice pack.
Layout/Appearance
Alignment
Border
Horizontal Spacing
Vertical Spacing
Size
Set Width 220
Set Height 291
Pixels
Percent
Insert As...
Regular Image
Image Map
Image Map
Edit Image Map...

FIGURE 15.4 Insert image control form.

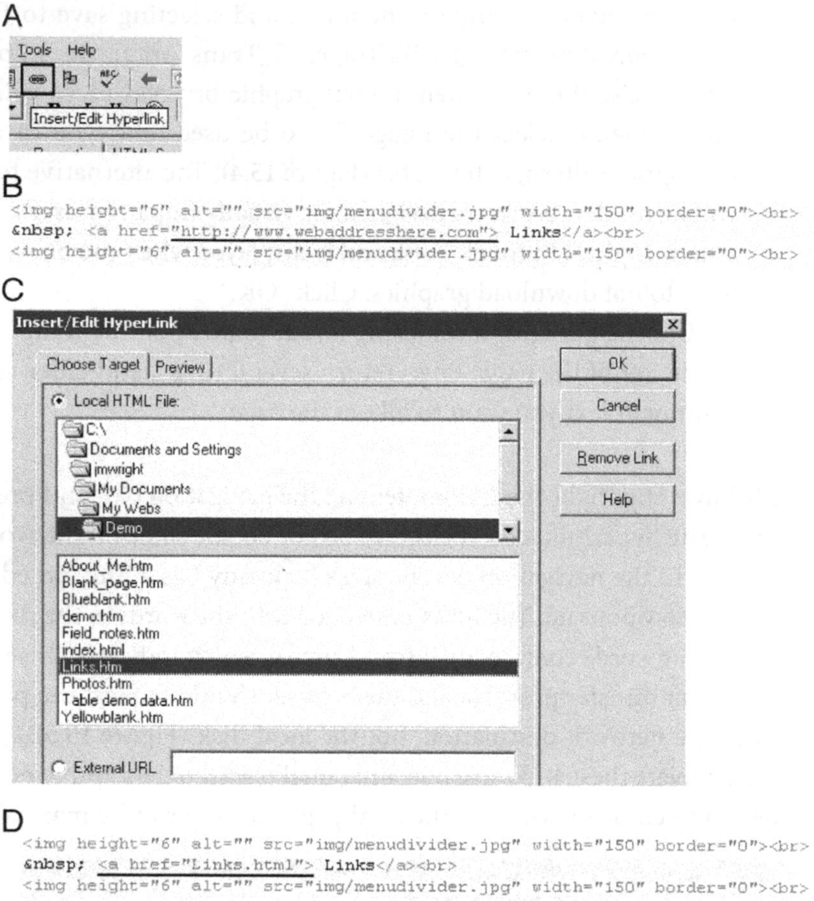
A
Tools Help
Insert/Edit Hyperlink
B
<img height="6" alt="" src="img/menudivider.jpg" width="150" border="0">
  <a href="http://www.webaddresshere.com"> Links</a>
<img height="6" alt="" src="img/menudivider.jpg" width="150" border="0">
C
Insert/Edit HyperLink
Choose Target
Preview
Local HTML File:
C:\
Documents and Settings
jmwright
My Documents
My Webs
Demo
About_Me.htm
Blank_page.htm
Blueblank.htm
demo.htm
Field_notes.htm
index.html
Links.htm
Photos.htm
Table demo data.htm
Yellowblank.htm
External URL
OK
Cancel
Remove Link
Help
D
<img height="6" alt="" src="img/menudivider.jpg" width="150" border="0">
  <a href="Links.html"> Links</a>
<img height="6" alt="" src="img/menudivider.jpg" width="150" border="0">

FIGURE 15.5 Editing a hyperlink. (A) Hyperlink menu icon, (B) template default text to be replaced, (C) Selecting the hyperlink target, (D) Changed HTML code pointing the link to a new location.

or through the menu. The HTML code shows the default text linked to the navigation button (Figure 15.5B).

Select the local link button and pick the `Links.html` file (Figure 15.5C). The preview tab can be used to bring up an image of the target page. Click “OK.” Examination of the HTML code shows that there now is a different anchor tag associated with the word links (Figure 15.5D). (Note: Links are not active in the editor mode; the preview tab needs to be selected for links to be active.) Save the page.

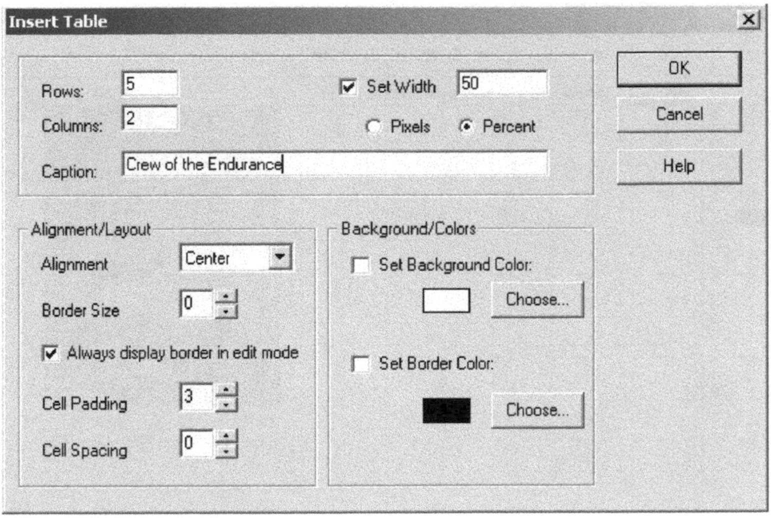
Insert Table
Rows: 5
Columns: 2
Caption: Crew of the Endurance
Set Width 50
Pixels
Percent
OK
Cancel
Help
Alignment/Layout
Alignment Center
Border Size 0
Always display border in edit mode
Cell Padding 3
Cell Spacing 0
Background/Colors
Set Background Color:
Choose...
Set Border Color:
Choose...

FIGURE 15.6 Table creation form.

9. **Creating and editing a table:** Tables are easily created through use of the table insert form (Figure 15.6). Here we specify five rows and two columns with a total width of 50% of the display to hold the listing of senior officers on the expedition. The caption is displayed above the table and is used by automated site-indexing software.

   The form produces the empty table shown in Figure 15.7. Highlight the caption text, and change the size to 4. Note that the caption display font has only seven sizes. These are not point sizes that would normally be used in a word processing document; these are HTML-specified sizes and are translated by the browser for display. The font specified is Tahoma; if the font is not installed on the client computer, the browser will select another typeface that is installed locally, which may in turn alter the point size used for the display. These changes can substantially alter the appearance and layout of a Web page. Enter crew member descriptions and names from the list in the Sample Data section.

10. **Adding audio:** Use Windows Explorer to locate `TADA.WAV` in the `system/media` folder. Copy the file to the PageBreeze sample directory in the `Program Files` directory. Locate the start of the page body in the HTML code. Add a blank line below it, and insert the following line: `<bgsound src="tada.wav">`; no closing tag is

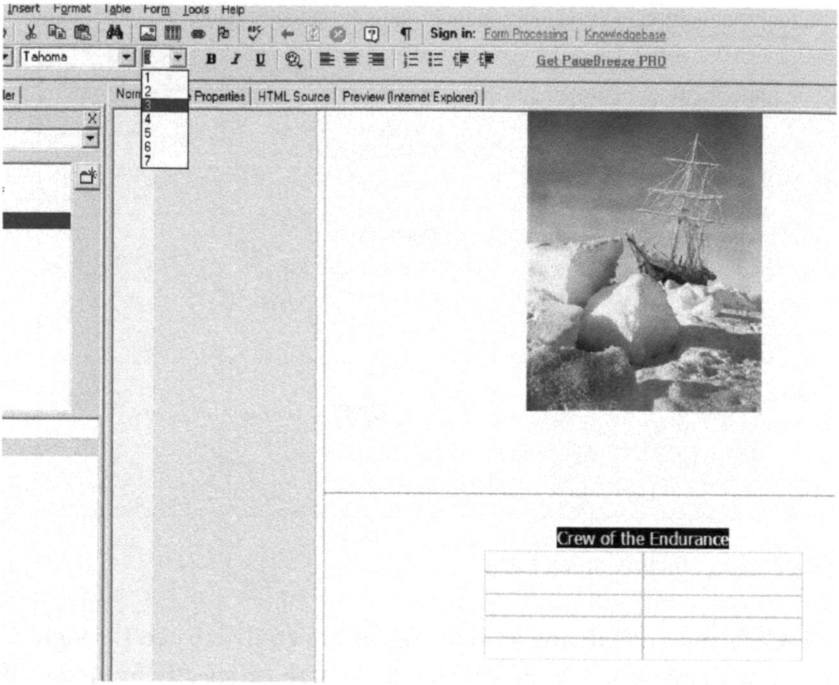
Insert Format Table Form Tools Help
Sign in: Form Processing | Knowledgebase
Tahoma
Get PageBreeze PRO
Properties | HTML Source | Preview (Internet Explorer)
Crew of the Endurance

FIGURE 15.7 Changing displayed text size.

necessary. This will play the sound file immediately before the page loads. Placing the line just before the closing body </body> tag will play the sound file after the page is loaded.

Part III Sample Data

Text:

Men wanted: for hazardous journey. Small wages, bitter cold, long months of complete darkness, constant danger, safe return doubtful. Honor and recognition in case of success. Ernest Shackleton.

Although widely attributed to Shackleton as a newspaper recruiting advertisement for the expedition, there is no evidence that it was ever run and appears to be an after-the-fact invention which has achieved the status of an urban legend.

Public domain image from Wikipedia article on the Imperial Trans-Antarctic Expedition.

http://en.wikipedia.org/wiki/Imperial_Trans-Antarctic_Expedition

Crew of the Endurance.

Ernest Shackleton	1st in Command
Frank Wild	2nd in Command
Frank Worsley	Captain of Endurance
Frank Hurley	Photographer
Hubert Hudson	Navigation Officer

SECTION 2 NUTS AND BOLTS OF BIOLOGICAL DATABASES AND THEIR CONSTRUCTION

Part I Introduction

This section describes what a database is, the various structures a database may have, and how data are searched and results retrieved. Important issues to consider prior to establishing a database are discussed. Selected available database management systems are briefly reviewed as well as some important online biological databases. The hands-on portion demonstrates how to establish an object-oriented database of existing data structures to create a searchable index of archived data. Lastly, data files are imported from existing spreadsheets and linked to facilitate information retrieval.

1. Database Basics

Fundamentally, a database is a collection of facts organized for efficient use. The basic unit of information is the record, which is a collection of one or more fields containing data. The organizational layout of a database is called the *schema*. There are a wide variety of database models with differing underlying schemas. Each was developed to address various data tracking and retrieval issues, and each has its own strengths and weaknesses. No one model solves all problems, and the various types of databases continue to be in use.

Databases may be static or dynamic. Dynamic data change and need to be updated on a regular basis, such as inventories of culture room supplies or laboratory chemicals. Static data seldom change and, if modified, the changes usually need to be tracked; laboratory notebooks, for example.

The collection of records is a database, but database management software (DBMS) is required to enter, locate, and extract information. The specifics of how the management software is implemented and how well it works for a particular purpose depends on the database schema.

Spreadsheet vs. database. There are fundamental differences between a spreadsheet and a database. Spreadsheet programs are designed to support general computation using cells to organize data and present results; the basic unit of information to be operated upon is the cell. Usually a row of cells is viewed by the user as a record of information about some object; however, spreadsheet software has no such interpretation of how the cells relate because it is designed to perform general computation.

In tabular format for presentation, a database record has the same appearance as a row of cells in a spreadsheet but, in a database, fields are linked by the software. Deleting a field deletes the field in all records and does not change the association of data within records. When a group of records is sorted by information in a single field, all the fields in the record remain associated.

In spreadsheets, deleting a cell (as opposed to clearing the information within a cell) will usually cause the remaining cells in the column to move up or cells in the row to move laterally to fill the vacated space. Another problem occurs when a column or region is selected for sorting; the other cells do not sort with them, unless they are also selected. Both issues break the perceived relationship of information between cells with potentially serious consequences.

Commercial spreadsheets have tools to provide database-like functionality because they are often used as databases, but the inherent differences remain. It may help to think of a database record as a page in a laboratory notebook, whereas a spreadsheet is a scrapbook collection of sticky notes.

Data vs. information. Data are simple facts. Information adds to knowledge as the result of organizing, processing, or transforming data. Information can change over time as new experiments are conducted. These issues can be confused and cause problems when implementing a laboratory database. Often, what is desired is a laboratory information management system similar to a laboratory notebook that integrates multiple sources of data and interpretation of results. This may include such items as pathology laboratory reports, patient history, genotype, biopsy pictures, microarray results, lists of genes altered in specific pathways, literature citations, and other items. Choice of the database schema and operating software will dictate how effectively these sources can be integrated.

2. Database Models

The database models commonly encountered are flat file, hierarchical, network, relational, object-oriented, and entity-relational. Normally, a user sees only the output of the database, and the database model behind it is neither obvious nor relevant. However, if you have reached a point where a formal database is necessary, you need to consider the strengths and weaknesses of the various options available. You also need to keep in mind that the various models are conceptual structures that require specifically written software to fully implement. Quite often, people start using one or more of the database structures without realizing it in an effort to cope with information overload.

Flat file data structure. A flat file is simply a text file containing all records and associated fields with a fixed file structure to indicate data fields and records. The FASTA file format used by NCBI (National Center for Biotechnology Information) is a flat file dataset. The ability to search and sort the information within the text file is limited to standard text editing tools unless time and effort are spent developing task-specific tools. For some standard flat file formats such as FASTA, tools are available, but are usually developed by and intended for use by professional computer staff. The FASTA format was developed to facilitate use of text-editing tools and scripting languages to search, edit, and manipulate sequence data.

Hierarchical structure. A hierarchical database is structured with a parent/child relationship from a source usually called the *root*. One parent may have multiple children, but children have only one parent. The data item is called a *leaf* or *node*, and the path back to the root is called a *branch*. Computer folder and file navigation is an example. Each drive is a root, and all files on that drive are linked back to the root through a hierarchical structure. When implemented in a formal software-driven database, algorithms called *tree searches* are used to locate and extract data. However, the data are forced into a structure that does not easily accommodate items that can be in multiple categories, e.g., proteins with multiple functions. The result is that single items are repeated on various branches in order to fit the schema. Each node is independent; if the information in one node is updated, the others are not. This leads to problems with multiple versions of information.

Hybrid models. Many laboratory datasets are organized as a hybrid model with flat files linked by a hierarchical navigation structure and can be quite useful for small datasets that do not change frequently (Figure 15.9). Users often have no specific software for searching and sorting beyond that available in the computer operating system. The drawbacks are that it is time consuming to locate, examine, and combine information from several files. Also, you may not recall the reason the dataset was

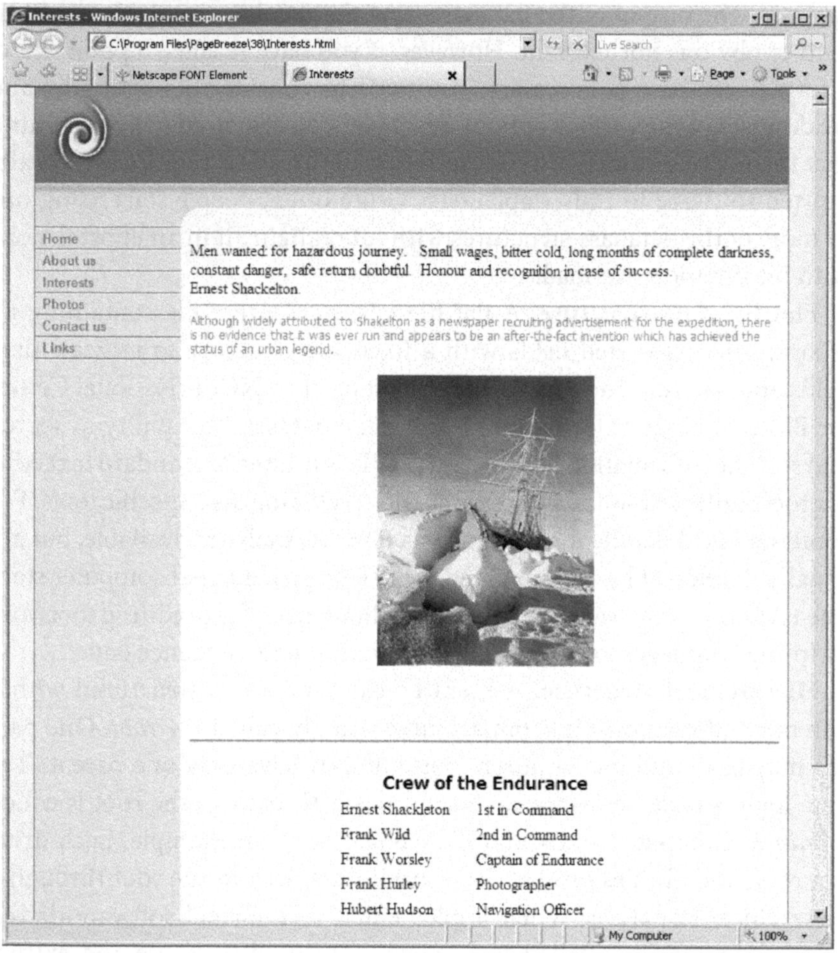
Interests - Windows Internet Explorer
C:\Program Files\PageBreeze\38\Interests.html
Live Search
Netscape FONT Element
Interests
Page
Tools
Home
About us
Interests
Photos
Contact us
Links
Men wanted: for hazardous journey. Small wages, bitter cold, long months of complete darkness, constant danger, safe return doubtful. Honour and recognition in case of success.
Ernest Shackelton.
Although widely attributed to Shakelton as a newspaper recruiting advertisement for the expedition, there is no evidence that it was ever run and appears to be an after-the-fact invention which has achieved the status of an urban legend.
Crew of the Endurance
Ernest Shackleton 1st in Command
Frank Wild 2nd in Command
Frank Worsley Captain of Endurance
Frank Hurley Photographer
Hubert Hudson Navigation Officer
My Computer
100%

FIGURE 15.8 Final form of the Web site displayed in IE7.

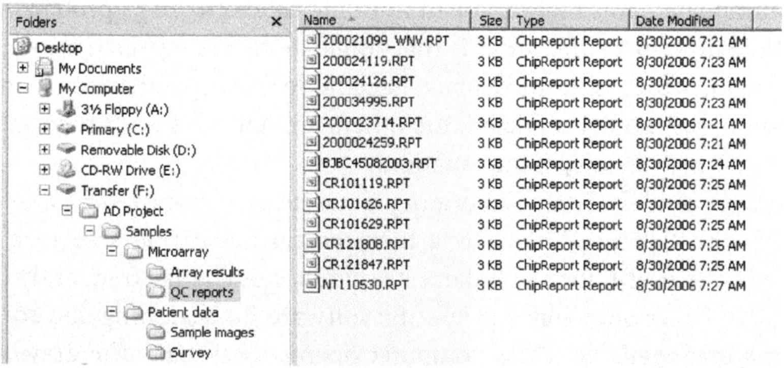
Folders
Desktop
My Documents
My Computer
3½ Floppy (A:)
Primary (C:)
Removable Disk (D:)
CD-RW Drive (E:)
Transfer (F:)
AD Project
Samples
Microarray
Array results
QC reports
Patient data
Sample images
Survey
Name
Size
Type
Date Modified
200021099_WNV.RPT 3 KB ChipReport Report 8/30/2006 7:21 AM
200024119.RPT 3 KB ChipReport Report 8/30/2006 7:23 AM
200024126.RPT 3 KB ChipReport Report 8/30/2006 7:23 AM
200034995.RPT 3 KB ChipReport Report 8/30/2006 7:23 AM
2000023714.RPT 3 KB ChipReport Report 8/30/2006 7:21 AM
2000024259.RPT 3 KB ChipReport Report 8/30/2006 7:21 AM
BJBC45082003.RPT 3 KB ChipReport Report 8/30/2006 7:24 AM
CR101119.RPT 3 KB ChipReport Report 8/30/2006 7:25 AM
CR101626.RPT 3 KB ChipReport Report 8/30/2006 7:25 AM
CR101629.RPT 3 KB ChipReport Report 8/30/2006 7:25 AM
CR121808.RPT 3 KB ChipReport Report 8/30/2006 7:25 AM
CR121847.RPT 3 KB ChipReport Report 8/30/2006 7:25 AM
NT110530.RPT 3 KB ChipReport Report 8/30/2006 7:27 AM

FIGURE 15.9 Example of a commonly used database structure with hierarchical organization of flat file datasets and photos.

placed in a particular branch; what is obvious today is less clear 2 years later, when research directions have shifted. Files may be inadvertently duplicated on another branch, resulting in different versions accumulating over a period of time.

Because this type of data organization is common in laboratories, some databases allow direct import of this structure to create an index that can be searched, sorted, and manipulated in the database.

Network model. A network database structure is similar to hierarchical, but there is not a strict one parent per child relationship between levels. One child level may have multiple parents, and the parents may be on different levels. This is called a *set structure*, where tables are related by an owner/member relationship; a single item may have multiple owners. As an example, CFTR in the GO (Gene Ontology) classification structure belongs in multiple functional categories: ATPase activity, ion channel activity, and PDZ domain binding. In a hierarchical model based on function, CFTR would be displayed in three branches; in a network model, it is displayed once with three owners. Although the data display may follow a network model, the actual database structure used to store, search, and retrieve information may be relational.

Efficient computational search and sorting in the network and hierarchical models is heavily dependent on hardware and physical location of records. Neither is easily adaptable to changing needs.

Relational model. Relational databases are good for large homogenous datasets where extraction of data is important. The relational database structure is derived from mathematical considerations and is organized to permit efficient use of set theory and predicate (true/false) logic to implement search and retrieval strategies. The name is derived from the relational calculus used to formulate queries. In a relational database data are organized by tables which are collections of records. Records are made up of linked fields containing data. Each record has a unique identifier called a *primary key*. The physical order of the records within a table or fields within a record is not relevant to retrieving information, thus removing some serious constraints of a network or hierarchical model. The standard method of managing a relational database, locating data, retrieving information, and organizing results is through use of SQL (Structured Query Language); however, other management methods can be used.

The strengths of the relational model lie in the linking of tables by common key fields. The software provides tools for creating an organizational structure, but it is the developer who must decide what tables are necessary. The process of breaking down large datasets and creating smaller data tables suitable for efficient use is called *normalization*. When properly constructed, each table addresses a single topic and contains a minimum of redundant information.

For example, there may be a table listing of protein records with sequence and function data in the fields, a table of species records with taxonomy data, and a table of tissues with morphological and functional descriptions. The records in each table have fields for keys from other tables; a record in one table is linked to a record in another table by use of the primary key. The DBMS can then search and associate data by use of a query in the form of SELECT - FROM – WHERE. For example, to extract a list of all tissues in which CFTR is expressed, a query would be constructed such as SELECT (all records) FROM (tissue table) WHERE (CFTR primary key is found). The search could be further narrowed by restricting the output to vertebrates.

Understanding what primary keys are and how records and tables should be laid out for efficient use is very important in creation of an efficient relational database. There are multiple implementations of relational databases and SQL DBMSs available both commercially and free. Although there are numerous technical papers and tutorials available online with each of the major database vendors, many of them assume the user has a technical or computer software background. For an introduction to relational databases written for a nontechnical audience, Hernandez [1] has a clear, minimally technical introduction with numerous examples. Even if you do not implement a relational database, understanding how they are organized and searched will be of substantial value in implementing object-oriented or other schema databases.

Object-oriented models. Object-oriented databases are well suited for complex datasets and work directly with object-oriented programming languages. An object-oriented database is based on objects that have properties. Although not technically correct, the terminology is sometimes used interchangeably with records and fields in user manuals and marketing information. One of the underlying concepts is the use of classes and inheritance to describe objects. This permits integration of multiple data types because an object in a class will be correctly interpreted. For example, an object in class `image` will be interpreted as a picture to be displayed and those in class `text` as text to be handled by a word processor; both image files and text files can be used as properties of an object without confusion.

Commercial object-oriented databases are a relatively recent innovation, and industry standards are still being developed. This can cause problems if moving data from one vendor's DBMS to another. Nonetheless, they can be extremely useful in managing laboratory data because of their use of intuitive object structure and ability to incorporate a wide variety of data sources, such as audio clips, photos, microarray result tables, text documents, etc., into a single object. The data structure may be very complex and not easily stored or searched in conventional databases.

The class structure is exemplified by example of an experiment, for which you may create a parent class `project` that has the basic properties of hypothesis and funding. Subclasses such as `animal` and `cell culture` are created that inherit the properties of the parent and add new ones specific to the subclass. For example, the class `cell culture` may have properties such as cell line, drug tested, drug concentration, incubation time, culture media, photos of the culture, and date. The class `cell culture` is used to create an object that holds the information for a specific experiment. All objects (experiments) in the class have the same basic properties but differ in details of the properties. This addition of properties appropriate to hold the data for an experiment can be extensive, but it may be better to create additional subclasses for clarity.

Entity-relational model. An entity-relational database combines the SQL query function of relational databases and the ability to store complex data formats of object-oriented databases in order to overcome limitations in storage, searching, and linking data from diverse sources. The NCBI database is an entity-relational model implemented in ASN1 (Abstract Syntactical Notation version 1) for describing and exchanging data even though the output in FASTA format is a flat file. There is a historical reason for this in that the original database was based on a flat file model, and users came to expect that format to be used for submission and retrieval. However, flat files cannot function efficiently with large datasets. The changeover was a substantial undertaking because it had to be made without interruption to the end user community. Currently, the data are being shifted from ASN1 format to XML because of limitations, evolving computer standards, and increasing use of the World Wide Web to distribute information (http://www.ncbi.nlm.nih.gov/IEB/ToolBox/XML/ncbixml.txt). This highlights the importance of considering the model, software selection, and evolving software standards when setting up a database.

Object-oriented features are starting to appear in some standard relational DBMS but they may not be marketed as entity-relational DBMS systems.

3. Additional Database Selection Considerations

Database applications usually fall into one of two categories: management of data collections or information analysis. Relational databases do extremely well with data collections when the schema is simple and the data types are few; the data may change, but relationships among data fields are stable. Examples are inventory control and microarray expression results. Object-oriented databases do well when there is need to navigate through and analyze large volumes of data in multiple formats.

Additional considerations include the overall size of the database and what existing data files need to be incorporated. Inclusion of large binary objects such as photos or confocal microscopy images can tremendously expand the physical size of the database. Relational databases do not handle images well without very specifically designed DBMS software. The issue is often avoided by including a path to the image rather than the image itself. However, this solution requires a very stable physical organization and extreme care in maintaining links when moving the database to a new computer.

Budget, availability of technical support staff, user training, and anticipated growth are other factors. What computing hardware and operating systems are available? What hardware and operating systems are required by the software? How many people need to access the data simultaneously? There may be a substantial investment in education, layout design, and training prior to becoming productive. Costs of commercial object-oriented databases are comparable to relational databases, and both models have free versions. There are no simple answers to all the issues but, in general, the more complex the dataset, the more difficult it is to implement as a relational database.

4. Databases on the Market

Coverage in this section is of the readily available, better-known products or interesting special-purpose database systems. There are far more database management programs available than mentioned here, many of which are intended for technically savvy users.

A. Relational Database Management Software A wide variety of relational model DBMS programs are available, each with differing file format characteristics, pricing, and utility. Vendors usually have conversion programs and utilities to translate the data file format of one system to that of another. Many third-party conversion utilities can be found at www.

download.com as well utilities that manipulate files and data in a variety of database formats. There are online user community support groups for the database programs which vary widely in quality and activity.

Some companies offer a free version with limited functionality as an introduction to their full-featured commercial products. Note that the DBMS usually does not have to be implemented on a server; a workstation may do fine for small laboratories. Most laboratories will do well with software intended for small business use. However, if substantial use is anticipated, server hardware and server operating system versions are recommended as well as consultation with a professional database developer. A few DBMS programs are now described:

Access (http://www.microsoft.com). Sold as a stand-alone product or included with Microsoft Office Professional, this is probably the easiest DBMS to implement for a laboratory that has MS Office installed, has few users, no need for simultaneous access, limited needs, and limited resources. The program can import directly from and export directly to Excel spreadsheets, and features a graphical user interface for designing SQL queries. The graphic interface makes the introduction to SQL a bit easier. As with other relational databases, knowledge of SQL is imperative to efficiently design and utilize the database. At a minimum, a spreadsheet can be converted to a database then manipulated in table mode, where it has the familiar appearance and some of the tools of a spreadsheet. The fictional Northwind Traders database, downloadable from Microsoft, provides numerous examples of common data search and retrieval functions. The demonstration database was developed for Access 2000 and can be easily converted to the Access 2003 standard.

Paradox (http://www.corel.com). Paradox is included in WordPerfect Office professional and the WordPerfect student and teacher edition. This is also a low-cost solution for those who have Corel corporate software installed. Paradox has its own programming language and structure; there are tools available should data transfer to another DBMS be needed. It also is a good introductory database program because its target users are business office workers.

MySQL (http://www.mysql.com). MySQL community edition is free and open-source, which means the source code is available. Online documentation is extensive but intended for a technology-oriented reader. Available for both Windows and Linux, but versions are specific for the processor utilized in the computer: Intel 32 or 64 bit, AMD 32 or 64 bit, PowerPC, and others.

Oracle (http://www.oracle.com). Oracle has been sold commercially since 1979 and offers several versions including large business, enterprise-class functionality. Oracle Database XE is a free version that can store up to 4 GB of user data, uses up to 1 GB RAM, and one CPU on the host machine. Both Windows and Linux versions are available.

SQL Server (http://www.microsoft.com). SQL Server is a Microsoft product. SQL Server Express is a free, limited version, for use by developers and is available for Windows XP and Vista; not recommended for novices. The DBMS includes a management console. MSDE (Microsoft Data Engine) is an earlier free version also intended for developers and has been integrated in many third-party programs to perform data management tasks. Runs on Windows operating systems.

SQLite (http://www.sqlite.org). SQLite is a public domain project where the DBMS is incorporated into another program instead of being a separate entity. The database exists as a single file that aids in transferring the database to another machine. Although free, it requires substantial programming knowledge to implement as intended. A stand-alone program called sqlite3 can be used to create a database, run queries, and manage the database. However, sqlite3 has a command-line interface and requires the user to be comfortable with database programming.

Iman (http://www.search-tech.com). Iman is a DBMS software designed specifically to manage images and runs on top of Microsoft SQL server. Data entry and management is Web based, so it is easily set up for use at multiple locations. This works well for managing and searching large libraries of digital images, with office staff performing most of the work. It is not intended for use with numeric datasets, but these can be created in SQL Server, which is required for Iman.

B. Object-Oriented and Object-Relational Database Management Software We now describe some popular DBMSs in this category:

Catalyzer (www.axiope.com). Marketed by Axiope, Catalyzer was specifically designed by researchers for biomedical research laboratory use. It runs on Windows, Mac OS X, and Linux. The program comes in two versions: one for desktop use and a server version for use as a central repository by the desktop installations. There is an evaluation version that allows creation of up to 100 objects without having to purchase a license. A Web browser is required for use with the server version; IE 6 and 7, Mozilla 1.4, Firefox 1.0, Netscape 7.0 and 7.1, Safari 2.0, and Opera 7.0 browsers are supported. There are modules that support direct inclusion and display of some common proprietary laboratory data files, including pClamp electrophysiology records, flow cytometry FCS files, MetaMorph images, Zeiss confocal, IPLab, and other imaging formats. The database is a single file that simplifies moving the store to a new computer.

OpenLink Virtuoso (http://www.openlinksw.com/virtuoso). OpenLink server is an object-relational model database server that combines multiple functions beyond just database management to include Web applications and file server functions, among others. The Virtuoso edition is open-source, comes with a free license, and runs on Windows, Linux, and Mac OS X. However, it is appropriate for advanced users, developers, and information technology professionals.

dBASE (http://www.dbase.com). dBASE was one of the early successful relational databases for microcomputers. Although originally a relational database, the current version is managed by an object-oriented programming language. dBASE did not use SQL; instead, it had a proprietary programming language for database management that has spawned a number of other programming languages referred to as xBase languages. The program is capable of directly connecting to SQL-based databases including MS SQL Server, Oracle, and Sybase. Runs on Windows operating systems.

Visual FoxPro (www.microsoft.com). FoxPro is usually regarded as a DBMS but actually is an xBase programming language used to write database applications. It is a popular database programming language among professionals.

5. Important Biological Databases

Online databases have become an important tool in molecular biology research. The annual January issue of *Nucleic Acids Research* is dedicated to online research databases; there were 858 in the 2006 update [2]. It is important to keep in mind that some databases are repositories of raw data with no guarantees of quality, whereas others may have extensive manual curation to ensure that only high-quality information is produced. Just as important as the databases are easy-to-use online tools to search and retrieve information.

NCBI (http://www.ncbi.nlm.nih.gov/) has one of the largest research sites on the Internet with a single-entry portal for search and navigation. At the time of this writing, there are twenty-one linked and searchable databases within the site. As mentioned earlier, it has an entity-relational structure written in ASN1. Many important bioinformatics databases are part of the NCBI family, including dbSNP, 3D Domains, RefSeq, Genomes, and proteins. In addition, databases from other sources, such as OMIM (Online Mendelian Inheritance in Man), have been incorporated into the NCBI site search capabilities. The GenBank database at NCBI, the DNA database of Japan, and EMBL (European Molecular Biology Laboratory) together form the International Nucleotide Sequence Database Collaboration with information exchanged among sites on a daily basis.

The Pfam (http://pfam.wustl.edu/) and GO databases have proved very useful in automated gene annotation. Protein sequences are compared against the Pfam database of sequence alignments of common protein families. This permits tentative assignment of function prior to laboratory bench work and is useful when large numbers of proteins are being investigated in genomewide scans. The database is accessible through Web sites hosted in several countries.

The GO database is a part of the Gene Ontology project (http://www.geneontology.org/), which provides a controlled vocabulary to describe gene and gene products. Ontology is a formal method of describing concepts and their relationships. The limited vocabulary and networked structure addresses the issue of several genes having multiple descriptive terms for the same process. Although a human would recognize that translation and protein synthesis probably have the same meaning when used by different authors, an automated computer search and retrieval system has no way to match the terms. Additionally, the database provides the level of confidence that can be ascribed to a descriptor, ranging from published laboratory-based work to automated annotation based on sequence similarity.

Manually curated and defined biological pathways are available through the KEGG (http://www.genome.jp/kegg/pathway.html) and BioCarta (http://www.biocarta.com/) databases. Automated links to these databases are now found in several microarray analysis programs in which a list of genes is evaluated for representatives occurring in specific pathways.

Part II Step-By-Step Tutorial

The tutorial has four goals in demonstrating basic data input and retrieval operations. They are as follows: how to create a searchable index of archived work, how to enter data, how to organize data for efficient use, and how to link objects by their properties in order to extract information. Data organization, called *normalization* in relational database work, is the key to creating an effective database. Data accumulation without organization is the computational equivalent of metastasis.

Catalyzer is used for demonstration purposes because it is easy to integrate existing datasets commonly found in laboratories, represents the newer object-oriented model technology toward which databases are moving, and requires the minimal amount of additional training to enter, locate, and extract data. An evaluation version can be used to work through the demonstration. The terms *object* and *property* are used interchangeably with *record* and *field* in the user manual, and they will be used in the same manner in this exercise.

1. Create a Catalog of Archived Data

Start by inserting a CD or other device with the data files to be cataloged. From the File drop-down menu, select Import>Folder then navigate to the drive and folder containing the data files, and click "import."

Check the options desired in the selection window. The imported structure can be limited to specific file types and exclude undesired files. In this case, we will generate small thumbnails of images for inclusion in the database.

The program creates a searchable index to the data files by importing the file structure, creating a thumbnail of images found, and extracting embedded image information (Figure 15.10). The generated thumbnail image, extracted image information hierarchical structure (pathway),

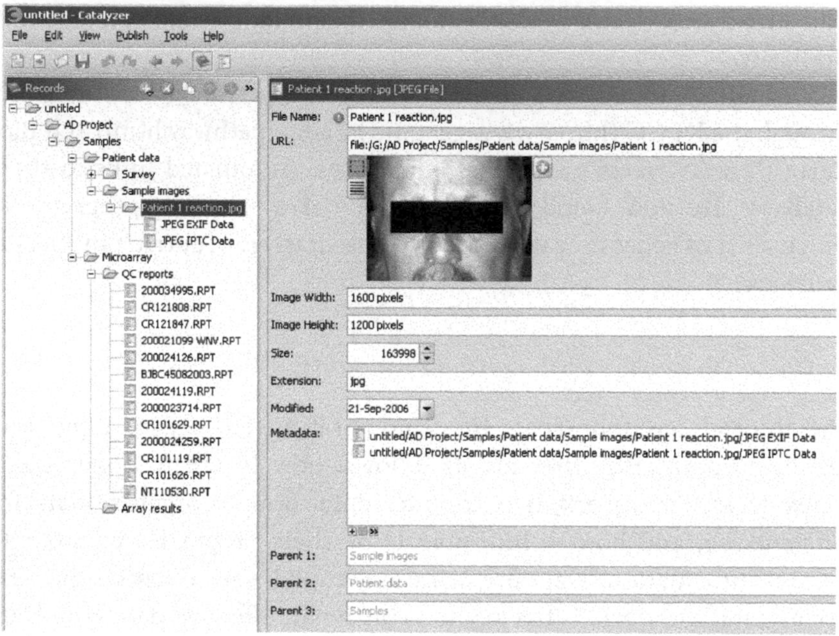
untitled - Catalyzer
File Edit View Publish Tools Help
Records
untitled
AD Project
Samples
Patient data
Survey
Sample images
Patient 1 reaction.jpg
JPEG EXIF Data
JPEG IPTC Data
Microarray
QC reports
200034995.RPT
CR121808.RPT
CR121847.RPT
200021099 WNV.RPT
200024126.RPT
BJBC45082003.RPT
200024119.RPT
2000023714.RPT
CR101629.RPT
2000024259.RPT
CR101119.RPT
CR101626.RPT
NT110530.RPT
Array results
Patient 1 reaction.jpg [JPEG File]
File Name: Patient 1 reaction.jpg
URL: file:/G:/AD Project/Samples/Patient data/Sample images/Patient 1 reaction.jpg
Image Width: 1600 pixels
Image Height: 1200 pixels
Size: 163998
Extension: jpg
Modified: 21-Sep-2006
Metadata: untitled/AD Project/Samples/Patient data/Sample images/Patient 1 reaction.jpg/JPEG EXIF Data
untitled/AD Project/Samples/Patient data/Sample images/Patient 1 reaction.jpg/JPEG IPTC Data
Parent 1: Sample images
Parent 2: Patient data
Parent 3: Samples

FIGURE 15.10 Database view of imported folder structure and images in the fictional archived AD project folder.

filenames, file size, file creation date, and other parameters are in the database; the data within the files are not. Note that the program recognized embedded data fields in .JPG files and extracted the information into EXIF and IPTC objects. This extracted data can be searched to locate images taken by a specific camera, exposure time, date, use of flash, or other parameters.

A. Sorting the Archived Data Catalog In the default display for objects within a container, switch to table view mode by clicking on the table view button in the lower right, outlined (Figure 15.11A) . Sort a column by clicking on the title field in a selected column (Figure 15.11B); a second click reverses the sort order. Despite having the appearance of a spreadsheet, all fields remain associated as a record even though only one field is selected for sorting.

B. Exporting the Catalog Data may be exported as a PDF file, as another catalog with only the selected information, or copied to the clipboard for use in other programs. In the table view, highlight the data you wish to

A

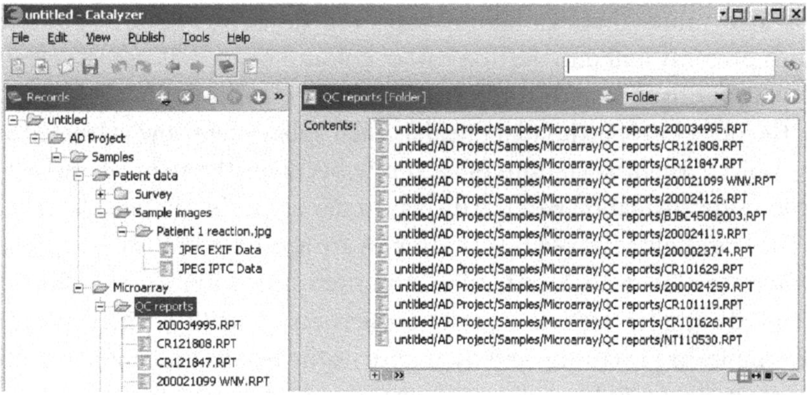
untitled - Catalyzer
File Edit View Publish Tools Help
Records
QC reports [Folder]
Folder
Contents:
untitled
AD Project
Samples
Patient data
Survey
Sample images
Patient 1 reaction.jpg
JPEG EXIF Data
JPEG IPTC Data
Microarray
QC reports
200034995.RPT
CR121808.RPT
CR121847.RPT
200021099 WNV.RPT
untitled/AD Project/Samples/Microarray/QC reports/200034995.RPT
untitled/AD Project/Samples/Microarray/QC reports/CR121808.RPT
untitled/AD Project/Samples/Microarray/QC reports/CR121847.RPT
untitled/AD Project/Samples/Microarray/QC reports/200021099 WNV.RPT
untitled/AD Project/Samples/Microarray/QC reports/200024126.RPT
untitled/AD Project/Samples/Microarray/QC reports/BJBC45082003.RPT
untitled/AD Project/Samples/Microarray/QC reports/200024119.RPT
untitled/AD Project/Samples/Microarray/QC reports/2000023714.RPT
untitled/AD Project/Samples/Microarray/QC reports/CR101629.RPT
untitled/AD Project/Samples/Microarray/QC reports/2000024259.RPT
untitled/AD Project/Samples/Microarray/QC reports/CR101119.RPT
untitled/AD Project/Samples/Microarray/QC reports/CR101626.RPT
untitled/AD Project/Samples/Microarray/QC reports/NT110530.RPT

B

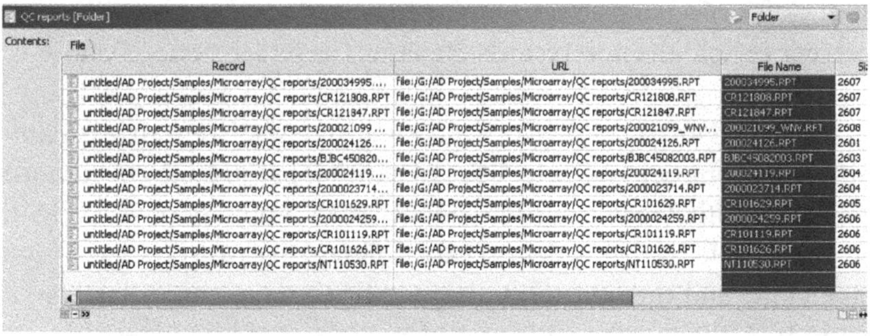
QC reports [Folder]
Folder
Contents:
File
Record
URL
File Name
2607
2607
2607
2608
2601
2603
2604
2604
2605
2606
2606
2606
2606

FIGURE 15.11 Manipulation of imported file structure in the database. (A) Use the highlighted icon to switch to table view of objects in a folder. This makes all data fields available. (B) Sorting the records based upon selection a single data field.

export. Right-click and select "copy to clipboard." The data can now be pasted into Excel or other programs. Each field becomes a column with the field label as a column header. In this instance there are separate fields for file name, path, size, file extension, and modification date.

Additional CDs, memory sticks, hard drives, network drives, or other accessible media containing archived data can be added to the database. The resulting dataset can be searched efficiently to locate files without having to search each CD or other device individually.

2. Data Entry

Because many laboratory records are already stored in Excel files, the simplest way to enter data is to cut and paste from an Excel spreadsheet.

Alternatively, an Excel worksheet can be used to create a file of comma-separated values that can be imported via the `file import` function. When properly laid out, the first row of column headers is used to create a template for the object (record), with the individual column labels becoming the object properties (data fields). Each subsequent row is treated as a new object. The first column of each row becomes the name displayed in the left side navigation bar; the displayed name can be changed later.

The workflow assumed is that samples are acquired in the clinic, sent to a laboratory for mRNA extraction, and then stored in a freezer. Portions of stored samples are later sent to a microarray facility, which prepares the sample for processing and then returns the results and quality control reports in data files labeled with their processing identifiers.

Select the folder where a new project will be located then right-click (Ctl-click for Mac) and select > add using class > folder; this will create a container to hold the new records. Rename the folder `Demo` Data.

A. Organize the Data This is a fictional dataset from an experiment designed to compare characteristics of skin biopsy samples from normal individuals with those from individuals with atopic dermatitis. In this case, the sample ID and patient ID are different because one was assigned by the clinic and the other was assigned by the microarray facility (Figure 15.12). Although the entire set of records could be entered at once, there are several issues that will cause problems later whether using either object-oriented or relational databases.

Inspection indicates there are three primary objects (tables, if creating a relational database): patients, samples, and microarrays. The first step is to reorganize the data to reflect that division using the patient ID as the common identifier in the resulting tables. Create an Excel worksheet containing information about the sample from Table 15.1 in the Data Section.

In Excel, select the data in the worksheet to be entered and copy it to the clipboard. In Catalyzer, with the `Demo Data` folder highlighted, use File> Import > Clipboard to import data from the clipboard. This enters the data from the clipboard and creates a subfolder named `Imported Data`. Rename the new folder to `Sample Data` for clarity (Figure 15.13A). Switching to the class view using the class creation icon, highlighted in Figure 15.13B, reveals that the program has also automatically created a class object named `Imported Data` (Figure 15.13C) for the imported dataset. The class can be renamed to match the renamed data folder. The properties of the object created by this class may also be modified in this

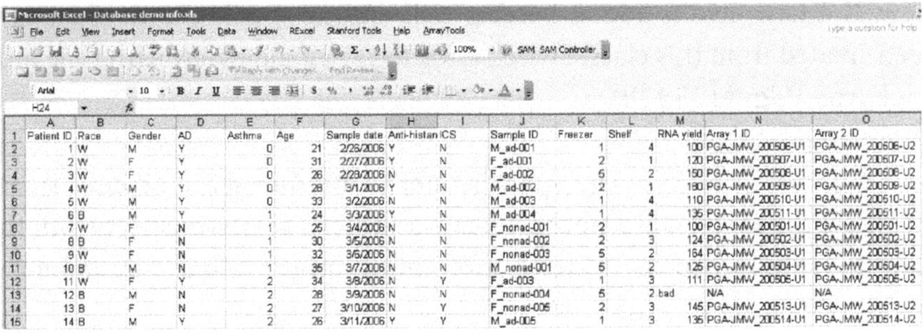

	Patient ID	Race	Gender	AD	Asthma	Age	Sample date	Anti-histan	ICS	Sample ID	Freezer	Shelf	RNA yield	Array 1 ID	Array 2 ID
2	1	W	M	Y	0	21	2/26/2006	Y	N	M_ad-001	1	4	100	PGA-JMW_200506-U1	PGA-JMW_200506-U2
3	2	W	F	Y	0	31	2/27/2006	Y	N	F_ad-001	2	1	120	PGA-JMW_200507-U1	PGA-JMW_200507-U2
4	3	W	F	Y	0	26	2/28/2006	N	N	F_ad-002	5	2	150	PGA-JMW_200508-U1	PGA-JMW_200508-U2
5	4	W	M	Y	0	28	3/1/2006	Y	N	M_ad-002	2	1	180	PGA-JMW_200509-U1	PGA-JMW_200509-U2
6	5	W	M	Y	0	33	3/2/2006	N	N	M_ad-003	1	4	110	PGA-JMW_200510-U1	PGA-JMW_200510-U2
7	6	B	M	Y	1	24	3/3/2006	Y	N	M_ad-004	1	4	136	PGA-JMW_200511-U1	PGA-JMW_200511-U2
8	7	W	F	N	1	25	3/4/2006	N	N	F_nonad-001	2	1	100	PGA-JMW_200501-U1	PGA-JMW_200501-U2
9	8	B	F	N	1	30	3/5/2006	N	N	F_nonad-002	2	3	124	PGA-JMW_200502-U1	PGA-JMW_200502-U2
10	9	W	F	N	1	32	3/6/2006	N	N	F_nonad-003	5	2	164	PGA-JMW_200503-U1	PGA-JMW_200503-U2
11	10	B	M	N	1	35	3/7/2006	N	N	M_nonad-001	5	2	125	PGA-JMW_200504-U1	PGA-JMW_200504-U2
12	11	W	F	Y	2	34	3/8/2006	N	Y	F_ad-003	1	3	111	PGA-JMW_200505-U1	PGA-JMW_200505-U2
13	12	B	M	N	2	28	3/9/2006	N	N	F_nonad-004	5	2	bad	N/A	N/A
14	13	B	F	N	2	27	3/10/2006	N	Y	F_nonad-005	2	3	145	PGA-JMW_200513-U1	PGA-JMW_200513-U2
15	14	B	M	Y	2	26	3/11/2006	Y	Y	M_ad-005	1	3	136	PGA-JMW_200514-U1	PGA-JMW_200514-U2

FIGURE 15.12 Spreadsheet of fictional experimental data.

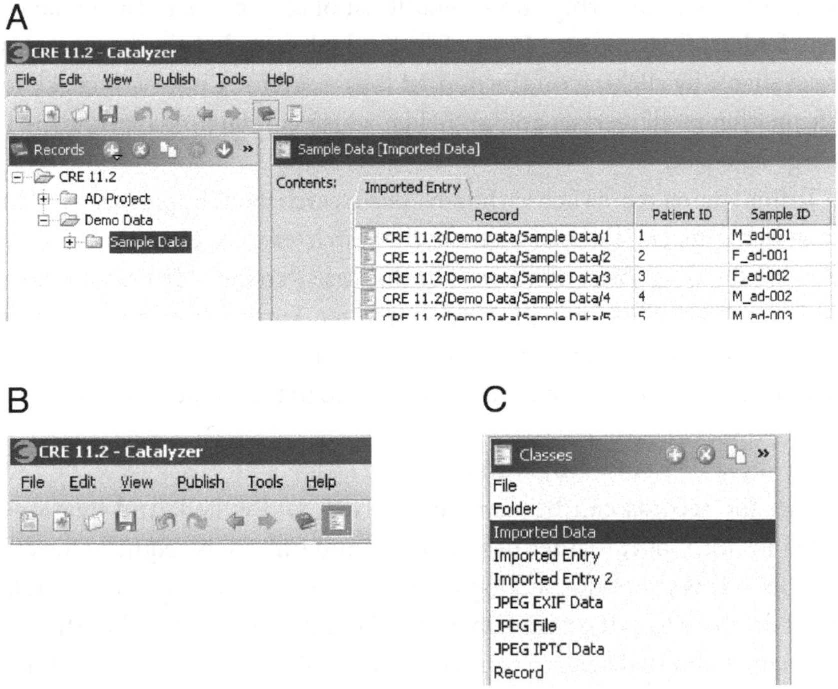
A
CRE 11.2 - Catalyzer
File Edit View Publish Tools Help
Records
CRE 11.2
AD Project
Demo Data
Sample Data
Sample Data [Imported Data]
Contents:
Imported Entry
Record
Patient ID
Sample ID
CRE 11.2/Demo Data/Sample Data/1
1
M_ad-001
CRE 11.2/Demo Data/Sample Data/2
2
F_ad-001
CRE 11.2/Demo Data/Sample Data/3
3
F_ad-002
CRE 11.2/Demo Data/Sample Data/4
4
M_ad-002
B
CRE 11.2 - Catalyzer
File Edit View Publish Tools Help
C
Classes
File
Folder
Imported Data
Imported Entry
Imported Entry 2
JPEG EXIF Data
JPEG File
JPEG IPTC Data
Record

FIGURE 15.13 Importing data from the clipboard automatically creates data folders and an object class for the data. (A) The Imported Data folder, created when data were entered via the clipboard, has been renamed to reduce confusion. (B) Location of the class creation icon to switch function from data manipulation to object property modification. (C) Imported Data class, also created when importing from the clipboard, is now visible and can be renamed to match the renamed data folder. The properties of the object created by this class may also be modified.

view. If a new field is added here, it will appear in all records that have been created from this class. Switch back to the catalog view using the blue book icon (Figure 15.13B).

B. Advanced Searches The records refer to another set of objects: freezers where the samples are stored. Create another Excel worksheet from the data in Table 15.2 in the Data Section and import the data. Rename the folder to "`freezer list`."

We will perform an advanced search to locate all –20°C freezers, and then link fields across records to have access to an inventory of individual freezer contents from the patient records. Select search from the menu or press Ctrl-F. This brings up a default list of all records in the database: files, folders, freezers, etc. Lists of individual records in a specific class are available by clicking on the named tabs. Searching now will check for information in all records and all fields, which can be time consuming in a large database.

To limit the search, switch to the advanced search mode by pressing the blue arrow in the upper right corner next to the search text box. Limit the search by selecting the class object of interest, in this case `Freezer`. The display automatically shows all records in the class freezer. Further limit the search by selecting the field `Temperature`, and then set the value to –20. This creates a list limited to freezers that have the value of –20 in the temperature field.

C. Create an Inventory by Linking Data Fields with Class Objects Although the records can be searched, sorted, and edited, and new ones added in this mode, further organization and linking is required for efficient use. This process is analogous to creating an SQL query in a relational database to extract information. The difference here is that the link is dynamic and updates are immediately visible in the results; in SQL, a query has to be run again to update the results.

Switch to class mode as previously mentioned. Select a record in the `sample data` folder. The class navigation pane shows the class for this object created when the data were first copied into the program; it can be renamed for clarity.

In the class design right side window, highlight the freezer field, and use the drop-down menu to change field type to `reference` (Figure 15.14A). There will be a warning message about potential `data loss`; click "continue." In the field conversion popup window, select the target class (Figure 15.14B) and list output (Figure 15.14C). In this case the class is `freezer`, and the list is `contents`.

In the sample data record, the `freezer` field now is an interactive link to the freezer containing the sample (Figure 15.15A). Without the active link, only the freezer number would be displayed. Clicking on the + sign in the `freezer` field opens a table view of all samples in freezer 1 (Figure 15.15B). The fields in the referenced sample records can be opened and edited from this window.

At this point we have created a live link to freezer contents from an individual patient record. Not only do we know which freezer the sample is in, but we also can find out what else is stored in the same freezer and go directly to information about other samples. If the freezer contents are

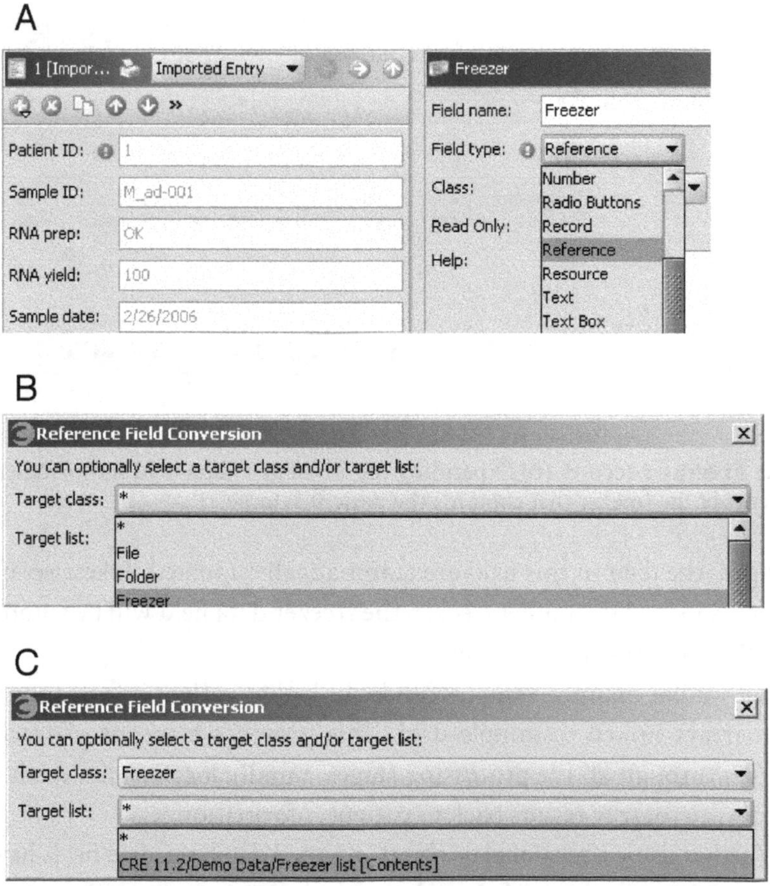
A
1 [Impor...
Imported Entry
Patient ID:
1
Sample ID:
M_ad-001
RNA prep:
OK
RNA yield:
100
Sample date:
2/26/2006
Freezer
Field name:
Freezer
Field type:
Reference
Class:
Read Only:
Help:
Number
Radio Buttons
Record
Reference
Resource
Text
Text Box
B
Reference Field Conversion
You can optionally select a target class and/or target list:
Target class:
*
Target list:
*
File
Folder
Freezer
C
Reference Field Conversion
You can optionally select a target class and/or target list:
Target class:
Freezer
Target list:
*
*
CRE 11.2/Demo Data/Freezer list [Contents]

FIGURE 15.14 Linking a field in a record to a reference. (A) Select the field name and change type to reference. (B) Select target class as `freezer`. (C) Select class output list as `freezer contents`.

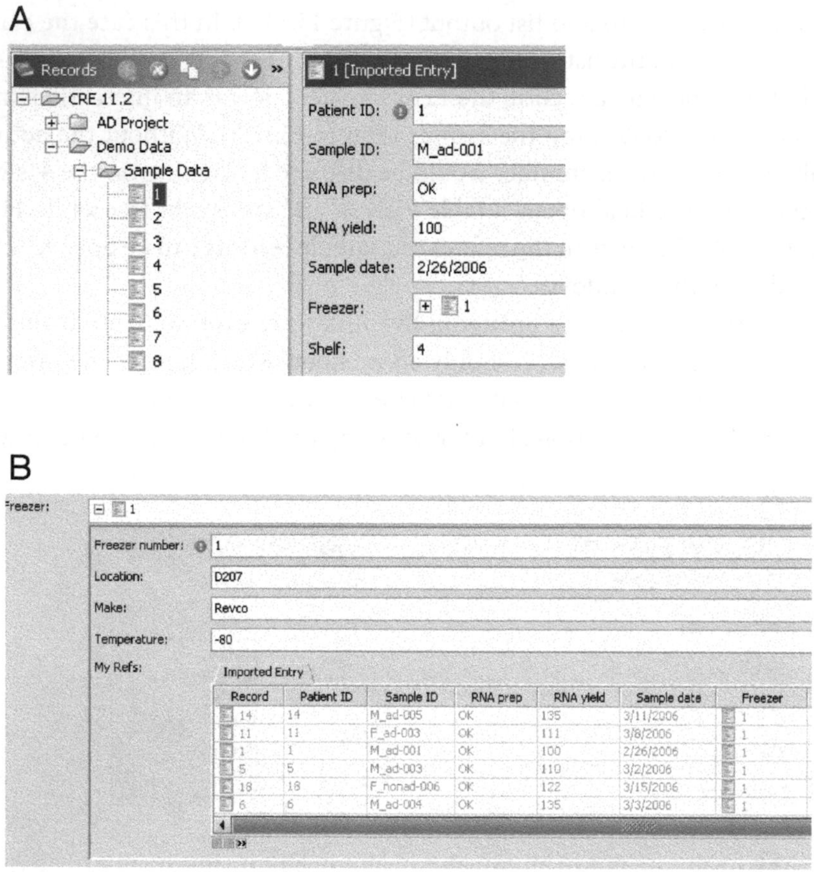
A
Records
CRE 11.2
AD Project
Demo Data
Sample Data
1
2
3
4
5
6
7
8
1 [Imported Entry]
Patient ID: 1
Sample ID: M_ad-001
RNA prep: OK
RNA yield: 100
Sample date: 2/26/2006
Freezer: 1
Shelf: 4
B
Freezer: 1
Freezer number: 1
Location: D207
Make: Revco
Temperature: -80
My Refs:
Imported Entry
Record Patient ID Sample ID RNA prep RNA yield Sample date Freezer
14 14 M_ad-005 OK 135 3/11/2006 1
11 11 F_ad-003 OK 111 3/8/2006 1
1 1 M_ad-001 OK 100 2/26/2006 1
5 5 M_ad-003 OK 110 3/2/2006 1
18 18 F_nonad-006 OK 122 3/15/2006 1
6 6 M_ad-004 OK 135 3/3/2006 1

FIGURE 15.15 (A) Patient record has a `freezer` data field that is an active link to the `freezer` record. (B) Expanding the `freezer` data field shows the entire contents of the freezer that contains the patient sample.

changed, the data in this field are automatically updated. Likewise, if the sample is moved to another freezer, the freezer data field will be automatically linked to the new location.

In a similar manner, samples can be linked to patient information and microarrays linked to sample data. This provides a means of tracking samples through all the processing stages, rapidly locating materials and linking microarray results back to patient information.

Whether using a relational or object-oriented database, data fields have to be linked to selected records in order to retrieve information. Organization and layout of basic elements, either in tables or as objects, is the key to successful use.

Part III Sample Data

Freezer Data (Table 15.1)
Sample Data (Table 15.2)

TABLE 15.1 Sample Data

Patient ID	Sample ID	RNA Preparation	RNA Yield	Sample Date	Freezer
1	M_ad-001	OK	100	02/26/2006	1
5	M_ad-003	OK	110	03/02/2006	1
6	M_ad-004	OK	135	03/03/2006	1
11	F_ad-003	OK	111	03/08/2006	1
14	M_ad-005	OK	135	03/11/2006	1
18	F_nonad-006	OK	122	03/15/2006	1

TABLE 15.2 Freezer Data

Freezer	Location	Make	Temperature
1	D207	Revco	-80
2	WBSB208	Revco	-80
3	D207	Sanyo	-20
4	D207	Revco	-20
5	BP216	Sanyo	-80

REFERENCES

1. Hernandez, M.J. *Database Design for Mere Mortals: A Hands-On Guide to Relational Database Design*, 2nd ed., Addison-Wesley, 2003.
2. Galperin, M.Y. The molecular biology database collection: 2006 update. *Nucl Acids Res* 34: D3–D5, 2006.

CHAPTER 16

Microsoft Excel and Access

Dmitry N. Grigoryev

CONTENTS

Recently developed gene expression platforms such as oligonucleotide and cDNA microarrays are powerful and extremely specific techniques for the identification of differentially regulated genes in a large variety of experimental models and human diseases. The search for new candidate genes of human diseases in global gene expression profiling using this high-throughput microarray technology has been widely employed. Microarray profiling is a robust tool in simultaneous identification of expression patterns of large groups of genes and elucidating the mechanisms underlying complex biological processes and diseases. This approach reveals responses of thousand of genes to a given challenge or condition. To cope with these overwhelming datasets, various computing algorithms were developed. The majority of these analytical programs and tools represents complex statistical packages and requires specific bioinformatics training of potential users. The biomedical scientists become dependent on the bioinformatics assistance, which reduces flexibility of the research and discovery processes. However, despite the complexity of the gene expression data, tools familiar to every biologist for data storage, manipulation, and analysis such as Microsoft Excel and Microsoft Access can be efficiently employed for the basic genomic analyses. While waiting for the detailed and finesse analysis from their bioinformaticians, biologists will be able to evaluate major trends of expressional changes and make an educational guess in what direction to proceed with their ongoing researches.

SECTION 1 MICROSOFT EXCEL

Part I Introduction

Several attempts to simplify gene expression analysis using Microsoft Excel were reported and successfully tested. However, these approaches are based on modifications and add-ins to the basic Microsoft Excel platform and introduced extra functionality that requires advanced knowledge of and experience with Microsoft Excel. The goal of this section, however, is to allow bioinformatics illiterates to conduct basic genomic analysis using standard Microsoft Excel functions such as sorting, filtering, and in-place calculations. The straightforward approach described in this section will allow researchers to conduct a basic gene expression analysis tailored to their immediate needs.

1. What Is Microsoft Excel?

Microsoft Excel is a spreadsheet program that allows creating and editing spreadsheets, which are used to store information in columns and rows that can then be organized and processed. Spreadsheets are designed to work with numeric and character entries that can be easily manipulated using multiple embedded functions. Mathematical calculations can be done on numerical entries and text modifications on characters entries in such a way that desired changes propagate throughout the spreadsheet automatically.

2. What Can Microsoft Excel Do?

In numerous tutorials, the description of the Microsoft Excel application is restricted to business-oriented tasks and focused on computations related to accounting and production of concise financial reports. The charting and text manipulation features make Microsoft Excel well suited to producing attractive and informative reports for business meetings. In this chapter we will employ the enormous calculating and data storage potential of this program for the analysis of large datasets generated during gene expression profiling. The standard Microsoft Excel spreadsheet can accommodate 65,536 entries and the largest Affymetrix Human U133 Plus 2 array contains 54,614 entries, which eliminates space concerns. We will demonstrate how the mathematical, statistical, and searching functions of Microsoft Excel would help us to retrieve biologically sensible information from meaningless rows of expression values.

3. Featured Functions in Microsoft Excel for Microarray Analysis

A. Open Data Sheet The variety of Microsoft Excel functions applicable for gene expression analysis will be introduced during progression of a sample microarray data processing. The single-channel microarray platform (microarrays that use one labeling dye) such as Affymetrix will be considered in this chapter. The researchers are usually provided with spreadsheet data in text (.txt) or excel (.xls) format. The standard outputs of these arrays contain various gene expression information including three key entries: unique probe identifier, gene expression signal, and detecting quality of this signal. The sample shorten version of common output for the Affymetrix platform is provided in Part III.

B. Log Transformation We will first generate the basic spreadsheet of gene expression information, which will contain only the key entries described earlier. Then expression data will be converted into log base two (log2) format. The results of microarray experiments are usually presented as fluorescent signal ratios of gene expression signals between control and treated samples.

We recommend processing these ratios in log format. In our example experiment in Part III, we will look at gene expression at Drugs A and B, and the results are relative expression levels compared to the control set as numerical 1. Assume Drug A upregulates a gene by two-fold and Drug B downregulates this gene by two-fold relative to control expression. The numerical ratio values are 2.0 and 0.5 for Drug A and B, respectively. Apparently we have the same magnitude of expressional changes in both drugs, but in an opposite direction. However using these numerical values for further calculation will generate a difference between Drug A (two-fold change or 2.0) and control (no change or 1) equal to +1.0, whereas that between control (1) and Drug B (0.5) equals −0.5. Thus, mathematical operations that use the difference between numerical values would lead to the conclusion that the two-fold upregulation was twice as significant as the two-fold downregulation, which is wrong. The log transformation eliminates this problem. When data from the previous example is log2-transformed, the data points become 0, 1.0, and −1.0 for control, Drug A, and Drug B, respectively. With these values, two-fold up- and two-fold down regulation are symmetric in relation to 0. For microarray analysis we recommend that you work with log2-transformed data using LOG function of Microsoft Excel and at the end of the analysis convert log2 data back to numeric values using power (^) function of the program.

C. Normalization Once data are log-transformed, the next step will be intraarray normalization. When analyzing multiple arrays, there is a need for data normalization, after calculating gene expression. The overall signal intensity can vary between arrays. This global variation is characterized by the overall shift in the intensity of expression signals. This nonbiological variation is multifactorial and depends on RNA quality, experimental and scanning procedures, sample labeling, and environmental factors such as light exposure. There are different algorithms available to normalize microarray data, and most microarray platforms reports already-normalized data. However, most normalization procedures are based on the whole array data including genes, expression of which is affected by experimental drugs. Therefore, it is preferable that these genes be excluded from the normalization procedure and only invariant between arrays genes utilized. Our basic tutorial will accept original Affymetrix normalized data and proceed with analysis without additional data normalization. The more complex process of global array normalization using invariant gene set with application of mean center array normalization will be described in the "Advanced" section of the tutorial.

D. Filtering Usually, only a fraction of genes is expressed in a given cell or tissue type. To simplify our analysis, we will exclude nonexpressors from further consideration using the filtering functions of Microsoft Excel. In our basic streamline analysis, we will consider only genes that were detected on all arrays. This approach will provide major information about global gene expression changes in response to a given drug. You have to realize that this strict filtering will leave out genes that were undetectable, for example, in control samples but become active after exposure to a certain drug. The more complex selective filtering, which recognizes these particular cases, will be described in the "Advanced" section of the tutorial.

E. Statistics Microsoft Excel provides a number of functions for statistical calculations, including ANOVA and *t*-test. Considering the overwhelming multiple comparison nature of array data where 20,000–50,000 thousand genes are tested simultaneously, the Microsoft Excel statistical package is not particularly suited for microarray analysis. However, it is efficient enough to identify general trends in transcriptional changes of implicated genes and suggest most probable gene candidates. An example of the basic gene expression analysis will be presented using the *t*-test function of the program.

F. Visualization In this section we will demonstrate how to generate color-coded expression tables and expression-pattern-based clusters using Microsoft Excel formatting and filtering functions. These techniques should be especially useful to laboratories that do not have access to specialized commercial visualization software. Huge columns of numbers are mind-numbing to most users. Microarray data are much easier to perceive as colored squares, where the color indicates whether a gene is up- or downregulated. We will use Microsoft Excel's `Conditional Formatting` function to highlight or replace expression ratios with a corresponding color, thereby transforming an array database into a color representation that will facilitate global assessment of gene expression patterns. We also will perform *clustering*, a powerful method for discovering patterns in array data. The goal of clustering is to subdivide genes in such a way that similar expression patterns fall in the same cluster. Gene expression clustering allows an open-ended exploration of the data, without getting lost among the thousands of individual genes. We will demonstrate how to cluster genes that show similar expression patterns across a number of samples, using the expression pattern of one gene as a template for the identification of its counterparts.

Part II Step-By-Step Tutorial

1. Basic Data Analysis Streamline

A. Retrieving Gene Expression Spreadsheet Researchers are usually provided with gene expression spreadsheets in text (.txt) or Excel (.xls) format, both of which can be opened using Microsoft Excel:

1. Start Microsoft Excel by clicking on `Start` -> `Programs` -> `Microsoft Excel` or by double-clicking on the `Microsoft Excel` icon on the desktop.

2. To open the gene expression spreadsheet click `File` -> `Open`. There are several shortcuts to call desired Microsoft Excel functions, which can significantly accelerate manipulation of large amounts of data. These shortcuts will be gradually introduced in this tutorial. The shortcut keys help provide an easier, and usually quicker, method of

navigating and using Microsoft Excel. Shortcut keys are commonly accessed by using the ALT, CTRL, and SHIFT keys in conjunction with a single letter (on IBM-compatible computers). The assigned shortcuts are listed in the corresponding menus (in our case under the `File` menu next to the `Open` function, we can see a description of assigned shortcuts for `Open` (modifier key plus single character; CTRL+O). In other words, CTRL+O is telling you to simultaneously press the CTRL key and the O key to activate the shortcut. Pressing CTRL+O would bypass the `File` function and perform the `Open` command directly. In addition to the assigned shortcuts, we can use ALT-associated shortcuts, which are alternatives to mouse clicking. You will note that the "F" in `File` has been underlined. This tells you that you can press the ALT key and F to access the `File` menu; then you can see that under `File` menu the `Open` has an underlined "O." Therefore, holding ALT and pressing F and then O will follow the mouse-clicking route and perform `File` and then `Open` function in succession. As you begin to work with shortcut keys, you will notice that your navigation throughout Microsoft Office will be dramatically accelerated.

(Note: Users outside the U.S. or users using a foreign copy of Microsoft applications may not be able to get all shortcut keys to perform the function listed in this tutorial.)

3. In the `Open` window, select your file and double-click on it.

4. If a gene expression spreadsheet was provided to you in raw text format (also called ASCII) and you cannot see it in the `Open` window, use `Files of type`: scroll down the menu of the `Open` window (alternatively use ALT+T) and then select `All files (*.*)` from the menu. Double-click on your file, and Microsoft Excel will automatically recognize it as a text file and start the `Import Wizard`.

5. Choose `Finish` and you will see the data placed in cells in a Microsoft Excel spreadsheet.

6. Convert this text file to a Microsoft Excel file with `File->Save as` or ALT+F+A and select `Microsoft Excel Workbook (*.xls)` from `Save as type:` pull-down menu of the dialog box `Save As`.

B. Formatting Data for Expression Analysis The analysis flow of a standard Affymetrix output, provided in Part III, which will be demonstrated in this tutorial, can be applied to other platforms as well:

1. Open `Affymetrix data source sample.xls` as described in Section 1. For our analysis we will need only three key entries: `Probe set ID`, `Signal`, and `Detection`. Therefore, we will remove all `Stat Pairs Used` columns and a `Description` column; the latter consumes a lot of computer memory and slows down the computing process.

2. Save `Affymetrix data source sample.xls` file as `Affymetrix expession.xls` using the `File->Save As` function.

(Tip: It is a good practice to save a file as a copy before you perform any modifications on it.)

3. Select column B `C1 Stat Pairs Used` by clicking on the title cell of this column, then holding down the CTRL key, select all other columns (...`Stat Pairs Used`) to be removed. To delete selected columns use `Edit->Delete` (Figure 16.1) or (ALT+E+D). Alterna-

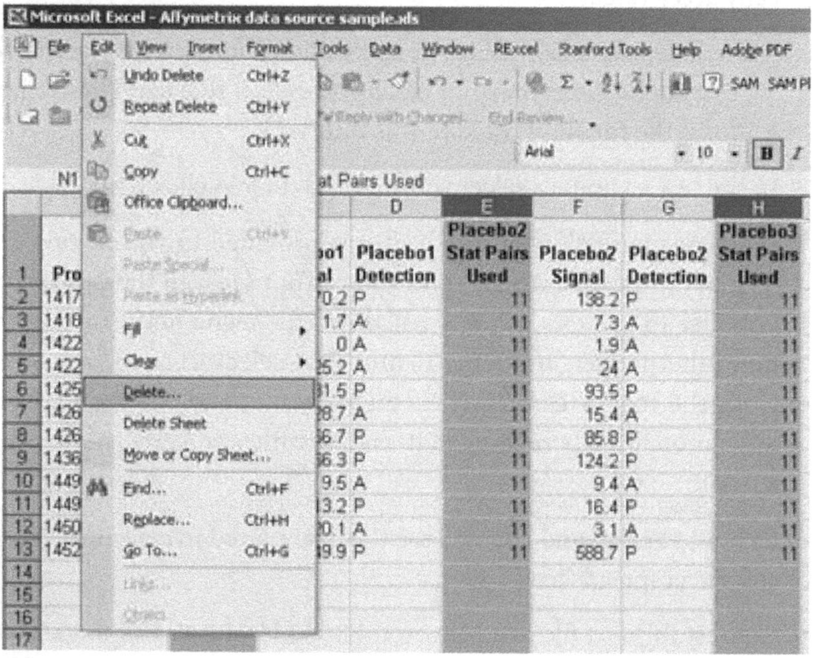
Microsoft Excel - Affymetrix data source sample.xls
Undo Delete Ctrl+Z
Repeat Delete Ctrl+Y
Cut Ctrl+X
Copy Ctrl+C
Office Clipboard...
Fill
Clear
Delete...
Delete Sheet
Move or Copy Sheet...
Find... Ctrl+F
Replace... Ctrl+H
Go To... Ctrl+G
Placebo1 Detection
Placebo2 Stat Pairs Used
Placebo2 Signal
Placebo2 Detection
Placebo3 Stat Pairs Used

FIGURE 16.1 Deleting multiple columns using the Microsoft Excel tool bar.

tively, while pointing on any of the selected columns right-click your mouse to display the popup menu, then select the `Delete` function (right-click ->`Delete`) (Figure 16.2).

Now, when we have only necessary information, we will log-transform expression signal values. To facilitate future computing, we will rearrange data in the spreadsheet by clustering `Signal` and `Detection` values.

4. Select all nine `Signal` columns as described earlier, and copy them using `Edit->Copy` or assigned function CTRL+C. Alternatively, you can use ALT+E+C or right-click->`Copy`.

5. Select the first empty column (in our case, T) and paste copied columns with `Edit->Paste` or CTRL+P. The alternative commands can be deduced from the previous pattern and are ALT+E+P or right-click->`Paste`.

6. Now we have to reselect original `Signal` columns and delete them. After completion of these manipulations, we will have all `Signal` values clustered in nine columns from K to S. Next, we will designate

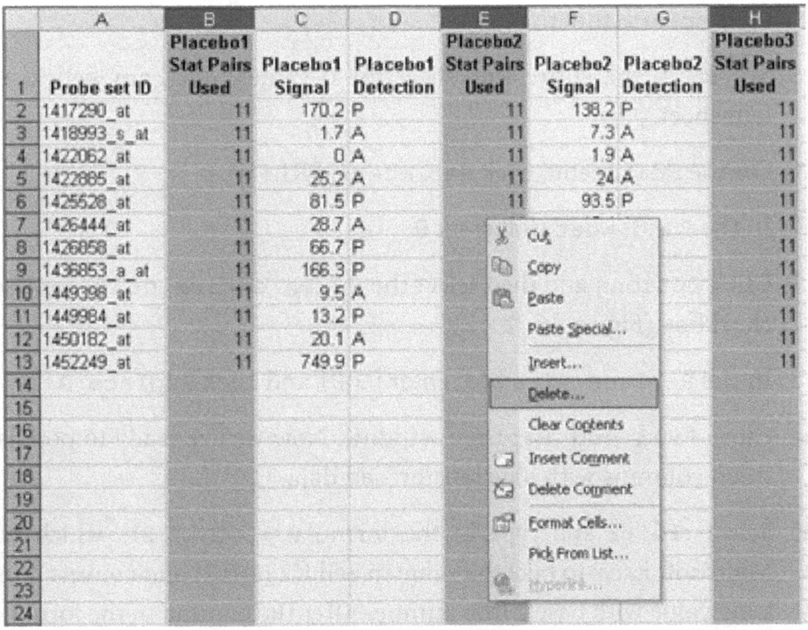

	A	B	C	D	E	F	G	H
		Placebo1			Placebo2			Placebo3
		Stat Pairs	Placebo1	Placebo1	Stat Pairs	Placebo2	Placebo2	Stat Pairs
1	Probe set ID	Used	Signal	Detection	Used	Signal	Detection	Used
2	1417290_at	11	170.2	P	11	138.2	P	11
3	1418993_s_at	11	1.7	A	11	7.3	A	11
4	1422062_at	11	0	A	11	1.9	A	11
5	1422865_at	11	25.2	A	11	24	A	11
6	1425528_at	11	81.5	P	11	93.5	P	11
7	1426444_at	11	28.7	A				11
8	1426858_at	11	66.7	P	✂ Cut			11
9	1436853_a_at	11	166.3	P				11
10	1449398_at	11	9.5	A	📋 Copy			11
11	1449984_at	11	13.2	P	📋 Paste			11
12	1450182_at	11	20.1	A	Paste Special...			11
13	1452249_at	11	749.9	P	Insert...			11
14								
15					Delete...			
16					Clear Contents			
17								
18					📄 Insert Comment			
19					📄 Delete Comment			
20					📄 Format Cells...			
21					Pick From List...			
22								
23					🔗 Hyperlink...			
24								

FIGURE 16.2 Deleting multiple columns using the Microsoft Excel popup window.

space for log-transformed data using column titles and text-modifying formulas.

7. Select cell T1, and type: `=K1&" log2"`. This equation tells Microsoft Excel to take the value in cell K1, which is a text entry "Placebo1 Signal" and add (ampersand sign, &) text "log2" in quotation marks. After you hit Enter, the content of T1 cell will be "Placebo1 Signal log2".

(Note: All entries starting with the equal sign are interpreted by Microsoft Excel as a formula. If you forget to put the equal sign, the program will see it as a regular entry.)

8. To title other columns accordingly, click on the T1 cell again and while holding the right mouse button, highlight eight first raw cells to the right (including AB column). To fill highlighted cells with titles, use the CTRL+R command (`Fill-to-the-Right` function). Microsoft Excel will move through cells from K1 to S1, applying the formula that we provided in the T1 cell to each highlighted cell.

Given that log transformation can be applied only to positive numbers, we will substitute all zero values in our datasheet with 0.001 using the `Replace` function.

9. To search the entire worksheet for 0 values, click any cell in the datasheet.
10. On the `Edit` menu, click `Replace` (CTRL+H).
11. In the `Find what` slot, enter 0.
12. Click `Options` and then select the `Match entire cell contest` checkbox (Figure 16.3).
13. In the `Replace with` box, enter 0.001 and click `Replace All`.
14. Close `Find and Replace` window. Now we are ready to populate titled columns with log-transformed data.
15. Select cell T2 and type in the formula: `=LOG(K2,2)`, which tells Microsoft Excel to take the value in cell K2 (170.2) and convert it into a log value with base 2 (the number after the comma in the formula).
16. Hit Enter. Select the entire area under new titles by clicking on the T2 cell and highlighting it by dragging the mouse pointer to cell AB13 while

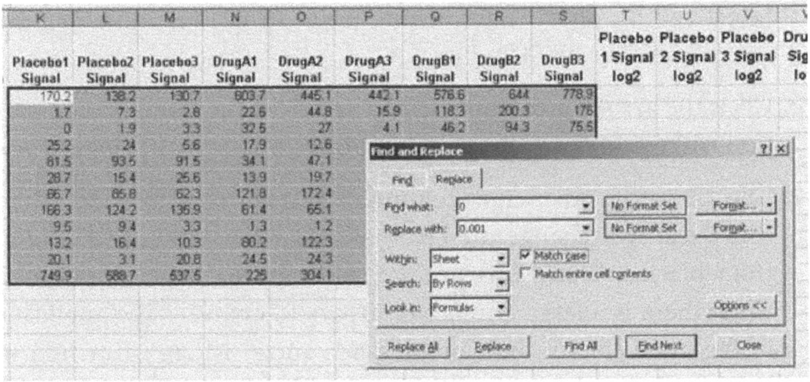

K	L	M	N	O	P	Q	R	S	T	U	V	
Placebo1 Signal	Placebo2 Signal	Placebo3 Signal	DrugA1 Signal	DrugA2 Signal	DrugA3 Signal	DrugB1 Signal	DrugB2 Signal	DrugB3 Signal	Placebo 1 Signal log2	Placebo 2 Signal log2	Placebo 3 Signal log2	Dru Sig lo
170.2	138.2	130.7	603.7	445.1	442.1	576.6	644	778.9				
1.7	7.3	2.8	22.6	44.8	15.9	118.3	200.3	176				
0	1.9	3.3	32.5	27	4.1	46.2	94.3	75.5				
25.2	24	5.6	17.9	12.6								
61.5	93.5	91.5	34.1	47.1								
28.7	15.4	25.6	13.9	19.7								
86.7	85.8	62.3	121.8	172.4								
166.3	124.2	135.9	61.4	65.1								
9.5	9.4	3.3	1.3	1.2								
13.2	16.4	10.3	80.2	122.3								
20.1	3.1	20.8	24.5	24.3								
749.3	588.7	537.5	225	304.1								

Find and Replace

Find | Replace

Find what: 0 — No Format Set — Format...

Replace with: 0.001 — No Format Set — Format...

Within: Sheet   ☑ Match case

Search: By Rows   ☐ Match entire cell contents

Look in: Formulas

Replace All | Replace | Find All | Find Next | Close

FIGURE 16.3 Application of the Microsoft Excel Find and Replace function to the selected area of a datasheet.

holding the right button. Alternatively, you can highlight the desired area by clicking on the T2 cell and then performing CTRL+SHIFT+End.

(Note: CTRL+Home and CTRL+End will select the very first or last cell in a spreadsheet, respectively.)

To fill the selected area, we have to use the Fill-Down (CTRL+D) and Fill-to-the-Right (CTRL+R) functions in succession in any order of these commands. This approach demonstrates the functionality of relative referencing in Microsoft Excel. A relative cell reference in a formula, such as T2, is based on the relative position of the cell that contains the formula and the cell the reference refers to (in our case, K2). If the position of the cell that contains the formula changes, the reference changes in the same direction. When we copy our formula down columns and then across rows, the reference will be automatically adjusted. For example, when we copy (fill down) formula =LOG(K2,2) in cell T2 to cell T3, the relative reference is automatically adjusted from K2 to K3.

C. Selecting Detectable Genes   To select genes that were expressed on all microarrays, we will apply filtering functions of Microsoft Excel. The Detection columns (B1–G1) contain information of transcript detectability by each probe set on an array and coded by Affymetrix as P for present, M for marginal, and A for absent. For our further analysis, we will select genes expression of which was classified by Affymetrix as "present (P)" on all arrays.

1. Select cell B1; on the `Data` menu, point to `Filter`, and then click `AutoFilter` or ALT+D+F+F.
2. Click the arrow in column B1 and select `P`. The `AutoFilter` function will retain only rows that contain `P` in column B1 (Figure 16.4).
3. Select `P` in the other eight columns. To track your filtering, Microsoft Excel will change the color of the arrow in filtered columns from black to blue. This `AutoFilter` approach becomes tedious when there is a large number of arrays to analyze. The application of the more efficient but complex `Advance Filter` function will be described in the section "Clustering." Now we will copy identified "present" genes to Sheet2 of this workbook.
4. Select `Probe set ID` column and all `Signal log2` columns and copy them (CTRL+C).
5. Click on the `Sheet2` tab in the right lower corner of the workbook, select the A1 cell in `Sheet2`, and paste (CTRL+P) the copied data. You should have six probe sets that detect their corresponding genes in all tested samples.

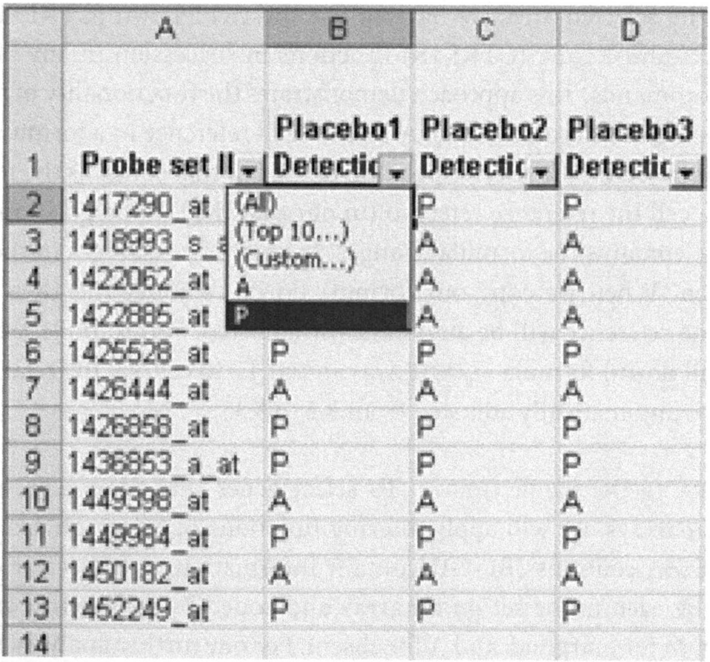

	A	B	C	D
1	Probe set II	Placebo1 Detectic	Placebo2 Detectic	Placebo3 Detectic
2	1417290_at	(All)	P	P
3	1418993_s_	(Top 10...) (Custom...)	A	A
4	1422062_at	A	A	A
5	1422885_at	P	A	A
6	1425528_at	P	P	P
7	1426444_at	A	A	A
8	1426858_at	P	P	P
9	1436853_a_at	P	P	P
10	1449398_at	A	A	A
11	1449984_at	P	P	P
12	1450182_at	A	A	A
13	1452249_at	P	P	P
14				

FIGURE 16.4 Application of the Microsoft Excel `AutoFilter` function.

6. You can rename `Sheet2` using the `Format` menu, point to `Sheet`, and then click `Rename`. Type the new name `Present` over the current (`Sheet2`) highlighted name.

D. Expression Analysis

1. Name columns K2 and M2 `Fold change DrugA` and `Fold Change DrugB`, respectively.

2. Name columns L2 and N2 `P value DrugA` and `P value DrugB`, respectively.

3. To calculate expression fold change for the first probe set 1417290_at select cell K2 and type: `=2^(AVERAGE(E2:G2)-AVERAGE(B2:D2))`.

   Microsoft Excel will subtract the average of control expression signals from the average of DrugA expression signals and convert the resulting log2 value to the numeric fold change.

   (Note: The subtraction of log values is an equivalent of division of numeric values.)

4. Repeat this step for Drug B (cell M2: `=2^(AVERAGE(H2:J2)-AVERAGE(B2:D2))`.

5. To access statistical tools, click `Data Analysis` in the `Tools` menu. If the `Data Analysis` is not available, ask your IT person to load the Analysis ToolPak add-in program.

6. In cell L2 type: `=TTEST(E2:G2,B2:D2,2,2)` where E2:G2 represent Drug A data set, B2:D2 represent Placebo data set, the third entry specifies the number of distribution tails (in our case it is a 2-tailed test), the fourth entry specifies the type of a test (1 = paired, 2 = two-sample equal variance (our case), 3 = two-sample unequal variance).

   Given that during log2 transformation the variability of expression values becomes equivalent, the two-sample equal variance type was selected.

7. Repeat this step for Drug B by typing in cell N2: `=TTEST(H2:J2,B2:D2,2,2)`.

8. Click on L2 cell, and select all cells under new titles using your mouse or shortcut CTRL+SHIFT+End.

9. Fill empty cells with CTRL+D.

Now we will select genes that were significantly (0.5 > Fold change > 2 and $P < 0.05$) affected by at least one of two tested drugs.

10. For Drug A select cell K1, and on the `Data` menu, point to `Filter` and then click `AutoFilter` or ALT+D+F+F.
11. Click the arrow in the column K1 and select `(Custom...)`.
12. In the box on the left, click the arrow and select `is greater than`.
13. In the box on the right, enter 2.
14. Add another criteria by clicking `Or` and repeat the previous step in blank boxes, but this time select `is less than`.
15. In the box on the right, enter 0.5 (Figure 16.5) and click `OK`.
16. Perform filtering for column L2 by selecting `is less than` and entering 0.05 in the box on the right.
17. Select `Probe set ID` and four last columns, copy them to a new (CTRL+N) workbook, and save it (CTRL+S) as `Candidate genes.xls`.
18. Return to `Affymetrix expression.xls` Present sheet; click the arrow in the column K1 and select `All`.

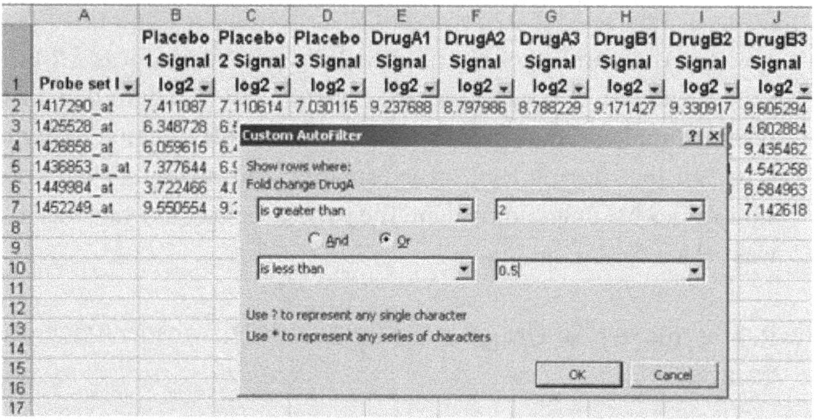
Custom AutoFilter
Show rows where:
Fold change DrugA
is greater than
2
And
Or
is less than
0.5
Use ? to represent any single character
Use * to represent any series of characters
OK
Cancel

FIGURE 16.5 User-defined filtering using (Custom…) feature of the Microsoft Excel `AutoFilter` function.

19. Repeat this step for column L1, and the original dataset will be restored.
20. Repeat the entire filtering procedure for Drug B, using columns M and N.
21. Click on cell A2, then holding SHIFT click on cell A7 and perform copy command (CTRL+C). Now your selection will be limited to a specific set of cells, which will allow you to paste selected `Probe set ID` entries under existing data in the `Candidate genes.xls` workbook.
22. Click on cell A6 in `Candidate genes.xls` and paste (CTRL+V) `Probe set IDs` from the Drug B gene list.
23. Repeat this step with data in columns K to N. Click on the cell K2, and holding SHIFT click on the cell N7, copy and paste into `Candidate genes.xls`, selecting cell B6.

These manipulations generated the candidate gene list for both drugs A and B. But there are most likely genes that were significantly affected by both drugs and, therefore, this list contains duplicated entries. To eliminate duplicates, we will filter for unique records using the `Advanced Filter` function.

24. Click cell A1 in the list.
25. On the `Data` menu point to `Filter` and then click `Advanced Filter`.
26. Click "Filter the list, in-place" and select the `Unique records only` checkbox, click `OK`.
27. To save unique candidate genes, perform CTRL+SHIFT+END -> CTRL+C, paste (CTRL+V) to Sheet2 of this workbook, and rename Sheet2 with Unique using `Rename` function as described earlier.

We will annotate these candidate genes in Section 2 using Microsoft Access.

E. Color-Coding Gene Expression Ratios First, we have to convert gene expression values into fold change ratios on a chip-to-chip basis:

1. Open (CTRL+O) `Color coding and clustering.xls` workbook provided in Part III, which represents familiar expression values in log2 format (`Expression values` spreadsheet).
2. Label columns K to P by copying titles form columns E to J, respectively.
3. Select cells K1 to P1 and replace `Signal log2` part of the title with `Fold change` as described in the following text.
4. On the `Edit` menu, click `Replace`. In the `Find what` box, enter `Signal log2`. In the `Replace with` box, enter `Fold change`.
5. Click `Replace All` and then `OK`.
6. In the cell K2 type formula: `=E2-AVERAGE($B2:$D2)`.

This formula tells Microsoft Excel to subtract average of log2 Placebos from individual gene expression value on each array for Drug A and Drug B.

7. Copy `Probe set ID` (column A) and six `Fold change` columns (K to P) to spreadsheet `Color coding` of this Microsoft Excel workbook.
8. Click on the first number containing cell B2, and select all numeric entries in `Color coding` spreadsheet with CTRL+SHIFT+END.
9. On the `Format` menu, click `Conditional Formatting`.
10. In the first slot of the `Conditional Formatting` dialog box, select `Cell Value Is`, select the comparison phrase `greater than`, and then type `1` in the third slot.

Remember that a 1-fold change in log2 format stands for a numeric 2-fold change.

11. Click the `Format` button in the `Conditional Formatting` dialog box, and in the `Format Cell` dialog box select the `Pattern` tab, choose red color, and click `OK`.
12. To add another condition, click the `Add` button in the `Conditional Formatting` dialog box.
13. Select the comparison phrase `less than`, and then type `-1` in the third slot. Click `Format`.

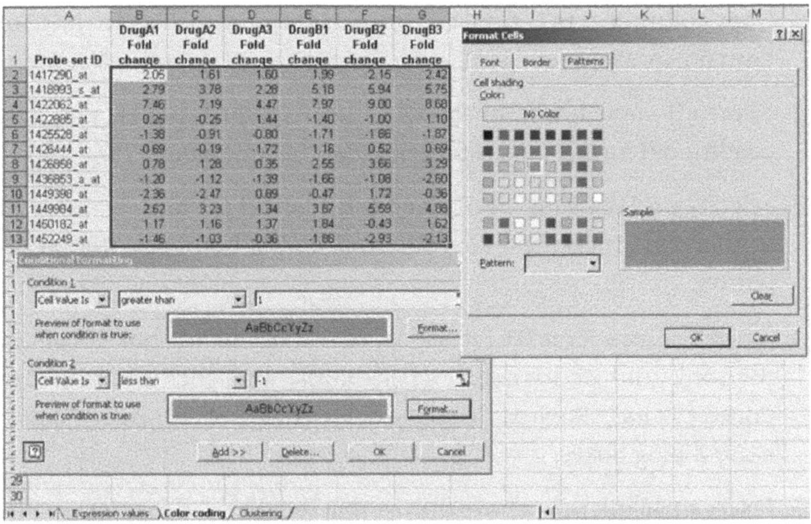

FIGURE 16.6 Visualization of the sought data points using `Format Cell` option of the Microsoft Excel `Conditional Formatting` function.

14. In the `Format Cell` dialog box, select `Pattern`, choose green color, and then click `OK` (Figure 16.6).

To facilitate segregation of genes that were similarly affected by both Drug A and Drug B, we will sort genes by their averaged expression values.

15. Type formula `=AVERAGE(B2:G2)` in cell H2.
16. Generate averages for all genes by selecting cells from H2 to H13 and filling down with CTRL+D.
17. Sort this datasheet by column H in descending order by clicking a cell in the column H and then the `Sort Descending` icon.

This manipulation will move genes that were mostly upregulated in both conditions upward and those that were mostly downregulated in both conditions downward.

F. Clustering We will build a cluster around 1449984_at probe set, which is our putative gene of interest. Our goal is to identify other genes with expression patterns similar to that of the 1449984_at probe set. We will set upper and lower fold change limits for our cluster using 2 logs deviation from the corresponding fold change values of the 1449984_at probe set.

1. Copy columns A to G from `Color coding` spreadsheet to the `Clustering` spreadsheet.
2. Select all copied cells in the `Clustering` spreadsheet, and delete conditional formatting as described in the following text.
3. Select `Conditional Formatting` from the `Format` menu and in the `Conditional Formatting` dialog box click the `Delete` button.
4. In the `Delete Conditional Formatting` dialog box select `Condition1` and `Condition2` checkboxes, and click `OK` in `Delete Conditional Formatting` and followed `Conditional Formatting` dialog boxes.
5. Insert six blank rows above the list that will be used for storing filtering criteria.
6. Select cells A1 to G6, and on the `Insert` menu click `Rows`.

The criteria containing rows must have their own column labels, so copy column titles from the seventh row and paste them in the first row. Given that we will filter our datasheet using two values (upper and lower cluster limits), we have to duplicate column titles in the filtering criteria area.

7. Paste another set of array titles (excluding `Probe set ID` title) to cells H1 to M1.
8. To identify upper limiting values of the cluster, type in the cell A4 title: `upper->` and in cell B4 type formula: `=B10+2`. This formula tells Microsoft Excel to add 2 logs (4 fold changes) to the expression value of 1449984_at probe set.
9. To identify lower limiting values, type in cell A5 title: `lower->`, and in cell B5 type formula: `=B10-2`.
10. Generate upper and lower value for all six arrays by highlighting the A4 to G5 area and filling highlighted cells with the CRTL+R command.

Now we will submit these numbers to filtering criteria.

11. Select cell B2 and type: `="<"&B4`. The resulting expression will tell Microsoft Excel to search for numbers that are below upper limits of the cluster.

12. Generate upper limit searching criteria for all six arrays by highlighting A2 to G2 area and filling highlighted cells with the CRTL+R command.

13. Select cell H2, and type: `=">"&B5`. The resulting expression will tell Excel to search for numbers that are above the lower limits of the cluster.

14. Generate lower limit searching criteria for all six arrays by highlighting the H2 to M2 area and executing the CRTL+R command.

(Note: The criteria range must be separated from your database by at least one blank row.)

15. Click cell A7 (it can be any cell inside the datasheet), then select `Filter->Advanced Filter` from the `Data` menu. The `Advanced Filter` dialog box will appear.

The `List range` in the dialog box is the area of the spreadsheet that Microsoft Excel determines as your database. On the spreadsheet, it will be outlined with dotted lines.

16. To indicate the criteria range, click in the `Criteria range` slot of the dialog box, then select (highlight) the A1 to M2 area that contains filtering criteria.

(Note: If the criteria range in the dialog box already contains cell references, highlight them prior to scrolling over our criteria range or type in correct cell addresses.)

We have the option of filtering the database "in place" or copying the filtered records to another location in the spreadsheet.

17. Leave `Filer the list, in-place` selected (Figure 16.7) and click `OK`.

Microsoft Excel identified three records that contain probe sets that are in the specified cluster range. To retrieve more genes, the limiting criteria should be relaxed by increasing deviation to 3 or 4 logs.

To view all records again, select `Filter->Show All` from the `Data` menu.

18. To perform graphical visualization, select filtering results to be graphed.

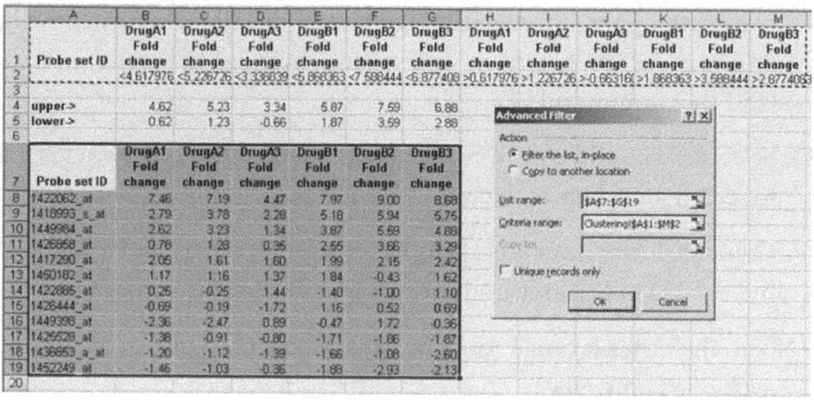

Probe set ID
DrugA1 Fold change | DrugA2 Fold change | DrugA3 Fold change | DrugB1 Fold change | DrugB2 Fold change | DrugB3 Fold change | DrugA1 Fold change | DrugA2 Fold change | DrugA3 Fold change | DrugB1 Fold change | DrugB2 Fold change | DrugB3 Fold change
<4.617976 <5.226726 <3.336839 <5.868363 <7.588444 <6.877408 >0.617976 >1.226726 >-0.66316(>1.868363 >3.588444 >2.877406
upper-> 4.62 5.23 3.34 5.87 7.59 6.88
lower-> 0.62 1.23 -0.66 1.87 3.59 2.88
Probe set ID | DrugA1 Fold change | DrugA2 Fold change | DrugA3 Fold change | DrugB1 Fold change | DrugB2 Fold change | DrugB3 Fold change
1422062_at 7.46 7.19 4.47 7.97 9.00 8.68
1418993_s_at 2.79 3.78 2.28 5.18 5.94 5.75
1449984_at 2.62 3.23 1.34 3.87 5.59 4.88
1426858_at 0.76 1.28 0.35 2.55 3.86 3.29
1417290_at 2.05 1.61 1.60 1.99 2.15 2.42
1460182_at 1.17 1.16 1.37 1.84 -0.43 1.62
1422885_at 0.25 -0.25 1.44 -1.40 -1.00 1.10
1426444_at -0.69 -0.19 -1.72 1.16 0.52 0.69
1449398_at -2.36 -2.47 0.89 -0.47 1.72 -0.36
1425528_at -1.38 -0.91 -0.80 -1.71 -1.86 -1.87
1436853_a_at -1.20 -1.12 -1.39 -1.66 -1.08 -2.60
1452249_at -1.46 -1.03 -0.36 -1.88 -2.93 -2.13
Advanced Filter
Action
Filter the list, in-place
Copy to another location
List range: $A$7:$G$19
Criteria range: Clustering!$A$1:$M$2
Copy to:
Unique records only
OK
Cancel

FIGURE 16.7 Application of the Microsoft Excel `Advanced Filter` function.

19. Put your cursor in cell A7, click hold the mouse button down, and drag to cell G11.
20. Click on the `Chart Wizard` button on the Microsoft Excel toolbar.
21. From the `Chart Wizard` dialog box that opens, select `Line` from the listing in the "Chart type" and the first template from the "Chart subtype"; then click `Finish` (Figure 16.8).

2. *Advanced Techniques*

A. Mean Center Array Normalization This array-normalizing approach is a simplified version of Z-transformation of array data and should be performed on log transformed data. The normalization is achieved by subtraction of the columnwise mean from the individual values in each row of this column, so that the mean value of the column becomes 0.

First, we will identify invariant probe sets:

1. Open file `Affymetrix expression.xls`.

Instead of filtering for present (`P`) detection as described in the "Selecting detectable genes" procedure, we will filter for "absent" (`A`) detection.

2. Apply `AutoFilter` function to all columns, and retain only rows that contain `A`.
3. Save identified absent genes that are representative of invariant data fraction to Sheet3 of this workbook.

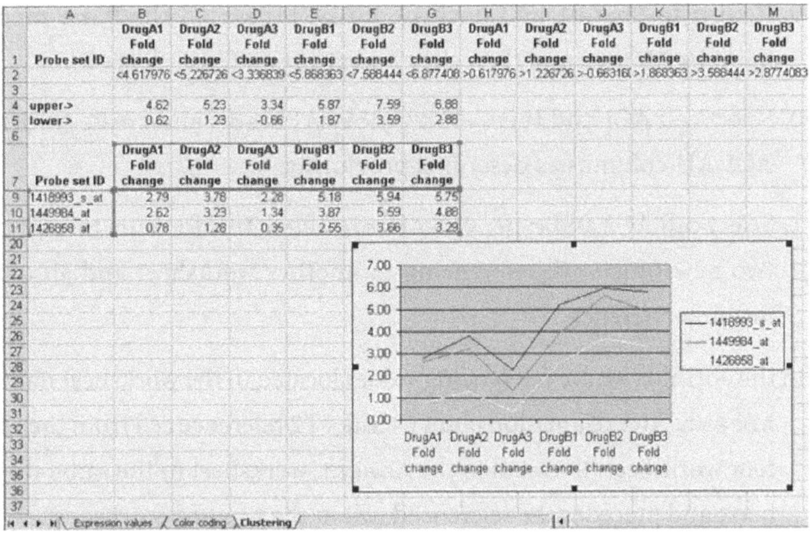

Probe set ID
DrugA1 Fold change
DrugA2 Fold change
DrugA3 Fold change
DrugB1 Fold change
DrugB2 Fold change
DrugB3 Fold change
upper->
lower->
1418993_s_at
1449984_at
1426858_at
Expression values
Color coding
Clustering

FIGURE 16.8 Creating linear graph for selected values using the Microsoft Excel `Chart Wizard`.

4. Select `Probe set ID` column and all `Signal log2` columns and copy them (CTRL+C).

5. Click on the `Sheet3` tab in the left lower corner of the workbook, select A1 cell in `Sheet3`, and paste (CTRL+P) copied data.

6. Rename `Sheet3` as `Absent`.

You should have four probe sets that did not detect their corresponding genes in any of tested samples, which characterize these genes as nonspecific for a given tissue. Therefore, hybridization signals of these probe sets represent nonspecific binding for all arrays and can serve as an invariant data set.

Now we will calculate mean value for each column.

7. In cell B6 type `=AVERAGE(B2:B5)` and hit Enter.

8. Select B6 to J6 and Fill-to-the-Right (CTRL+R).

Now you can evaluate the technical variability of array brightness, where `Placebo3` is the dimmest and `DrugB3` is the brightest array in the experiment.

9. Return to Sheet1 and uncheck AutoFilter: on the Data menu, point to Filter, and then click AutoFilter (ALT+D+F+F).
10. Select cell AC1 and type: =T1&" normalized", and create titles for AC1-AJ1 columns as described previously.
11. Select cell AC2 and type: =T2-Absent!B$6. This formula introduces two new functions: referencing to another worksheet and absolute formula reference.

In this formula we are referencing values located in the worksheet named Absent. An exclamation point (!) links T2 reference cell from the current worksheet (Sheet1) to the Absent worksheet in the same workbook and precedes the referenced cell in the Absent worksheet.

Contrary to relative referencing described in the log transformation paragraph, an absolute cell reference in a formula, such as $B$6, always refers to a cell in a specific location. If the position of the cell that contains the formula changes, the absolute reference remains the same. However, each column in our case is represented by its own mean value located in the sixth row of the Absent worksheet; therefore, we will apply mixed referencing. A mixed reference has either an absolute column and relative row or absolute row and relative column. An absolute row reference takes the form B$6, which allows changes in column address but keeps the row location constant.

12. Apply the normalizing formula to all nine columns.

B. Expanded Selection of Detectable Genes In our basic streamline analysis we identified detectable genes that were present (P) on all arrays. This approach ignored genes that were unaffected by one condition but become detectable in another. In this section we will identify all genes that were detectable at any given experimental condition. We will define genes that are detectable on two out of three arrays under any experimental conditions.

1. Copy `Probe set ID`, `Detection`, and `Normalized` columns (A-J, AC-AK) in `Sheet1` of workbook `Affymetrix expression.xls` to new workbook (CTRL+N or by clicking on `New` icon in the Microsoft Excel toolbar); and save it (CTRL+S) as `Normalized expression.xls`.
2. In the `Detection` columns we will replace A and M with 0, and P with 1 using the `Replace` function.
3. Select cell B2 and, holding SHIFT, select cell J13.
4. On the `Edit` menu, click `Replace` (CTRL+H).
5. In the `Find what` box, enter P.
6. Click `Options` and then select `Match entire cell contest` checkbox.
7. In the `Replace with` slot, enter 1 and click `Replace All`.
8. Replace M and A with 0.
9. Close the `Find and Replace` window.
10. Name columns T, U, and V `Placebo detection`, `DrugA detection`, and `DrugB detection`, respectively.
11. In cells T2, U2, and V2 type: `=INT(AVERAGE(B2:D2)/(2/3))`, `=INT(AVERAGE(E2:G2)/(2/3))`, `=INT(AVERAGE(H2:J2)/(2/3))`, respectively. This formula will assign 1 to a probe set that was present on 2 or 3 arrays ($\geq 2/3$) and 0 to probe sets that were present on less than 2 arrays ($<2/3$), where an integer function (INT) reports numbers rounded down to an integer.
12. Fill the corresponding columns down (CTRL+D).
13. Name column W `Overall detection`.
14. In cell W2 type: `=SUM(T2:V2)` and fill down the column.

Now we will filter out probe sets with overall detection equal to 0.

15. Select cell W1, and on the `Data` menu point to `Filter`; then click `AutoFilter` or ALT+D+F+F.
16. Click the arrow in the column W1 and select `(Custom...)`.
17. In the box on the left, click the arrow and select "does not equal".

18. In the box on the right, enter 0 and click `OK`.
19. Copy (CTRL+C) `Probe set ID` column and nine "Normalized" columns (K-S) to `Sheet2` and rename `Sheet2` as `Present genes`.

This approach identified eight detectable genes compared to six genes that were detected using the "present on all arrays" approach in `Affymetrix expression.xls`.

Part III Sample Data

`Affymetrix data source sample.xls` represents shortened (12 probe sets) example of standard (12000–54000 probe sets) Affymetrix array output.

`Affymetrix expression example.xls` represents the final outcome of all manipulation described in the tutorial for `Affymetrix expression.xls`.

`Candidate genes example.xls` demonstrates final selection of candidate genes generated by analyses described in the tutorial.

`Color coding and clustering.xls` contains basic expression data for color coding and clustering exercises.

`Color coding and clustering example.xls` represents the final outcome of all manipulation described in the tutorial.

`Normalized expression example.xls` represents the final outcome of all manipulation described in the advanced section of Microsoft Excel tutorial for `Normalized expression.xls`.

SECTION 2 MICROSOFT ACCESS

Part I Introduction

DNA microarray experiments provide expression information for thousand of genes, including well-known, newly identified, and unannotated genes. This flood of information can overwhelm biologists and make extraction of useful information a challenging task. Lack of sufficient annotation often becomes rate-limiting as sifting through available databases and biomedical literature sources is a monotonous, laborious, and time-consuming task. To accelerate this process, Microsoft Access-based tools and macros were developed and successfully applied for gene annotation. However, these tools require installation of additional scripts and macros. In this chapter we will show how to annotate selected genes using the basic functions of Microsoft Access. The application of these simple functions allows batch querying of multiple genomic databases and obtaining up-to-date annotation for candidate genes of interest.

1. What Is Microsoft Access?

Microsoft Access is a database management system that functions in the Windows environment and allows one to create and process data in a relational database. In simple terms a relational database is an organized collection of tabular data that can be linked (related) using similar fields between multiple tables such as gene symbols or accession numbers. The flexible Microsoft Access database management system provides researchers with tools they need to organize gene expression data tailored to specific objectives. Microsoft Access also provides a user-friendly interface that allows users to manipulate data in a graphical environment that is less intimidating for general users such as biologists.

2. What Can Microsoft Access Do?

Microsoft Access supports the framework for storing and analyzing large amounts of information in a database in a matter of minutes. Tables comprise the fundamental building blocks of the Microsoft Access database and are extremely similar to Microsoft Excel spreadsheets. Microsoft Access allows researchers to add, modify, or delete data from the database, ask questions (queries) about the data stored in the database, and produce reports summarizing selected contents. The report-building wizards provide the capability to quickly produce summaries of the data contained in several tables tailored to researcher's objectives.

3. Featured Functions in Microsoft Access for Biologists

Microsoft Access offers one of the simplest database management functions. Although Microsoft Access is a powerful stand-alone application, we will use it as an auxiliary tool for selection of common entries from multiple gene expression and annotation datasheets and filtering large textual entries for gene ontology. We also will demonstrate applicability of basic Microsoft Access functions for cross-referencing gene lists for creation of Venn diagrams. The Venn diagram is a useful visualization tool to indicate the extent of overlaps and differences between multiple gene lists generated during microarray analyses. The flexibility of the Microsoft Access filtering function allows identification of not only numbers of shared genes or genes specific to a particular condition but also description of these genes.

A. Queries The Microsoft Access query function will be used to display information from a combination of tables. For example, suppose you have one table that lists the mouse probe sets that detected significant changes in expression pattern of their target genes. A second table contains descriptive information about all mouse probe sets. By creating a query based on both tables, you could retrieve a listing of each significantly changed probe set along with the name of a gene and its biological function. We will also use the query function for cross-linking orthologous genes. It is common in genomic research that human genes are better annotated than rodent, canine, or other species. To enrich, for example, mouse annotation, one can link mouse genes to their corresponding human orthologues, which is easily done with Microsoft Access query function.

B. Filtering Large Entries The Microsoft Excel filtering function can handle approximately 300 characters and spaces per cell in a spreadsheet. However, the majority of the gene ontology descriptions go well beyond 300 characters. Therefore, the search for a specific gene ontology using the Microsoft Excel filtering function would be incomplete. The Microsoft Access filtering function does not have this limitation and will generate a complete list of queried ontologies.

Part II Step-By-Step Tutorial

1. Starting Microsoft Access

1. Double-click on the Microsoft Access icon on the desktop or click on `Start` -> `Programs` -> `Microsoft Access`. If a dialog box is automatically displayed with options to create a new database or open an existing one, select `Candidate gene annotation.mdb` database and then click `OK`.

2. If the dialog box is not displayed open provided in Part III database manually with CTRL+O, then in the `Look in` field of the `Open` dialog box select `Candidate gene annotation.mdb` file and click `Open` button.

3. Under `Table` object you will see four tables: `Candidate genes`, `Human annotation`, `Mouse annotation`, and `Human-Mouse orthologues`. `Candidate genes` table represents six candidate

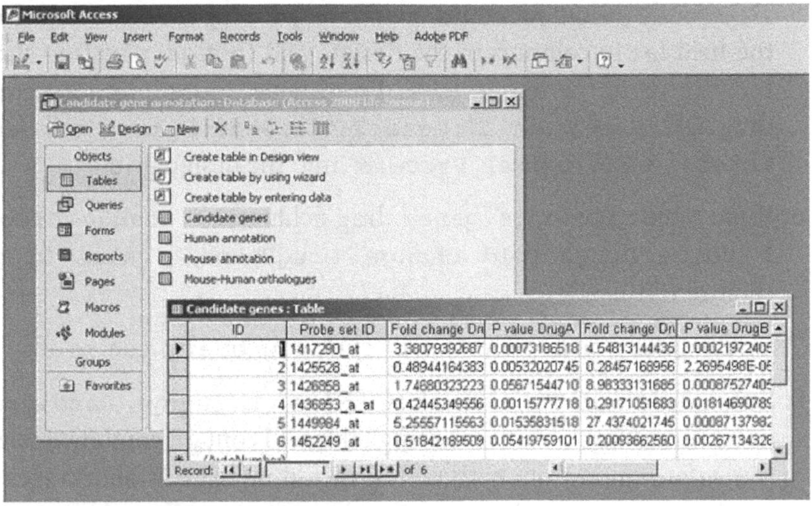
Microsoft Access
File Edit View Insert Format Records Tools Window Help Adobe PDF
Open Design New
Objects
Tables
Queries
Forms
Reports
Pages
Macros
Modules
Groups
Favorites
Create table in Design view
Create table by using wizard
Create table by entering data
Candidate genes
Human annotation
Mouse annotation
Mouse-Human orthologues
Candidate genes : Table
ID
Probe set ID
Fold change Dr
P value DrugA
Fold change Dr
P value DrugB
1 1417290_at 3.38079392687 0.00073186518 4.54813144435
2 1425528_at 0.48944164383 0.00532020745 0.28457168956
3 1426858_at 1.74680323223 0.05671544710 8.98333131685
4 1436853_a_at 0.42445349556 0.00115777718 0.29171051683
5 1449984_at 5.25557115563 0.01535831518 27.4374021745
6 1452249_at 0.51842189509 0.05419759101 0.20093662560
Record: 1 of 6

FIGURE 16.9 Locating tabular data in the Microsoft Access environment.

genes previously identified in Microsoft Excel section and can be opened by double-clicking on the table title (Figure 16.9).

2. Assigning Probe Sets to Their Target Genes

1. In the `Database` window click `Queries` under `Objects`, and then click `New` on the `Database` window toolbar.

2. In the `New Query` dialog box, select `Design View`, and then click `OK`. Alternatively, you can double-click on `Create query in Design view`.

3. In the `Show Table` dialog box, select `Candidate genes` and `Mouse annotation` tables by double-clicking the name of each table and Microsoft Access will add these tables to the query; and then click `Close`.

4. In `Query` design view (top panel), drag a `Probe set ID` field from `Candidate genes` table to the `Probe set ID` field in `Mouse annotation` table, and the join line should appear. This line between fields of two different tables tells Microsoft Access how the data in one table is related to the data in the other. With this type of join, Microsoft Access selects records from both tables only when the values in the joined fields are equal (in our case, the same probe set IDs).

5. To generate the output of this query, we will drag selected fields from the field list to columns in the design grid (bottom panel) to show these fields in the output table. From table `Mouse annotation`, drag fields `Probe Set ID`, `Gene Title`, `Gene Symbol`, and `Gene Ontology Biological Process` into the design grid.

6. From table `Candidate genes` drag fields `Fold change DrugA`, `P value DrugA`, `Fold change DrugB`, and `P value DrugB` (Figure 16.10).

7. Under `Query` menu select `Run` or click `Run query` icon.

8. Save query with `File->Save` (CTRL+S) as `Query Candidate gene annotation`. The resulting table will contain combined selective information from both tables, including expression changes of candidate genes and their functional annotation. Close the query.

3. *Assigning Probe Sets to Their Orthologues in Another Species*

1. Open `Candidate gene annotation.mdb`. In the `Database` window click `Queries` under `Objects`, and then double-click on `Create query in Design view`.

2. In the `Show Table` dialog box double-click on `Human annotation` and `Mouse-Human orthologues`; two tables should appear in the query design view.

3. Click `Queries` tab in the `Show Table` dialog box and double-click on `Query Candidate gene annotation` listing. Close `Show Table` dialog box.

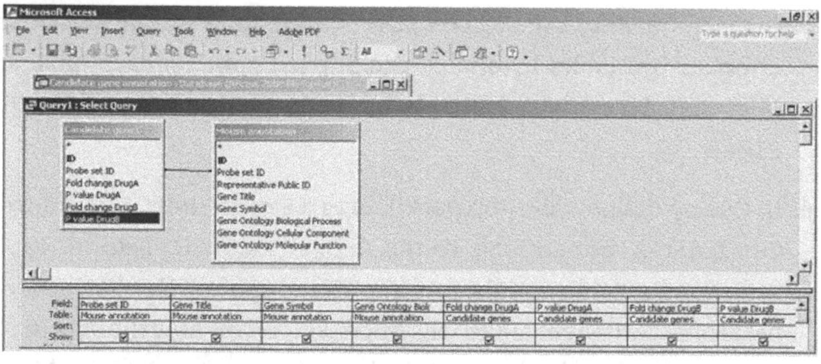

FIGURE 16.10 Building up a simple, two-component query using the Microsoft Access `Design View` panels.

4. Connect `Probe set ID` field from the `Query Candidate gene annotation` table with the `Mouse Probe Set` field in `Mouse-Human orthologues` table. Connect `Probe set ID` field from the `Human annotation` table with the `Human Probe Set` field in the `Mouse-Human orthologues` table.

5. Starting with the `Probe set ID` field in `Query Candidate gene annotation` table, select multiple entries by holding SHIFT and clicking on the last field `P value DrugB`. Drag the whole selection down into the first column of design grid (bottom panel).

6. Drag `Probe set ID` and `Gene Ontology Biological Process` fields from the `Human annotation` table into the second and fifth column of the design grid, respectively (Figure 16.11).

7. Run query and save resulting table as `Query Mouse-Human ontology`.

The resulting table will contain set of mouse-human orthologous candidate genes that are annotated using functional information from both species.

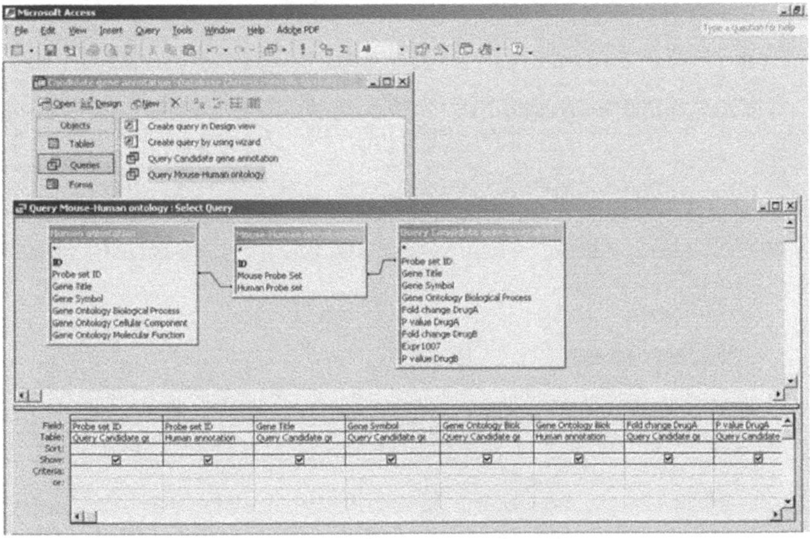
Microsoft Access
Query Mouse-Human ontology : Select Query
Probe set ID
Gene Title
Gene Symbol
Gene Ontology Biological Process
Mouse Probe Set
Human Probe set
Fold change DrugA
P value DrugA
Fold change DrugB
P value DrugB
Field:
Table:
Sort:
Show:
Criteria:
or:

FIGURE 16.11 Building up a complex, three-component query using the Microsoft Access `Design View` panels.

4. Filtering for Large Entries

1. Open `Mouse annotation` table.
2. On the Microsoft Access `Records` menu, point to `Filter`, and then click `Advanced Filter/Sort`.
3. Add to the design grid the `Gene Ontology Biological Process` field by dragging it from the `Mouse annotation` table.
4. In the "Criteria:" cell for the "Field:" `Gene Ontology Biological Process` enter `*transport*`. The asterisk (*) represents the wild card character that is commonly used in computing to substitute for any other character, number, or text. This expression tells Microsoft Access to search for any entry that contains the combination of letters *transport*. Therefore, all gene ontologies related to any kind of molecular transport will be retrieved (Figure 16.12).
5. Apply the filter with `Record->Apply Filter/Sort` or by clicking `Apply Filter` on the toolbar. Three genes with transport-related gene ontology are identified.

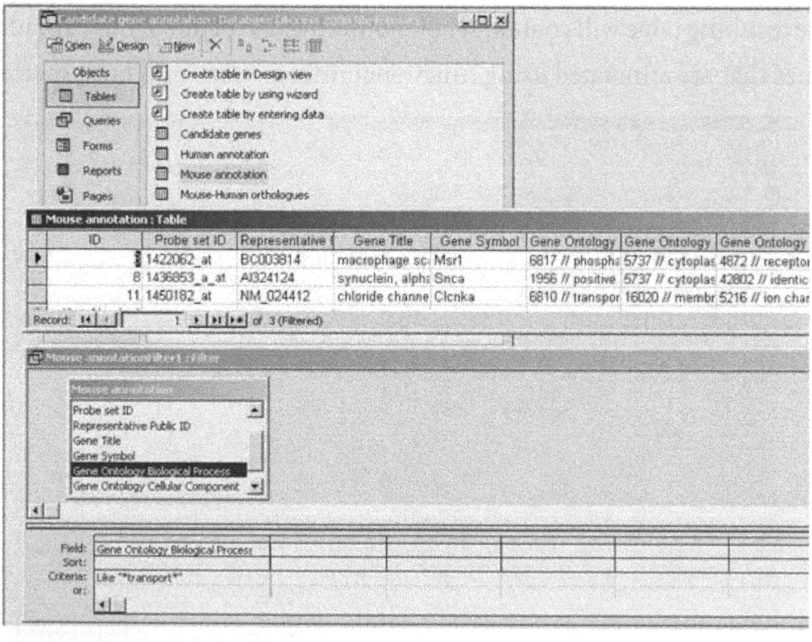
Objects
Tables
Queries
Forms
Reports
Pages
Create table in Design view
Create table by using wizard
Create table by entering data
Candidate genes
Human annotation
Mouse annotation
Mouse-Human orthologues
Mouse annotation : Table
ID
Probe set ID
Gene Title
Gene Symbol
1422062_at
BC003814
macrophage sc
Msr1
1436853_a_at
AI324124
synuclein, alpha
Snca
1450182_at
NM_024412
chloride channe
Clcnka
Probe set ID
Representative Public ID
Gene Title
Gene Symbol
Gene Ontology Biological Process
Gene Ontology Cellular Component
Field:
Sort:
Criteria:
or:
Gene Ontology Biological Process
Like "*transport*"

FIGURE 16.12 Searching for oversized entries using the Microsoft Access `Filter` function.

6. To reproduce this filtering approach in Microsoft Excel, open `Mouse annotation.xls` file.

7. Select cell E1 (Gene Ontology Biological Processes), on the `Data` menu, point to `Filter` and then click `AutoFilter` or ALT+D+F+F.

8. Click the arrow in the column E1 and select `(Custom…)`.

9. In the box on the left, click the arrow and select `contains`.

10. In the box on the right, type `transport` and click `OK`.

This approach identified only two genes related to molecular transport (the *Snca* gene identified by Microsoft Access is missing because letter combination *transport* is located beyond 300 characters and spaces (at 734-742 to be precise) in ontology annotation for this genes and therefore cannot be accessed by the filtering function of Microsoft Excel.

5. Cross-Referencing for Venn Diagram Construction

We will generate a basic two-component Venn diagram and identify corresponding gene lists using candidate genes for Drug A (three genes) and Drug B (five genes) produced in Section 1 with more stringent significance set at $p < 0.01$.

1. Open the Microsoft Access database `Venn diagram.mdb`, which is provided in Part III.

2. In the `Database` window click `Queries` under `Objects` listings, and then double-click on `Create query in Design view`.

3. Put tables `Candidates Drug A` and `Candidates Drug B` into the query (top panel) by double-clicking the name of each table in the `Show Table` dialog box, and then click `Close`.

4. In query design view (top panel), connect `Probe set ID` field from `Candidates Drug A` table to the `Probe set ID` field in the `Candidates Drug B` table.

5. From table `Candidates Drug A` drag field `Probe Set ID` into the first column and from table `Candidates Drug B` drag field `Fold Change B` into the second column of the design grid (bottom panel) to show these fields in the output table.

6. Under `Query` menu select `Run` or click `Run query` icon.

Two genes are identified as affected by both drugs. Therefore, in a Venn diagram the area of shared genes will have two genes; as a result, Drug-A- and Drug-B-specific areas will be represented by one and three genes, respectively.

Now we will identify these drug-specific genes.

7. On the Microsoft Access tool bar click `View -> Design view`. The query design will reappear.
8. Click on the intertable link with the right button and in the appeared dialog box select `Join Properties` (Figure 16.13).
9. In the `Joint Properties` dialog box select radio button "2: Include ALL records from 'Candidates Drug A' and only those records from 'Candidates Drug B' where the joined fields are equal" and click `OK`.
10. After you run this query, the resulting table will have all `Probe set ID` for genes affected by Drug A and `Probe set ID` affected by both drugs.

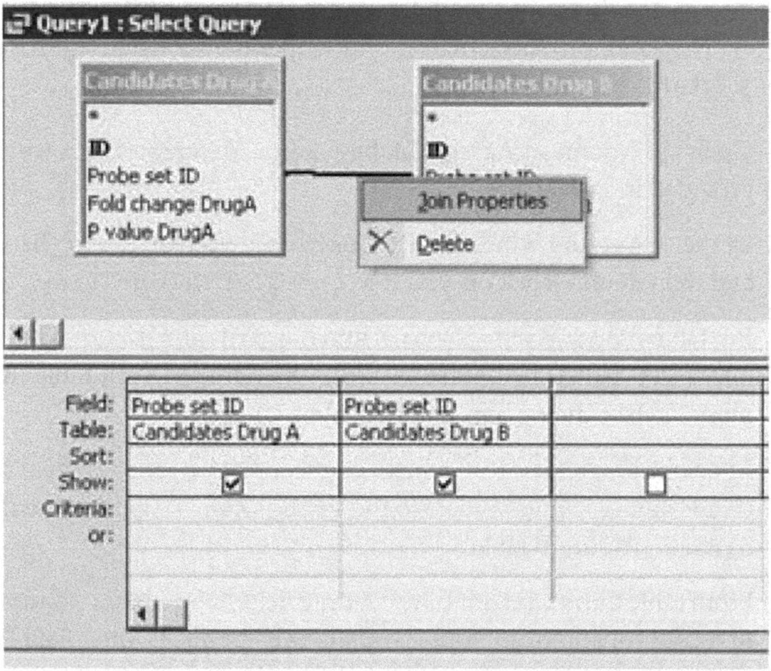
Query1 : Select Query
*
ID
Probe set ID
Fold change DrugA
P value DrugA
*
ID
Join Properties
Delete
Field:
Table:
Sort:
Show:
Criteria:
or:
Probe set ID
Candidates Drug A
Probe set ID
Candidates Drug B

FIGURE 16.13 Manipulating with query output by modifying table joining default settings using the Microsoft Access `Join properties` function.

Query1 : Select Query

Candidates Drug A.Probe set ID	Candidates Drug B.Probe set ID
1417290_at	1417290_at
1425528_at	1425528_at
1436853_a_at	

FIGURE 16.14 Tabular output of a query from Figure 16.13.

The `Probe set ID` field for probe set 1436853_a_at in Drug B column is empty (Figure 16.14), which identifies this gene as specific for Drug A.

11. Repeat the previous steps, but now select in the `Joint Properties` dialog box radio button "3: Include ALL records from 'Candidates Drug B' and only those records from 'Candidates Drug A' where the joined fields are equal" and click `OK`.
12. Run the query.

Now we have identified three genes in the Drug B list that have blank cells in the Drug A column. On the lines of the previous conclusion, these genes are specific for Drug B.

Part III Sample Data

`Candidate gene annotation example.mdb` and `Venn diagram example.mdb` contain queries generated during tutorial using `Candidate gene annotation.mdb` and `Venn diagram.mdb`, respectively.

`Mouse annotation.xls` represents an annotation table for twelve probe sets in the sample Affymetrix array from Section 1.

SECTION 3 INTERCHANGE OF DATA BETWEEN MICROSOFT EXCEL AND ACCESS

Microsoft Excel is tightly integrated with Microsoft Access, which makes data transfer between these applications an easy task. Microsoft Access accepts tabular information in multiple formats and automatically generates its own program compatible tables. Given that Microsoft Access tables are extremely similar to Microsoft Excel spreadsheets, the cross-program conversion of data is a straightforward task. Once data are imported into Microsoft Access, researchers will be able to manipulate data inside the Microsoft Access database and easily export the resulting tables back to Microsoft Excel as program-compatible spreadsheets.

Part I Introduction

1. Import

Microsoft Access provides functions for using data from an external data source. You can import the data into a new Microsoft Access table, which is a way to convert data from a different format (in our case, a Microsoft Excel spreadsheet) and copy it into Microsoft Access. Imported data represent a copy of the Microsoft Excel spreadsheet information in a Microsoft Access database, with the source table not being altered during the importing process. Therefore, updating genomic information in your Microsoft Excel spreadsheet will not affect content of its copy in the Microsoft Access database. Although there is a function that allows one to link Microsoft Excel and Microsoft Access tables, we would recommend the simple reimporting of updated genome annotation files upon their availability.

2. Export

Microsoft Access objects are probably the easiest objects to export to Microsoft Excel because, as they are created with the same application, these objects are accordingly formatted and recognizable. The `Export` function will allow quick transformation of a Microsoft Access table into the familiar environment of the regular Microsoft Excel workbook.

Part II Step-By-Step Tutorial

1. Double-click on the Microsoft Access icon on the desktop or perform `Start -> Programs -> Microsoft Access`.
2. If a dialog box is automatically displayed with options to create a new database, select `Blank Access database` and then click `OK`.
3. If a dialog box is not displayed, create a new database with CTRL+N, then on the `New File` tab, which appears on the right of your screen click `Blank Database`.
4. Use `File name:` slot to name database `Candidates.mdb` and specify location using `File New Database` wizard.
5. Click `Create` to start defining our database.

1. Importing Tables

We will import `Candidates Drug A.xls` and `Candidates Drug B.xls` files provided in Part III of this section.

1. To import the `Candidates Drug A.xls` spreadsheet: on the `File` menu, point to `Get External Data`, and then click `Import`.

2. In the `Import` dialog box, in the `Files of type` slot, click the arrow and select `Microsoft Excel (*.xls)`.

3. Click the arrow to the right of the `Look in` box, select the drive and folder where `Candidates Drug A.xls` is located, and then double-click its icon.

4. In the `Import Spreadsheet Wizard` dialog box select worksheet `Sheet1` (you can import from only one spreadsheet within a workbook at the time).

5. Click `Next`, select `First Row Contains Column Headings` checkbox and click `Finish`.

6. Click the `OK` button in the report of successful importing that will appear (Figure 16.15).

7. Click once on the new Microsoft Access table with default name `Sheet1` and rename this table as `Candidates Drug A`.

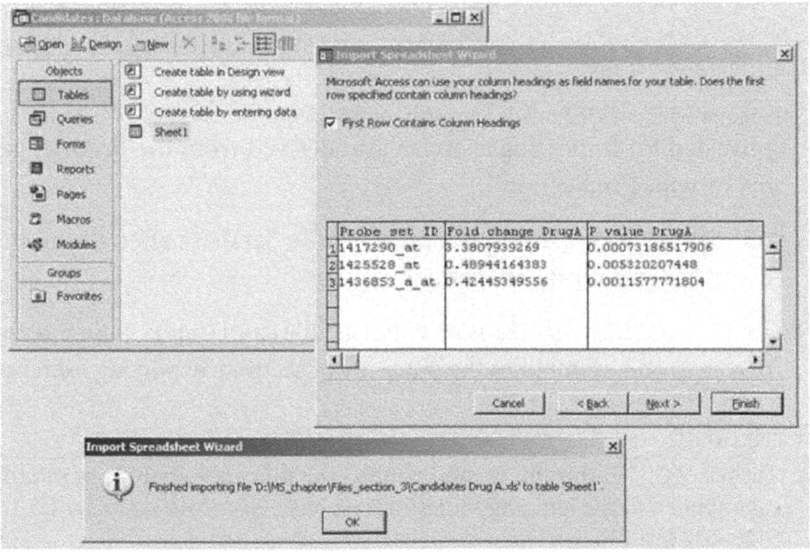
Open Design New
Objects
Tables
Queries
Forms
Reports
Pages
Macros
Modules
Groups
Favorites
Create table in Design view
Create table by using wizard
Create table by entering data
Sheet1
Import Spreadsheet Wizard
Microsoft Access can use your column headings as field names for your table. Does the first row specified contain column headings?
First Row Contains Column Headings
Probe set ID
Fold change DrugA
P value DrugA
1417290_at
3.3807939269
0.00073186517906
1425528_at
0.48944164383
0.005320207448
1436853_a_at
0.42445349556
0.0011577771804
Cancel
< Back
Next >
Finish
Import Spreadsheet Wizard
Finished importing file 'D:\MS_chapter\Files_section_3\Candidates Drug A.xls' to table 'Sheet1'.
OK

FIGURE 16.15 Importing Microsoft Excel table using the Microsoft Access `Get External Data` function.

8. Repeat these steps to import file `Candidates Drug B.xls` and name it `Candidates Drug B`.

Now this database should be an exact copy of `Venn diagram.mdb` described in Section 2. Query these tables for common `Probe Set ID` and name resulting table `Query common candidates`.

2. Exporting Queries

Now we will convert this Microsoft Access table back into Microsoft Excel.

1. On the `File` menu, click `Export`. In the `Save as type:` slot select `Microsoft Excel 97-2002 (.xls)`.
2. Click the arrow to the right of the `Save in:` box, and select the drive and folder to save to. Name the exported file `Common candidates` and click `Export All`.

(Note: Given that some gene ontology fields are pretty long do not select the "Save formatted" check box, it will truncate all output fields to 255 characters and spaces.)

Now we have familiar Microsoft Excel workbook environment for genes that were affected by both drugs.

Part III Sample Data

`Candidates Drug A.xls` and `Candidates Drug B.xls` files provided for importing exercise and derived from the original gene list provided in Section 1.

`Candidates example.mdb` represents the database after all the steps described in tutorial are completed.

`Common candidates.xls` is an exported file from `Candidates example.mdb` and exemplifies the outcome of accurate exporting exercise.

REFERENCES

1. Tusher, V.G., Tibshirani, R., and Chu, G. Significance analysis of microarrays applied to the ionizing radiation response. *Proc Natl Acad Sci USA* 98, 5116–5121, 2001.

2. Schageman, J.J., Basit, M., Gallardo, T.D., Garner, H.R., and Shohet, R.V. MarC-V: a spreadsheet-based tool for analysis, normalization, and visualization of single cDNA microarray experiments. *Biotechniques* 32, 338–340, 342, 344, 2002.
3. Conway, T., Kraus, B., Tucker,D.L., Smalley, D.J., Dorman, A.F., and McKibben, L. DNA array analysis in a Microsoft Windows environment. *Biotechniques* 32, 110, 112–114, 116, 118–119, 2002.
4. Breitkreutz, B.J., Jorgensen, P., Breitkreutz, A., and Tyers, M. AFM 4.0: a toolbox for DNA microarray analysis. *Genome Biol* 2, SOFTWARE0001, 2001.
5. Beckers, G.J. and Conrath, U. Microarray data analysis made easy. *Trends Plant Sci* 11, 322–323, 2006.
6. Li, C. and Hung Wong, W. Model-based analysis of oligonucleotide arrays: model validation, design issues and standard error application. *Genome Biol* 2, RESEARCH0032, 2001.
7. Li, C. and Wong, W.H. Model-based analysis of oligonucleotide arrays: expression index computation and outlier detection. *Proc Natl Acad Sci USA* 98, 31–36, 2001.
8. Eisen, M.B., Spellman, P.T., Brown, P.O., and Botstein, D. Cluster analysis and display of genome-wide expression patterns. *Proc Natl Acad Sci USA* 95, 14863–14868, 1998.
9. Cheadle, C., Cho-Chung, Y.S., Becker, K.G., and Vawter, M.P. Application of z-score transformation to Affymetrix data. *Appl Bioinf* 2, 209–217, 2003.
10. Brown, A., McKie, M., van Heyningen, V., and Prosser, J. The human PAX6 mutation database. *Nucl Acids Res* 26, 259–264, 1998.

Selected Websites

CHAPTER 1 GENOME ANALYSIS

UCSC Genome Browser Home	http://genome.ucsc.edu
NCBI-BLAST	http://www.ncbi.nlm.nih.gov/BLAST
Ensembl Genome Browser	http://www.ensemble.org
ECR Browser	http://ecrbrowser.dcode.org
Argo Genome Browser	http://www.broad.mit.edu/annotation/argo

CHAPTER 2 TWO COMMON DNA ANALYSIS TOOLS

REBASE	http://rebase.neb.com
NEB Cutter	http://tools.neb.com/NEBcutter2/index.php
Primer3	http://frodo.wi.mit.edu/cgi-bin/primer3/primer3_www.cgi
Restriction/Mapper	http://arbl.cvmbs.colostate.edu/molkit/mapper/index.html
	http://www.restrictionmapper.org/

CHAPTER 3 PHYLOGENETICS ANALYSIS

Clustalw	http://www.ebi.ac.uk/clustalw/
3D-Coffee	http://igs-server.cnrs-mrs.fr/Tcoffee/tcoffee_cgi/index.cgi
MUSCLE	http://www.drive5.com/muscle/
PROBCONS	http://probcons.stanford.edu/
MAFFT	http://timpani.genome.ad.jp/mafft/server/

CHAPTER 4 SNP AND HAPLOTYPE ANALYSES

dbSNP	http://www.ncbi.nlm.nih.gov/projects/SNP
HapMap	http://www.hapmap.org/
Haploview	http://www.broad.mit.edu/mpg/haploview/
PharmGKB	http://www.pharmgkb.org/
Seattle SNPs	http://pga.gs.washington.edu/

CHAPTER 5 GENE EXPRESSION PROFILING BY MICROARRAY

Affymetrix	http://affymetrix.com/index.affx
GEO	http://www.ncbi.nlm.nih.gov/geo/
GeneSpring	http://www.genespring.com
MeV	http://www.tm4.org/mev.html
SAM	http://www-stat.stanford.edu/~tibs/SAM/

CHAPTER 6 GENE EXPRESSION PROFILING BY SAGE

SAGEnet	http://www.sagenet.org
NCBI-SAGE	http://www.ncbi.nlm.nih.gov/sage
NCI-SAGE	http://cgap.nci.nih.gov/SAGE
NCBI-GEO	http://www.ncbi.nlm.nih.gov/geo

CHAPTER 7 REGULATION OF GENE EXPRESSION

NCBI-LocusLink	http://www.ncbi.nlm.nih.gov/LocusLink/
EASED	http://eased.bioinf.mdc-berlin.de/
RNA Editing Web Site	http://dna.kdna.ucla.edu/rna/index.aspx

CHAPTER 8 MICRORNOMA GENOME-WIDE PROFILING BY MICROARRAY

miRBase:	http://microrna.sanger.ac.uk/sequences/
RNAdb:	http://research.imb.uq.edu.au/rnadb/
MIAME:	http://www.ebi.ac.uk/miamexpress/
miRGen:	http://www.diana.pcbi.upenn.edu/miRGen.html.

CHAPTER 9 SIRNA

siRNA at Whitehead:	http://jura.wi.mit.edu/bioc/siRNAext/
siDirect:	http://design.RNAi.jp/
CGAP RNAi at NCI:	http://cgap.nci.nih.gov/RNAi
HuSiDa:	http://www.hnman-siRNA-database.net
Protein Lounge siRNA database:	http://www.proteinlounge.com/sirna

CHAPTER 10 PROTEOMIC DATA ANALYSIS

Software for proteomics and genomics research	http://www.nonlinear.com
Matrix Science – Home	http://www.matrixscience.com
Pubchem	http://pubchem.ncbi.nlm.nih.gov/omssa/

CHAPTER 11 PROTEIN SEQUENCE ANALYSIS

RCSB PDB	http://www.rcsb.org/pdb/home/home.do
PredictProtein	http://www.predictprotein.org
BioInfoBank	http://bioinfo.pl/meta/livebench.pl
InterProScan	http://www.ebi.ac.uk/InterProScan/

CHAPTER 12 PROTEIN FUNCTION ANALYSIS

Uniprotein	http://www.uniprot.org/
Ex-PASY	http://www.expasy.org/tools/
EBI site	http://www.ebi.ac.uk/services/
CBS Prediction Servers	http://www.cbs.dtu.dk/services/

CHAPTER 13 FUNCTIONAL ANNOTATION OF PROTEINS IN MURINE MODELS

The Jackson Laboratory	http://www.jax.org.
Mouse Genome Informatics	http://www.informatics.jax.org
Charles River Laboratories	http://www.criver.com/
IGTC	http://www.genetrap.org/
Baygenomics	http://baygenomics.ucsf.edu/

CHAPTER 14 APPLICATION OF PROGRAMMING LANGUAGES IN BIOLOGY

Perl website:	http://www.cpan.org
R website:	http://cran.r-project.org
Biojava website:	http://www.biojava.org
Bioperl website:	http://bioperl.open-bio.org
BioConductor website:	http://www.bioconductor.org

CHAPTER 15 WEBSITE AND DATABASE DESIGN

Wiki-Database	http://en.wikipedia.org/wiki/Database
HTML tags	http://www.w3schools.com
Software download	http://www.download.com
HTML and CSS tutorials	http://www.htmldog.com

CHAPTER 16 MICROSOFT EXCEL AND ACCESS

Microsoft Corp.	http://office.microsoft.com/en-us/default.aspx
Microsoft Excel	http://www.exceltrainingsite.com/
Microsoft Access	http://www.accesstrainingsite.com/

Glossary

Accession number: This refers to the unique GenBank identifier a sequence has been assigned. This number can be used to search Genbank records for a specific sequence.

Affymetrix: Affymetrix was founded by Stephen P.A. Fodor, Ph.D. and others in the late 1980s with the revolutionary idea to use semi-conductor manufacturing techniques to create GeneChips (an Affymetrix trademark) or generically DNA microarrays.

α-value: The nominal probability (set by the investigator) of making a type 1 error.

Algorithm: any well-defined procedure describing how to accomplish a particular task.

ALT: the Alt (alternative) key on the IBM computer keyboard used to change the function other keys.

Alternative promoter: an alternative region from which transcripts of a gene originate. The existence of multiple transcripts for a single gene that differ in their 50 termini reflects the presence of alternative promoters.

Alternative splicing: alternative splicing means some pre-mRNAs can be spliced in more than one way, generating alternative mRNAs.

Amplicon: a DNA fragment product of polymerase chain reaction

Analysis of variance: A statistical test for determining differences in mean values between two or more groups.

Argonaute proteins: endonucleases of the RISC complex capable of degrading the target mRNA strand whose sequence is complementary to that of the siRNA guide strand.

Bayesian probability: The probability of a proposition being true, which is conditional on the observed data.

Bioconductor: Bioconductor is an open source and open development software project to provide tools for the analysis of SNP and transcriptional profiling data (SAGE, microarrays or Affymetrix chips) and the integration of genomic meta data.

Biojava: Biojava is an open source project dedicated to providing Java tools for processing biological data. It includes objects for manipulating sequences, file parsers, CORBA interoperability, DAS, dynamic programming, and simple statistical routines.

Bioperl: Bioperl is a collection of reusable Perl scripts and modules that can be used to develop complex bioinformatics applications. It contains a rich set of scripts or modules for sequence manipulation, accessing of databases using a range of data formats, and parsing of the results of various molecular biology programs including Blast, clustalw, TCoffee, genscan, ESTscan and HMMER.

BLAST: a method to ascertain sequence similarity.

Blat: a fast sequence alignment tool similar to BLAST.

Bonferroni correction: A family-wise error rate (FWER) control procedure that sets the -value level for each test and strongly controls the FWER for any dependency structure among the tests.

Bootstrap analysis: A form of computer-intensive resampling-based inference. Pseudo-data sets are created by sampling from the observed data with replacement (that is, after a case is resampled, it is returned to the original data and can, potentially, be drawn again).

Case: In a microarray experiment, a case is the biological unit under study; for example, one soybean, one mouse or one human.

cDNA: complementary DeoxyriboNucleic Acid. Single-stranded DNA that is complementary to messenger RNA or DNA that has been synthesized from messenger RNA by reverse transcriptase.

ChIP-on-chip: is a microarray-based technique for understanding gene regulation in disease. It uses chromatin immuno-precipitation (ChIP) to discover how regulatory proteins interact with the genome of living cells.

Chromosome: Chromosome refers to the structure in the cell composed of a very long molecule of DNA and associated proteins called Histones.

Conditional knockout gene: to delete a gene in a particular organ, cell type, or stage of development.

CTRL: the Ctrl (control) key on the IBM computer keyboard used to change the function other keys.

Dicer: an RNAse III ribonuclease that cleaves double-stranded RNA (dsRNA) and pre-microRNA into short double-stranded siRNA.

DNA Microinjection: DNA injected into the pronucleus of a fertilized ovum. Following injection, DNA would incorporate into the genome of the cell.

Domains: Compact, globular regions of proteins that are the basic units of tertiary structure

Dscam: The Drosophila melanogaster Down syndrome cell adhesion molecule (Dscam) gene: this gene encodes an axon guidance receptor and can generate 38,016 different isoforms via the alternative splicing of 95 variable exons. Dscam contains 10 immunoglobulin (Ig), six Fibronectin type III, a transmembrane (TM), and cytoplasmic domains.

EMBL (The European Molecular Biology Laboratory): Major research center coordinating molecular biology research. It includes sites in Grenoble, Hamburg, Heidelberg, Hinxton, and Monterotondo.

Ensembl: a joint project between EMBL-EBI and the Sanger Centre to develop a software system which produces and maintains automatic annotation on eukaryotic genomes.

Expressed Sequence Tags ESTs: partial, single-pass sequences from either end of a cDNA clone

False-discovery rate (FDR): The expected proportion of rejected null hypotheses that are false positives. When no null hypotheses are rejected, FDR is taken to be zero.

5' splice site: the exon-intron boundary at the 5' end of the intron

Fold change: A metric for comparing a gene's mRNA-expression level between two distinct experimental conditions. Its arithmetic definition differs between investigators.

Format: to create or edit the layout of a document; to change a document so it will fit onto a different type of page or to prepare a mass storage medium for initial use, erasing any existing data in the process in computing.

GenBank: the NIH genetic sequence database; an annotated collection of all publicly available DNA sequences

Gene Expression Omnibus (GEO): the largest fully public repository for high-throughput molecular abundance data, primarily gene expression data at the National Center for Biotechnology Information (NCBI).

Gene Ontology: A way of describing gene products in terms of their associated biological processes, cellular components and molecular functions in a species-independent manner.

Gene trapping: a method based on the random integration of a gene-trap vector into the mouse genome. A promoterless reporter gene following a splice acceptor will produce a fusion transcript between the trapped gene and the reporter gene when the vector inserts into an intron. This allows the identification of the trapped genes easily by 5' rapid amplification of cDNA ends and also to investigate both the in vitro and in vivo expression patterns of trapped genes.

Gene-expression profiling: Determination of the level of expression of hundreds or thousand of genes through the use of microarrays. Total RNA extracted from the test tissue or cells and labeled with a fluorescent dye is tested for its ability to hybridize to the spotted nucleic acids.

Genelist: A group of genes/proteins with some common property, such as putative interaction with miRNAs or same expression profiles. Are generated by target prediction programs or by calculations performed by GeneSpring or other bioinformatics tools.

Genespring software: GeneSpring is a powerful analysis tool that analyzes the scanned microarray data by assigning experiment parameters and interpretation to filter genes for differential expression and cluster to identify similar regulated groups.

Genetic code: rules by which information encoded in genetic material is translated into proteins. It defines the relationship between trinucleotides (codons) and amino acids

Genome Browser: a tool which collates all relevant genomic sequence information in one location and provides a rapid, reliable and simultaneous display of any requested portion of genomes at any scale in a graphical design.

Genome: genetic information of an organism

Haplotype: a combination of alleles at different markers along the same chromosome that are inherited as a unit.

Haplotype tagging: refers to methods to select minimal number of SNPs that uniquely identify common haplotypes (>5% in frequency).

Hierarchical clustering technique: A computational method that groups genes (or samples) into small clusters and then group these clusters into increasingly higher level clusters. As a result, a dendrogram (i.e., tree) of connectivity emerges.

HUGO (The Human Genome Organization): is the international organization of scientists involved in human genetics. Established in 1989 by a collection of the world's leading human geneticists, the primary ethos of the Human Genome Organization is to promote and sustain international collaboration in the field of human genetics.

IBM: International Business Machines corporation. American computer technology corporation headquartered in Armonk, New York. The company is one of the few information technology companies with a continuous history dating back to the 19th century; it was founded in 1888. IBM manufactures and sells computer hardware, software, infrastructure services, hosting services, and consulting services in areas ranging from mainframe computers to nanotechnology.

Intersection-union tests: Multicomponent tests in which the compound null hypothesis consists of the union of two or more component null hypotheses.

Isoschizomers, restriction endonucleases that recognize the same restriction site

Linkage disequilibrium (LD): a term used in the study of population genetics for the non-random association of alleles at two or more loci.

Markup language: originally developed by the publishing industry to communicate page layout information between the editor, writer and printer. The text is accompanied by embedded codes which indicate specifics of styling such as font, point size, italic, paragraph indent, line spacing, etc. Specific code systems have been developed which extend the concept to data interchange among various computer systems.

MiRNoma: The full spectrum of microRNAs expressed in a particular cell type.

Multiple sequence alignment (MSA): to align more than two sequences at a time in a given query set. MSA is often used in identifying conserved sequence regions across a group of sequences hypothesized to be evolutionarily related.

Normalization: The process by which microarray spot intensities are adjusted to take into account the variability across different experiments and platforms.

Null hypothesis: The hypothesis that is being tested in a statistical test. Typically in a microarray setting it is the hypothesis that states: there is no difference between gene-expression levels across groups or conditions.

Object oriented database: a database based upon objects which have properties that are variable and can used to identify individual objects. Objects are defined by classes; this permits integration and manipulation of many data types including non-numeric data e.g. class – person: properties - name, height, weight, appearance (photo), DNA sequence. Well suited for working with complex data structures.

Open reading frame (ORF): a portion of DNA that begins with an initiation codon (ATG) and ends with a nonsense/stop codon (TAG, TAA, TGG). An open reading frame has the potential to encode a polypeptide beginning with methionine

Overfitting: This occurs when an excessively complex model with too many parameters is developed from a small sample of 'training' data. The model fits those data well, but does so by capitalizing on chance variations and, therefore, will fit a fresh set 'test' data poorly.

Parameter: A quantity (for example, mean) that characterizes some aspect of a (usually theoretically infinite) population.

Perl: Perl stands for Practical Extraction Report Language and is a programming language designed to handle a variety of system administrator functions. It provides comprehensive string handling functions and is widely used to write Web server programs for tasks such as automatically updating user accounts and newsgroup postings, processing removal requests, synchronizing databases and generating reports. Perl has also been adapted to non-UNIX platforms. Perl is one of the most popular languages used by Biologists.

Permutation test: A statistical hypothesis test in which some elements of the data are permuted (shuffled) to create multiple new pseudo-data sets. One then evaluates whether a statistic quantifying departure from the null hypothesis is greater in the observed data than a large proportion of the corresponding statistics calculated on the multiple pseudo-data sets.

Phamacogenetics: the study of genetic variants and how these variants relate to interindividual response to drug therapy.

Phylogenetics: the study of evolutionary relatedness among various groups of organisms.

Phylogeny (or phylogenesis): the origin and evolution of a set of organisms, usually a set of species.

Plasmode: A real (not computer simulated) data set for which the true structure is known and is used as a way of testing a proposed analytical method.

Polylinker: a very short segment of artificial DNA that harbors restriction enzyme recognition sites

Polymerase chain reaction (PCR): a highly specific in vitro exponential amplification of the target DNA using a thermostable polymerase

Posterior probability: The Bayesian probability that a hypothesis is correct, which is conditional on the observed data.

Power: This is classically defined as the probability of rejecting a null hypothesis that is false. However, power has been defined in several ways for microarray studies.

Prediction analysis of microarrays (PAM): A statistical technique that identifies a subgroup of genes that best characterizes a predefined class and uses this gene set to predict the class of new samples.

Primary structure of protein: The sequence of amino acids in a polypeptide chain;

Primer: oligonucleotide that binds to complementary target sequences and is extended during the synthesis of DNA by DNA polymerase

Probability-based algorithm: For mass spectrometry database search, the probability-based algorithms model to some extent the peptide fragmentation process and calculate the probability that a particular peptide sequence produced the observed spectrum by chance.

Programming Language: Programming Language are a series of instructions written by a programmer according to a given set of rules or conventions ("syntax"). Programming language instructions are converted into programs in language specific to a particular operating system so that the computer can interpret and carry out the instructions. Some common programming languages are Perl, JAVA, C, and C++.

Promoter: the genomic sequence immediately upstream of the transcriptional start site defined by the 5' end of an mRNA. It is this region that is presumed to bind the transacting factors required to transcribe the gene.

Protein annotation: Refers to information associated to an amino-acid sequence. Besides annotation relevant to the protein structure and function, it could also include references and cross-references to related data sources, for instance.

Protein function: In the narrow sense, protein function refers to the molecular function that a protein performs based on its biochemical properties. In a broader sense, protein function also refers to the biological role in which a protein is involved - which can be both on the level of biological processes as on the phenotypic level.

Protein microarray: In a protein microarray, capture molecules are immobilized in a very small area, and probed for various biochemical activities. There are two general types of protein microarrays: analytical microarrays and functional protein microarrays.

Protein sequence analysis: Refers to the prediction of protein features such as biochemical properties, post-translational modifications and the presence of structural or functional domains.

Post-translational modifications: Refers to specific amino-acid modifications that occur during the late steps of protein synthesis and performed by enzymatic mechanisms. In addition to specific proteolytic cleavages, more than 350 naturally occurring post-translational modifications have been identified to date.

Protein-protein interaction: Refers to the association of protein molecules and the study of these associations from a structural, biochemical and network perspective.

Proteomics: Proteomics is an emerging scientific field that involves the identification, characterization, and quantification of proteins in cells, tissues or body fluids.

p-value: The probability, were the null hypothesis true, of obtaining results that are as discrepant or more discrepant from those expected under the null hypothesis than those actually obtained.

Real time PCR (q-RT-PCR): a PCR method in which the amount of PCR produced is monitored in real time, during each cycle

Recombinant DNA: artificial DNA made by combining two or more different strands of DNA from the same of different organisms.

Relational database: a database organized to efficiently utilize set theory and predicate (true/false) logic to search, sort and retrieve data. This structure works well with large amounts of numeric data in limited formats.

Restriction enzyme: endonuclease that cleaves DNA as part of the defense mechanism against foreign DNA

Restriction mapping: the characterization of double-stranded DNA that is based on the location of the restriction endonucleases cleavage sites

Restriction site: the DNA sequence (usually 4 - 6 bases) cleaved by restriction enzymes. They usually have dyad symmetry (palindromic sequences)

RNA splicing: is a process that removes introns and joins exons in a primary transcript.

RNAi: RNAi is a short form of RNA interference. It is a mechanism in eukaryotic cells by which short fragments of double-stranded ribonucleic acid (dsRNA) interfere with the expression of a particular gene whose sequence is complementary to the dsRNA.

RNA-induced silencing complex (RISC): a multi-protein siRNA complex which cleaves (incoming viral) dsRNA and binds the antisense RNA strand to a protein which seeks out the complementary strand. When it finds the complementary strand, it activates RNAse activity and cleaves the RNA.

Sampling variation: The variability in statistics that occurs among random samples from the same population and is due solely to the process of random sampling.

Schema: In computer databases it is the underlying organizational structure or model by which data are organized. The organizational structure dictates the type and efficiency of algorithms that can be used to search, sort and retrieve data.

Secondary structure of protein: The regular arrangement of amino acids, such as α-helix and β sheet, within localized regions of a polypeptide chain;

Selection bias: This occurs when the prediction accuracy of a rule is estimated using cases that had some role in the derivation of the rule. It is an upward bias — that is, one that overestimates the predictive accuracy.

SHIFT: the Shift key on the IBM computer keyboard is a modifier key that used to type capital letters and other alternate "upper" characters. There are typically two Shift keys, on the left and right sides of a keyboard.

Significance analysis of microarrays (SAM): A statistical method used in microarray analyses that calculates a score for each gene and thus identifies genes with a statistically significant association with an outcome variable such as transfection with a specific miRNAs.

Silent mutagenesis: change in nucleotide sequence that preserves the encoded protein sequence

Single Nucleotide Polymorphism (SNP): single-base variations in a DNA sequence. For example, two sequenced DNA from different individuals, AAGCCTA to AAGCTTA, contain a difference in a single nucleotide. In this case we say that there are two alleles, C and T.

siRNA: siRNA is the short double-stranded RNA, called small interfering RNAs (siRNAs), also known as short interfering RNA or silencing RNA. siRNAs have a well defined structure: a short (usually 21-nt) double-strand of RNA (dsRNA) with 2-nt 3' overhangs on either end. They are underlying RNA interference.

Spliceosome: Protein-RNA complex that removes introns from eukaryotic nuclear RNAs. Splicing is catalyzed by spliceosome. Spliceosome consists of many small nuclear RNA and associated proteins.

SQL: Structured Query Language, a computer language designed for retrieving data from a relational database. It also has functions to create a database, modify the database schema and modify data; thus, it is primary control center for the database administrator. Although SQL has been defined by the American National Standards Institute, companies may have custom implementations of the language specific for their product.

SR proteins: are Serine / Arginine -residue proteins which are involved in regulating and selecting splice sites in eukaryotic mRNA.

SWISS-PROT: a curated protein sequence database which strives to provide a high level of annotation (such as the description of the function of a protein, its domains structure, post-translational modifications, variants, etc.), a minimal level of redundancy and high level of integration with other databases.

Table Browser: a tool provides text-based access to the genome assemblies and annotation data stored in the Genome Browser database.

Tag: A SAGE tag is a 14-nucleotide sequence that has been found within an mRNA. The relative abundance of a particular SAGE tag within a pool of tags gives some indication of the level of expression of the gene(s) containing that tag.

Targeted insertion: to insert the DNA into embryonic stem cells and selecting for cells with homologous recombinants.

Tertiary structure of protein: The three dimensional folding of a polypeptide chain that gives the protein its functional form;

3' splice site: the exon-intron boundary at the 3' end of the intron

Transformation: The application of a specific mathematical function so that data are changed into a different form. Often, the new form of the data satisfies assumptions of statistical tests. The most common transformation in microarray studies is $\log_2$.

Transgenic mouse: a mouse that has had foreign DNA introduced into one or more of its cells artificially.

t-tests: Statistical tests that are used to determine a statistically significant difference between two groups by looking at differences between two independent means.

Two-dimensional differential in gel electrophoresis or 2D DIGE: 2D-DIGE is a fairly recent improvement of the 2DE technology. Prior to gel electrophoresis, the proteins from different disease states or experimental treatments are separately labeled with different fluorescent dyes which are matched with mass and charge and each has a different emission wavelength. The labeled samples are then combined and subjected to 2DE.

Type 1 error: A false positive or the rejection of a true null hypothesis; for example, declaring a gene to be differentially expressed when it is not.

Type 2 error: A false negative, or failing to reject a false null hypothesis; for example, not declaring a gene to be differentially expressed when it is.

U133: Gene Chip based on Indigene build 133.

Venn: John Venn (August 4, 1834 – April 4, 1923) was a British logician and philosopher, who is famous for conceiving the Venn diagrams, which are used in many fields, including set theory, probability, logic, statistics, and c

Index

C

D

Q

R

S

W

X

Y

Z

Milton Keynes UK
Ingram Content Group UK Ltd.
UKHW021933071024
449327UK00022B/1789

9 780367 388751